Canadian Mathematical S
Société mathématique du

Editors-in-Chief
Rédacteurs-en-chef
K. Dilcher
K. Taylor

Advisory Board
Comité consultatif
P. Borwein
R. Kane
S. Shen

For other titles published in this series, go to
www.springer.com/series/4318

Michael J. Jacobson, Jr.
Hugh C. Williams

Solving the Pell Equation

 Springer

Michael J. Jacobson, Jr.
Department of Computer Science
University of Calgary
2500 University Drive NW
Calgary AB T2N 1N4
Canada
jacobs@cpsc.ucalgary.ca

Hugh C. Williams
Department of Mathematics and Statistics
University of Calgary
2500 University Drive NW
Calgary AB T2N 1N4
Canada
williams@math.ucalgary.ca

Editors-in-Chief
Rédacteurs-en-chef

Karl Dilcher
K. Taylor
Department of Mathematics and Statistics
Dalhousie University
Halifax, Nova Scotia B3H 3J5
Canada
cbs-editors@cms.math.ca

ISBN 978-1-4419-2747-7   e-ISBN 978-0-387-84923-2
DOI 10.1007/978-0-387-84923-2

Mathematics Subject Classification (2000): 11D09 11A55

Printed on acid-free paper

9 8 7 6 5 4 3 2 1

springer.com

To the children: Amanda, Alexa, Hannah, Graeme, and Sarah.

Preface

Let D be a positive non-square integer. The misnamed Pell equation is an expression of the form

$$T^2 - DU^2 = 1 \,, \tag{0.1}$$

where T and U are constrained to be integers. For example, if $D = 13$, then $T = 649$ and $U = 180$ is a solution of (0.1). This very simple Diophantine equation seems to have been known to mathematicians for over 2000 years. Indeed, there is very strong evidence that it was known to Archimedes, as the Cattle Problem, attributed to him in antiquity, makes very clear. Even today, research involving this equation continues to be very active; at least 150 articles dealing with it in various contexts have appeared within the last decade. One of the main reasons for this interest is that the equation has a habit of popping up in a variety of surprising settings; it is also of great importance in solving the general second-degree Diophantine equation in two unknowns:

$$ax^2 + bxy + cy^2 + dx + ey + f = 0 \,.$$

Furthermore, the problem of solving (0.1) is connected to that of determining the regulator, an important invariant of a real quadratic number field, and to solving the discrete logarithm problem in such structures. Today, this latter problem is of interest to cryptographers.

Such is the interest in the Pell equation that at least three books have been devoted to it:

- H. Konen, *Geschichte der Gleichung* $t^2 - Du^2 = 1$, Leipzig, 1901.
- E. E. Whitford, *The Pell Equation*, College of the City of New York, New York, 1912
- Edward J. Barbeau, *Pell's Equation*, Springer, 2003

The first two have been out of print for a long time and are currently very difficult to find. Also, because they are quite old, they do not deal with the modern theory of the equation. Much has been learned since 1912. The last book, according to its author, "is a focused exercise book in algebra" and is intended to motivate college students to develop an appreciation of mathematical technique. As such, it succeeds very well, but there is no attempt in the book to explore the deeper aspects of this equation, nor was that its author's intent.

It is well known that for any positive non-square integer D, (0.1) has an infinitude of solutions, which can be easily expressed in terms of the *fundamental solution* t, u, where $t, u > 0$. For example, the fundamental solution of (0.1) for $D = 7$ is $t = 8$ and $u = 3$. If $D = 1620$, then $t = 161$, but if $D = 1621$, then t is a number of 76 digits! As we will see in Chapter 13, there are even more extreme examples of this phenomenon for larger values of D. It is this puzzle of finding the fundamental solution that we refer to as the problem of solving the Pell equation. It was very likely investigated by the ancients, but it was not until the early 7th century AD that the Indian mathematician Brahmagupta discovered an ad hoc method of solving this problem. Unfortunately, his method and its more deterministic successors, which make use of the theory of continued fractions, cannot conveniently be used when D becomes large, say in excess of 15 digits.

The purpose of this book is to provide a comprehensive discussion of how to find the fundamental solution and, in particular, to describe methods for doing this that have been developed since 1972 for large values of D. As much of this material is scattered rather widely throughout the literature, this will be the first book to discuss this subject in any detail. The principal component of our enquiry will be computational techniques, but in order to derive these, it will be necessary to develop the required theory. In doing this, we will explore a great variety of different topics in number theory, some indication of which may be found by examining the table of contents.

As our approach to the Pell equation is largely computational, we assume that the reader is at least vaguely familiar with the basic precepts of measuring the computational complexity of an algorithm. We will use the terms "complexity," "time complexity," and just plain "runtime" or "time" interchangeably to describe the efficiency (the number of bit operations needed) by a particular computational technique. Thus, for example, we might say that a particular algorithm executes in time complexity $O(f(n))$, where f is some function and n is the input length, or we might say that it completes its computation in time $O(f(n))$. We may also have to measure the maximum number of bits required by an algorithm in order for it to execute. We call this the *space* or *space complexity* of the algorithm.

In order to solve (0.1) for large D, it has been discovered that it is easier first to evaluate the regulator R of the associated real quadratic number field $\mathbb{Q}(\sqrt{D})$. Nevertheless, the problem of computing R can still be very difficult, particularly when the value of the *radicand* D becomes very large ($> 10^{25}$). (Clearly, the actual value of the regulator can never be computed because it is a transcendental number; we are content to produce a rational number (often an integer) R' which is within 1 of the actual value.) The best method currently available for computing R' is Buchmann's subexponential method.

Unfortunately, the correctness of the value of R' produced by this technique is conditional on a generalized Riemann hypothesis, for which there is as yet no proof. The best unconditional algorithm (the value of R' is unconditional, not the running time) for computing the regulator of a real quadratic field is Lenstra's $O(D^{1/5+\epsilon})$ Las Vegas algorithm.

In this book we will discuss all of the above techniques and ultimately describe a rigorous method for verifying the regulator produced by the subexponential algorithm. This technique is of complexity $O(D^{1/6+\epsilon})$ and is unconditional, once we have a candidate for R'. It has been used to verify a 33-digit R' for a field with a 65-digit value of D. In addition, these methods can be extended to the problem of determining rigorously for real quadratic fields of large radicand whether or not a given ideal is principal. This, as we will point out, is of great importance in solving certain Diophantine equations. We will also describe some rather surprising applications of this material to cryptography.

Most of these techniques rely on estimations of certain irrational quantities; thus, in order to establish our results rigorously, it is essential that we have provable upper bounds on the errors that result from our use of these approximations. We provide a complete discussion of this and the associated algorithms, but, unfortunately, certain aspects of this are necessarily very technical and, frankly, rather wearisome. In order to facilitate a relatively smooth flow of this material, we have relegated the greater portion of the more tedious minutiae required by this investigation to an appendix.

We use different modes of presentation of the algorithms discussed in this book. The most formal of these include a name, such as NUCOMP or WNEAR, and a detailed listing of the pseudocode for the algorithm. This is usually provided for the basic algorithms, which are frequently employed in the latter part of the book. Several of these can be found in the aforementioned appendix. Some of the other algorithms, which in this formal format would be far too long, are described in pseudocode, which is less detailed. This is particularly the case for the index-calculus techniques described in Chapter 13. Finally, we sometimes simply describe certain processes rather informally as a simple sequence of steps which involve the use of the more formally presented algorithms that have been described previously. For example, this is the case for the technique of rigorously verifying the value of R' mentioned in Chapter 15.

We wish to emphasize here that this book is not intended to be used as a textbook; its focus is much too narrow, and although we do include a number of examples, we provide no exercises. It could, however, be used as supplementary reading for students enrolled in a second course in number theory. The intended primary audience is number theorists, both professional and amateur, and students, but as we discuss a number of cryptographic applications of the material that we develop in the book, a possible secondary readership would be that of mathematical cryptographers at about the same level as the primary readership. The subject matter should be accessible to anyone with an undergraduate knowledge of elementary number theory, abstract algebra, and analysis. We have provided many references and notes for those who may wish to follow up on various topics, but in spite of the size of the Reference section, we must point out that it should not be regarded as complete. We have mostly included citations to work which is relevant to our theme of deriving methods for solving (0.1), and we sincerely hope that we have not though ignorance or inadvertence omitted any important contributions.

We had two principal objectives in writing this book. One was to provide a relatively gentle introduction for senior undergraduates, and others with the same level of preparedness, to the delights of algebraic number theory through the medium of a mathematical object that has fascinated people since the time of Archimedes. Our other goal was to detail the enormous progress that has been made, since Shanks' discovery in 1972 of what he termed the *infrastructure* of an ideal class, on the development of efficient algorithms for performing arithmetic in quadratic number fields. What we are able to do today is most remarkable; it certainly surprises us.

Acknowledgements

The idea of writing this book was conceived about two decades ago when it became apparent that a conditional, subexponential algorithm could be used to solve the Pell equation. During the time that has elapsed, many changes were made to the original concept of this book. Only during the last 2 years, however, did we think that the state of research on this topic had stabilized to the point that we were able to complete this work. It is important to emphasize that it is not possible to write a book such as this in isolation. Over the years, many individuals have made contributions to this work either in providing advice, ideas, or encouragement. We wish to acknowledge, in particular, Mark Bauer, Mike Bennett, Andrew Booker, Richard Brent, Henri Cohen, Wayne Eberly, Mark Giesbrecht, Andrew Granville, Robbert de Haan, Safuat Hamdy, Hendrik Lenstra, Jr., Stephane Louboutin, Richard Lukes, Keith Matthews, Markus Maurer, Richard Mollin, Stefan Neis, Roger Patterson, Alper Ozdamar, Sachar Paulus, Michael Pohst, Alf van der Poorten, Shantha Ramachandran, Rei Safavi-Naini, Renate Scheidler, Arthur Schmidt, Jon Sorensen, Andreas Stein, Arne Storjohann, Edlyn Teske, Patrick Theobald, Ulrich Vollmer, Gary Walsh, and Jim White.

We also wish to single out some individuals for special thanks. The first of these is Karl Dilcher, who in his capacity as one of the Editors in Chief of the Canadian Mathematical Society's *Books in Mathematics* series solicited this work. It is fair to say that neither of us would have even considered writing, let alone completing, this book without his continued support and encouragement. It is difficult for us to express the extent of our gratitude to Johannes Buchman. He has acted as mentor, contributor, and friend to this project since its conception. He also made available, before its publication, a copy of his book with Ulrich Vollmer entitled *Binary Quadratic Forms: An Algorithmic Approach*. As this book covers in part some of the material of this volume, we are most appreciative of this gesture. It allowed us to produce a more focused work which presents some of the same material from a different point of view and which we hope will be seen as complementary to his. We extend our considerable thanks also to John P. Robertson for his eagerness in asking to be involved as a proofreader for this book from the very beginning and for his enormous competence in carrying out this duty. His efforts in this regard have resulted in a much better book. What errors remain, and it is inevitable that there will be some, are totally the responsibility of the authors.

To our former student, Reg Sawilla, we wish to express our thanks for the considerable programming effort that has gone into testing the many algorithms that appear in this work. We are also most indebted to our current student, Alan Silvester, for completing with great competence the enormous task of entering and editing this work on the computer. This should be seen in the light of our almost completely indecipherable handwriting, particularly that of the second author.

There are also a number of institutions that we wish to acknowledge. First and foremost of these is the Alberta Informatics Circle of Research Excellence (iCORE). Their generous support of our efforts has provided us with precious time and several much needed opportunities to interact with many scholars in the preparation of this work. We also thank Alberta Ingenuity and the Canadian Foundation for Innovation (CFI) for providing us with the funds needed to acquire the computing machinery that was used so often in producing the results and testing the routines in this work. In this connection, we want to thank Marc Wrubleski for his considerable efforts in keeping the machines working efficiently. Of course, we are also most grateful to the National Science and Engineering Research Council (NSERC) for their continued support of our research. We are also indebted to the libraries of several universities for providing us access to their collections. In no particular order these are the University of Calgary, the University of Toronto, the University of Sydney, Australian National University, and the University of Illinois at Urbana-Champaign. The second author would like to express his gratitude to the Fields Institute and to the Department of Computing Sciences at Macquarie University for providing him sanctuaries in which much of the work needed for his contribution to this work could be undertaken free of the inevitable interruptions that occur in his own institution.

Finally, we wish to acknowledge the contribution of our respective families to this project. During the time needed to complete this work, we have not been as attentive to them as we should have been, and we deeply appreciate this sacrifice on their part.

Calgary, AB,   *Michael J. Jacobson, Jr.*
June, 2008   *Hugh C. Williams*

Contents

List of Symbols

We list below, in the order in which they appear in the book, several of the symbols that we frequently use. However, the reader is cautioned that occasionally the same symbol may be used to represent two or more different objects. For example, the symbol (a, b), usually used to denote the greatest common divisor of a and b, is also used to denote the pair a, b. In these cases, the exact meaning of the symbol should be clear through context.

1

Introduction

1.1 Diophantine Equations

A Diophantine equation is an indeterminate equation whose unknowns are only allowed to assume integral* or sometimes rational values. The study of such equations goes back to the ancients; indeed, they are named after Diophantus of Alexandria (c. 200–284 AD) in honour of his work on them.[1]. However, it is most likely that the Greek mathematicians were investigating their properties much earlier than this. To take a simple example, consider the equation

$$x^2 + y^2 = z^2 \,, \tag{1.1}$$

where we constrain a solution (x, y, z) to be a triple of integers.[2] Every student of high school geometry is familiar with the solution $(3, 4, 5)$ and some are even made aware of the additional solutions $(5, 12, 13)$ and $(8, 15, 17)$. In fact, as we shall see below, there exists an infinitude of distinct integral solutions of (1.1) for which $(x, y, z) = 1$.

There are a number of questions that can be asked concerning any particular Diophantine equation.

We might only be interested in whether any solutions exist. For example, the simple Diophantine equation

$$x^2 + y^2 = 3$$

can have no solution in integers, nor can

$$x^2 - 3y^2 = 2 \,.$$

Sometimes a certain Diophantine equation has one or more trivial (obvious) solutions, and we are interested in whether it has any non-trivial solutions. Consider the celebrated equation of Fermat[3]

---

* We will use the term "integer" here to refer to a rational integer, an element of $\mathbb{Z}$.

$$x^n + y^n = z^n \qquad (n > 2) \,.$$

Clearly this has some trivial solutions where $xyz = 0$, but it was not until recently that it was finally shown by Taylor and Wiles that it has no non-trivial solutions.[4]

Some Diophantine equations have a few non-trivial solutions and then have no more solutions. Back in 1942, Ljunggren[5] showed that the only solutions in positive integers of

$$x^2 - 2y^4 = -1$$

are $(x, y) = (1, 1), (239, 13)$. In 1844 Catalan conjectured that the only consecutive powers except the trivial 0 and 1 are 8 and 9; that is, the Diophantine equation

$$x^p - y^q = 1 \qquad (|x| \neq 1; p, q > 1)$$

has only the solutions $(x, y, p, q) = (\pm 3, 2, 2, 3)$. There is an enormous literature on this conjecture, which was finally proved by Mihăilescu[6] in 2002. As a third example we mention the Ramanujan-Nagell equation

$$2^n - 7 = x^2 \,.$$

In 1913, Ramanujan[7] asked if this had any other solution other than those for which $n = 3, 4, 5, 7, 15$. This was answered in the negative by Nagell[8] in 1948.

We might also be interested in whether there are only a finite number of solutions or whether there are infinitely many. In the latter case, can we characterize all solutions? We will now show that (1.1) has an infinitude of solutions and we will show how to characterize all of them.

We begin by observing that if (1.1) is soluble in integers x, y, z, then if d divides any two of x, y, or z, then d must divide the third. Also, if (x, y, z) is a solution of (1.1), then so is (dx, dy, dz) for any integer d. We may therefore only consider *primitive* solutions of (1.1); these are solutions for which $(x, y, z) = 1$. Thus, one of x or y must be odd and the other even; for if they were both odd, then $x^2 + y^2 \equiv 2 \pmod 4$, but z^2 can only be 1 or 0 $\pmod 4$. With no loss of generality we assume that $2 \mid y$ and write

$$y^2 = z^2 - x^2 = (z - x)(z + x) \,.$$

If we put $g = (z - x, z + x)$, then $g \mid 2z$ and $g \mid 2x$, which means that $g \mid (2z, 2x)$ or $g \mid 2(z, x)$. Since (x, y, z) is a primitive solution, we have $(z, x) = 1$; hence, $g \mid 2$. Since $x \equiv z \equiv 1 \pmod 2$, we have $g = 2$. It follows that

$$\left(\frac{y}{2}\right)^2 = \left(\frac{z - x}{2}\right)\left(\frac{z + x}{2}\right) \,,$$

where $((z - x)/2, (z + x)/2) = 1$. Thus, both $(z - x)/2$ and $(z + x)/2$ must be perfect squares. We put $(z - x)/2 = m^2$ and $(z + x)/2 = n^2$ and find that $y = 2mn$, $z = m^2 + n^2$, and $x = n^2 - m^2$, where m and n must have different parity. To verify that this parametric representation of a solution to (1.1) is valid, we simply note the identity

$$(n^2 - m^2) + (2mn)^2 = (m^2 + n^2)^2 . \qquad (1.2)$$

We can now characterize all of the solutions of (1.1) as those given by the form $(d(n^2 - m^2), 2dmn, d(m^2 + n^2))$, where d, m, and n are integers and m and n have the opposite parity. Putting $d = 1, n = 2$, and $m = 1$, we get $(x, y, z) = (3, 4, 5)$.

The Diophantine equation (1.1), sometimes called the Pythagorean equation, is an example of one that is particularly easy to solve; however, it is important to point out that many Diophantine equations are very difficult to solve. Up to the early years of the 20th century there were few, if any, generally applicable techniques that might be successfully employed for this purpose. The study seemed to be little more than a grab bag of mathematical tricks that might or (more likely) might not be useful in solving a particular equation. This situation has changed profoundly during the last half-century, and there are now a number of very deep and powerful methods that can be employed to solve a wide variety of Diophantine equations.[9] However, it must be emphasized that it is a mathematical fact that there cannot exist any general algorithm which can be used for solving all Diophantine equations.[10]

1.2 The Pell Equation

Let us now return to (1.1), but with a small change. We will try to solve, for a given integer a, the Diophantine equation

$$x^2 + ay^2 = z^2 . \qquad (1.3)$$

We note that the following generalization of (1.2) holds:

$$(m^2 - an^2)^2 + a(2mn)^2 = (m^2 + an^2)^2 ;$$

thus, the equation has an infinitude of solutions, but are all of the solutions (x, y, z) characterized by the parametric triple $(d(m^2 - an^2), 2dmn, d(m^2 + an^2))$, where d, m, and n are integers? While we have shown that this is the case for $a = 1$, this is not the case generally. For, consider the simple equation

$$x^2 - 21y^2 = z^2 ,$$

and note that $(5, 1, 2)$ is a solution. This is clearly not included in the set of solutions mentioned earlier.

We will therefore approach this equation in another way. There is no loss of generality in assuming that $a < 0$ and $-a$ is not a perfect square (this would lead us back to the problem of solving (1.1)). We will denote $-a$ by D.

Consider the identity

$$(x^2 - Dy^2)(p^2 - Dq^2) = (xp + Dyq)^2 - D(xq + yp)^2 \, . \qquad (1.4)$$

Thus, if

$$p^2 - Dy^2 = 1 \, , \qquad (1.5)$$

then if (x, y, z) is a solution of

$$x^2 - Dy^2 = z^2 \, , \qquad (1.6)$$

so is

$$(xp + Dyq, xq + yp, z) \, ;$$

that is, we get another solution of (1.6) from a given one, provided we can solve (1.5).

The Diophantine equation

$$T^2 - DU^2 = 1 \qquad (1.7)$$

is called the Pell equation. As a simple example consider

$$T^2 - 7U^2 = 1 \, ,$$

which has the solutions $(\pm 1, 0)$ (trivial solutions), $(\pm 8, \pm 3)$, $(\pm 127, \pm 48)$, etc. This deceptively simple looking Diophantine equation has been the object of study by mathematicians for over two millennia. It is named after John Pell because of an error in attribution by Euler[11] to a method of solving it in "Wallis's works." This was most likely the result of a cursory reading by Euler of Wallis' *Algebra*. As noted by several authorities, most recently Weil,[12] Pell's name occurs frequently in *Algebra*, but never in connection with the Pell equation. In fact, it seems most likely that the method referred to by Euler for solving (1.7) is a technique that Wallis credits to Lord Brouncker. In spite of ample evidence attesting to Euler's carelessness,[13] there have even been relatively recent efforts made to connect[14] Pell with (1.7). This seems to have begun with a misunderstanding of a remark of Hankel,[15] who actually stated, in speaking of the Pell equation, "Pell has done it no other service than to set it forth again in a much read work." The "much read work" is the English translation[16] of the *Teutschen Algebra* of Rahn. However, careful examinations of this work by Konen,[17] Wertheim,[18] and Eneström[19] did not result in the discovery of any mention of (1.7). There can be little doubt that much of this book, particularly pages 100–192, were due to Pell,[20] yet the only mention of anything even resembling the Pell equation in it is the equation

$$x = 12y^2 - z^2 \qquad (1.8)$$

on page 143. This persuaded Whitford[21] that Pell had some acquaintance with (1.7) and seems also to have served as the reason that Pell's biography[22] suggests this may have been the case. However, a thorough inspection of the context in which (1.8) arises reveals that it is to be used to find x after values for y and z have been selected. This, then, can scarcely be regarded as the Pell equation. Thus, there is no evidence whatsoever linking Pell with (1.7). Nevertheless, as Weil[23] asserts, the "traditional designation [of (1.7)] as 'Pell's equation' is unambiguous and convenient." Consequently, it is the term used throughout this work for (1.7), even though it is both historically wrong and unjust to those early individuals who did make important contributions to its study.

The Pell equation has a habit of appearing in a variety of settings, some quite unexpected.[24] Consider the simple problem of finding integers that are both triangular and square. A triangular number simply counts the number of points in a grid in which the first row contains a single point and each subsequent row contains one more point than the previous. Thus, the triangular numbers are $1, 3, 6, 10, 15$, etc. and are given by the formula $x(x+1)/2$. We are therefore searching for those integers x such that $x(x+1)/2 = y^2$ for some integer y. This means that

$$4x^2 + 4x = 8y^2$$

or

$$(2x+1)^2 = 8y^2 + 1 , \tag{1.9}$$

a Pell equation with $D = 8$. In fact, the smallest positive value of x ($\neq 1$) satisfying (1.9) is 8, yielding the square triangular number 36. In fact, as we shall see below, there exists an infinitude of such numbers.

Another simple problem is that of finding integral Pythagorean triangles whose non-hypotenuse sides differ by 1. We already know that these sides must be given by $m^2 - n^2$ and $2mn$ for integral values of m and n. Thus, we must find such m and n for which

$$m^2 - n^2 - 2mn = \pm 1 .$$

This can be rewritten as

$$(m - n)^2 - 2n^2 = \pm 1 ,$$

an example of a Pell equation and an analogous equation with $T^2 - DU^2 = -1$ with $D = 2$. Putting $m - n = 3$ and $n = 2$, we get the triangle with sides $21, 20, 29$.

As another, less simple example, we mention the surprising occurrence of the Pell equation in Lehmer's[25] parameterization of solutions of the cubic Diophantine equation $x^3 + y^3 + z^3 = 1$.

We will now show that (1.7) always has a non-trivial ($U \neq 0$) solution when D is positive and not a perfect square. We first require a simple lemma.

Lemma 1.1. *Let s be any positive integer. Integers t and u always exist such that*

$$|t - u\sqrt{D}| < \frac{1}{s} \leq \frac{1}{|u|} .$$

Proof. For each integer u such that $0 \le u \le s$, put $t = \lceil u\sqrt{D} \rceil$. Then for each such (t, u) pair we have

$$0 < t - u\sqrt{D} < 1 .$$

If we divide the interval between 0 and 1 into s subintervals, each of length $1/s$, we see by the box principle that two of the above $s+1$ pairs, say (t_1, u_1) and (t_2, u_2), must be such that $t_1 - u_1\sqrt{D}$ and $t_2 - u_2\sqrt{D}$ lie within the same interval. Since $u_1 \ne u_2$, we see that $t_1 - u_1\sqrt{D}$ and $t_2 - u_2\sqrt{D}$ are distinct and

$$\frac{-1}{s} < t_1 - u_1\sqrt{D} - (t_2 - u_2\sqrt{D}) < \frac{1}{s}$$

or

$$|t_1 - t_2 - (u_1 - u_2)\sqrt{D}| < \frac{1}{s} .$$

Also, since $|u_1 - u_2| \le s$, we get

$$|t_1 - t_2 - (u_1 - u_2)\sqrt{D}| < \frac{1}{s} \le \frac{1}{|u_1 - u_2|} .$$

$\square$

Corollary 1.2. *There exists an infinitude of pairs of integers (t, u) such that*

$$|t - u\sqrt{D}| < \frac{1}{|u|} .$$

Proof. Suppose there exists only a finite set S of such pairs. Then there must exist some minimal integer M such that

$$\frac{1}{M} < \min\{t - u\sqrt{D} : (t, u) \in S\} .$$

By the lemma there must exist integers t' and u' such that

$$|t' - u'\sqrt{D}| < \min\left\{ \frac{1}{M}, \frac{1}{|u'|} \right\} .$$

Since $|t' - u'\sqrt{D}| < 1/|u'|$, we have $(t', u') \in S$. However, since

$$|t' - u'\sqrt{D}| < \frac{1}{M} < \min\{t - u\sqrt{D} : (t, u) \in S\} ,$$

this is impossible. $\square$

We will now use S to denote the infinite set of all pairs (t, u) such that

$$|t - u\sqrt{D}| < \frac{1}{|u|} .$$

Theorem 1.3. *The Pell equation always has at least one non-trivial solution.*

Proof. If $(t, u) \in S$, then

$$|t + u\sqrt{D}| \leq |t - u\sqrt{D}| + |2u\sqrt{D}| < \frac{1}{|u|} + 2|u|\sqrt{D} .$$

Hence,

$$|t^2 - Du^2| = |t - Du||t + u\sqrt{D}| < \frac{1}{|u|} \left(\frac{1}{|u|} + 2|u|\sqrt{D} \right)$$

$$= \frac{1}{u^2} + 2\sqrt{D}$$

$$\leq 1 + 2\sqrt{D} .$$

Thus, for all $(t, u) \in S$, we have $|t^2 - Du^2| < 1 + 2\sqrt{D}$. Since $1 + 2\sqrt{D}$ is fixed, we must, again by the box principle, have an infinitude of pairs $(t, u) \in S$ such that

$$t^2 - Du^2 = k$$

for some fixed $k \in \mathbb{Z}$ with $|k| < 1 + 2\sqrt{D}$. It must also be the case that there exists an infinitude of these pairs for which both the t values and the u values are the same modulo k. Let (t_1, u_1) and (t_2, u_2) be two such pairs where $t_1 \neq \pm t_2$ and $u_1 \neq \pm u_2$. By (1.4) we see that

$$(t_1 t_2 - Du_1 u_2)^2 - D(t_1 u_2 - t_2 u_1)^2 = k^2 .$$

Now, $t_1 u_2 - t_2 u_1 \equiv 0 \pmod{k}$; hence, $t_1 t_2 - Du_1 u_2 \equiv 0 \pmod{k}$ and

$$\left(\frac{t_1 t_2 - Du_1 u_2}{k} \right)^2 - D \left(\frac{t_1 u_2 - t_2 u_1}{k} \right)^2 = 1 .$$

Since $(t_1 t_2 - Du_1 u_2)/k, (t_1 u_2 - t_2 u_1)/k \in \mathbb{Z}$, we have a non-trivial solution of (1.7) as long as $t_1 u_2 - t_2 u_1 \neq 0$. However, if $t_1 u_2 - t_2 u_1 = 0$, then $t_1 t_2 - Du_1 u_2 = \pm k$, and these are two equations which can be simultaneously satisfied only if $t_1 = \pm t_2$ and $u_1 = \pm u_2$, possibilities that we have already excluded. □

Theorem 1.3 is a most remarkable result because if, for example, we started to conduct an exhaustive search for the smallest value of T for which

$$T^2 - 1621U^2 = 1 ,$$

we would go a very long way before we found such a T with $U \neq 0$. Indeed, we would likely become convinced that there is no such value of T; this is because the least such value for T is a number of 76 decimal digits. Nevertheless, we know that this equation does have a non-trivial solution, but the proof of Theorem 1.3 gives us no information as to how to determine T and U. Much of the rest of this book will be devoted to developing results that provide this kind of information.

1.3 Representation of All Solutions

We have show that (1.7) always has at least one non-trivial solution. In this section we will characterize all of the non-trivial solutions of the Pell equation. Indeed, as we will need slightly more general results later, we will examine the solutions of the Diophantine equation

$$X^2 - DY^2 = 4\sigma, \qquad \sigma \in \{-1, 1\}. \tag{1.10}$$

Evidently, if X, Y, and σ satisfy (1.10), then $X \equiv DY \pmod{2}$. Also, in the case of $X \equiv DY \equiv 0 \pmod{2}$ and $\sigma = 1$, there are two possible subcases. When $X \equiv Y \equiv 0 \pmod{2}$, we get (1.7) with $T = X/2, U = Y/2$. When $X \equiv 0 \pmod{2}$ and $Y \equiv 1 \pmod{2}$, then we must have $D \equiv 0 \pmod{4}$. We then see that if we put $T = X/2$ and $U = Y$, we again obtain (1.7) on replacing D by $D/4$. Of course, by Theorem 1.3, (1.10) always has a non-trivial solution. We will now show that it has an infinitude of non-trivial solutions. We will use (X, Y, σ) to denote any integral solution of (1.10).

Theorem 1.4. *If (x_1, y_1, σ_1) and (x_2, y_2, σ_2) are solutions of (1.10) where we do not have $x_1 = \eta x_2$, $y_1 = -\eta y_2$ for $\eta \in \{-1, 1\}$, then (x_3, y_3, σ_3) is a solution of (1.10), where*

$$x_3 = \frac{x_1 x_2 + D y_1 y_2}{2}, \qquad y_3 = \frac{x_1 y_2 + x_2 y_1}{2}, \qquad \sigma_3 = \sigma_1 \sigma_2.$$

Proof. By (1.4) it suffices to show that

$$x_1 y_2 + x_2 y_1 \equiv x_1 x_2 + D y_1 y_2 \equiv 0 \pmod{2}$$

and $y_3 \neq 0$. The first of these follows easily from the observations that $x_1 \equiv Dy_1$, $x_2 \equiv Dy_2 \pmod{2}$. Also, if $y_3 = 0$, then since $x_1 = -x_2 y_1/y_2$ and $x_3^2 = 4\sigma_1\sigma_2$, we get $x_3 = \pm 2$. If we put $\eta = x_1/x_2$, then $y_1/y_2 = -\eta$ and we find from $x_3 = \pm 2$, that $\eta(x_2^2 - Dy_2^2) = \pm 4$. Since $x_2^2 - Dy_2^2 = 4\sigma_2$, we see that $|\eta| = 1$, a case we have excluded. □

Remark 1.5. If x_1, x_2, x_3, y_1, y_2, and y_3 are defined as in Theorem 1.4, then if $\lambda_1 = (x_1 + y_1\sqrt{D})/2$, and $\lambda_2 = (x_2 + y_2\sqrt{D})/2$, we have

$$\lambda_1 \lambda_2 = \frac{x_3 + y_3\sqrt{D}}{2}.$$

Thus, if (x_1, y_1, σ_1) is any solution of (1.10), we can produce an infinitude of solutions of (1.10) as $(x_n, y_n, \sigma_1{}^n)$ $(n = 1, 2, 3, \dots)$, where

$$\frac{x_n + y_n\sqrt{D}}{2} = \lambda_1{}^n.$$

These must all be distinct because if $(x_m, y_m, \sigma_1{}^m) = (x_n, y_n, \sigma_1{}^n)$ for $n > m$, then $\lambda_1{}^n = \lambda_1{}^m$ and $\lambda_1{}^{n-m} = 1$. However, this means that $y_{n-m} = 0$, which is impossible by Theorem 1.4.

We will next show how all the solutions of (1.10) can be generated. We will need some preliminary results.

Lemma 1.6. *If (x, y, σ) is a solution of (1.10), then $x + y\sqrt{D} > 2$ if and only if $x > 0, y > 0$.*

Proof. Certainly, if $x, y > 0$, then $x + y\sqrt{D} \geq 1 + \sqrt{D} > 2$. Suppose $x + y\sqrt{D} > 2$. Since

$$(x + y\sqrt{D})(x - y\sqrt{D}) = 4\sigma ,$$

we get

$$\frac{|x - y\sqrt{D}|}{2} = \frac{2}{x + y\sqrt{D}} < 1 .$$

Hence, $-2 < x - y\sqrt{D} < 2$. Since $x + y\sqrt{D} > 2$, it follows that $x, y > 0$. $\square$

Lemma 1.7. *If (x, y, σ) is a solution of (1.10) and $x, y > 0$, then $2yD \geq 8$. If $x > 0$ and $y > 1$, then $2x + 1 + (2y - 1)D > 8$.*

Proof. Clearly both inequalities hold for $D \geq 8$. Suppose $D < 8$; since D is not a square we can only have $D = 2, 3, 5, 6, 7$. If $D = 2, 3, 6, 7$, then $2 \mid x$ and $2 \mid y$, which means that $x \geq 2, y \geq 2$ and both inequalities are satisfied. If $D = 5$, then $2yD \geq 10 > 8$, and if $x \geq 1, y > 1$, then $2x + 1 + (2y - 1)D > 8$. $\square$

Theorem 1.8. *Suppose (x_1, y_1, σ_1) and (x_2, y_2, σ_2) are solutions of (1.10). If $x_1, x_2, y_1, y_2 > 0$, we have*

$$x_2 + y_2\sqrt{D} > x_1 + y_1\sqrt{D}$$

if and only if $x_2 > x_1$ and $y_2 \geq y_1$.

Proof. It is evident that if $x_2 > x_1$ and $y_2 \geq y_1$, then $x_2 + y_2\sqrt{D} > x_1 + y_1\sqrt{D}$. Suppose now that $x_2 + y_2\sqrt{D} > x_1 + y_1\sqrt{D}$. We distinguish two cases.

CASE 1: $x_1 - y_1\sqrt{D} > 0$:

In this case we have

$$\frac{x_1 - y_1\sqrt{D}}{2} = \frac{2}{x_1 + y_1\sqrt{D}} > \frac{2}{x_2 + y_2\sqrt{D}} = \frac{|x_2 - y_2\sqrt{D}|}{2} .$$

Hence,

$$\frac{y_1\sqrt{D} - x_1}{2} < \frac{x_2 - y_2\sqrt{D}}{2} < \frac{x_1 - y_1\sqrt{D}}{2} .$$

Since $x_2 + y_2\sqrt{D} > x_1 + y_1\sqrt{D}$, we see that

$$-y_2\sqrt{D} = \frac{-x_2 - y_2\sqrt{D}}{2} + \frac{x_2 - y_2\sqrt{D}}{2}$$

$$< \frac{-x_1 - y_1\sqrt{D}}{2} + \frac{x_1 - y_1\sqrt{D}}{2} = -y_1\sqrt{D}$$

and $y_2 > y_1$. Since $y_2 \geq y_1 + 1$, we have $Dy_2{}^2 > Dy_1{}^2 + 2Dy_1$. It follows that

$$x_2{}^2 = Dy_2{}^2 + 4\sigma_2 > Dy_1{}^2 + 2Dy_1 + 4\sigma_2 = x_1{}^2 - 4\sigma_1 + 4\sigma_2 + 2Dy_1$$
$$\geq x_1{}^2 + 2Dy_1 - 8 \geq x_1{}^2$$

and $x_2 > x_1$.

CASE 2: $x_1 - y_1\sqrt{D} < 0$:
    We have

$$\frac{y_1\sqrt{D} - x_1}{2} = \frac{2}{x_1 + y_1\sqrt{D}} > \frac{2}{x_2 + y_2\sqrt{D}} = \frac{|x_2 - y_2\sqrt{D}|}{2} ,$$

which means that

$$x_2 = \frac{x_2 - y_2\sqrt{D}}{2} + \frac{x_2 + y_2\sqrt{D}}{2} > \frac{x_1 - y_1\sqrt{D}}{2} + \frac{x_1 + y_1\sqrt{D}}{2} = x_1 .$$

If $y_2 < y_1$, then $y_2 \leq y_1 - 1$ and

$$4\sigma_2 \geq (x_1 + 1)^2 - D(y_1 - 1)^2 = 4\sigma_1 + 2x_1 + 1 + (2y_1 - 1)D .$$

Since $y_2 > 0$, we have $y_1 > 1$ and, by Lemma 1.7,

$$2x_1 + 1 + (2y_1 - 1)D > 8 \geq 4\sigma_2 - 4\sigma_1 \geq 2x_1 + 1 + (2y_1 - 1)D ,$$

a contradiction. Hence, $y_2 \geq y_1$.

$\square$

We know that (1.10) has a solution $(x, y, 1)$, where $x \equiv y \equiv 0 \pmod 2$, and without any loss of generality we may assume that $x, y > 0$. Now, suppose we have another solution (x_1, y_1, σ_1) of (1.10) with $x_1, y_1 > 0$ and $x_1 + y_1\sqrt{D} < x + y\sqrt{D}$. By Theorem 1.8, we know that $x_1 < x$ and $y_1 \leq y$. These bounds dictate that there can only be a finite number of possibilities for (x_1, y_1, σ_1) and therefore only a finite number of possibilities for $x_1 + y_1\sqrt{D}$. We can therefore define a unique solution (x_1, y_1, σ_1) of (1.10) for which $x_1 + y_1\sqrt{D}$ exceeds 2 and is least. We call this solution of (1.10) the *fundamental solution* and we put $\epsilon = (x_1 + y_1\sqrt{D})/2$.

Theorem 1.9. *If (x', y', σ') is any solution of (1.10), then*

$$\eta = (x' + y'\sqrt{D})/2 = \pm\epsilon^n$$

for some $n \in \mathbb{Z}$.

Proof. Since

$$\left(\frac{x' - y'\sqrt{D}}{2}\right)\left(\frac{x' + y'\sqrt{D}}{2}\right) = \sigma' ,$$

we see that one and only one of η, $-\eta$, η^{-1}, or $-\eta^{-1}$ is greater than 1. Denote this by γ $(= (|x'| + |y'|\sqrt{D})/2$ by Lemma 1.6). Since $\gamma > 1$ and $\epsilon > 1$, there must exist some non-negative $n \in \mathbb{Z}$ such that

$$\epsilon^n \leq \gamma < \epsilon^{n+1} .$$

If $\gamma = \epsilon^n$, we are done because $n \neq 0$ and $\eta \in \{\gamma, -\gamma, \gamma^{-1}, -\gamma^{-1}\}$. If $\gamma \neq \epsilon^n$, then

$$1 < \gamma\epsilon^{-n} < \epsilon .$$

Since $\epsilon(x_1 - y_1\sqrt{D})/2 = \sigma_1$, we have

$$\epsilon^{-n} = \sigma_1{}^n \left(\frac{x_1 - y_1\sqrt{D}}{2} \right)^n .$$

Since $(x_1, -y_1, \sigma_1)$ is a solution of (1.10), it follows from Remark 1.5

$$\lambda = \gamma\epsilon^{-n} = \frac{x_2 + y_2\sqrt{D}}{2}$$

for some $x_2, y_2 \in \mathbb{Z}$ with $x_2 \equiv Dy_2 \pmod{2}$. Also, since $\lambda > 1$, by Lemma 1.6 we have $x_2, y_2 > 0$ and by Theorem 1.4 we have

$$x_2{}^2 - Dy_2{}^2 = 4\sigma'\sigma_1{}^n .$$

Hence, $(x_2, y_2, \sigma'\sigma_1{}^n)$ is a solution of (1.10) and $1 < \lambda < \epsilon$. By selection of ϵ, this is impossible.   □

Let (t, u) be that solution of the Pell equation (1.7) for which $t, u > 0$ and $t + u\sqrt{D}$ is least. We call this the *fundamental solution* of (1.7).

Corollary 1.10. *Let (t, u) be the fundamental solution of the Pell equation. If*

$$T^2 - DU^2 = 1 \quad and \quad T, U > 0 ,$$

then

$$T + U\sqrt{D} = (t + u\sqrt{D})^n$$

for some positive integer n.

Proof. By the theorem we know that

$$t + u\sqrt{D} = \epsilon^k \tag{1.11}$$

and

$$T + U\sqrt{D} = \epsilon^m$$

for $k, m \in \mathbb{Z}$. Since (t, u) is the fundamental solution of (1.7), we must have $m > k > 0$. Let

$$m = nk + r ,$$

where $n > 0$, $0 \leq r < k$. Then,

$$T + U\sqrt{D} = (t + u\sqrt{D})^n \epsilon^r$$

and

$$1 \leq \epsilon^r = (T + U\sqrt{D})(t - u\sqrt{D})^n \ .$$

Since $\epsilon^r = (x' + y'\sqrt{D})/2$, for some $x', y' \in \mathbb{Z}$, we see by the preceding equation, that if $r > 0$, we must have $x' \equiv y' \equiv 0 \pmod{2}$ and

$$\left(\frac{x'}{2}\right)^2 - D\left(\frac{y'}{2}\right)^2 = 1 \ .$$

Since $r < k$, we have $\epsilon^r < \epsilon^k = t + u\sqrt{D}$, but this contradicts the definition of the solution (t, u). Hence, $r = 0$ and

$$T + U\sqrt{D} = (t + u\sqrt{D})^n \ .$$

$\square$

Remark 1.11. Notice that if (T, U) is any solution of the Pell equation, then

$$T + U\sqrt{D} = \pm(t + u\sqrt{D})^n$$

for some $n \in \mathbb{Z}$ and proper choice of sign.

Thus, we can characterize all of the solutions of the Pell equation or its more general form (1.10) once we know the corresponding fundamental solution.

We should emphasize that it is not always possible to solve (1.10) for certain preselected values of σ or $X \pmod{2}$. For example, there is no integer solution of $X^2 - DY^2 = -4$ when $D \equiv -1 \pmod{4}$. Also, there is no integer solution of $X^2 - DY^2 = 4$ for $X \equiv Y \equiv 1 \pmod{2}$ when $D = 11$. Thus, when we speak of a fundamental solution of (1.10), we do not necessarily know the value of σ_1 and $x_1 \pmod{2}$ a priori. However, as we have seen, there is always a fundamental solution of (1.10) as long as we do not preselect σ and $X \pmod{2}$.

We next attack the problem of determining k in (1.11). We note that if $2 \mid x_1$ and $\sigma_1 = 1$, then $2 \mid y_1$, or if $2 \nmid y_1$, then $4 \mid D$. In the first case, $\epsilon = t + u\sqrt{D}$; in the second, it is easy to see that $t + u\sqrt{D} = \epsilon^2$. If $2 \mid x_1$ and $\sigma_1 = -1$, then $t + u\sqrt{D} \neq \epsilon$, but since $\sigma_1{}^2 = 1$, we have $t + u\sqrt{D} = \epsilon^2$. If $2 \nmid x_1$, then $2 \nmid y_1$, and if we define (x_i, y_i) by

$$\frac{x_i + y_i\sqrt{D}}{2} = \epsilon^i \ ,$$

we see that $2 \nmid y_2$ $(y_2 = x_1 y_1)$ and $2 \nmid y_4$ $(y_4 = x_2 y_2, \ x_2 = x_1^2 - 2\sigma_1)$. However, since

$$x_3 = (x_1{}^3 + 3y_1{}^2 x_1 D)/2 = 2x_1(\sigma_1 + y_1{}^2 D) \, ,$$
$$y_3 = (3x_1{}^2 y_1 + y_1{}^3 D)/2 = 2y_1(x_1{}^2 - \sigma_1) \, ,$$

we see that $(x_3/2, y_3/2)$ is a solution of

$$T^2 - DU^2 = \sigma_1 \, .$$

Thus, if $\sigma_1 = 1$, then $k = 3$; if $\sigma_1 = -1$, then $k \neq 4$ because $2 \nmid y_4$ and $k \neq 5$ because $x_5{}^2 - Dy_5{}^2 = -4$. Hence, since $2 \mid x_6$ and $2 \mid y_6$ and $\sigma_1{}^6 = 1$, we see that $k = 6$ when $\sigma_1 = -1$ and $2 \nmid x_1$. We summarize these results in Table 1.1.

Table 1.1. Value of k in (1.11)

$x_1 \pmod 2$	$y_1 \pmod 2$	σ_1	k
0	0	1	1
0	1	1	2
0	—	-1	2
1	—	1	3
1	—	-1	6

We remark that it is easy to determine σ_1 when we know the values of $x_1, y_1 \pmod 8$ and that of $D \pmod{16}$.

1.4 The Lucas Functions

In this section we will briefly discuss the Lucas functions. We will not provide proofs of their many properties, as these can easily be found in the literature.[26] These functions are useful in characterizing all of the solutions of (1.7) and (1.10) once a fundamental solution is known.

We let P and Q be coprime integers and let α and β denote the zeros of $x^2 - Px + Q$. The Lucas functions are defined as follows:

$$u_n = u_n(P, Q) = (\alpha^n - \beta^n)/(\alpha - \beta) \, ,$$
$$v_n = v_n(P, Q) = \alpha^n + \beta^n \, .$$

The properties of these functions have been investigated for well over a century, and as a result, a great deal of literature concerning them has accumulated.[27] We will assume here that $d = (\alpha - \beta)^2 = P^2 - 4Q \neq 0$. Note that

$$v_n^2 - du_n^2 = 4Q^n \, .$$

If we put $\alpha = \epsilon = (x_1 + y_1\sqrt{D})/2$ and $\beta = (x_1 - y_1\sqrt{D})/2$, then $P = \alpha + \beta = x_1$ and $Q = \sigma_1$. Also, $\alpha - \beta = y_1\sqrt{D}$ and $d = y_1{}^2 D$. If we define x_n and y_n by

$$\frac{x_n + y_n\sqrt{D}}{2} = \epsilon^n\,,$$

then

$$x_n = \alpha^n + \beta^n = v_n(x_1, \sigma_1)\,,$$

$$y_n = \frac{y_1(\alpha^n - \beta^n)}{\alpha - \beta} = y_1 u_n(x_1, \sigma_1)\,.$$

Also, as noticed by Lehmer,[28] if we put $\alpha = t + u\sqrt{D}$, $\beta = t - u\sqrt{D}$, then for $T_n + U_n\sqrt{D} = (t + u\sqrt{D})^n$, we get

$$2T_n = v_n(2t, 1)\,,$$

$$U_n = u u_n(2t, 1)\,.$$

Thus, the successive solutions to (1.7) or (1.10) can be easily characterized in terms of v_n and u_n and many properties of x_n and y_n or T_n and U_n can be derived from those of the Lucas functions. In particular, there are several results concerning the divisibility properties of u_n that will be useful in the sequel.

We begin by noting that if $n \mid m$, then $u_n \mid u_m$; that is, the sequence $\{u_n\}$ is a divisibility sequence.

Definition 1.12. *Let $n = 2^\nu m$, where $2 \nmid m$. Then if $4 \mid k$ or $k \equiv 1 \pmod 4$, the Kronecker symbol $(\frac{k}{n})$ is defined as*

$$\left(\frac{k}{n}\right) = \left(\frac{k}{2}\right)^\nu \left(\frac{k}{m}\right)\,,$$

where $(\frac{k}{m})$ is the Jacobi symbol and

$$\left(\frac{k}{2}\right) = \begin{cases} 0 & when \quad 4 \mid k \\ 1 & when \quad k \equiv 1 \pmod 8 \\ -1 & when \quad k \equiv 5 \pmod 8\,. \end{cases}$$

The *Law of Apparition* concerning $\{u_n\}$ states that if p is any prime such that $p \nmid Q$, then $p \mid u_{p-\delta}$, where δ is the Kronecker symbol $(\frac{k}{n})$. The *Law of Repetition* concerning $\{u_n\}$ states that if $p^u \parallel u_n$, then $p^{u+\nu} \mid u_{p^\nu n}$, and if $p^u \neq 2$, then $p^{u+\nu} \parallel u_{p^\nu n}$ when $p \nmid n$. Thus, if $(m, Q) = 1$, we see that there is some n such that $m \mid u_n$. We call the least positive value of n the *rank of apparition* of m and denote this by $\omega(m)$, or ω.

It is known that $m \mid u_n$ for some n if and only if $\omega(m) \mid n$. Thus, by combining these results we see that if $(m, Q) = 1$, then

$$\omega(m) \mid \phi_d(m) \, ,$$

where we define $\phi_d(m)$ by

$$\phi_d(m) = \prod_{p^\alpha \| m} p^{\alpha-1} \left(p - \left(\frac{d}{p} \right) \right) .$$

Indeed,

$$\omega(m) \mid \operatorname*{lcm}_{p^\alpha \| m} \left[p^{\alpha-1} \left(p - \left(\frac{d}{p} \right) \right) \right] . \tag{1.12}$$

Notice that if d is a perfect square, then $\phi_d(m)$ is Euler's totient function $\phi(m)$ when $(m, d) = 1$.

By using these results, we can show that when $(m, Q) = 1$, the value of $\omega(m)/m$ can be bounded.

Theorem 1.13. *If* $(m, Q) = 1$, *then*

$$\omega(m) \le \left(\frac{2\omega(2)}{3} \right) m \le 2m \, .$$

Proof. Let $m = 2^k n_1 n_2$, where $(2, n_1 n_2) = 1$, $(n_2, d) = 1$, and any prime divisor of n_1 must divide d. We have

$$\omega(m) \mid \operatorname{lcm}\left[\omega(2^k), \omega(n_1), \omega(n_2) \right]$$

and, therefore,

$$\omega(m) \le \omega(2^k) \omega(n_1) \omega(n_2) \, . \tag{1.13}$$

If $p \mid d$, then it is easy to prove that $p \mid u_p$; hence, by the Law of Repetition, $\omega(n_1) \mid n_1$; also, $\omega(2^k) \mid 2^{k-1} \omega(2)$. If $n_2 = p^j$ for some prime p, then

$$\omega(n_2) \le p^{j-1}(p+1) \le \frac{4}{3} p^j = \frac{4}{3} n_2 \, .$$

If n_2 is not a prime power, then by (1.12),

$$\omega(n_2) \le \frac{\phi_d(n_2)}{2^{t-1}} \, ,$$

where $t \, (\ge 2)$ is the number of distinct primes dividing n_2. Since[29]

$$\phi_d(n_2)/2^{t-1} < n_2 \, ,$$

the result follows from (1.13).   □

Now, suppose that $D = f^2 D_0$, where D_0 has no integer square divisor. Let (t, u) be the fundamental solution of

$$T^2 - D_0 U^2 = 1 \tag{1.14}$$

and put $g = (f, u)$. Let (t', u') be the fundamental solution of

$$T^2 - DU^2 = 1 .$$

Since this equation can also be written as

$$T^2 - D_0(fU)^2 = 1 ,$$

we see that

$$t' + fu'\sqrt{D_0} = (t + u\sqrt{D_0})^n = T_n + U_n\sqrt{D_0}$$

for some minimal $n \in \mathbb{Z}^{>0}$; that is,

$$fu' = U_n = uu_n(2t, 1)$$

or $f/g \mid u_n(2t, 1)$. This means that $n = \omega(f/g)$, $t' = t_n$ and $u' = u_n/f$. Since all of the possible values for n must be divisors of an easily computed integer $\phi_d(f/g)$, it is not a difficult problem to determine (t', u') once (t, u) is known, provided we have the complete factorization of f. Thus, the problem of finding a fundamental solution of (1.7) can be reduced to that of finding a fundamental solution of (1.14) when we have a complete factorization of f.

Notes and References

[1] Very little is known about the life of Diophantus Even the dates given here are to some degree conjectural and have been disputed. See [Kno93]. Diophantus' work, *Arithmetica*, can be found in [Hea64].

[2] This equation has a very long history which may extend back to Pythagoras. See [Dic19, Vol. II, p. 165].

[3] This problem is posed in a marginal note in Fermat's copy of Diophantus' *Arithmetica*. Fermat claimed to have a proof that this equation has only trivial solutions but was unable to include it because the margin was too small to contain it. The note itself is believed to have been written about 1630. Few today believe that Fermat had a correct proof.

[4] See [Wil95a] and [TW95]. For a lively discussion of the proof of this result, see [vdP96].

[5] See [Lju42], [ST91], and [Che94].

[6] [Mih03] and [Mih04]. For a discussion of this result, see [Bil04] and [Bil05].

[7] [Ram00], p. 327.

[8] [Nag48] and [Nag61].

[9] See, for example, [Sma98], [Spr93], and [ST86].

[10] This is essentially the substance of Hilbert's Tenth Problem: Given any Diophantine equation, devise a process according to which it can be decided in a finite number of operations whether the equation is solvable in integers. This problem was solved in the negative by Matiyasevich in 1970. For a full account of this, the reader is referred to [Mat93].

[11] [Eul43].

[12] [Wei84], p. 174.

[13] [Whi12], p. 59 footnotes.

[14] See, for example, [Sco38], p. 268.

[15] [Han65], p. 203.

[16] [Rah68].

[17] [Kon01], pp. 33–34, footnote 1.

[18] [Wer02].

[19] [Ene02].

[20] See [Scr74].

[21] [Whi12], p. 2.

[22] [ACL91], pp. 1973–1975.

[23] [Wei84], p. 174.

[24] Several of these are mentioned in Chapter 8 of [Mor69].

[25] [Leh56].

[26] Much information concerning the properties of these functions, together with proofs, can be found in [Rib96] and [Wil98].

[27] See [Wil98].

[28] [Leh26b] and [Leh28].

[29] See, for example, Lemma 4.3.11 of [Wil98].

Early History of the Pell Equation

2.1 The Cattle Problem of Archimedes

This chapter is devoted to various aspects of the history of the Pell equation before the work of Lagrange. As this topic has already been dealt with in some detail by Konen,[1] Whitford,[2] and Dickson,[3] our discussion here will be brief. We will concentrate on providing a more modern historical perspective and a somewhat different presentation of this material than that given in these earlier works.

In 1773, the poet and literary critic Gotthold Ephraim Lessing (1729–1781) published[4] a Greek epigram which he had edited from an Arabic manuscript in the Herzog-August Library in Wolfenbüttel in northern Germany. The text of this epigram consists of a heading, followed by a poem of 44 lines made up of 22 elegiac distichs, a scholium giving a (false) solution, and a lengthy analysis of the problem by Chr. Leiste. There has been some controversy[5] concerning the exact translation of the heading, but it seems that Fraser's version,[6] given belown, is about as accurate as can be expected.

> A problem which Archimedes set in epigrammatic form and sent to those interested in these matters in Alexandria, in the letter addressed to Eratosthenes of Cyrene.

The most frequently cited translation of the problem itself is that of Thomas.[7]

> If thou art diligent and wise, O stranger, compute the number of cattle of the Sun, who once upon a time grazed on the fields of the Thrinacian isle of Sicily, divided into four herds of different colours, one milk white, another a glossy black, the third yellow and the last dappled. In each herd were bulls, mighty in number according to these proportions: Understand, stranger, that the white bulls were equal to a half and a third of the black together with the whole of the yellow, while the black were equal to the fourth part of the dappled and a fifth, together with, once more, the whole of the yellow. Observe

further that the remaining bulls, the dappled, were equal to a sixth part of the white and a seventh, together with all the yellow. These were the proportions of the cows: The white were precisely equal to the third part and a fourth of the whole herd of the black; while the black were equal to the fourth part once more of the dappled and with it a fifth part, when all, including the bulls went to pasture together. Now the dappled in four parts[8] were equal in number to a fifth part and a sixth of the yellow herd. Finally the yellow were in number equal to a sixth part and a seventh of the white herd. If thou canst accurately tell, O stranger, the number of cattle of the Sun, giving separately the number of well-fed bulls and again the number of females according to each colour, thou wouldst not be called unskilled or ignorant of numbers, but not yet shall thou be numbered among the wise. But come, understand also all these conditions regarding the cows of the Sun. When the white bulls mingled their number with the black, they stood firm, equal in depth and breadth, and the plains of Thrinacia, stretching far in all ways, were filled with their multitude. Again, when the yellow and the dappled bulls were gathered into one herd they stood in such a manner that their number, beginning from one, grew slowly greater till it completed a triangular figure, there being no bulls of other colours in their midst nor none of them lacking. If thou art able, O stranger, to find out all these things and gather them together in your mind, giving all the relations, thou shalt depart crowned with glory and knowing that thou hast been adjudged perfect in this species of wisdom.

Recently, a charming translation[9] by Hillion and Lenstra has appeared, which possesses much of the light-hearted spirit of the original.

This problem is referred to in a scholium to Plato's Charmides as being called the Cattle Problem by Archimedes. It may also have been mentioned in some work of Cicero.[10] Since Krumbiegel's[11] criticism of this work in 1880, it has been customary to regard the problem, now called the Cattle Problem, as most likely having originated with Archimedes (c. 287–212 BC), but the poem itself as a Hellenistic fabrication. However, Fraser[12] has argued very convincingly that we should also accept Archimedes as the author of the poetical form of the problem, and there seeems to be no good reason to dispute this judgement.

The problem is to find the numbers W, X, Y, and Z of the white, black, dappled, and yellow bulls, respectively and the numbers w, x, y, and z of the cows of corresponding colours. We can now write the equations which these quantities satisfy as

$$W = \left(\frac{1}{2} + \frac{1}{3}\right) X + Z , \qquad (2.1)$$

$$X = \left(\frac{1}{4} + \frac{1}{5}\right) Y + Z \,, \tag{2.2}$$

$$Y = \left(\frac{1}{6} + \frac{1}{7}\right) W + Z \,, \tag{2.3}$$

$$w = \left(\frac{1}{3} + \frac{1}{4}\right) (X + x) \,, \tag{2.4}$$

$$x = \left(\frac{1}{4} + \frac{1}{5}\right) (Y + y) \,, \tag{2.5}$$

$$y = \left(\frac{1}{5} + \frac{1}{6}\right) (Z + z) \,, \tag{2.6}$$

$$z = \left(\frac{1}{6} + \frac{1}{7}\right) (W + w) \,, \tag{2.7}$$

$$W + X = \square \,, \tag{2.8}$$

$$Y + Z = \triangle \,. \tag{2.9}$$

Leiste[13] found integral solutions of (2.1), (2.2), and (2.3) as

$$Y = 1580m \,, \quad Z = 891m \,, \quad W = 2226m \,, \quad X = 1602m \,, \tag{2.10}$$

where m is an integer parameter. (This is simply linear algebra.) He then went on to find solutions to (2.1)–(2.7) for the unknowns that were all 20 times larger than they might be. However,[14] if we multiply (2.4) by 4800, (2.5) by 2800, (2.6) by 1260, and (2.7) by 462 and add, we get

$$4657w = 2800X + 1260Y + 462Z + 143W \,.$$

By using (2.10), we find that $m = 4657n$, for an integer parameter n. From this and (2.4)–(2.7), we find that

$$\begin{aligned}
W &= 10366482n \,, & X &= 7460514n \,, \\
Y &= 7358060n \,, & Z &= 4149387n \,, \\
w &= 7206360n \,, & x &= 4893246n \,, \\
y &= 3515820n \,, & z &= 5439213n \,.
\end{aligned} \tag{2.11}$$

Since the coefficients of n have greatest common divisor 1, (2.11) represents all of the possible solutions of (2.1)–(2.7). As mentioned earlier, Leiste gave a solution with $n = 20$ and the scholium,[15] with no explanation, gives a solution for $n = 80$. Neither of these satisfies (2.8) or (2.9).

It remains to consider (2.8) and (2.9). Since $W + X$ must be a square and

$$W + X = 4 \times 957 \times 4657n \,,$$

we must have $n = 957 \cdot 4657U^2 = 4456749U^2$. Also, $Y + Z = V(V + 1)/2$ means that

$$T^2 = 8(Y + Z) + 1 = DU^2 + 1 ,$$

where $T = 2V + 1$ and

$$D = 410286423278424 . \tag{2.12}$$

Thus, in order to solve the Cattle Problem, we must solve the Pell equation (1.7) of Chapter 1 with D given by (2.12). We will discuss in the next chapter (Example 3.10) how this problem can be solved.

There has been some dispute[16] about the exact wording of the Cattle Problem, but no significant changes to it have been met with acceptance by modern scholars. Some doubt has been expressed concerning whether the second part of the problem actually reduces to a Pell equation. This has to do with whether to interpret the text of the problem as asking for $W + X$ to be an integral square or whether the bulls when packed together should fill a square. As a bull is longer than it is broad, the latter reading would simply ask that $W + X$ should be a rectangular number. This problem is called Wurm's problem,[17] as it was solved by him to produce a solution where

$$W + X = 1409076 \cdot 1485583 .$$

This suggests, then, that the ratio of the length to the breadth of the bulls would be $1485583/1409076$, which is rather close to 1. As the authors of this book come from the cattle-producing province of Alberta, we are able to attest that we have seen many bulls, but never a bull with these proportions, and it is unlikely that the bulls in Sicily ever had such proportions either. Indeed, as Dijksterhuis[18] has noted, this apparently simplifying assumption is nothing of the sort, because if we assume that the ratio of the length to the breadth of a bull is λ, a rational number, then the condition that $W + X$ be a square becomes $W + X$ is λ times a square, and the supposed simplification of the problem is lost. Moreover, as Archimedes was far too good a mathematician not to include in his statement of the problem all of the values needed to solve it, and no value for λ is provided, we must assume that his intent was that the second part of the problem should reduce to what we now call a Pell equation. Although it is only implicit, as far as is currently known, the Cattle Problem represents the earliest mention in history of a Pell equation.

We are left with a number of questions concerning this remarkable work. For example, what caused Archimedes to devise it in the first place? Hultsch[19] has provided a very clever explanation for this. Apollonius of Perga (c. 262– c. 190 BC) in his *Easy Delivery* produced a better approximation to π than that of Archimedes in his earlier *Measurement of the Circle*, and it seems that part of Apollonius's motivation for doing this was to exhibit his superior skill in this sort of numerical manipulation. Certainly, he must have performed more difficult multiplications than those mentioned in the *Measurement of the Circle*. Another work of Apollonius concerning the multiplication of large numbers, preserved within the *Synogoge* or *Collection* of Pappus (c. 290–c. 350

AD), although inspired by Archimedes's *Sand-reckoner*, also seems to imply some criticism of Archimedes' methods. Thus, it does not seem unreasonable for Archimedes to have responded by issuing the Cattle Problem as a challenge to Apollonius and others; for, as we shall see in §3.3, solving the second part of it involves the manipulation of enormous numbers. This supposition is to some extent supported by the *Cattle Problem's* lightly satirical tone, which is particularly evident in the mockery displayed in the last lines of the epigram, which Fraser[20] translates as:

> If thou findest out these things, and layest them to mind, giving all the measures of the numbers, go victorious in glory and know in truth that thou hast been judged consummate in this wisdom at least.

From a mathematician's perspective, the tone of this provides us with the best reason to reject the Wurm hypothesis mentioned earlier: his solution is just too simple to derive. Of course, as Dijksterhuis[21] rightly points out, it is impossible to verify these suppositions, but it is interesting that, as Apollonius spent most of his career in Alexandria, he might very likely have been there during the time that the letter containing the problem was sent to Eratosthenes (276–194 BC). Knorr[22] has made the interesting suggestion that Eratosthenes had composed the first part of the problem and that Archimedes had responded by sending it back to him with the addition of the second, more difficult part. This suggestion, however, does not seem to have found much support among scholars, as most seem to accept Krumbiegel's earlier judgement[23] that "there is no ground whatever in the poem for... a division of authorship."

The problem appears to owe some of its inspiration to Homer; for, in Book XII, lines 127–139, of the *Odyssey*, the poet wrote[24]:

> Your next landfall will be the island of Thrinacie, where the Sun-god pastures his large herds and well-fed sheep. There are seven herds of cattle and as many flocks of beautiful sheep, with fifty head in each.

The Greek word "thrinacian" means three-cornered and was used to designate the three-cornered Island of Sicily,[25] where Archimedes lived. Notice that there also seems to be a computation problem in Homer's lines. Any educated Greek of the time would have recognized this Homeric allusion in the *Cattle Problem*.

There is also another important question concerning this problem: Could Archimedes himself solve it? Given our discussion of its solution in §3.3, the answer must be no. Although the basic idea of how to go about solving it had been demonstrated by Amthor[26] as early as 1880, it was not until the advent of modern computing devices that it was possible to compute the enormous numbers representing the size of the various herds. Indeed, as late as 1964, Beiler[27] could write concerning this problem that "stupendous feats of calculation have been performed and the answers have not yet been completely computed nor is it likely that they ever will be." The more important question, as noted by Vardi,[28] is: Did Archimedes know that it had a solution? As we will see in the next section, this could be the case, but we will likely never know for certain. One thing, however, must be borne in mind. In our modern society, with its very sophisticated mathematics and computers, it is easy to lose sight of what a remarkable piece of work this is. Given its date of composition and the state of mathematics (as far as we currently understand it) at this time, it must be regarded as a work of considerable genius. Who else, but Archimedes, could have posed it? Moreover, the poem with its lighter side also contributes something to our understanding of this extraordinary man. In this regard, we can do no better than to conclude this section with a quote of Fraser.[29]

> The poem... helps us to gain a picture of Archimedes as one who, for all his extraordinary pre-eminence in his abstract and theoretical world, possessed a warm and lively human sympathy, and this side of his character is worthy of emphasis no less than the superlative tributes to his mathematical genius.

2.2 Further Contributions of the Greeks

The first explicit mention of a Pell equation seems to occur in the work of Theon of Smyrna. (c. 130 AD)[30] If we put $s_1 = 1$ and $d_1 = 1$ and compute

$$s_{n+1} = s_n + d_n, \quad d_{n+1} = 2s_n + d_n \quad (n = 1, 2, 3, \ldots),$$

then

$$d_n^2 - 2s_n^2 = (-1)^n. \tag{2.13}$$

Of course, Theon does not use the modern notation that we are employing here, nor did he provide a proof of (2.13), being content instead to simply verify it for the first few cases. Some further light was shed on these observations much later by the neoplantonist philosopher Proclus[31] (412–485 AD). He referred to an identity, which in our notation would be expressed as

$$(2x + y)^2 + y^2 = 2x^2 + 2(x + y)^2 \tag{2.14}$$

and appears to appeal to Proposition 10 in Book II of Euclid's *Elements* for a proof. If we rewrite the identity, we get

$$(2x + y)^2 - 2(x + y)^2 = -(y^2 - 2x^2),$$

which does provide a proof of (2.13), although Proclus does not say this. Most mathematical historians agree that both Theon and Proclus appear to be drawing on a much earlier Pythagorean source for this material. What is remarkable about these side and diagonal numbers is that they suggest that the Pythagoreans used the values of d_n/s_n as a means of producing ever better rational approximations of $\sqrt{2}$. As the early Greek mathematicians were interested in the problem of irrationality, it is possible that the existence of this infinite sequence approaching, but never reaching, the value of $\sqrt{2}$ might have been used in producing an early (but incorrect) proof of the irrationality of this quantity.

In fact, it is possible to use (2.13) to produce a correct proof of the irrationality of $\sqrt{2}$. For, if we assume that $\sqrt{2}$ is rational, then $\sqrt{2} = a/b$ for some $a, b \in \mathbb{Z}^{>0}$. Hence, we can rewrite (2.13) as

$$bd_n + as_n = \frac{b^2}{|bd_n - as_n|} .$$

Since $bd_n - as_n \neq 0$ (otherwise (2.13) could not hold), we have $|bd_n - as_n| \geq 1$ and

$$0 < bd_n + as_n < b^2 . \tag{2.15}$$

As d_n and s_n increase beyond any limit, we see that (2.15) is impossible for all $n \in \mathbb{Z}^{>0}$. While this proof seems very simple to us, it is by no means likely that the Pythagoreans would have discovered it.

Thus, it appears that the early Greeks knew how to produce solutions of (1.7) when $D = 2$. It is difficult to say with any certainty that they extended the idea of side and diagonal numbers any further, but if[32] we put $D = 3$ and define $s_1 = 1$ and $d_1 = 2$,

$$s_{n+1} = s_n + d_n, \quad d_{n+1} = 3s_n + d_n \quad (n = 1, 2, 3, \dots) ,$$

we get

$$\frac{d_n}{s_n} = \frac{2}{1}, \frac{5}{3}, \frac{7}{4}, \frac{19}{11}, \frac{26}{15}, \frac{71}{41}, \frac{97}{56}, \frac{265}{153}, \frac{362}{209}, \frac{989}{571}, \frac{1351}{780}, \dots \tag{2.16}$$

as $n = 1, 2, 3, \dots, 11, \dots$. These are exactly the convergents in the simple continued fraction expansion (see §3.2) of $\sqrt{3}$. Furthermore, in the *Measurement of the Circle*, Archimedes[33] introduces with no explanation the inequality

$$\frac{265}{153} < \sqrt{3} < \frac{1351}{780} . \tag{2.17}$$

Note that both of the bounds used in this occur in (2.16). However, there are several other methods[34] by which Archimedes might have discovered (2.17). What does seem to be clear is that the Greeks were in possession of some techniques that allowed them to find good rational approximations to $\sqrt{n}$ (and other irrationals) for certain integral values of n. As will be demonstrated in Chapter 3, simple continued fractions can be used to produce the best rational approximations to a given irrational. Could the Greeks have been aware, at least on some level, of these objects? The answer is yes. In Proposition 2 of Book X of Euclid's *Elements* we have:

> If, when the less of two unequal magnitudes is continually subtracted in turn from the greater, that which is left never measures the one before it, the magnitudes will be incommensurable.

The process Euclid (c. 325–265 BC) is describing here is called anthyphairesis, and it has become the subject of considerable scrutiny by modern historians of early Greek mathematics. Anything like a full discussion of this is well beyond the scope of this book, and we refer the interested reader to the excellent books of Knorr[35] and Fowler[36] for a fascinating treatment of this subject. We will be content here with a few simple observations.

Euclid's understanding of a magnitude is what we might call a line segment and is distinct from what he understood by a number (integer). If we have two line segments A and B, we will write $A < B$ to denote that the line segment A is shorter than the line segment B. Now, suppose we are given two line segments L_0 and L_1, where $L_1 < L_0$. We apply the anthyphairesis process to L_1 and L_0; that is, we subtract L_1 from L_0 a certain number of times, say q_0 times, until we get a remaining line segment $L_2 < L_1$. We then repeat the procedure with L_2 and L_1, etc. We will get the following sequence of equations, where the q values are all positive integers:

$$L_0 = q_0 L_1 + L_2 \quad (L_2 < L_1) ,$$
$$L_1 = q_1 L_2 + L_3 \quad (L_3 < L_2) ,$$
$$\vdots$$
$$L_i = q_i L_{i+1} + L_{i+2} \quad (L_{i+2} < L_{i+1}) ,$$
$$\vdots$$

If this process does not terminate (no L_n ever "measures" L_{n-1}; i.e., no length of any L_{n+1} is ever 0), then L_0 and L_1 are not commensurable or, in more modern parlance, L_0/L_1 is irrational. If we examine this process from a modern perspective and put

$$\phi_i = \frac{L_i}{L_{i+1}} \quad (i = 0, 1, 2, \dots) ,$$

then

$$0 < \phi_i - q_i = \frac{L_{i+2}}{L_{i+1}} < 1 .$$

Thus, $q_i = \lfloor \phi_i \rfloor$ and

$$\phi_{i+1} = (\phi_i - q_i)^{-1} > 1 \quad (i = 0, 1, 2, \dots) ; \tag{2.18}$$

that is, the anthyphairesis of L_0/L_1 is given by

$$\frac{L_0}{L_1} = \phi_0 = [q_0, q_1, q_2, \dots, q_i, \dots] ,$$

the simple continued fraction expansion of ϕ_0 (see §3.2). We call the q_i ($i = 1, 2, \dots$) the *partial quotients* in this representation.

We know that there were several instances in which the Greeks might have employed this process, both geometrically and arithmetically.[37] This is corroborated by early references to the ancient's (5th-4th century BC) understanding that magnitudes are in proportion to each other if they have the same anthyphairesis (same sequence of partial quotients). Indeed, this seems to have formed the basis of their concept of proportion. Concerning this, Knorr[38] states:

> We can conceive of only one reason for the ancients' invention of the anthyphairetic definition of proportion: to extend the formal numerical definition so that proportions of incommensurable magnitudes may be included.

We also know that the early Greek mathematicians were very interested in the problem of incommensurability; in particular, they seem to have spent a lot of effort in demonstrating the possible incommensurability of line segments whose ratio is $\sqrt{n}/\sqrt{m}$, where m and n are positive integers,[39] and they could construct geometrically such line segments. It is not unreasonable to assume in their earliest investigations into this that they might have employed the anthyphairetic process to such line segments. This certainly seems to be what is behind parts of Books II, X, and XIII of the *Elements*. The main problem in doing this, as Fowler[40] has observed, would be the difficulty that they would face in determining the partial quotients that would be needed to express the anthyphairesis of $\sqrt{n}/\sqrt{m}$. This is simply because their arithmetic procedures would not permit the easy manipulation of the decimal numbers that would result. Fowler[41] has provided a possible and plausible solution to this problem by making use of concepts that would be known to the ancients. The basis of his procedure is what he calls the Parmenides Proposition (PP), which we give below as Proposition 2.1. A form of this result appears in Plato's *Parmenides* and was very likely known to the Greeks of Plato's time (427–347 BC). Certainly, it appears in the much later *Collection* of Pappus and could easily be derived from results[42] in Books VII or V of the *Elements*. We give this proposition next.

Proposition 2.1 (The Parmenides Proposition). *Let $A, B, C, D \in \mathbb{Z}^{>0}$. If $A/B < C/D$, then*

$$\frac{A}{B} < \frac{A+C}{B+D} < \frac{C}{D} \, .$$

Now, suppose ϕ is any real number and

$$\frac{A}{B} < \phi < \frac{C}{D} \, ,$$

where $A, B, C, D \in \mathbb{Z}^{>0}$. We have $\phi B - A > 0$ and $C - \phi D > 0$; hence, $(\phi B - A)/(C - \phi D) > 0$, and, consequently, there exist positive integers p and p' such that $p > (\phi B - A)/(C - \phi D)$ and $p' > (C - \phi D)/(\phi B - A)$. This means that

$$\frac{pC + A}{pD + B} > \phi \text{ and } \frac{p'A + C}{p'B + D} < \phi \, .$$

These observations lead us to the following simple algorithm, proposed by Fowler, for finding rational approximations to ϕ.

Algorithm 2.1:

Input: Suppose ϕ, A, B, C, D are defined as above, and

$$\frac{A}{B} < \phi < \frac{C}{D} \ .$$

1: Compute $R = (A + C)/(B + D)$. We now have two cases.
2: **case 1:** $R > \phi$
3:     Apply PP repreatedly to find a q so that

$$\frac{A}{B} < \frac{(q+1)A + C}{(q+1)B + D} < \phi < \frac{qA + C}{qB + D} < \frac{C}{D} \ .$$

4:     Return q, $C' = qA + C$, and $D' = qB + D$. Note that

$$\frac{A}{B} < \phi < \frac{C'}{D'} < \frac{C}{D} \ .$$

5: **end case**
6: **case 2:** $R < \phi$
7:     Apply PP repeatedly to find a q so that

$$\frac{A}{B} < \frac{A + qC}{B + qD} < \phi < \frac{A + (q+1)C}{B + (q+1)D} < \frac{C}{D} \ .$$

8:     Return q, $A' = A + qC$, and $B' = B + qD$. Note that

$$\frac{A}{B} < \frac{A'}{B'} < \phi < \frac{C}{D} \ .$$

9: **end case**

When this algorithm is applied repeatedly, the cases will strictly alternate; that is, if a given iteration falls under Case 1, then the next iteration will fall under Case 2, and vice versa.

In Case 1 we can compute q directly from

$$q = \left\lfloor \frac{C - \phi D}{\phi B - A} \right\rfloor$$

and in Case 2 from

$$q = \left\lfloor \frac{\phi B - A}{C - \phi D} \right\rfloor \ .$$

Suppose we consider the simple case of $\phi = \sqrt{n}/1 = \sqrt{n}$ for some non-square positive integer n. We begin with

$$\frac{\lfloor \phi \rfloor}{1} < \phi < \frac{\lfloor \phi \rfloor + 1}{1} \,. \tag{2.19}$$

If $\phi_0 = \phi$, $q_0 = \lfloor \phi_0 \rfloor$, and $R = \lfloor \phi \rfloor + 1/2 < \phi$, then $q_1 = 1/(\phi - q_0) = 1$; but if $R > \phi$, then we can apply Algorithm 2.1 to (2.19) to obtain

$$\frac{\lfloor \phi \rfloor}{1} < \phi < \frac{q\lfloor \phi \rfloor + \lfloor \phi \rfloor + 1}{q + 1} \,,$$

where

$$q = \left\lfloor \frac{\lfloor \phi \rfloor + 1 - \phi}{\phi - \lfloor \phi \rfloor} \right\rfloor = q_1 - 1 \,.$$

Thus, if $R < \phi$, we already have, by (2.19),

$$\frac{q_0}{1} < \phi < \frac{q_1 q_0 + 1}{q_1} \,,$$

and if $R > \phi$, we get

$$\frac{q_0}{1} < \phi < \frac{q_1 q_0 + 1}{q_1}$$

after the application of Algorithm 2.1 to (2.19).

To proceed further with our analysis we will need a result which is proved in §3.1. If we put $A_{-2} = 0$, $A_{-1} = 1$, $B_{-2} = 1$, and $B_{-1} = 0$ and define subsequent values for A_i and B_i by the recursive formulas (3.4), then by (3.9) we have

$$\phi_{i+1} = \frac{A_{i-1} - \phi B_{i-1}}{\phi B_i - A_i} \,. \tag{2.20}$$

By our previous remarks we may assume that we have, after a possible application of Algorithm 2.1 to (2.19),

$$\frac{A_0}{B_0} < \sqrt{n} < \frac{A_1}{B_1} \,.$$

Also, it is easy to see that if for some $i \geq 1$,

$$\frac{A_{i-1}}{B_{i-1}} < \sqrt{n} < \frac{A_i}{B_i} \,, \tag{2.21}$$

then since $\phi_{i+1} > 1$, we must have $R = (A_i + A_{i-1})/(B_i + B_{i-1}) < \sqrt{n}$. Thus, on applying Algorithm 2.1 to (2.21), we get

$$q = \left\lfloor \frac{A_{i-1} - \phi B_{i-1}}{\phi B_i - A_i} \right\rfloor = \lfloor \phi_{i+1} \rfloor = q_{i+1}$$

by (2.18) and (2.20). Also,

$$\frac{A_{i+1}}{B_{i+1}} < \sqrt{n} < \frac{A_i}{B_i} \,.$$

Similarly, if

$$\frac{A_i}{B_i} < \sqrt{n} < \frac{A_{i-1}}{B_{i-1}} ,$$

we find after the application of Algorithm 2.1 (here $R > \phi = \sqrt{n}$) that $q_{i+1} = q$
and

$$\frac{A_i}{B_i} < \sqrt{n} < \frac{A_{i+1}}{B_{i+1}} .$$

By induction (a process of deduction not likely known to the early Greek mathematicians), this procedure of repeated application of Algorithm 2.1 will produce the anthyphairesis of $\sqrt{n}/1 = [q_0, q_1, q_2, \ldots]$.

Algorithm 2.1 is evidently a very simple process that anyone with knowledge of the PP could, for example, apply successively to $\sqrt{n}/1$ in the manner that we have described above. It is highly unlikely, of course, that the Greeks of the time would have been able to prove formally that this procedure would produce the anthyphairesis of $\sqrt{n}/1$ as we have done here, but they could easily have computed the successive convergents A_i/B_i to $\sqrt{n}$ and discovered their anthyphairesis to be $[q_0, q_1, q_2, \ldots, q_i]$ ($i = 0, 1, 2, \ldots$). As they would have known by construction of the convergents that the value of $\sqrt{n}$ is always bounded above and below by two successive convergents, they would likely conclude (correctly) that the anthyphairesis $\sqrt{n}/1$ is $[q_0, q_1, q_2, \ldots]$. For small values of n, they would notice the periodic structure of $[q_0, q_1, q_2, \ldots]$ and perhaps, as Fowler[43] suggests, be able to prove geometrically that their conjectured anthyphairesis is correct. The difficulty of checking the inequalities that occur in Algorithm 2.1 would be much diminished because $\phi = \sqrt{n}$; hence, all that would be needed in each case is the determination of whether or not some rational number a/b exceeded $\sqrt{n}$. This, of course, is possible simply by checking the value of the integer $a^2 - nb^2$. During the process of checking these values, the Greeks would have discovered that if this process is carried out far enough for a given n, they would get $A_i^2 - nB_i^2 = 1$ (see §3.3) for perhaps several values of i, and thereby find solutions to the Pell equation for $D = n$. While this would not have been their original objective, they would nevertheless have been struck by the discovery, just as the Pythagoreans were in the case of $n = 2$.

Of course, this is conjectural, and it is possible to develop other plausible processes whereby the ancients might have been able to find good rational approximations to $\sqrt{n}$, but it fits very well with what we have been able to deduce from the few tantalizing grains of information that have survived time's winnowing. Certainly, the Greeks must have been able to perform some calculations like these, at least for small values of n. For example, if Archimedes had applied this to $\sqrt{27}$ (a better choice than $\sqrt{3}$ for his purpose[44]), he would have found that the first few convergents are $5/1, 26/5, 265/51$, and $1351/260$ and that

$$\frac{265}{51} < \sqrt{27} < \frac{1351}{260} ,$$

which, on dividing by 3, yields (2.17). Indeed, $1351^2 - 3 \cdot 780^2 = 1$. Thus, it is reasonable to infer that an expert calculator like Archimedes had some knowledge about how to solve the Pell equation for small values of D, at least. Possibly these investigations prompted him to believe that the Pell equation is always solvable, but that when D is large, this is a very difficult problem. This would explain his thinking in setting the Cattle Problem.

One other place where the Pell equation is explicitly mentioned by the Greeks is in Diophantus' *Arithmetica*. In Sections 9 and 11 of Book V, he solved (1.7) for $D = 26$ and $D = 30$, respectively. While this might cause us to think that the later Greeks had found a technique for solving the Pell equation, it is important to realize that the method given would, in general, only find rational solutions to the Pell equation, not integral ones. Diophantus also showed in a lemma in Section 14 of Book VI how one could find, given rationals x and y and integers D and r, a second rational solution to $x^2 - Dy^2 = r^2$. The concentration in the *Arithmetica* on techniques that only produce rational solutions to Diophantine equations strongly suggests that the later Greeks were either not able or not interested in producing integral solutions. Tannery[45] suggested that possibly Diophantus might have considered such problems, particularly the Pell equation, in the then lost seven books of the *Arithmetica*; however, although more recent research[46] has revealed some of these lost books, there is still no evidence that Diophantus ever considered the problem of finding only integral solutions. This, then, represents the very unsatisfactory state of our knowledge concerning the ancients' contributions to the study of the Pell equation.

2.3 The Indian Mathematicians

The situation is much different when we consider the achievements of the Indian mathematicians of the early to late middle ages.[47] As early as the 5th century AD, Aryabhata I (b. 476 AD) had developed a method for solving the linear Diophantine equation

$$ax - by = c \tag{2.22}$$

for integers x and y, given positive integers a, b, and c. Aryabhata's original problem was to find an integer n which on being divided by a given integer a leaves a given remainder of r_1 and on division by a given b leaves a remainder r_2. On putting $c = |r_1 - r_2|$, this problem reduces to making either $(ax + c)/b$ or $(by + c)/a$ a positive integer according to whether $r_1 > r_2$ or $r_2 > r_1$. Aryabhata then goes on to describe a solution technique, called the *kuttaka* (pulverizer), which is a variant of the now standard method of solving this problem by making use of the continued fraction expansion of a/b (see §3.2). It is often assumed by number theorists that the Greeks must have found a method of solving (2.22). Indeed, no less of an authority than Thomas Heath[48] seems to have believed this.

Thus, the solution of the equation $ax - by = c$, given by Aryabhata... is an easy development from Euclid's method of finding the greatest common measure or proving by that process that two numbers have no common factor (Eucl. VII. 1, 2, X. 2, 3), and it would be strange if the Greeks had not taken this step.

It would not be strange, however, if the Greeks had no interest in the problem. We have seen that the earlier Greeks were concerned with finding rational approximations to irrationals, but the problem of finding a rational approximation to a rational like a/b, would likely not have been regarded as a problem at all. The later Greeks seemed to be interested only in rational solutions of Diophantine equations, and this explains why Diophantus never dealt with (2.22). In any event, what is true is that we have no evidence at all that any of the Greek mathematicians made the slightest contribution to the problem of solving (2.22) for integers x and y.

In 628, Brahmagupta (598–670) was the first to discover our identity (1.4); that is, if

$$A^2 - DB^2 = Q \tag{2.23}$$

and

$$P^2 - DR^2 = S , \tag{2.24}$$

then

$$(AP + DBR)^2 - D(AR + BP)^2 = QS . \tag{2.25}$$

Today we call this process of multiplying two quadratic forms to yield a third quadratic form *composition*, but the Indian mathematicians referred to it as *samasa*.

If we have $Q = S = \pm 2$, $A = P$, and $B = R$, then $T = (A^2 + DB^2)/2 = A^2 - (\pm 1)$, $U = AB$ is a solution of (1.7). Brahmagupta discovered this result together with those in Table 1.1 and this enabled him to solve the Pell equation whenever he had any solution (A, B) of

$$A^2 - DB^2 = -1 , \pm 2 , \pm 4 . \tag{2.26}$$

However, he could do more than this: He developed an *ad hoc* way of solving the Pell equation. For example,[49] consider the equation $x^2 - 92y^2 = 1$, about which Brahmagupta declared, "[a person solving this problem] within a year [is] a mathematician." He first notes that $10^2 - 92 = 8$ and then composes this with itself to obtain $192^2 - 92 \cdot 20^2 = 64$. After dividing this equation by 64, he gets $24^2 - 92(5/2)^2 = 1$, and on composing this latter equation with itself, he obtains $1151^2 - 92 \cdot 120^2 = 1$. Brahmagupta also realized that by using this composition principle he could produce many more solutions to the Pell equation, once he had one solution.

However, the crowning achievement of Indian mathematics with respect to the Pell equation was the development of the cyclic method for solving it. The technique, described by Bhaskara II (1114–1185) in 1150 AD, and its history are well described by Selenius[50] and the interested reader should consult this work for further details and references. We will only sketch, with additional information, a variant (there are several) of the algorithm here.

We will assume that $Q, A, B \in \mathbb{Z}$ and that $(A, B) = 1$ in (2.23); this means that $(B, Q) = 1$. As the technique for solving (2.22) was known, the step of finding an integer P such that $Q \mid BP + A$ could be easily achieved by the kuttaka process. It follows that since $(B, Q) = 1$, we must have $Q \mid P^2 - D$ and $Q \mid AP + DB$. By putting $R = 1$ in (2.24), we see from (2.25) that

$$\left(\frac{AP + DB}{Q}\right)^2 - D\left(\frac{A + BP}{Q}\right)^2 = \frac{P^2 - D}{Q}. \qquad (2.27)$$

From this simple observation we can develop the cyclic method for solving the Pell equation.

Given integers n, A_{n-1}, B_{n-1}, Q_n, and P_n where $(A_{n-1}, B_{n-1}) = 1$ such that

$$\left|A_{n-1}^2 - DB_{n-1}^2\right| = Q_n ,$$

find by the kuttaka process a positive[51] integer P_{n+1} such that $|P_{n+1}^2 - D|$ is minimal and $Q_n \mid (P_{n+1}B_{n-1} + A_{n-1})$. Put $Q_{n+1} = |P_{n+1}^2 - D|/Q_n$,

$$A_n = \frac{A_{n-1}P_{n+1} + DB_{n-1}}{Q_n} , \qquad B_n = \frac{B_{n-1}P_{n+1} + A_{n-1}}{Q_n} . \qquad (2.28)$$

By (2.27) we get

$$\left|A_n^2 - DB_n^2\right| = Q_{n+1} , \qquad (2.29)$$

and $(A_n, B_n) = 1$. The latter result follows easily by observing that $|A_nB_{n-1} - B_nA_{n-1}| = 1$. The method terminates when, for some n, $Q_{n+1} = 1, 2, 4$ because, as we have explained above, Brahmagupta had already shown how to solve the Pell equation once any solution of (2.26) is known.

Consider the example of $D = 67$. We begin with $n = 0$, $A_{-1} = 1$, $B_{-1} = 0$, $Q_0 = 1$, and $P_0 = 0$. We now summarize in Table 2.1 the solution of the Pell equation by this process, called the *cakravala* (the circle or cyclic method) by the Indians.

Table 2.1. Cakravala for $D = 67$

n	P_n	Q_n	A_{n-1}	B_{n-1}	P_{n+1} (mod Q_n)
0	0	1	1	0	1
1	8	3	8	1	1
2	7	6	41	5	5
3	5	7	90	11	2
4	9	2	221	27	

Since $221^2 - 67 \cdot 27^2 = -2$, we get $T = 221^2 + 1 = 48842$, $U = 27 \cdot 221 = 5967$ as a solution of the Pell equation $T^2 - 67U^2 = 1$. Concerning this technique, Hankel[52] stated, "It is beyond all praise; it is certainly the finest thing that was achieved in the theory of numbers before Lagrange." Unfortunately, the Indians did not provide a proof that the cyclic method would always work. They were content, it seems, in the empirical knowledge that it always seemed to do so, and they used it to solve the Pell equation for $D = 61, 67, 97, 103$. It was not until the late 1930s that a proof that the cyclic method would always produce a value of $Q_i = 1$ was produced by Ayyangar.[53] He noted that this process could be represented as the expansion of $\sqrt{D}$ into a type of semiregular continued fraction which would always be periodic.

We note that if (as is certainly the case for $n = 0$)

$$P_n B_{n-1} \equiv A_{n-1} \pmod{Q_n} ,$$

then, by (2.28),

$$P_{n+1} B_n - A_n = B_{n-1}(P_{n+1}^2 - D)/Q_n$$
$$\equiv 0 \pmod{Q_{n+1}} .$$

Thus, by induction we may assume that $Q_n \mid (P_n B_{n-1} - A_{n-1})$. Since $Q_n \mid (P_{n+1} B_{n-1} + A_{n-1})$ by construction and $(Q_n, B_{n-1}) \mid (A_{n-1}, B_{n-1})$, we get $(Q_n, B_{n-1}) = 1$ and

$$P_{n+1} \equiv -P_n \pmod{Q_n} .$$

Hence,

$$P_{n+1} = q_n Q_n - P_n \qquad (2.30)$$

for some $q_n \in \mathbb{Z}$. If we now begin with $n = 0$ and define

$$\phi_i = \frac{P_i + \sqrt{D}}{Q_i} (> 0) \quad (i = 0, 1, 2, \dots) ,$$

$$\eta_{i+1} = \text{sign}(P_{i+1}^2 - D) ,$$

we get

$$\phi_{i+1} = \frac{P_{i+1} + \sqrt{D}}{Q_{i+1}} = \frac{\eta_{i+1} Q_i}{\sqrt{D} - P_{i+1}} .$$

By (2.30),

$$\frac{\sqrt{D} - P_{i+1}}{Q_i} = \phi_i - q_i ;$$

hence,

$$\phi_{i+1} = \frac{\eta_{i+1}}{\phi_i - q_i} . \qquad (2.31)$$

We now investigate the problem of the value of q_n.

Theorem 2.2. *If we put $q = \lfloor (P_n + \sqrt{D})/Q_n \rfloor$, then $0 < q \leq q_n \leq q + 1$.*

Proof. Put $P = qQ_n - P_n$, $P' = (q+1)Q_n - P_n$ and note that $P \equiv P' \equiv P_{n+1} \pmod{Q_n}$. By definition of q, we have $P < \sqrt{D}$ and $P' > \sqrt{D}$.

If $q_n < q$, then

$$0 < P_{n+1} = q_n Q_n - P_n < P < \sqrt{D}\,.$$

Hence, $|D - P_{n+1}^2| = D - P_{n+1}^2$, $|D - P^2| = D - P^2$. Since $P_{n+1} < P$, we get $D - P_{n+1}^2 > D - P^2$, which is impossible by selection of P_{n+1}.

If $q_n > q + 1$, then

$$P_{n+1} = q_n Q_n - P_n > P' > \sqrt{D}\,.$$

In this case, $|D - P_{n+1}^2| = P_{n+1}^2 - D$, $|D - P'^2| = P'^2 - D$, and $P_{n+1}^2 - D > P'^2 - D$, which is also impossible. $\square$

By Theorem 2.2 and (2.31), we see that

$$\phi_{i+1} > 1 \quad (i = 0, 1, 2, \dots)\,.$$

This means that the expression (2.31) can be used to give us

$$\sqrt{D} = q_0 + \cfrac{\eta_1}{q_1 + \cfrac{\eta_2}{q_2 + \cfrac{\eta_3}{q_3 + \ddots}}}\,, \tag{2.32}$$

a semiregular[54] continued fraction expansion of $\sqrt{D}$.

A number of misconceptions continue to circulate concerning the cyclic method. One of these is that it was rediscovered by Lagrange. This, as Selenius has pointed out, is not the case. Lagrange made use of simple continued fractions, which would not necessarily be the same as the semiregular continued fractions implicitly employed by the cyclic method. Often the algorithm is attributed to Bhaskara II, but as mentioned by Shankar Shukla,[55] Bhaskara made no claim to being the originator of the method, and as Jayadeva, who worked in the 10th century or earlier, had discovered a variant of the technique, it seems that it must have been developed much earlier than the time of Bhaskara. Finally, there is the belief, perhaps due to Tannery,[56] that the cyclic method derives from Greek influences. There seems, in spite of Tannery's analysis, to be little solid evidence in support of this. The simple fact is that, as mentioned earlier, we do not really know what the Greeks knew about the Pell equation. What we do know, however, is that the Indian methods display a history of steady development and refinement up to and including the discovery of the cyclic method, and this very strongly suggests that Hankel's[57] position that the Indians evolved the technique by themselves is the correct one.

2.4 Fermat and His Successors

The story of the Pell equation resumes with the challenge[58] issued in 1657 to Frénicle in particular and mathematicians in general by Fermat. Fermat had most likely, through his research, come to recognize the fundamental nature of the Pell equation. He asks for a proof of the following statement:

> Given any [positive] number [D] whatever that is not a square, there are also given an infinite number of squares such that, if the square is multiplied into the given number and unity is added to the product, the result is a square.

It next requests a general rule by which solutions of the problem could be determined and, as examples, asks for solutions when $D = 109, 149, 433$.

The story of how the second part of this challenge was answered by Brouncker and Wallis has been very well told by Weil[59] and Mahoney[60] and needs no elaboration here. Instead, we will content ourselves with giving a somewhat different account from that provided by Weil[61] concerning Brouncker's technique for solving the Pell equation. We emphasize that, although Brouncker's method is equivalent to what we will describe, he did not think about it in quite this way.

Let $P, Q, R \in \mathbb{Z}$, where $Q \neq 0$,

$$P^2 - QR = D > 0 ,$$

and D is not an integral square. Put

$$F(X, Y) = QX^2 - 2PXY + RY^2 \tag{2.33}$$

and let ρ and ρ' denote the zeros of $F(x, 1)$. Since D is not a square, we know that $\rho, \rho' \notin \mathbb{Q}$. Brouncker seems to have used the following result, although he provides no proof of it.

Proposition 2.3. *Suppose* $\rho > 1$ *and* $\rho' < 0$. *If* $F(X, Y) = 1$, *where* $X, Y \in \mathbb{Z}$ *and* $X > Y > 1$, *then* $\lfloor \rho \rfloor < X/Y < \lfloor \rho \rfloor + 1$.

Proof. Since $F(X, Y) = 1$, we may assume that $X = qY + Z$, where $0 < Z < Y$. Also,

$$|Q||X - \rho'Y||X - \rho Y| = 1 . \tag{2.34}$$

Since $\rho' < 0$, we get $|X - \rho'Y| = X - \rho'Y > X > 1$. Also, $X - \rho Y = (q - \rho)Y + Z$; thus, if $q - \rho < -1$, then $X - \rho Y < -Y + Z \leq -1$, and if $q - \rho > 0$, then $X - \rho Y > Z \geq 1$. In either case, $|X - \rho Y| > 1$, which is impossible by (2.34). It follows that $\rho - 1 < q < \rho$ or $q = \lfloor \rho \rfloor$. □

If we substitute $X = qY + Z$ in (2.33), we get

$$F'(Y, Z) = Q'Y^2 - 2P'YZ + R'Z^2 ,$$

where $Q' = q^2Q - 2qP + R$, $P' = P - qQ$, $R' = Q$, and

$$P'^2 - Q'R' = D. \tag{2.35}$$

It is easy to show that

$$\frac{P' - \sqrt{D}}{Q'} = \frac{1}{\frac{P+\sqrt{D}}{Q} - q}, \qquad \frac{P' + \sqrt{D}}{Q'} = \frac{1}{\frac{P-\sqrt{D}}{Q} - q}.$$

Thus, if τ and τ' are the zeros of $F'(x,1)$, then $\tau = 1/(\rho - q)$, $\tau' = 1/(\rho' - q)$. If $\rho > 1$, $\rho' < 0$, and $q = \lfloor \rho \rfloor$, then $\tau > 1$, $\tau' < 0$.

With these preliminary observations, we can now go on to describe Brouncker's very ingenious technique. We suppose T, U is a solution of $T^2 - DU^2 = 1$ and put $Q_0 = 1$, $P_0 = 0$, $R_0 = -D$, $X_0 = T$, and $X_1 = U$. We have $F_0(X_0, X_1) = Q_0 X_0^2 - 2P_0 X_0 X_1 + R_0 X_1^2 = 1$ and $\rho_0 = \sqrt{D}$, $\rho_0' = -\sqrt{D}$ are the zeros of $F_0(x,1)$. Putting $q_0 = \lfloor \rho_0 \rfloor$ and substituting $q_0 X_1 + X_2$ for X_0 in $F_0(X_0, X_1)$ we get $F_1(X_1, X_2) = 1$ $(0 < X_2 < X_1)$. Here,

$$Q_1 = q_0^2 Q_0 - 2q_0 P_0 + R_0, \qquad P_1 = P_0 - q_0 Q_0, \qquad R_1 = Q_0.$$

We put $\rho_1 = 1/(\rho_0 - q_0)$, $q_1 = \lfloor \rho_1 \rfloor$, and $X_1 = q_1 X_2 + X_3$ $(0 < X_3 < X_2)$ and compute $F_2(X_2, X_3)$ $(= 1)$, etc. In fact, if $F_i(X_i, X_{i+1}) = 1$ $(0 < X_{i+1} < X_i)$, we put

$$\rho_i = \frac{1}{\rho_{i-1} - q_{i-1}}, \tag{2.36}$$

$q_i = \lfloor \rho_i \rfloor$, and

$$X_i = q_i X_{i+1} + X_{i+2} \tag{2.37}$$

in F_i to obtain $F_{i+1}(X_{i+1}, X_{i+2}) = 1$ with

$$Q_{i+1} = (P_{i+1}^2 - D)/Q_i, \qquad P_{i+1} = P_i - q_i Q_i, \qquad R_{i+1} = Q_i,$$

by (2.35).

As the sequence $\{X_i\}$ is a strictly decreasing (for increasing i) sequence of positive integers, this process must come to a halt with $X_j = 1$, $X_{j+1} = 0$ for some $j \geq 0$. To find T and U, all that is necessary is to proceed backward using (2.37) once all the values of $q_0, q_1, q_2, \ldots, q_{j-1}$ have been determined.

We will now exemplify[62] the process for the case of

$$T^2 - 13U^2 = 1. \tag{2.38}$$

Here,

$$F_0(X_0, X_1) = X_0^2 - 13X_1^2, \qquad q_0 = \left\lfloor \sqrt{13} \right\rfloor = 3;$$

$$F_1(X_1, X_2) = -4X_1^2 + 6X_1 X_2 + X_2^2, \qquad q_1 = \left\lfloor \frac{3 + \sqrt{13}}{4} \right\rfloor = 1;$$

$$F_2(X_2, X_3) = 3X_2^2 - 2X_2X_3 - 4X_3^2 \,, \qquad q_2 = \left\lfloor \frac{1 + \sqrt{13}}{3} \right\rfloor = 1 \,;$$

$$F_3(X_3, X_4) = -3X_3^2 + 4X_3X_4 + 3X_4^2 \,, \qquad q_3 = \left\lfloor \frac{2 + \sqrt{13}}{3} \right\rfloor = 1 \,;$$

$$F_4(X_4, X_5) = 4X_4^2 - 2X_4X_5 - 3X_5^2 \,, \qquad q_4 = \left\lfloor \frac{1 + \sqrt{13}}{4} \right\rfloor = 1 \,;$$

$$F_5(X_5, X_6) = -X_5^2 + 6X_5X_6 + 4X_6^2 \,, \qquad q_5 = \left\lfloor \frac{3 + \sqrt{13}}{1} \right\rfloor = 6 \,;$$

$$F_6(X_6, X_7) = 4X_6^2 - 6X_6X_7 - X_7^2 \,, \qquad q_6 = \left\lfloor \frac{3 + \sqrt{13}}{4} \right\rfloor = 1 \,;$$

$$F_7(X_7, X_8) = -3X_7^2 + 2X_7X_8 + 4X_8^2 \,, \qquad q_7 = \left\lfloor \frac{1 + \sqrt{13}}{3} \right\rfloor = 1 \,;$$

$$F_8(X_8, X_9) = 3X_8^2 - 4X_8X_9 - 3X_9^2 \,, \qquad q_8 = \left\lfloor \frac{2 + \sqrt{13}}{3} \right\rfloor = 1 \,;$$

$$F_9(X_9, X_{10}) = -4X_9^2 + 2X_9X_{10} + 3X_{10}^2 \,, \qquad q_9 = \left\lfloor \frac{1 + \sqrt{13}}{4} \right\rfloor = 1 \,;$$

$$F_{10}(X_{10}, X_{11}) = X_{10}^2 - 6X_{10}X_{11} - 4X_{11}^2 \,.$$

We observe that $F_{10}(X_{10}, X_{11}) = 1$ can be easily achieved with $X_{10} = 1$ and $X_{11} = 0$. We can now find

$$
\begin{array}{llll}
X_9 = q_9 X_{10} + X_{11} = 1 \,, & X_8 = q_8 X_9 + X_{10} = 2 \,, & X_7 = 3 \,, \\
X_6 = 5 \,, & X_5 = 33 \,, & X_4 = 38 \,, \\
X_3 = 71 \,, & X_2 = 109 \,, & X_1 = 180 \,, \\
X_0 = 649 \,, & &
\end{array}
$$

and put $T = 649$, $U = 180$ as a solution of (2.38).

Brouncker used his method to find solutions of several difficult Pell equations, including $x^2 - 433y^2 = 1$. This was a major feat of calculation, as the value of y is a number of 19 digits. However, neither he nor Wallis nor Frénicle was able to provide a proof that the Pell equation could always be solved (non-trivially) for any positive non-square value of D. Fermat[63] took notice of this and stated that he had such a proof "by means of *descente* duly and appropriately applied." Unfortunately, Fermat provided no further information concerning his proof than this. Hofmann[64] and, with greater success, Weil[65] have attempted to reconstruct what Fermat's method might have been. While we may never really know what this was, it is nevertheless very likely that Fermat did have a proof. The fact that he selected 109, 149, and

433 for values of D as challenge examples is particularly suggestive because the corresponding Pell equations have large values of t and u.

The method of Brouncker was modified and extended by Euler, who realized that, as is apparent from (2.36), continued fractions could be used to provide an efficient algorithm for solving the Pell equation. However, even through he had devised all of the important tools, he just fell short of proving that his method would work for any non-square D. As mentioned earlier, the development of such a technique was first done by Lagrange in a rather clumsy work, which he later improved. For further information on this particularly interesting part of mathematical history, the reader is referred to Weil's book. In the next chapter we will describe Lagrange's method of using simple continued fractions to solve the Pell equation.[66]

Notes and References

[1] [Kon01].

[2] [Whi12].

[3] [Dic19], Vol. II, Ch. 12.

[4] [Les73], pp. 421–446. A more accessible source for some of this is [Les97], p. 100.

[5] [Fra72], Vol. II, p. 587, note 243.

[6] [Fra72], Vol. I, p. 409.

[7] [Tho80], Vol. II, p. 203–206.

[8] That is, a fifth and sixth of both of the males and of the females.

[9] [Arc99] and [Len02].

[10] [Kru80], p. 124; [Fra72], Vol. II, p. 588, note 245; [Dij87], p. 398, note 2.

[11] [Kru80].

[12] [Fra72], Vol. I, pp. 407–408.

[13] See [Dic19], Vol. II, pp. 342–343.

[14] See [Amt80], p. 155ff or [Hea12], p. 320.

[15] For the original Greek version of the scholium and a French translation, see [Arc71], pp. 171–173.

[16] See, for example, [Sch93] and [Wat95].

[17] [Wur30]. For a more easily accessible version, see [Hea12], pp. 319–323.

[18] [Dij87], p. 399, note 3.

[19] [Pau96b], II. 1, p. 534–535; [Hea12], p. xxxv.

[20] [Fra72], Vol. I, p. 409.

[21] [Dij87], p. 399; [Fra72], Vol. II, p. 590, note 256.

[22] [Kno86], p. 295.

[23] [Kru80], p. 124. See also [Fra72], Vol. II, p. 590, note 257.

[24] [Hom46], Book XII, lines 127–130, p. 192.

[25] [Str61], 6.2.1; [Thu92], Book VI, 2.

[26] [Amt80].

[27] [Bei64], p. 249.

[28] [Var98].

[29] [Fra72], p. 409.

[30] For a lengthy treatment of this, see [Kno75b], Ch. II. A translation of the relevant work of Theon is in [Fow87], p. 58.

[31] [Fow87], pp. 101–102.

[32] [Kno75a], p. 137.

[33] [Hea12], pp. 91–98.

[34] See [Hea12], lxxx–xcix; [Kno75a], pp. 136–139.

[35] [Kno75b].

[36] [Fow87].

[37] [Kno75b], pp. 255–261; [Fow87], Ch. 2.

[38] [Kno75b], p. 258.

[39] [Hea81], Vol. I, pp. 202–212; [vdW54], pp. 165–179.

[40] [Fow87], p. 45.

[41] [Fow87], Section 2.3(b).

[42] [Kno75a], p. 138; [Fow87], pp. 42–44.

[43] [Fow87], Ch. 3.

[44] [Fow87], p. 50, pp. 54–55.

[45] [Tan84].

[46] [Ses82], Part I.

[47] Two useful sources for this material are [DS62] and [Sri67].

[48] [Hea64], p. 281.

[49] [Col17], p. 363.

[50] [Sel63] and [Sel75].

[51] This is never explicitly stated, but it seems to be implicit in the *kuttaka* process that would be used to find P_{n+1}.

[52] [Han65], p. 202.

[53] See [Ayy40].

[54] See §38 of Vol. I of [Per57].

[55] [Shu54], p. 1 and p. 20.

[56] [Tan37], p. 240ff.

[57] [Han65], pp. 203–204.

[58] [Fer12], pp. 333–335. An English version can be found in [Hea64], pp. 285–286.

[59] [Wei84].

[60] [Mah94].

[61] [Wei84], pp. 92–97.

[62] This example of Brouncker's can be found in [Fer12], Vol. III, p. 480. It is also reprinted in [Whi12], pp. 53–55.

[63] [Fer12], p. 433.

[64] [Hof94].

[65] [Wei79]; [Wei84], Section XIII.

[66] For another perspective on this see Edwards [Edw05], pp. 65–112.

3

Continued Fractions

3.1 General Continued Fractions

While the techniques of the Indian mathematicians and those of Brouncker, Euler, and Lagrange for solving the Pell equation are different to some degree, they can all be unified by considering the theory of what are called semiregular continued fractions.

If we have two sequences of integers $\{a_n\}$ for $n \geq 1$ and $\{q_n\}$ for $n \geq 0$ and some complex number ϕ, we put $\phi_0 = \phi$ and define

$$\phi_{j+1} = \frac{a_{j+1}}{\phi_j - q_j} .$$

Then a general continued fraction[1] expansion of ϕ can be given as

$$\phi_0 = q_0 + \cfrac{a_1}{q_1 + \cfrac{a_2}{q_2 + \cfrac{a_3}{\ddots \cfrac{}{q_{i-1} + \cfrac{a_i}{q_i + \cfrac{a_{i+1}}{\phi_{i+1}}}}}}} \tag{3.1}$$

In the case where ϕ is irrational, (3.1) is said to be *semiregular*[2] if the following hold:

1. $|a_i| = 1 \; (i \geq 1)$.
2. $q_i \geq 1, q_i + a_{i+1} \geq 1 \; (i \geq 1)$.
3. $q_i + a_{i+1} \geq 2$ infinitely often.

For example, if ϕ is real and $q_n = \lfloor \phi_n + 1/2 \rfloor$, the nearest integer to ϕ_n, and sign $a_{n+1} = \text{sign}(\phi_n - q_n) \; (n \geq 0)$, we have what is called the *nearest integer continued fraction* expansion of ϕ.

For our purposes,[3] however, it will only be necessary to consider expressions of the form (3.1), where $a_n = 1$ $(n \geq 1)$. In this case we let $q_0, q_1, q_2, \ldots, q_i, \ldots$ be any given sequence of integers (*partial quotients*) and let ϕ $(= \phi_0)$ be any given complex number. If we define

$$\phi_{j+1} = \frac{1}{\phi_j - q_j} \quad (j = 0, 1, 2, \ldots, i), \tag{3.2}$$

then we can express ϕ_0 as the *continued fraction*

$$\phi_0 = q_0 + \cfrac{1}{q_1 + \cfrac{1}{q_2 + \cfrac{1}{\ddots \quad q_{i-1} + \cfrac{1}{q_i + \cfrac{1}{\phi_{i+1}}}}}}.$$

We denote this by

$$\phi_0 = \langle q_0, q_1, \ldots, q_i, \phi_{i+1} \rangle \,,$$

where ϕ_{i+1} is called a *complete quotient*. We will now develop some simple properties of these continued fractions. As its use has become more-or-less standard in computational number theory, we will, for the most part, make use of Perron's[4] notation in what follows.

We put $A_{-2} = 0, A_{-1} = 1, B_{-2} = 1$, and $B_{-1} = 0$ and define

$$\begin{cases} A_{j+1} = q_{j+1} A_j + A_{j-1} \\ B_{j+1} = q_{j+1} B_j + B_{j-1} \end{cases} \quad (j = -1, 0, 1, \ldots) \,. \tag{3.3}$$

It is easy to establish by induction that

$$A_j B_{j-1} - B_j A_{j-1} = (-1)^{j-1} \,. \tag{3.4}$$

We also see by induction that

$$\frac{A_i}{B_i} = \langle q_0, q_1, \ldots, q_i \rangle, \quad \frac{A_i}{A_{i-1}} = \langle q_i, q_{i-1}, \ldots, q_1, q_0 \rangle \,,$$
$$\frac{B_i}{B_{i-1}} = \langle q_i, q_{i-1}, \ldots, q_1 \rangle \,. \tag{3.5}$$

We call A_i/B_i the *ith convergent* of a continued fraction (3.2) and note that

$$\phi_0 = \frac{\phi_{i+1} A_i + A_{i-1}}{\phi_{i+1} B_i + B_{i-1}} \tag{3.6}$$

or

$$\phi_{i+1} = -\frac{\phi_0 B_{i-1} - A_{i-1}}{\phi_0 B_i - A_i}. \tag{3.7}$$

We also observe that $\phi^{-1} = \langle 0, \phi \rangle$. Thus, if A/B ($A, B \neq 0$) is a convergent of the continued fraction (3.2), then B/A will be a convergent of the continued fraction $\langle 0, q_0, q_1, \ldots, q_i, \phi_{i+1} \rangle = \phi^{-1}$.

We will now restrict the possible values of ϕ_0 to

$$\phi_0 = \frac{P|Q| + \sqrt{Q^2 D}}{Q|Q|},$$

where $P, D, Q \in \mathbb{Z}$, $\sqrt{D} \notin \mathbb{Q}$, and $Q \mid D - P^2$. This latter condition is not as restrictive as it may seem, as we can always replace D by $Q^2 D$, P by $|Q|P$, and Q by $|Q|Q$. Since $\phi_0 = (P|Q| + \sqrt{Q^2 D})/Q|Q|$ and $Q^2 \mid DQ^2 - Q^2 P^2$, we have the necessary condition.

Before going any further, we will need some simple observations concerning numbers of the form $a + b\sqrt{D}$, where $D, a, b, \in \mathbb{Q}$ (the rationals) and $\sqrt{D} \notin \mathbb{Q}$. If $\alpha = a + b\sqrt{D}$, we define its *conjugate* $\overline{\alpha}$ to be $a - b\sqrt{D}$ and its *norm* $N(\alpha)$ to be $\alpha\overline{\alpha}$. Note that $N(\alpha) \in \mathbb{Q}$. Also, if $\alpha = a + b\sqrt{D}$ and $\beta = c + d\sqrt{D}$, $a, b, c, d \in \mathbb{Q}$, then $\overline{\alpha\beta} = \overline{\alpha}\overline{\beta}$, and since $1/\alpha = \overline{\alpha}/N(\alpha)$, we get $\overline{(\alpha/\beta)} = \overline{\alpha}/\overline{\beta}$. Also, $N(\alpha\beta) = N(\alpha)N(\beta)$.

Put $P_0 = P$, $Q_0 = Q$, and $G_j = Q_0 A_j - P_0 B_j$. We have

$$Q_0 \left[A_j - \left(\frac{P_0 - \sqrt{D}}{Q_0} \right) B_j \right] = G_j + \sqrt{D} B_j \tag{3.8}$$

and

$$G_j B_{j-1} - B_j G_{j-1} = Q_0 (-1)^{j-1} \tag{3.9}$$

by (3.4). Also by (3.7) and (3.8) we see that

$$\phi_{i+1} = -\frac{G_{i-1} - \sqrt{D} B_{i-1}}{G_i - \sqrt{D} B_i}. \tag{3.10}$$

By using (3.2) and induction, we find that

$$\phi_j = \frac{P_j + \sqrt{D}}{Q_j},$$

where $P_j, Q_j \in \mathbb{Z}$, $Q_j \mid D - P_j^2$, and

$$P_{j+1} = q_j Q_j - P_j, \quad Q_{j+1} = \frac{D - P_{j+1}^2}{Q_j} \tag{3.11}$$

($j = 0, 1, 2, \ldots, i$). If we now refer to (3.10) and equate rational and irrational parts, we get

$$\begin{cases} DB_i = P_{i+1}G_i + Q_{i+1}G_{i-1} \,, \\ G_i = P_{i+1}B_i + Q_{i+1}B_{i-1} \,. \end{cases} \qquad (3.12)$$

From these results, (3.9), and (3.3), it is easy to deduce that

$$\begin{cases} (-1)^{i+1}Q_{i+1} = (G_i{}^2 - DB_i{}^2)/Q_0 \,, \\ (-1)^i P_{i+1} = (G_iG_{i-1} - DB_iB_{i-1})/Q_0 \end{cases} \qquad (3.13)$$

and

$$\begin{cases} G_i = (DB_{i-1} + P_{i+1}G_{i-1})/Q_i \,, \\ B_i = (G_{i-1} + P_{i+1}B_{i-1})/Q_i \,. \end{cases} \qquad (3.14)$$

If we define

$$\theta_{j+2} = A_j - B_j\bar{\phi}_0 = \frac{G_j + \sqrt{D}B_j}{Q_0} \qquad (j = -2,-1,0,1,\dots) \,,$$

we get

$$\theta_0 = \frac{\sqrt{D} - P_0}{Q_0} = -\bar{\phi}_0 \,, \quad \theta_1 = 1 \,, \quad \theta_{j+1} = q_j\theta_j + \theta_{j-1} \,. \qquad (3.15)$$

By (3.10),

$$\phi_j = \frac{-\bar{\theta}_j}{\bar{\theta}_{j+1}} \quad \text{and} \quad -\frac{1}{\bar{\phi}_j} = \frac{\theta_{j+1}}{\theta_j} \,.$$

Putting

$$\psi_j = -\frac{1}{\bar{\phi}_j} = \frac{P_j + \sqrt{D}}{Q_{j-1}} \,,$$

we get

$$\theta_{j+1} = \psi_j\theta_j \,. \qquad (3.16)$$

Hence, since $\psi_k = Q_k\phi_k/Q_{k-1}$, we see from (3.16) that

$$\theta_{j+1} = \prod_{k=1}^{j} \psi_k = \frac{Q_j}{Q_0} \prod_{k=1}^{j} \phi_k \quad (j \geq 1) \,. \qquad (3.17)$$

Note that

$$N(\theta_{j+1}) = (-1)^j \frac{Q_j}{Q_0} \,. \qquad (3.18)$$

From this we get

$$\begin{aligned} \theta_{j+1}{}^{-1} &= (-1)^j \frac{Q_0\bar{\theta}_{j+1}}{Q_j} \\ &= (-1)^j \frac{G_{j-1} - \sqrt{D}B_{j-1}}{Q_j} \\ &= (-1)^j (B_{j-2} + B_{j-1}\bar{\phi}_j) \end{aligned} \qquad (3.19)$$

by (3.12).

Notice that all of these results are dependent only on the value of ϕ_0 and the numbers in the sequence $\{q_n\}$. Suppose ϕ_0 is real. In the special case that $q_1, q_2, \ldots, q_i \geq 1$ and $\phi_{i+1} > 1$, we say that the continued fraction is *regular* or *simple* (SCF) and denote this by

$$\phi_0 = [q_0, q_1, \ldots, q_i, \phi_{i+1}] .$$

3.2 Simple Continued Fractions

In order to have $\phi_{i+1} > 1$, we must have $(\phi_i - q_i)^{-1} > 1$ or $0 < \phi_i - q_i < 1$. It follows that $q_i = \lfloor \phi_i \rfloor$. Since $q_i \geq 1$, we can argue that $q_{i+1} = \lfloor \phi_{i+1} \rfloor$, etc. Hence, $q_j = \lfloor \phi_j \rfloor$ for $j = 0, 1, 2, \ldots, i$. Simple continued fractions[5] will play a very important role in much of what is to follow. We will begin by considering the SCF of a rational number K/L, where $K, L \in \mathbb{Z}$ and $L > 0$. We put $R_{-2} = K$ and $R_{-1} = L$ and define

$$q_j = \lfloor R_{j-2}/R_{j-1} \rfloor , \tag{3.20}$$

$$R_{j-2} = q_j R_{j-1} + R_j \tag{3.21}$$

$(0 < R_j < R_{j-1}; \; j = 0, 1, 2, \ldots, n - 1)$. This is simply the Euclidean[6] algorithm; we ultimately find $R_n = 0$ for some n, and, therefore,

$$\frac{K}{L} = [q_0, q_1, \ldots, q_n] .$$

Also, as is well known, $(K, L) = R_{n-1}$.

It is easy to establish by induction that for this continued fraction we have

$$(-1)^{j+1} R_j = L A_j - K B_j \tag{3.22}$$

and

$$L = B_j R_{j-1} + B_{j-1} R_j . \tag{3.23}$$

Also, if $C_j = (-1)^{j-1} B_j$, then it is clear that

$$C_j = C_{j-2} - q_j C_{j-1} . \tag{3.24}$$

As is well known, this algorithm can be extended to the problem of solving the linear Diophantine equation

$$ax + by = c , \tag{3.25}$$

where $a, b,$ and c are given integers and we want integer values for x and y. We first note that in order for (3.25) to have integer solutions, it is necessary that $d = (a, b)$ be a divisor of c. We therefore make this assumption and put $\bar{a} = a/d$, $\bar{b} = b/d$, and $\bar{c} = c/d$. If we put $K = a$ and $L = b$ (we may assume with no loss of generality that $b > 0$ because its sign can be absorbed by y), we get

$$\frac{\bar{a}}{\bar{b}} = \frac{a}{b} = [q_0, q_1, \ldots, q_n] = \frac{A_n}{B_n} .$$

By (3.4) we have

$$A_n B_{n-1} - B_n A_{n-1} = (-1)^{n-1} ;$$

hence, $(A_n, B_n) = 1$, and since $(\bar{a}, \bar{b}) = 1$, $B_n > 0$, and $\bar{b} > 0$, we must have $\bar{a} = A_n$ and $\bar{b} = B_n$. Hence,

$$\bar{a} B_{n-1} - \bar{b} A_{n-1} = (-1)^{n-1}$$

from which we see that

$$\bar{a}(-1)^{n-1} \bar{c} B_{n-1} + \bar{b}(-1)^n \bar{c} A_{n-1} = \bar{c}$$

or

$$a(-1)^{n-1} \bar{c} B_{n-1} + b(-1)^n \bar{c} A_{n-1} = c .$$

Thus, a particular solution of (3.25) is given by

$$x_0 = (-1)^{n-1} \bar{c} B_{n-1} , \quad y_0 = (-1)^n \bar{c} A_{n-1} .$$

To find all solutions of (3.25) we assume that x, y is any solution. We must have

$$ax + by = c = ax_0 + by_0 .$$

It follows that

$$\bar{a}(x - x_0) = -\bar{b}(y - y_0) .$$

Since $(\bar{a}, \bar{b}) = 1$, we must have $x - x_0 = \bar{b}t$ and $y - y_0 = -\bar{a}t$ for some integer parameter t. It is easy to verify that $x = x_0 + \bar{b}t$, $y = y_0 - \bar{a}t$ satisfies (3.25) for any integer t. Thus, we can characterize all solutions of (3.25) by $x = x_0 + \bar{b}t$, $y = y_0 - \bar{a}t$, where (x_0, y_0) is any particular solution and t is any integer.

Of course, to get some idea of how quickly these algorithms execute, we will need to be able to bound the value of n above. A result usually attributed to Lamé[7] states that if $|a|, b < N$, then[8]

$$n < \frac{\log(\sqrt{5}N)}{\log \tau} - 1 , \quad \text{where } \tau = \frac{1 + \sqrt{5}}{2} .$$

Hence, we know that these algorithms will execute in polynomial (in $\log N$) time. In fact, a careful analysis of the runtime for this algorithm, taking into consideration the fact that the numbers involved become smaller as the algorithm proceeds, allows us to assert that it executes in $O((\log N)^2)$ bit operations.[9] The Knuth-Schönhage greatest common divisor (GCD) algorithm[10] is based on fast Fourier transform (FFT) multiplication[11] and is asymptotically the fastest GCD algorithm known, executing in running time $O(n \log^2 n \log \log n)$, where $n = \log N$. Unfortunately, this algorithm can be practical only when the inputs are sufficiently large for the FFT multiplication technique to be efficient. In 2004, Stehlé and Zimmermann[12] presented a binary-type recursive GCD algorithm with the same complexity of the Knuth-Schönage algorithm. While this new algorithm is much easier to implement and more practical, its inputs still need to be about 25,000 digits for it to be comparable in efficiency to the k-ary algorithm of Jebelean[13] and Sorenson[14] as implemented in the GNU MP (multiprecision) integer arithmetic library.[15]

Lehmer[16] pointed out that for large a and b it is possible to modify the Euclidean algorithm to make it run faster. Briefly put, the idea is to work with only single-precision numbers in the divide steps rather than multiprecision numbers. This is possible because in order to compute $\lfloor R_{i-2}/R_{i-1} \rfloor$, all that is usually needed are the leading digits of R_{i-2} and R_{i-1}. The actual technique is somewhat more complicated than this, but it does improve the execution time of the algorithm.[17] It is remarkable that methods for finding the GCD of two integers have been resistant to any significant speed-up over the technique of the basic Euclidean algorithm.[18] Indeed, no one has even discovered a way of effectively parallelizing this or any other GCD algorithm.[19]

We will now derive some features of the SCF of a real irrational number ϕ. As we will need this result later, we mention that by (3.17) and the fact that $\phi_i \geq 1$ $(i \geq 1)$, we get

$$\frac{\theta_{j+1}}{Q_j} > \frac{\theta_{i+1}}{Q_i} \tag{3.26}$$

for $j > i \geq 0$. We note that because ϕ_i is irrational for all $i \geq 0$, the sequence of partial quotients $\{q_i\}$ must be infinite. Since by (3.6)

$$\phi = [q_0, q_1, \ldots, q_i, \phi_{i+1}] = \frac{\phi_{i+1}A_i + A_{i-1}}{\phi_{i+1}B_i + B_{i-1}},$$

we see that

$$\phi - \frac{A_i}{B_i} = \frac{\phi_{i+1}A_i + A_{i-1}}{\phi_{i+1}B_i + B_{i-1}} - \frac{A_i}{B_i} = \frac{B_i A_{i-1} - A_i B_{i-1}}{B_i(\phi_{i+1}B_i + B_{i-1})}$$

$$= \frac{(-1)^i}{B_i(\phi_{i+1}B_i + B_{i-1})} \tag{3.27}$$

by (3.2).

At this point we make a number of observations concerning the sequence $\{B_j\}$ $(j = -2, -1, 0, \ldots)$. We have $B_{-2} = 1$, $B_{-1} = 0$, $B_0 = 1$, and $B_1 = q_1(\geq 1)$. Since

$$B_j = q_j B_{j-1} + B_{j-2}$$

and $q_j \geq 1$ for $j \geq 1$, we have $B_j \geq 1$ $(j \geq 0)$. Also,

$$B_j > B_{j-1} \quad (j \geq 2). \tag{3.28}$$

Hence,

$$B_j \geq B_{j-1} \quad (j \geq 0) \,.$$

We also have

$$\phi_{j+1} B_j + B_{j-1} > q_{i+1} B_j + B_{j-1} = B_{j+1} \quad (j \geq 0) \,.$$

From this and (3.27) it follows that

$$\left| \phi - \frac{A_i}{B_i} \right| < \frac{1}{B_i B_{i+1}} \quad (i \geq 0) \,. \tag{3.29}$$

Thus, as mentioned in the previous chapter, the convergents of the SCF expansion of ϕ provide very good rational approximations to the value of ϕ. We will now show that as i increases, the convergents A_i/B_i provide successively better approximations of ϕ.

We note that in a SCF, we have $q_i + 1 > \phi_i$ $(i \geq 0)$; hence,

$$\phi_i B_{i-1} + B_{i-2} < (q_i + 1) B_{i-1} + B_{i-2} = q_i B_{i-1} + B_{i-2} + B_{i-1}$$
$$= B_i + B_{i-1} \leq q_{i+1} B_i + B_{i-1} = B_{i+1} \quad (i \geq 0) \,.$$

It follows from (3.27) that

$$\left| \phi - \frac{A_{i-1}}{B_{i-1}} \right| = \frac{1}{B_{i-1}(\phi_i B_{i-1} + B_{i-2})} > \frac{1}{B_{i-1} B_{i+1}} \quad (i \geq 1)$$

and

$$|B_{i-1}\phi - A_{i-1}| > \frac{1}{B_{i+1}} > |\phi B_i - A_i| \quad (i \geq 1) \tag{3.30}$$

by (3.29). We next see that

$$\left| \phi - \frac{A_i}{B_i} \right| = \frac{1}{B_i} |\phi B_i - A_i| < \frac{1}{B_i} |\phi B_{i-1} - A_{i-1}|$$
$$\leq \frac{1}{B_{i-1}} |\phi B_{i-1} - A_{i-1}| = \left| \phi - \frac{A_{i-1}}{B_{i-1}} \right| \quad (i \geq 1) \,.$$

Thus,

$$\left| \phi - \frac{A_i}{B_i} \right| < \left| \phi - \frac{A_{i-1}}{B_{i-1}} \right| \quad \text{for} \quad i \geq 1 \,. \tag{3.31}$$

From (3.28) and (3.29), we see that

$$\phi = \lim_{i \to \infty} \frac{A_i}{B_i} \,.$$

Thus, any irrational number ϕ is uniquely expressible, through the process described at the beginning of this section, as an infinite SCF; that is, we can write ϕ as

$$\phi = [q_0, q_1, q_2, \ldots] ,$$

where the sequence $\{q_i\}$ is infinite. Furthermore, any such continued fraction determined by an infinite sequence of partial quotients $\{q_i\}$ which are all positive for $i > 0$ must represent an irrational number.

We now prove the following useful theorem.

Theorem 3.1. *If a and b are integers, $b > 0$, and $|\phi b - a| < |\phi B_i - A_i|$ for some $i \geq 0$, then $b \geq B_{i+1}$.*

Proof. Suppose that $|\phi b - a| < |\phi B_i - A_i|$ for some $i \geq 0$ and $b < B_{i+1}$. We will show that this leads to a contradiction. By (3.4) there must exist integers x and y such that

$$xB_i + yB_{i+1} = b ,$$

$$xA_i + yA_{i+1} = a .$$

If $x = 0$, then $b = yB_{i+1}$, which means that $y > 0$ and $b \geq B_{i+1}$, a contradiction. If $y = 0$, then $b = xB_i$ and $a = xA_i$; hence, $x \neq 0$ and

$$|\phi b - a| = |\phi x B_i - x A_i| = |x||\phi B_i - A_i| \geq |\phi B_i - A_i| ,$$

another contradiction.

We next show that x and y must have opposite signs. If $y < 0$, then since $xB_i = b - yB_{i+1}$, we must have $x > 0$. If $y > 0$, then our assumption that $b < B_{i+1}$ means that $b < yB_{i+1}$ and, therefore, $xB_i < 0$ and $x < 0$. Since $\phi_{i+2} > 0$, we see by (3.7) that $\phi B_i - A_i$ and $\phi B_{i+1} - A_{i+1}$ have opposite signs; hence, both $x(\phi B_i - A_i)$ and $y(\phi B_{i+1} - A_{i+1})$ must have the same sign. Now, as

$$\phi b - a = x(\phi B_i - A_i) + y(\phi B_{i+1} - A_{i+1}) ,$$

it follows that

$$|\phi b - a| = |x(\phi B_i - A_i)| + |y(\phi B_{i+1} - A_{i+1})|$$
$$> |x(\phi B_i - A_i)| = |x||\phi B_i - A_i| \geq |\phi B_i - A_i| ,$$

a contradiction. □

We can now use this theorem to prove a most important result[20] concerning the convergents of the SCF of ϕ.

Theorem 3.2. *Suppose that $a, b \in \mathbb{Z}$ and $b > 0$. If $|\phi - a/b| < 1/2b^2$, then a/b must be a convergent in the SCF expansion of ϕ.*

Proof. Since $b > 0$, by (3.28) there must exist some integer $i \geq 0$ such that

$$B_i \leq b < B_{i+1} .$$

By Theorem 3.1, we must have

$$\frac{1}{2b} > |\phi b - a| \geq |\phi B_i - A_i| .$$

Clearly, if $a/b \neq A_i/B_i$, then $|bA_i - aB_i| \geq 1$. Hence,

$$\frac{1}{bB_i} \leq \frac{|bA_i - aB_i|}{bB_i} = \left| \frac{A_i}{B_i} - \frac{a}{b} \right|$$

$$\leq \frac{|\phi B_i - A_i|}{B_i} + \left| \phi - \frac{a}{b} \right|$$

$$< \frac{|\phi b - a|}{B_i} + \frac{1}{2b^2}$$

$$< \frac{1}{2bB_i} + \frac{1}{2b^2} .$$

However, this means that $b < B_i$, contradicting the selection of B_i.   □

We see, then, that not only do the convergents of the SCF expansion of a real number ϕ provide very good rational approximations of ϕ, but if we have a very good rational approximation to ϕ, it must be such a convergent. We will see how this is applicable to the problem of solving the Pell equation in the next section.

To get some idea of how we will proceed, we conclude this section with the following result.

Theorem 3.3. *If* $x, y, D, n \in \mathbb{Z}$; $D, x, y > 0$; $\sqrt{D} \notin \mathbb{Q}$; $|n| < \sqrt{D}$; *and*

$$x^2 - Dy^2 = n ,$$

then x/y *is a convergent in the SCF expansion of* $\sqrt{D}$.

Proof. We first consider the case of $n > 0$. Here we have $x - \sqrt{D}y > 0$ and $x + \sqrt{D}y > 2\sqrt{D}y$. Hence,

$$0 < x - \sqrt{D}y = \frac{n}{x + \sqrt{D}y} < \frac{\sqrt{D}}{2\sqrt{D}y} = \frac{1}{2y} ,$$

$$\left| \frac{x}{y} - \sqrt{D} \right| < \frac{1}{2y^2}$$

and the result holds by Theorem 3.2.

When $n < 0$, we get

$$Dy^2 - x^2 = |n|$$

and

$$y^2 - \frac{1}{D}x^2 = \frac{|n|}{D}$$

with $y > (1/\sqrt{D})x$. Since

$$y - \frac{1}{\sqrt{D}}x = \frac{|n|}{D(y + x/\sqrt{D})} < \frac{|n|}{2D(x/\sqrt{D})} < \frac{1}{2x} ,$$

we get

$$\left| \frac{y}{x} - \frac{1}{\sqrt{D}} \right| < \frac{1}{2x^2} \, .$$

Thus, y/x must be a convergent in the SCF expansion of $1/\sqrt{D}$ and the theorem follows.   □

3.3 Simple Continued Fractions of Quadratic Irrationals

In this section we will develop some of the properties of SCF expansions of real quadratic irrationals. Such numbers can be represented as $\phi = (P + \sqrt{D})/Q$, where $P, Q, D \in \mathbb{Z}$, $D > 0$, $\sqrt{D} \notin \mathbb{Q}$, and $Q \mid D - P^2$. We first require a very simple result.

Proposition 3.4. *If ϕ is a quadratic irrational defined as above and if for some j in the SCF expansion of ϕ the complete quotient ϕ_j (> 1) satisfies $\overline{\phi}_j < 0$, then $-1 < \overline{\phi}_{j+1} < 0$.*

Proof. Since

$$\phi_{j+1} = \frac{1}{\phi_j - q_j} \, ,$$

we have

$$\overline{\phi}_{j+1} = \frac{1}{\overline{\phi}_j - q_j} \, .$$

The result follows easily by noting that $q_j = \lfloor \phi_j \rfloor \geq 1$ and $\overline{\phi}_j < 0$.   □

From this we can easily prove some useful results concerning the integers P_i and Q_i for $i \geq j$. If $\phi_j > 1$ and $\overline{\phi}_j < 0$, then, by induction, $\phi_i > 1$ and $\overline{\phi}_i < 0$ for all $i \geq j$. This means that $\phi_i = (P_i + \sqrt{D})/Q_i > 1$ and $-\overline{\phi}_i = (\sqrt{D} - P_i)/Q_i > 0$. By adding these we get $2\sqrt{D}/Q_i > 1$ or $0 < Q_i < 2\sqrt{D}$. Also, we see that $P_i < \sqrt{D}$. If $i > j$, then $\overline{\phi}_i = (P_i - \sqrt{D})/Q_i > -1$, and since $\phi_i > 1$, we get $2P_i/Q_i > 0$ or $P_i > 0$.

Thus, if $i > j$, we have

$$\begin{cases} 0 < Q_i < 2\sqrt{D} \, , \\ 0 < P_i < \sqrt{D} \, . \end{cases} \tag{3.32}$$

We also observe that since $\psi_i = (P_i + \sqrt{D})/Q_{i-1}$, we have

$$\psi_i < \frac{2\sqrt{D}}{Q_{i-1}} \, , \tag{3.33}$$

and since

$$\psi_i = q_{i-1} + \frac{\sqrt{D} - P_{i-1}}{Q_{i-1}} = q_{i-1} + \left(\frac{\sqrt{D} + P_{i-1}}{Q_i}\right)^{-1},$$

we get

$$\psi_i > 1 + \frac{Q_i}{2\sqrt{D}}. \tag{3.34}$$

We now address the question of whether we ever find some j for which $\overline{\phi}_j < 0$.

Theorem 3.5. *If in the SCF expansion of the quadratic irrational $\phi = (P + \sqrt{D})/Q$ we get*

$$B_{j-2}B_{j-1} > \frac{|Q|}{2\sqrt{D}},$$

then $\overline{\phi}_j < 0$.

Proof. Since the B_i increase without bound, there must be some j for which $B_{j-2}B_{j-1} > |Q|/(2\sqrt{D})$. From (3.19), we have

$$\overline{\phi}_j = \frac{-B_{j-2} + (-1)^j \theta_{j+1}^{-1}}{B_{j-1}}, \tag{3.35}$$

where $\theta_{j+1} = A_{j-1} - B_{j-1}\overline{\phi}$. Now,

$$(-1)^j \theta_{j+1} = (-1)^j(A_{j-1} - B_{j-1}\phi) + (-1)^j B_{j-1}(\phi - \overline{\phi})$$
$$= (-1)^j(A_{j-1} - B_{j-1}\phi) + (-1)^j B_{j-1}(2\sqrt{D}/Q).$$

By (3.29) we know that $|A_{j-1} - B_{j-1}\phi| < 1/B_j$. Also, as observed in the proof of Theorem 3.1, we know that for $i = 0, 1, 2, \ldots$, the quantities $A_{i-1} - B_{i-1}\phi$ and $A_i - B_i\phi$ have opposite signs. Since $A_0 - B_0\phi = q_0 - \phi_0$ is negative, it follows that $\text{sign}(A_{j-1} - B_{j-1}\phi) = (-1)^j$.

Clearly, if $(-1)^j \theta_{j+1} < 0$, then $\overline{\phi}_j < 0$ by (3.35). Suppose $(-1)^j \theta_{j+1} > 0$. If $(-1)^j Q < 0$, then

$$\frac{1}{B_{j-2}} \geq \frac{1}{B_j} > |A_{j-1} - B_{j-1}\phi| > B_{j-1}\left(\frac{2\sqrt{D}}{|Q|}\right),$$

but since $B_{j-2}B_{j-1} > |Q|/(2\sqrt{D})$, this is a contradiction. If $(-1)^j Q > 0$, then

$$(-1)^j \theta_{j+1} > B_{j-1}\frac{2\sqrt{D}}{|Q|} > \frac{1}{B_{j-2}};$$

hence, $\overline{\phi}_j < 0$ by (3.35). $\qquad\square$

We also have the following results which will be useful later.

Theorem 3.6. *If $Q_0 > 0$ and $\overline{\phi}_{m-1} > 0$ ($m \in \mathbb{Z}^{>0}$), then $|\theta_m| \le 1$.*

Proof. This result is trivially true for $m = 1$. If $m = 2$, then since $\theta_2 = \psi_1$, $\psi_1 = -1/\overline{\phi}_1$, and $\phi_1 = 1/(\phi_0 - q_0)$, we see that $\psi_1 < 0$ and $\psi_1 = q_0 - \overline{\phi}_0 > \phi_0 - \overline{\phi}_0 - 1 = 2\sqrt{D}/Q_0 - 1 > -1$. Hence, $|\theta_2| < 1$. If $m \ge 3$, then by (3.35) we have
$$\overline{\phi}_{m-1} = \frac{-B_{m-3} + (-1)^{m+1}\theta_m^{-1}}{B_{m-2}} > 0 \,.$$
It follows that
$$(-1)^{m+1}\theta_m^{-1} > B_{m-3} \ge 1 \,.$$
Thus, $|\theta_m^{-1}| > 1$ and $|\theta_m| < 1$.   □

Note that if t is the least positive integer such that $\overline{\phi}_t < 0$, then since $0 < Q_t < 2\sqrt{D}$, $|P_t| < \sqrt{D}$, and $Q_t Q_{t-1} = D - P_t^2$, we either have
$$0 < Q_t < \sqrt{D} \text{ or } 0 < Q_{t-1} < \sqrt{D} \,.$$
Thus, there must exist some least positive integer m such that
$$0 < Q_{m-1} < \sqrt{D}$$
and $m - 1 \le t$.

Theorem 3.7. *If $Q_0 > 0$ and m (≥ 1) is defined as above, then $|\theta_m| < 2$.*

Proof. If $m = 1$, the result is trivial. If $m = 2$, then since $Q_0 > \sqrt{D}$ and $D - P_1^2 = Q_0 Q_1 > 0$, we get $|P_1| < \sqrt{D}$. Also, $\theta_2 = \psi_1 = (P_1 + \sqrt{D})/Q_0$; hence, $|\theta_2| \le 2\sqrt{D}/Q_0 < 2$. If $m \ge 3$ and $\overline{\phi}_{m-1} > 0$, then $|\theta_m| < 1$ by Theorem 3.6. If $m \ge 3$, $\overline{\phi}_{m-1} < 0$ and $\overline{\phi}_{m-2} < 0$, we get $t \le m - 2$, which is impossible because $t \ge m - 1$. Hence, $\overline{\phi}_{m-2} > 0$ and $|\theta_{m-1}| \le 1$. Since $\overline{\phi}_{m-1} < 0$, we get $|P_{m-1}| < \sqrt{D}$ and $Q_{m-2} > 0$. Thus, $Q_{m-2} > \sqrt{D}$, which means that
$$|\psi_{m-1}| = \frac{|P_{m-1} + \sqrt{D}|}{Q_{m-2}} < 2$$
and $|\theta_m| = |\psi_{m-1}\theta_{m-1}| < 2$.   □

From this result we know that $\overline{\phi}_j < 0$ for some j. Also, by (3.32), once this happens we also know that the values of P_i and Q_i are bounded for all $i > j$. This means that there can only be a finite number of distinct complete quotients ϕ_i for $i > j$. Thus, as there exist infinitely many complete quotients, for some minimal k and p (> 0), we must have
$$\phi_{k+p} = \phi_k \,.$$
However, since $q_i = \lfloor \phi_i \rfloor$, we see by (3.2) and induction that

$$\phi_{m+p} = \phi_m, \quad q_{m+p} = q_m,$$

for all $m \geq k$; that is, the infinite sequence of complete (and partial) quotients in the SCF expansion of any quadratic irrational must ultimately be periodic. (Also, if the SCF expansion of any irrational ϕ is periodic, then ϕ must be a quadratic irrational.) We often represent this by the notation

$$\phi = [q_0, q_1, \ldots, q_{k-1}; \overline{q_k, q_{k+1}, \ldots, q_{k+p-1}}],$$

where the bar is placed over the periodic part of the sequence of partial quotients. The ordered set of partial quotients $\{q_0, q_1, \ldots, q_{k-1}\}$ is called the *preperiod* and the ordered set $\{q_k, q_{k+1}, \ldots, q_{k+p-1}\}$ is called the *period* of the SCF expansion. The value of k is called the *length of the preperiod* and the value of p is called the *length of the period*. Certain semiregular continued fraction expansions of quadratic irrationals are also periodic. For example, Hurwitz[21] showed that this is the case for both the nearest integer continued fraction and the singular continued fraction.

Of particular interest are those real quadratic irrationals ϕ which have no preperiod ($k = 0$). Such continued fractions are said to be *purely periodic*. If ϕ has a purely periodic expansion, then

$$\phi = [\overline{q_0, q_1, \ldots, q_n}]$$

and $p = n + 1$. Hence, ϕ can be written as

$$\phi = [q_0, q_1, \ldots, q_n, \phi],$$

and by (3.6),

$$\phi = \frac{\phi A_n + A_{n-1}}{\phi B_n + B_{n-1}}.$$

It follows that if

$$F(x) = x^2 B_n + x(B_{n-1} - A_n) - A_{n-1},$$

then $F(\phi) = 0$ and the only other zero of $F(x)$ is $\overline{\phi}$.

Now, because ϕ has a purely periodic expansion, we must have $q_0 \geq 1$ because $q_0 = q_{n+1} \geq 1$. Thus, $\phi > 1$. We note, however, that $F(0) = -A_{n-1} < 0$ and $F(-1) = B_n - B_{n-1} + A_n - A_{n-1} > 0$; hence, $F(x)$ has a zero between -1 and 0. This can only be $\overline{\phi}$. Hence,

$$-1 < \overline{\phi} < 0, \phi > 1. \tag{3.36}$$

Thus, if ϕ is a quadratic irrational with a purely periodic SCF, it satisfies (3.36).

Now, suppose $\phi \ (= \phi_0)$ satisfies (3.36); then we know by Proposition 3.4 that ϕ_i satisfies (3.36) for all $i \geq 0$. We must have some j and $k \ (k > j)$ such that $\phi_j = \phi_k$, but by (3.2), this means that

$$q_{j-1} - q_{k-1} = \overline{\phi}_{k-1} - \overline{\phi}_{j-1} \, .$$

Since $|\overline{\phi}_{k-1} - \overline{\phi}_{j-1}| < 1$, we can only have $q_{j-1} = q_{k-1}$ and $\phi_{j-1} = \phi_{k-1}$ by (3.2). By induction, we will finally get $\phi = \phi_0 = \phi_{k-j}$. Thus, we have proved the following theorem.

Theorem 3.8. *The SCF expansion of the real quadratic irrational ϕ is purely periodic if and only if ϕ satisfies (3.36).*

As we mentioned in §1.3, the problem of determining the fundamental solution t, u of the Pell equation can be reduced to the problem of determining ϵ, the fundamental solution of (1.10). We will now show how ϵ can be found by making use of continued fractions. We point out that the technique about to be described will find ϵ for $D \in S = \{2, 3, 5, 8, 12\}$, but it is more convenient simply to present these values of ϵ in Table 3.1 and assume that $D \notin S$ in what follows.

Table 3.1. ϵ for $D \leq 5$

D	ϵ	σ_1
2, 8	$1 + \sqrt{2}$	-1
3, 12	$2 + \sqrt{3}$	1
5	$\dfrac{1 + \sqrt{5}}{2}$	-1

We first note that if (1.10) is solvable for $X \equiv DY \equiv 1 \pmod 2$, then $D \equiv 1 \pmod 4$. For a given D, we define s and q as follows:

$$s = \begin{cases} 2 & \text{if } 4 \mid D \text{ or } D \equiv 1 \pmod 4 \\ 1 & \text{otherwise} , \end{cases}$$

$$q = \begin{cases} 0 & \text{if } D \not\equiv 1 \pmod 4 \\ 1 & \text{if } D \equiv 1 \pmod 4 . \end{cases}$$

We put

$$\delta = \frac{q + \sqrt{D}}{s} > 1$$

and observe that $N(\delta) \in \mathbb{Z}$ and $\epsilon = w_1 + \delta z_1$, where

$$w_1 = \frac{x_1 - qy_1}{2}, \quad z_1 = \frac{sy_1}{2}$$

are both integers. We next see that (1.10) becomes

$$\epsilon \overline{\epsilon} = \sigma_1$$

or
$$(w_1 + \delta z_1)(w_1 + \overline{\delta} z_1) = \sigma_1 .$$
Since $\delta + \overline{\delta} = 2q/s = q$, we can write this as
$$(w_1 + \delta z_1)(w_1 + qz_1 - \delta z_1) = \sigma_1 ;$$
hence,
$$|w_1 + qz_1 - \delta z_1| = |w_1 + \delta z_1|^{-1} . \tag{3.37}$$
Since $x_1^2 - Dy_1^2 = 4\sigma_1$, we have $x_1^2 \geq Dy_1^2 - 4 > 5y_1^2 - 4 \geq y_1^2$. It follows that $x_1 > (s-1)y_1$ and $w_1, z_1 > 0$. Also, since $\delta > 2$, we have $\epsilon = w_1 + \delta z_1 > 2z_1$. From this we can deduce the following inequality from (3.37):
$$\left| \frac{w_1 + qz_1}{z_1} - \delta \right| < \frac{1}{2z_1^2} .$$
By Theorem 3.2, we see that $(w_1 + qz_1)/z_1$ must be a convergent in the SCF expansion of δ; that is, if
$$\delta = \phi_0 = [q_0, q_1, \ldots, q_i, \phi_{i+1}] ,$$
$Q_0 = s$, and $P_0 = q$, then
$$\frac{w_1 + qz_1}{z_1} = \frac{A_n}{B_n}$$
for some minimal n (≥ 0). Since $(w_1, z_1) = 1$, we must have $z_1 = B_n$, $w_1 + qz_1 = A_n$, from which we see that $w_1 = A_n - qB_n$ and $sw_1 + qz_1 = sA_n - qB_n = G_n$. By (3.13) we get
$$\sigma_1 = \frac{(-1)^{n+1} Q_{n+1}}{s} .$$
Since $\phi_0 = \delta > 1$ and $\overline{\delta} < 0$, we have (3.32) satisfied for all $i \geq 0$. Hence, $\sigma_1 = (-1)^{n+1}$ and $Q_{n+1} = s$. Since $Q_{n+1} \mid D - P_{n+1}^2$, we have $P_{n+1} \equiv -q \pmod{s}$. If we put $P_{n+1} = sS - q$, then because $0 < (\sqrt{D} - P_{n+1})/s < 1$, we get $S = \lfloor \delta \rfloor = q_0$ and $P_{n+1} = sq_0 - q = P_1 \neq q = P_0$ because $\delta > 2$. Hence, $q_{n+1} = 2q_0 - q$ and $P_{n+2} = sq_{n+1} - P_{n+1} = sq_0 - q = P_1$. Thus, since $P_{n+2} = P_1$ and $Q_{n+2} = (D - P_{n+2}^2)/Q_{n+1} = (D - P_1^2)/Q_0 = Q_1$, we see that the period length of the SCF expansion of δ is $n + 1$ and the preperiod consists of only q_0. We therefore have
$$\delta = [q_0; \overline{q_1, q_2, \ldots, q_p}] ,$$
where $p = n + 1$. Also,
$$\epsilon = \frac{G_{p-1} + \sqrt{D} B_{p-1}}{s} = \theta_{p+1} .$$
The minimality condition on ϵ guarantees that there cannot be a period of smaller length than p.

We now have the following algorithm for finding ϵ. Put $P_0 = q$, $Q_0 = s$, $q_0 = \lfloor (\sqrt{D} + q)/s \rfloor$, $B_{-1} = 0$, $B_0 = 1$, $G_{-1} = s$ and $G_0 = sq_0 - q$. Use the recurrences

$$P_{i+1} = q_i Q_i - P_i ,$$

$$Q_{i+1} = \frac{D - P_{i+1}^2}{Q_i} ,$$

$$q_{i+1} = \left\lfloor \frac{P_{i+1} + \sqrt{D}}{Q_i} \right\rfloor ,$$

$$G_{i+1} = q_{i+1} G_i + G_{i-1} ,$$

$$B_{i+1} = q_{i+1} B_i + B_{i-1} \quad \text{(from (3.3))}$$

for $i = 0, 1, 2, \ldots$ until we find the least positive p for which $Q_p = s$. Then

$$\epsilon = \frac{G_{p-1} + \sqrt{D} B_{p-1}}{s} \quad \text{and} \quad \sigma_1 = (-1)^p .$$

The determination of $\lfloor (P_{i+1} + \sqrt{D})/Q_i \rfloor$ might be considered troublesome because $\sqrt{D}$ is not an integer, but it is easy to show that for any real number ρ and any positive integer m that $\lfloor \rho/m \rfloor = \lfloor \lfloor \rho \rfloor/m \rfloor$. Hence, we have

$$\left\lfloor \frac{P_i + \sqrt{D}}{Q_i} \right\rfloor = \left\lfloor \frac{P_i + \lfloor \sqrt{D} \rfloor}{Q_i} \right\rfloor .$$

Example 3.9. Find ϵ when $D = 46$.
Here $D \not\equiv 1 \pmod 4$ and $q = 0$, $s = 1$, $\phi = \delta = \sqrt{46}$, and $\lfloor \sqrt{46} \rfloor = 6$. We begin with $P_0 = 0$, $Q_0 = 1$, $q_0 = 6$, $B_{-1} = 0$, $B_0 = 1$, $G_{-1} = 1$, and $G_0 = 6$. We use the formulas above to produce Table 3.2.

Table 3.2. Computing ϵ when $D = 46$

n	P_n	Q_n	q_n	G_n	B_n	n	P_n	Q_n	q_n	G_n	B_n
1	6	10	1	7	1	7	6	5	2	2150	317
2	4	3	3	27	4	8	4	6	1	3147	464
3	5	7	1	34	5	9	2	7	1	5297	781
4	2	6	1	61	9	10	5	3	3	19038	2807
5	4	5	2	156	23	11	4	10	1	24335	3588
6	6	2	6	997	147	12	6	1 (= s)			

Hence, in this case we get $p = 12$, $\sigma_1 = 1$, and

$$\epsilon = 24335 + 3588\sqrt{46} .$$

Also, by the results in Table 1.1 we know that $t + u\sqrt{46} = \epsilon$, so $t = 24335$ and $u = 3588$.

Of course, it is clear that we can rewrite the Pell equation as

$$(2T)^2 - (4D)U^2 = 4 .$$

If we replace D by $4D$, we get $s = 2, q = 0$, and $\delta = \sqrt{D}$. Thus, we can find the solution (t, u) of the Pell equation by simply developing the continued fraction expansion of $\sqrt{D}$ until we find $Q_p = 1$. However, there are cases in which, for a given D, it is more convenient to solve for ϵ than to find (t, u) as described by this process. We illustrate this in the next example.

Example 3.10. Find ϵ when $D = 541$.
In this case, we have $D \equiv 1 \pmod 4$ and $\delta = (1+\sqrt{541})/2$. Here, $\lfloor \sqrt{541} \rfloor = 23$ and $P_0 = 1$, $Q_0 = 2$, $q_0 = 12$, $B_{-1} = 0$, $B_0 = 1$, $G_{-2} = 2$, and $G_0 = 23$. We now produce the values in Table 3.3.

Table 3.3. Computing ϵ when $D = 541$

n	P_n	Q_n	q_n	G_n	B_n
1	23	6	7	163	7
2	19	30	1	186	8
3	11	14	2	535	23
4	17	18	2	1256	54
5	19	10	4	5559	239
6	21	10	4	23492	1010
7	19	18	2	52543	2259
8	17	14	2	128578	5528
9	11	30	1	181121	7787
10	19	6	7	1396425	60037
11	23	2 $(= s)$			

Here, we have $p = 11$, $\sigma = (-1)^{11} = -1$, and

$$\epsilon = \frac{1396425 + 60037\sqrt{541}}{2} .$$

Referring to Table 1.1, we see that

$$t + u\sqrt{541} = \epsilon^6 ;$$

hence, the fundamental solution of the Pell equation $T^2 - 541, U^2 = 1$ is given by

$$t = 3707453360023867028800645599667005001$$

and

$$u = 1593958697212701100771871387138775196900 \, .$$

If we were to solve this equation by employing the SCF expansion of $\sqrt{541}$, we would get a fundamental solution of $T^2 - DU^2 = -1$ and $p = 39$.

We conclude this series of examples with a discussion of the Cattle Problem.

Example 3.11. The Cattle Problem.
As mentioned in Chapter 2, the Cattle Problem reduces to that of solving the Pell equation

$$T^2 - DU^2 = 1 \tag{3.38}$$

for $D = 410286423278424$. We could solve this by finding the SCF expansion[22] of $\sqrt{D}$, but as the period length[23] is 203254, this would take a great deal of effort. In 1880, Amthor[24] noted that

$$D = (9314)^2 4729494 \, .$$

Thus, he could make use of ideas similar to those presented in §1.4 to solve (3.38) by first solving

$$T^2 - D_0 U^2 = 1 \, ,$$

where $D_0 = 4729494$. The period length of the SCF expansion of $\sqrt{D_0}$ is only 92, and for this equation we get

$$t = 109931986732829734979866232821433543901088049$$

and

$$u = 50549485234315033074477819735540408986340 \, .$$

We now need to find n such that the fundamental solution (t', u') of (3.38) is given by

$$t' + u'\sqrt{D} = (t + u\sqrt{D_0})^n \, .$$

We know that $f = 9314$ and $n = \omega(f/g)$, where $g = (f, u) = 2$. Thus, $n = \omega(4657)$. Since 4657 is a prime and $(D_0/4657) = -1$, we know that $\omega(4657) \mid 4658$. In fact, Amthor found that $\omega(4657) = 2329$. Hence,

$$t' + u'\sqrt{D} = (t + u\sqrt{D_0})^{2329} \, .$$

Amthor was unable to complete the solution of the Cattle Problem because the value of u' is approximately 1.86×10^{103265}. The total number of cattle was finally computed[25] in 1965 by making use of a computer; it was later published by Nelson[26] in 12 pages of fine print. Vardi[27] has given a very elegant representation of the total number of cattle as

$$\left[\frac{25194541}{184119142} \left(\begin{matrix} 1099319867328297349798662328214335439010880 49 \\ + 50549485234315033074477819735540408986340\sqrt{4729494} \end{matrix} \right)^{4658} \right]$$

and Lenstra[28] has presented a short table which describes all possible solutions for each herd of cattle. The least possible number of cattle on the Thrinacian isle is approximately 7.76×10^{206544}, an enormous figure, which Archimedes could not possibly have known.

The SCF method of solving the Pell equation had been pretty well developed by the time of Euler and Lagrange. Subsequent to its development, several authors, starting with Legendre in 1798, produced tables of values of t and u for values of D in certain ranges. These tables were described by Lehmer.[29] This work culminated in an unpublished table of Lehmer[30] in 1926 which dealt with all non-square D in the range $1700 < D \le 2000$. At this point, (1.7) had been solved for all positive non-square $D < 2000$. Considering that all of these tables had been produced by hand calculation, it is not difficult to see why this was about as much as could be done for several years. In 1941 and more extensively in 1955, Patz[31] produced tables of the SCF of $\sqrt{D}$ for all non-square values of D such that $1 < D < 10000$, but he did not compute t and u for each D. Finally, in 1961 Kortum and McNeil[32] used a computer to produce a very large table, giving, among other things, the SCF of $\sqrt{D}$ and the least solution of $T^2 - DU^2 = \pm 1$ for all non-square D such that $1 < D < 10000$. For reasons that will become clear later, no further tables of Pell equations have ever been published. It should, however, be mentioned that before the advent of computers, several authors attempted to solve certain Pell equations for values of D in excess of 2000. One of the more inpressive of these is the solution for $D = 9817$ by Martin[33] in 1877. Not only is his solution correct, but the value of t is a number of 97 decimal digits.

At this point it may appear to the reader that the problem of solving the Pell equation has been solved. This is certainly the case for $D < 10000$, but there are many questions left outstanding. For example, how large does p get for a certain size of D? This is important to know in order to determine the computational complexity of solving (1.7) by the continued fraction method. Also, associated with this problem is that of bounding the size of t and u for a given D. This is important for giving us information about just how likely we can actually solve (1.7) for large values of D. Are there faster ways than the continued fraction method for solving (1.7)? For what size of D can we expect to be able to solve the Pell equation? We will attempt to answer all of these questions and more in the subsequent chapters of this book. However, just to motivate this discussion further, we mention that if

$$D = 990676090995853870156271607886,$$

the value of t is a number consisting of over 2×10^{15} decimal digits, whereas the solutions of the Cattle Problem are numbers of just over 2×10^5 digits.

3.4 Some Special Results

Because of the importance of simple continued fractions in solving the Pell equation, we will provide some further results concerning these objects. We first mention that the development of the SCF of any quadratic irrational $\phi = (P + \sqrt{D})/Q$ can be made more efficient by the use of what has become known as Tenner's[34] modification to the expansion technique given earlier. We first define $Q_{-1} = (D - P^2)/Q$. We let R_{i-1} be the remainder on dividing the integer $P_{i-1} + \lfloor \sqrt{D} \rfloor$ by Q_{i-1}. Since

$$P_{i-1} + \lfloor \sqrt{D} \rfloor = q_{i-1}Q_{i-1} + R_{i-1} \, ,$$

we get

$$P_i = q_{i-1}Q_{i-1} - P_{i-1} = \lfloor \sqrt{D} \rfloor - R_{i-1} \, .$$

Also,

$$
\begin{aligned}
Q_i &= \frac{D - P_i^2}{Q_{i-1}} = \frac{D - (q_{i-1}Q_{i-1} - P_{i-1})^2}{Q_{i-1}} \\
&= \frac{(D - P_{i-1}{}^2)}{Q_{i-1}} - q_{i-1}{}^2 Q_{i-1} + 2P_{i-1}q_{i-1} \\
&= Q_{i-2} - q_{i-1}(P_i - P_{i-1}) \quad (i \geq 1) \, .
\end{aligned}
$$

Thus, we can now find ϕ_i from ϕ_{i-1} by dividing $P_{i-1} + \lfloor \sqrt{D} \rfloor$ by Q_{i-1} and obtaining the quotient q_{i-1} and remainder R_{i-1}. Then

$$P_i = \lfloor \sqrt{D} \rfloor - R_{i-1}$$

and

$$Q_i = Q_{i-2} - q_{i-1}(P_i - P_{i-1}) \, .$$

As most computers provide both the quotient and the remainder on a simple divide instruction, this means that (ignoring the cost of a subtraction) we replace a process for finding ϕ_i from ϕ_{i-1} which requires 2 divides, 1 multiply, and 1 squaring by one that requires only 1 divide and 1 multiply (usually by a small number; see the latter part of this section).

We have seen that in the SCF expansion of ϕ_0, it is a simple matter to determine ϕ_i from ϕ_{i-1}; however, if $\phi_0 > 1$ and $\overline{\phi}_0 < 0$, then we can also determine ϕ_{i-1}, given only ϕ_i ($i > 1$). We do this by noting that

$$Q_{i-1} = \frac{D - P_i^2}{Q_i}$$

and

$$
\begin{aligned}
\frac{P_i + \sqrt{D}}{Q_{i-1}} &= \frac{q_{i-1}Q_{i-1} - P_{i-1} + \sqrt{D}}{Q_{i-1}} \\
&= q_{i-1} + \frac{\sqrt{D} - P_{i-1}}{Q_{i-1}} \, .
\end{aligned}
$$

Since, by Proposition 3.4, $0 < -\overline{\phi}_{i-1} = (\sqrt{D} - P_{i-1})/Q_{i-1} < 1$ $(i > 1)$, we get

$$q_{i-1} = \left\lfloor \frac{P_i + \lfloor\sqrt{D}\rfloor}{Q_{i-1}} \right\rfloor \quad (i > 1) . \tag{3.39}$$

With this information, we find P_{i-1} by

$$P_{i-1} = q_{i-1}Q_{i-1} - P_i .$$

The Tenner variant of this process is to let R'_{i-1} be the remainder on dividing $P_i + \lfloor\sqrt{D}\rfloor$ by Q_{i-1}. Then by using reasoning similar to that used earlier, we get

$$Q_{i-1} = Q_{i+1} - q_i(P_i - P_{i+1}) ,$$

$$q_{i-1} = \left\lfloor \frac{P_i + \lfloor\sqrt{D}\rfloor}{Q_{i-1}} \right\rfloor ,$$

$$P_{i-1} = \lfloor\sqrt{D}\rfloor - R'_{i-1} .$$

We now discuss some properties of the SCF of special values of $\phi = (P + \sqrt{D})/Q$, where

1. $Q \mid 2P$
2. $\overline{\phi} < \lfloor\phi\rfloor - 1$.

Notice that both δ and $\sqrt{D}$ satisfy these conditions. In Examples 3.9 and 3.10, we saw that the values of Q_i, P_i, and q_i $(i = 1, 2, \ldots, p)$ are symmetric around $\lfloor p/2 \rfloor$. This is a consequence of the following theorem.

Theorem 3.12. *In the SCF expansion of ϕ satisfying properties (1) and (2) above, we have*

$$Q_{p-i} = Q_i ,$$
$$P_{p-i} = P_{i+1}$$

for $0 \le i < p$, where p is the period length. Also, if $0 < i < p$, then

$$q_{p-i} = q_i .$$

Proof. Put $P_0 = P$, $Q_0 = Q$. We know that $\phi_1 > 1$, and since $\overline{\phi}_0 < \lfloor\phi_0\rfloor - 1$, we know that $-1 < \overline{\phi}_1 < 0$; hence, the SCF expansion of ϕ_1 is purely periodic by Theorem 3.8 and (3.39) must hold. It follows that for the period length p, we have $\phi_{p+i} = \phi_i$ for $i \ge 1$ and

$$\phi = [q_0, \overline{q_1, q_2, \ldots, q_p}] .$$

Since $\phi_{p+1} = \phi_1$, we get $P_{p+1} = P_1$, $Q_{p+1} = Q_1$. Now,

$$Q_{p+1}Q_p = D - P_{p+1}^2 = D - P_1^2 = Q_1Q_0 ;$$

hence, $Q_p = Q_0$. Also, since $P_1 = q_0 Q_0 - P_0$ and $P_{p+1} = q_p Q_p - P_p$, we have $P_0 \equiv P_p \pmod{Q_0}$. However, we know by property (1) that $P_0 \equiv -P_0 \pmod{Q_0}$ and, therefore, $P_0 \equiv -P_p \pmod{Q_0}$. Let $P_p = mQ_0 - P_0$. Since $-1 < \overline{\phi}_p < 0$, we get $m = \lfloor (P_0 + \sqrt{D})/Q_0 \rfloor = q_0$ and $P_p = q_0 Q_0 - P_0 = P_1$. We now have $P_p = P_1$ and $Q_p = Q_0$. From this we see that the theorem is true for $p = 1$. Suppose $p > 1$. Since

$$Q_{p-1} = \frac{D - P_p{}^2}{Q_p} = \frac{D - P_1{}^2}{Q_0} = Q_1 \,,$$

we find by (3.39) that

$$q_{p-1} = \left\lfloor \frac{P_p + \sqrt{D}}{Q_{p-1}} \right\rfloor = \left\lfloor \frac{P_1 + \sqrt{D}}{Q_1} \right\rfloor = q_1$$

and

$$P_{p-1} = q_{p-1} Q_{p-1} - P_p = q_1 Q_1 - P_1 = P_2 \,.$$

Thus, the result is true for $i = 0, 1$. Suppose next that $Q_{p-i} = Q_i$, $P_{p-i} = P_{i+1}$, and $q_{p-i} = q_i$ for some i such that $0 < i < p$. Then

$$Q_{p-i-1} = \frac{D - P_{p-i}{}^2}{Q_{i-1}} = \frac{D - P_{i+1}{}^2}{Q_i} = Q_{i+1} \,.$$

If $i + 1 < p$, then, by (3.39),

$$q_{p-i-1} = \left\lfloor \frac{P_{p-i} + \sqrt{D}}{Q_{p-i-1}} \right\rfloor = \left\lfloor \frac{P_{i+1} + \sqrt{D}}{Q_{i+1}} \right\rfloor = q_{i+1}$$

and

$$\begin{aligned} P_{p-i-1} &= q_{p-i-1} Q_{p-i-1} - P_{p-i} \\ &= q_{i+1} Q_{i+1} - P_{i+1} \\ &= P_{i+2} \,. \end{aligned}$$

Thus, the theorem follows by induction on i.   □

Theorem 3.13. *Let p be the period length of the SCF expansion of ϕ satisfying conditions (1) and (2) above. If $p = 2t$, then $P_t = P_{t+1}$ and*

$$G_{p-1} = G_t B_{t-1} + G_{t-1} B_{t-2} \,,$$
$$B_{p-1} = B_{t-1}(B_t + B_{t-2}) \,;$$

if $p = 2t + 1$, then $Q_t = Q_{t+1}$ and

$$G_{p-1} = G_t B_t + G_{t-1} B_{t-1} \,,$$
$$B_{p-1} = B_t{}^2 + B_{t-1}{}^2 \,.$$

Proof. If $p = 2t$, then by Theorem 3.12,

$$\frac{A_{p-1}}{B_{p-1}} = [q_0, q_1, \ldots, q_{t-1}, q_t, q_{t-1}, \ldots, q_1] .$$

By (3.5), we have

$$[q_t, q_{t-1}, \ldots, q_1] = \frac{B_t}{B_{t-1}} ,$$

and by (3.6),

$$\frac{A_{p-1}}{B_{p-1}} = \frac{(B_t/B_{t-1})A_{t-1} + A_{t-2}}{(B_t/B_{t-1})B_{t-1} + B_{t-2}} = \frac{B_t A_{t-1} + B_{t-1}A_{t-2}}{B_t B_{t-1} + B_{t-1}B_{t-2}} .$$

If u is any prime such that $u \mid B_t A_{t-1} + B_{t-1}A_{t-2}$ and $u \mid B_t B_{t-1} + B_{t-1}B_{t-2}$, then, by (3.4), it is easy to see that $u \mid B_t$ and $u \mid B_{t-1}$, which by (3.4) is impossible. Hence,

$$A_{p-1} = B_t A_{t-1} + B_{t-1}A_{t-2} = A_t B_{t-1} + A_{t-1}B_{t-2} \quad \text{by (3.4)}$$

and

$$B_{p-1} = B_{t-1}(B_t + B_{t-2}) .$$

From the definition of G_i, it is easy to show via the above two results that

$$G_{p-1} = G_t B_{t-1} + G_{t-1}B_{t-2} .$$

If $p = 2t + 1$, then

$$\frac{A_{p-1}}{B_{p-1}} = [q_0, q_1, q_2, \ldots, q_{t-1}, q_t, q_t, q_{t-1}, \ldots, q_1]$$

$$= \frac{(B_t/B_{t-1})A_t + A_{t-1}}{(B_t/B_{t-1})B_t + B_{t-1}} = \frac{A_t B_t + A_{t-1}B_{t-1}}{B_t^2 + B_{t-1}^2} .$$

By using (3.4), it is easy to show that $(B_t^2 + B_{t-1}^2, A_t B_t + A_{t-1}B_{t-1}) = 1$ and, therefore,

$$A_{p-1} = A_t B_t + A_{t-1}B_{t-1} ,$$
$$B_{p-1} = B_t^2 + B_{t-1}^2 .$$

We can then easily deduce from the definition of G_i that

$$G_{p-1} = G_t B_t + G_{t-1}B_{t-1} .$$

$\square$

In the next theorem we will require the following lemma.

Lemma 3.14. *If in the SCF expansion of ϕ we have $P_k = P_{k+1}$ $(k \geq 0)$, then*

$$G_{k-1}{}^2 + DB_{k-1}{}^2 = Q_k(G_kB_{k-1} + G_{k-1}B_{k-2})$$
$$2G_{k-1} = Q_k(B_k + B_{k-2}) \,.$$

If $Q_k = Q_{k+1}$ $(k \geq 0)$, then

$$G_{k-1}G_k + DB_kB_{k-1} = Q_{k+1}(G_kB_k + G_{k-1}B_{k-1}) \,,$$
$$G_kB_{k-1} + G_{k-1}B_k = Q_{k+1}(B_n{}^2 + B_{k-1}{}^2) \,.$$

Proof. These can be easily verified by making use of (3.12), (3.14), and (3.3). For example, to verify that $G_{k-1}G_k + DB_kB_{k-1} = Q_{k+1}(G_kB_k + G_{k-1}B_{k-1})$ when $Q_k = Q_{k+1}$, we note that by (3.12) and (3.14),

$$\begin{aligned}
G_{k-1}G_k + DB_kB_{k-1} &= G_k(B_kQ_k - P_{k+1}B_{k-1}) \\
&\quad + B_{k-1}(P_{k+1}G_k + Q_{k+1}G_{k-1}) \\
&= Q_kG_kB_k + Q_{k+1}G_{k-1}B_{k-1} \\
&= Q_{k+1}(G_kB_k + G_{k-1}B_{k-1}) \,.
\end{aligned}$$

$\square$

Theorem 3.15. *If in the SCF expansion of δ we get $P_k = P_{k+1}$, for some minimal $k \geq 1$, then $p = 2k$; if we get $Q_k = Q_{k+1}$ for some minimal $k \geq 0$, then $p = 2k + 1$.*

Proof. By the symmetry properties in Theorem 3.12, we may assume that $k \leq \lfloor p/2 \rfloor = t$. If $k \neq t$, then $k < t$ and $k + 2 \leq t + 1$. By (3.34) and (3.17), we note that

$$\theta_{k+1} < \theta_{t+1} \,.$$

Also, since $|\overline{\psi}_i| = 1/\phi_i < 1$ $(i = 0, 1, 2, \ldots)$, we get, by (3.17),

$$1 \geq |\overline{\theta}_{k+1}| > |\overline{\theta}_{k+2}| \geq |\overline{\theta}_{t+1}| > |\overline{\theta}_{t+2}| \,;$$

hence,

$$1 \leq |\overline{\theta}_{k+1}|^{-1} < |\overline{\theta}_{k+2}|^{-1} \leq |\overline{\theta}_{t+1}|^{-1} < |\overline{\theta}_{t+2}|^{-1} \,.$$

It follows that $\theta_{k+1}/|\overline{\theta}_{k+1}|, \theta_{k+1}/|\overline{\theta}_{k+2}| > 1$ and each is less than either $\theta_{t+1}/|\overline{\theta}_{t+1}|$ or $\theta_{t+1}/|\overline{\theta}_{t+2}|$. By the definition of θ_j, (3.19), Theorem 3.13, and Lemma 3.14, we see that

$$\epsilon = \begin{cases} \theta_{t+1}/|\overline{\theta}_{t+1}| & \text{when } P_t = P_{t+1} \\ \theta_{t+1}/|\overline{\theta}_{t+2}| & \text{when } Q_t = Q_{t+1} \,. \end{cases}$$

However, if $P_k = P_{k+1}$, then $|N(\theta_{k+1}/|\overline{\theta}_{k+1}|)| = 1$, which by Lemma 3.14 means that we have a solution to (1.10) for $\sigma = N(\theta_{k+1}/|\overline{\theta}_{k+1}|)$. If $Q_k = Q_{k+1}$, then by (3.18), we get

$$|N(\theta_{k+1}/|\overline{\theta}_{k+2}|)| = |N(\theta_{k+1})/N(\theta_{k+2})| = Q_k/Q_{k+1} = 1$$

and we have a solution to (1.10) for $\sigma = N(\theta_{k+1}/|\overline{\theta}_{k+2}|)$.

In either case we get a solution (x, y, σ) for (1.10) such that $1 < (x + y\sqrt{D})/2 < \epsilon$. By definition of ϵ, this is impossible.   □

This means that we have a faster way of finding ϵ than by going though the entire period of the SCF of δ. We need only go as far as t where either $P_t = P_{t+1}$ or $Q_t = Q_{t+1}$. If we return to Example 3.9, we see that $P_6 = P_7$; hence, $p = 12$ and $t = 6$. We get

$$G_{11} = 23 \cdot 997 + 156 \cdot 9 = 24335 \ ,$$
$$B_{11} = 23(147 + 9) = 3588 \ .$$

In Example 3.10, we see that $Q_5 = Q_6$ and $p = 11$, $t = 5$. Here,

$$G_{10} = 5559 \cdot 239 + 1256 \cdot 54 = 1396425 \ ,$$
$$B_{10} = 239^2 + 54^2 = 60037 \ .$$

It is possible to deduce by elementary reasoning[35] that the period length of the SCF expansion of δ (or $\sqrt{D}$) is bounded by

$$p = O(\sqrt{D}\log D) \ .$$

By using results in §5.3 and §9.5 it is possible to produce a more precise bound,[36] but in order to do so we will need some preliminary results. We begin by examining the well-known sequence $\{F_n\}$ of Fibonacci, where $F_0 = 0, F_1 = 1$, and $F_{j+1} = F_j + F_{j-1}$ $(j = 1, 2, \dots)$.

Now, as is well known, F_i can be represented in closed form by the Binet formula $F_i = (\tau^i - \overline{\tau}^i)/\sqrt{5}$, where $\tau = (1 + \sqrt{5})/2$. Since $\tau + \overline{\tau} = 1$ and $\tau - \overline{\tau} = \sqrt{5}$, we have $\tau^2 - \overline{\tau}^2 = \sqrt{5}$. Also, since $\overline{\tau} = -\tau^{-1}$, we have $-1 < \overline{\tau} < 0$; hence, $0 < \overline{\tau}^{2i} \leq \overline{\tau}^2 < \tau^2$,

$$\tau^2 - (-1)^{i-1}\overline{\tau}^{2i} \geq \sqrt{5} \ ,$$

and

$$\tau^{i+1} - \overline{\tau}^{i+1} \geq \sqrt{5}\tau^{i-1} \quad (i \geq 1) \ .$$

Thus,

$$F_{i+1} \geq \tau^{i-1} \ .$$

We now establish the following simple proposition.

Proposition 3.16. *Consider the SCF expansion of δ and let θ_j be defined as in §3.1. We have*

$$\theta_{m+i} > F_{i+1}\theta_m \ ,$$

where $i, m \geq 1$.

Proof. Since $\psi_m > 1$ (see (3.34)), we see that the result holds for $i = 1$ ($F_2 = 1$). Now,

$$\psi_{m+1}\psi_m = \psi_m \frac{P_{m+1} + \sqrt{D}}{Q_m}$$

$$= \psi_m \frac{q_m Q_m - P_m + \sqrt{D}}{Q_m}$$

$$= \psi_m q_m + \frac{(\sqrt{D} - P_m)(\sqrt{D} + P_m)}{Q_m Q_{m-1}}$$

$$= \psi_m q_m + 1 .$$

Since $m \geq 1$, we have $q_m \geq 1$ and $\psi_m \psi_{m+1} > 2$; thus, the result holds for $i = 2$. That it holds for all $i \geq 3$ follows by using

$$\theta_j = q_{j-2}\theta_{j-1} + \theta_{j-2} \geq \theta_{j-1} + \theta_{j-2}$$

and mathematical induction. □

If we now put $i = n - 1$ and $m = 1$ in Proposition 3.16, we get

$$\theta_n > F_n \geq \tau^{n-2} ;$$

hence,

$$\log \theta_n > (n - 2) \log \tau$$

or

$$n < \frac{\log \theta_n}{\log \tau} + 2 . \tag{3.40}$$

Since $1/\log \tau < 2.078087$, we get, on putting $n = p + 1$,

$$p < 2.078087 \log \theta_{p+1} + 1 .$$

To get a better idea of how p and $\log \theta_{p+1}$ usually relate, we appeal to a result from the metrical theory of continued fractions.[37] We will now review some results from this theory that will be of importance to us later. If

$$\phi = [q_0, q_1, \ldots, q_j, \phi_{j+1}] ,$$

then $t = \phi - q_0 = [0, q_1, \ldots, q_j, \phi_{j+1}]$. Since $0 \leq t < 1$, we may confine ourselves to the study of the SCF of t for $0 \leq t < 1$. We use $\phi_{j+1}(t)$ to denote ϕ_{j+1} above.

Theorem 3.17 (Khintchine-Lévy[38]). *For almost all real t such that $0 \leq t < 1$, we have*

$$\lim_{n \to \infty} \left(\prod_{i=1}^{n} \phi_i \right)^{1/n} = e^\lambda ,$$

where $\phi_i = \phi_i(t)$ and $\lambda = \pi^2/(12 \log 2) \approx 1.186569$.

From (3.17) and the fact that Q_i is bounded in the SCF expansion of $t = \delta - \lfloor \delta \rfloor$, we see that this means that we would expect that

$$\log \theta_{p+1} \approx p\lambda , \qquad (3.41)$$

or

$$p \approx \frac{1}{\lambda} \log \theta_{p+1} \approx 0.842766 \log \theta_{p+1} .$$

For example, if $D = 26437680473689$, the value of p in the SCF expansion[39] of δ is 18334815 and $\log \theta_{p+1} = 21737796.43$. In this case we get

$$\frac{p}{\log \theta_{p+1}} = 0.84345 .$$

However, it must be emphasized that Theorem 3.17 is not true for all real t. Indeed, there are many values of D for which it is not true for $\delta - q_0$. For example, consider $D = M^2 + 1$, where $2 \nmid M$. In this case it is easy to see that $\delta = \sqrt{D}$ and $\phi_i = M + \sqrt{D}$ $(i = 1, 2, \dots)$. Hence,

$$\lim_{n \to \infty} \left(\prod_{i=1}^{n} \phi_i \right)^{1/n} = M + \sqrt{D} \neq e^\lambda .$$

Another useful result from the metrical theory is the Gauss-Kuz'min theorem.

Theorem 3.18 (Gauss-Kuz'min[40]). *As before, we let $0 \leq t < 1$ and let $y \geq 1$ be a given real number. If $\Pr(\phi_{j+1}(t) > y)$ denotes the probability that $\phi_{j+1}(t) > y$ $(j \geq 0)$, then*

$$\Pr(\phi_{j+1}(t) > y) = \frac{\log(1 + 1/y)}{\log 2} + O(q^j) ,$$

where q is some real number such that $0 < q < 0.76$.

If we put $y = 10$, we see that we would expect in the SCF of any t that $\phi_j > 10$ for about 13.75% of the j values. In other words, most of the partial quotients in a SCF tend to be small. For a given integer k, this result can be used[41] to determine $\Pr(q_{j+1} = k)$ $(j \geq 0)$, the probability that any partial quotient q_{j+1} is equal to k. This is given by

$$\Pr(q_{j+1} = k) = \frac{\log(1 + 1/k(k+2))}{\log 2} + O\left(\frac{q^j}{k(k+1)}\right) .$$

We provide some approximate values of these probabilities in Table 3.4. Notice that we expect that over 58% of all the partial quotients[42] are either 1 or 2. Also, $\Pr(q_n = r)$ and $\Pr(q_{n+m} = s)$ are only weakly dependent, as it can be proved[43] that

$$\Pr(q_n = r \text{ and } q_{n+m} = s) = \Pr(q_n = r) \Pr(q_{n+m} = s) (1 + O(q^m)) ,$$

where $0 < q < 1$.

Table 3.4.

k	$\Pr(q_{i+1} = k)$
1	0.415037
2	0.169925
3	0.093109
4	0.058894
5	0.040642

Notes and References

[1]The best reference for this is still [Per57]. However, there are a number of other sources that are extremely useful, especially for the analytic properties of continued fractions. In this category, we mention [Wal48] and [LW92]. Also, in [Bre80] there is a historical account of much of this material.

[2][Per57] Vol. I, §38. At the beginning of this section it is shown that if $\phi_i > 1$ (for all $i > 0$), then (3.1) must be semiregular.

[3]This is a more general presentation than that for the usual simple continued fraction, but it will be found to be useful for improving certain algorithms.

[4][Per57], Vol. I.

[5]There is a considerable literature concerning simple continued fractions. We mention only a few sources here such as [Old63], [RS92], and [Per57]. It should be pointed out that many elementary textbooks on number theory (such as [NZM91], Ch. 7; [Dav60], Ch. 4; [Sta70], Ch. 7) contain a chapter on simple continued fractions.

[6]This is probably one of the oldest algorithms known. It appears in Euclid's *Elements* around 300 BC as Proposition 2 in Book VII. See [Hea56].

[7]However, see [Sha94].

[8]See [Knu98], §4.5.3.

[9]See, for example, [BS96a], p. 70.

[10][Sch71]. See also [AHU74].

[11][SS71]. See also [AHU74] and [Knu98].

[12][SZ04].

[13][Jeb93].

[14][Sor04a].

[15][Gra07].

[16][Leh38]. See also [Jeb95]. The complete algorithm can be found in [JSW06a] as Algorithm 3.1.

[17]See [Sor95].

[18]See, for example, [Sor04b], [Sor94], [Web95], and [Bsh99].

[19]See [CG90] and the discussion in Note 4.10 on p. 96 of [BS96a].

[20]This result has been generalized to $|\phi - a/b| < c/b^2$ for any real positive c by Worley [Wor81].

[21][Hur89].

[22]This was attempted by Meyer in 1867 (see [Dic19], Vol. II, p. 344), but he gave up after he had computed 240 steps in the continued fraction.

[23][Len02].

[24][Amt80].

[25][WGZ65].

[26][Nel81].

[27][Var98].

[28][Len02].

[29][Leh41].

[30][Leh26a].

[31][Pat41] and [Pat55].

[32][KM61]. See also Shanks' interesting review of this work in [Sha62].

[33][Mar77b].

[34]Tenner's algorithm can be found on p. 372 of Vol. II of [Dic19]. It should be emphasized that only the formula for P_i is due to Tenner. The formula for Q_i seems to go back to at least 1941, as is was mentioned in [Pat41], p. xiii, formula (3).

[35]See [RS92], p. 50.

[36]Some indication of how this can be done is given in [SSW76] and [Wil81]. Some numerical testing of this is provided in [Wil81] and [PW85].

[37]For sources on this material, see [RS92], Ch. V, or [Khi97] or [Knu98].

[38][Khi36] and [Lév36]. Khintchine demonstrated that the limit exists independently of t almost everywhere and Lévy at about the same time also determined this, together with the value of the limiting value λ.

[39]See [Sha74].

[40]See [Knu98] or Chapter V of [RS92] A proof of this is provided in [RS92], p. 152.

[41][RS92], p. 156ff.

[42]However, this is a probabilistic result and need not be true in any given instance. For example, if we return to the SCF expansion of $\sqrt{D}$ for $D = M^2 + 1$ ($2 \nmid M$), we find that $q_0 = M$ and $q_i = M + [\sqrt{D}] = 2M$ for all $i > 0$.

[43][RS92], p. 159.

4

Quadratic Number Fields

4.1 Algebraic Numbers

In this chapter we will show how the Pell equation has an important connection to the theory of real quadratic number fields. In order to do this, it will first be necessary to develop some theory of these structures. Several of the results in this section will be given without proof because they are standard elementary results which can be found in several texts.[1] We will begin with the definition of an algebraic number

Definition 4.1. *A complex number* α *is called* algebraic *if it is a zero of some polynomial* $f(x) \in \mathbb{Q}[x]$.

A very important property of an algebraic number is provided in the following result.

Theorem 4.2. *An algebraic number* α *is the zero of a unique irreducible (over* $\mathbb{Q}$*) monic polynomial* $g(x) \in \mathbb{Q}[x]$. *Furthermore, if* $h(x) \in \mathbb{Q}[x]$ *and* $h(\alpha) = 0$, *then* $g(x)$ *divides* $h(x)$ *in* $\mathbb{Q}[x]$.

Definition 4.3. *The* minimal polynomial *of an algebraic number* α *is the* $g(x)$ *described in Theorem 4.2. The* degree *of* α *is the degree of its minimal polynomial.*

We will require a useful result concerning monic polynomials with integer coefficients.

Theorem 4.4 (Gauss' Lemma[2]). *If a monic polynomial* $f(x) \in \mathbb{Z}[x]$ *factors into the product of two monic polynomials* $g(x), h(x) \in \mathbb{Q}[x]$, *then* $g(x), h(x) \in \mathbb{Z}[x]$.

We now need the concept of an algebraic integer.

Definition 4.5. *An algebraic number* α *is an* algebraic integer *if* α *is the zero of a monic polynomial* $f(x) \in \mathbb{Z}[x]$.

Note that if $m \in \mathbb{Z}$, then m is the zero of $f(x) = x - m$; hence, m is an algebraic integer. We now show that the only elements of $\mathbb{Q}$ which can be algebraic integers are those in $\mathbb{Z}$.

Theorem 4.6. *If $\alpha \in \mathbb{Q}$ and α is an algebraic integer, then $\alpha \in \mathbb{Z}$.*

Proof. If $\alpha \in \mathbb{Q}$, then $\alpha = a/b$, where $a, b \in \mathbb{Z}$, $b \neq 0$, and $(a, b) = 1$. Since α is a zero of a monic polynomial $f(x) = x^n + c_1 x^{n-1} + c_2 x^{n-2} + \cdots + c_n$ with $c_1, c_2, \ldots, c_n \in \mathbb{Z}$, we get

$$\left(\frac{a}{b}\right)^n + c_1 \left(\frac{a}{b}\right)^{n-1} + c_2 \left(\frac{a}{b}\right)^{n-2} + \cdots + c_n = 0$$

or

$$a^n + c_1 b a^{n-1} + c_2 b^2 a^{n-2} + \cdots + c_n b^n = 0 .$$

It follows that $b \mid a^n$. Since $(a, b) = 1$, we can only have $b = \pm 1$ and therefore $\alpha \in \mathbb{Z}$. □

We see, then, that the term *algebraic integer* in Definition 4.5 is a generalization of the common use of the term *integer*. We will use the term *rational integer* (or just integer) to distinguish the elements of $\mathbb{Z}$ from the irrational algebraic integers such as $\sqrt{-2}$ and $(1 + \sqrt{5})/2$.

Theorem 4.7. *The minimal polynomial of an algebraic integer is monic with integer coefficients.*

Proof. Clearly, we may assume that the minimal polynomial is monic. What we need to establish is that the remaining coefficients of this polynomial are (rational) integers. Suppose that the algebraic integer α is a zero of a monic $f(x) \in \mathbb{Z}[x]$ and let the minimal polynomial of α be $g(x)$. By Theorem 4.2, we know that $f(x) = g(x)h(x)$, where $g(x), h(x) \in \mathbb{Q}[x]$, and $h(x)$ is monic. By Theorem 4.4, we have $g(x) \in \mathbb{Z}[x]$. □

It is evident that if we let A be the set of all algebraic numbers, then A is a field. Furthermore, the set of all algebraic integers I is a ring contained in A.

Definition 4.8. *An* algebraic number field *is any subfield of A.*

For example, if α is an algebraic number, then

$$\mathbb{K} = \left\{ \frac{f(\alpha)}{g(\alpha)} : f(x), g(x) \in \mathbb{Q}[x]; g(\alpha) \neq 0 \right\}$$

is a subfield of A. We denote such a field by $\mathbb{Q}(\alpha)$, the *extension field of $\mathbb{Q}$ by adjoining α*. Also, if I' is the set of algebraic integers in $\mathbb{Q}(\alpha)$, then I' must be a subring of I and therefore is a subring of $\mathbb{Q}(\alpha)$.

Theorem 4.9. *If α is an algebraic number of degree n, then every element $\beta \in \mathbb{Q}(\alpha)$ can be written uniquely as*

$$\beta = a_0 + a_1\alpha + a_2\alpha^2 + \cdots + a_{n-1}\alpha^{n-1} ,$$

where $a_i \in \mathbb{Q}$ $(i = 0, 1, 2, \ldots, n - 1)$.

The study of algebraic number fields has resulted in the development of what is called algebraic number theory, a very beautiful and deep collection of results, many of which are both fascinating and surprising.[3] As we will see, there are still many open questions concerning algebraic number theory, even in the simple instance of it that we will investigate.[4] Indeed, as our interest here concerns the Pell equation, we will confine our attention to the case of $\mathbb{Q}(\alpha)$ where α is of degree 2. These fields are known as *quadratic number fields*, or simply *quadratic fields*. Thus, we may assume that α is the zero of an irreducible polynomial $f(x) = ax^2 + bx + c$, where $a, b, c \in \mathbb{Z}$. Hence,

$$\alpha = \frac{-b \pm \sqrt{b^2 - 4ac}}{2a} .$$

If we put $D = b^2 - 4ac \in \mathbb{Z}$, we get $\mathbb{K} = \mathbb{Q}(\alpha) = \mathbb{Q}(\sqrt{D})$. From this point forward, we will use $\mathbb{K}$ to denote $\mathbb{Q}(\sqrt{D})$. We call D the *radicand* of $\mathbb{K}$. Notice that $\sqrt{D} \notin \mathbb{Q}$ because $f(x)$ is irreducible. Also, if $D = f^2 D_0$, where D_0 is squarefree, then $\mathbb{K} = \mathbb{Q}(\sqrt{D_0})$. Thus, either D_0 or any square multiple of it can be the radicand of $\mathbb{K}$.

If $\beta \in \mathbb{K}$, we have $\beta = (c_1 + c_2\sqrt{D})/c_3$, where $c_1, c_2, c_3 \in \mathbb{Z}$, $c_3 \neq 0$. We have previously defined $\overline{\beta} = (c_1 - c_2\sqrt{D})/c_3$ and $N(\beta) = \beta\overline{\beta}$; we now introduce the *trace* of β, denoted by $T(\beta)$, as $\beta + \overline{\beta}$. Note that

$$\beta^2 = T(\beta)\beta - N(\beta) . \tag{4.1}$$

Let β $(\notin \mathbb{Z})$ be an algebraic integer of $\mathbb{K}$; then β must be a zero of a monic irreducible polynomial $g(x)$ of degree 2 over $\mathbb{Q}$. Since β is a zero of $f(x) = x^2 - T(\beta)x + N(\beta)$, we must have $g(x) \mid f(x)$. It follows that $g(x) = f(x)$ and $T(\beta), N(\beta) \in \mathbb{Z}$.

Put

$$r = \begin{cases} 1 & \text{when } D_0 \not\equiv 1 \pmod 4 \\ 2 & \text{when } D_0 \equiv 1 \pmod 4 \end{cases}$$

and define $\omega_0 = (r - 1 + \sqrt{D_0})/r$. Notice that if we are only given D instead of D_0, we can easily evaluate r. If $2^\alpha \parallel D$, then $r = 1$ when α is odd. If α is even, then $D_0 \equiv D/2^\alpha \pmod 4$. We have $T(\omega_0) = 2(r-1)/r = r - 1$ and $N(\omega_0) = \omega_0\overline{\omega_0} = ((r-1)^2 - D_0)/r^2 \in \mathbb{Z}$. We are now able to completely characterize the algebraic integers of $\mathbb{K}$.

Theorem 4.10. β *is an algebraic integer of* $\mathbb{K} = \mathbb{Q}(\sqrt{D_0})$ *if and only if* $\beta = x + y\omega_0$, *where* $x, y, \in \mathbb{Z}$.

Proof. If $\beta = x + y\omega_0$, then $T(\beta) = \beta + \bar{\beta} = 2x + yT(\omega_0) \in \mathbb{Z}$ and $N(\beta) = x^2 + xyT(\omega_0) + y^2N(\omega_0) \in \mathbb{Z}$. Hence, by (4.1), β is an algebraic integer of $\mathbb{K}$. We next suppose that β is an algebraic integer of $\mathbb{K}$. We have

$$\beta = \frac{c_1 + c_2\sqrt{D_0}}{c_3} \quad (c_1, c_2, c_3 \in \mathbb{Z}, c_3 > 0) ,$$

and we may assume with no loss of generality that $(c_1, c_2, c_3) = 1$. Also, $T(\beta) = 2c_1/c_3 \in \mathbb{Z}$ and $N(\beta) = (c_1{}^2 - c_2{}^2D_0)/c_3{}^2 \in \mathbb{Z}$. If p is any prime such that $p \mid c_1$ and $p \mid c_3$, then since $N(\beta) \in \mathbb{Z}$, we get $p^2 \mid c_2{}^2D_0$. Since D_0 is squarefree, it follows that $p \mid c_2$, a contradiction; hence, $(c_1, c_3) = 1$. Since $T(\beta) \in \mathbb{Z}$, we can only have $c_3 \mid 2$. There are two possible cases.

CASE 1: $c_3 = 2$:

Here, $c_1{}^2 - c_2{}^2D_0 \equiv 0 \pmod{4}$. If $2 \mid c_2$, then $2 \mid c_1$, a contradiction. Thus, c_2 is odd and c_1 is odd, which means that $D_0 \equiv 1 \pmod{4}$ and $\beta = (c_1 + c_2\sqrt{D_0})/2$, $c_1 \equiv c_2 \equiv 1 \mod 2$. We have

$$\beta = \frac{c_1 - c_2}{2} + c_2 \left(\frac{1 + \sqrt{D_0}}{2}\right) = x + y\omega_0 \quad (x, y \in \mathbb{Z}) .$$

CASE 2: $c_3 = 1$:

Here,

$$\beta = c_1 + c_2\sqrt{D_0} = c_1 - c_2 + 2c_2 \left(\frac{1 + \sqrt{D_0}}{2}\right) = x + y\omega_0 \quad (x, y \in \mathbb{Z}) .$$

$\square$

4.2 Modules and Orders of $\mathbb{K}$

The structure of the set of algebraic integers of $\mathbb{K}$ is that of a module.[5]

Definition 4.11. *Let A be any additive abelian group. We say that $\mathcal{M}$ is a $(\mathbb{Z}\text{-})$module of A if $\mathcal{M}$ is an additive abelian subgroup of A.*

If $\mathcal{L}$ and $\mathcal{M}$ are modules and $\mathcal{L} \subseteq \mathcal{M}$, then since both $\mathcal{L}$ and $\mathcal{M}$ are additive abelian groups, we can produce the group of cosets (or quotient group) of $\mathcal{M}$ modulo $\mathcal{L}$ and we denote this by $\mathcal{M}/\mathcal{L}$. When $\mathcal{M}/\mathcal{L}$ is finite, we call the number of distinct cosets, denoted by $|\mathcal{M}/\mathcal{L}|$, the *index* of $\mathcal{L}$ in $\mathcal{M}$.

Proposition 4.12. *If $\mathcal{M}, \mathcal{L}$, and $\mathcal{K}$ are modules and $\mathcal{K} \subseteq \mathcal{L} \subseteq \mathcal{M}$, and if $\mathcal{M}/\mathcal{K}$ is finite, we have*

$$|\mathcal{M}/\mathcal{K}| \geq |\mathcal{M}/\mathcal{L}| .$$

If $\mathcal{K}$ is properly contained in $\mathcal{L}$, then

$$|\mathcal{M}/\mathcal{K}| > |\mathcal{M}/\mathcal{L}| .$$

Proof. If $\alpha, \beta \in \mathcal{M}$ and $\alpha - \beta \notin \mathcal{L}$, then $\alpha - \beta \notin \mathcal{K}$. This means that if μ is a coset representative for a coset in $\mathcal{M}/\mathcal{L}$, then μ is also a coset representative for a coset in $\mathcal{M}/\mathcal{K}$ and the first result follows. If $n = |\mathcal{M}/\mathcal{K}| = |\mathcal{M}/\mathcal{L}|$, we may assume that $\mathcal{M}/\mathcal{K}$ and $\mathcal{M}/\mathcal{L}$ have precisely the same coset leaders: $\mu_1, \mu_2, \ldots, \mu_n$. We may also assume with no loss of generality that $\mu_1 = 0$. If L properly contains $\mathcal{K}$, then there must exist some $\lambda \in \mathcal{L}$ such that $\lambda \notin \mathcal{K}$. Since $\lambda \in \mathcal{K} + \mu_i$ for some i $(1 \leq i \leq n)$, we have $\lambda \in \mathcal{L} + \mu_i$, but since $\mathcal{L} + \mu_i$ and $\mathcal{L}$ are disjoint, this is impossible. $\qquad\square$

Definition 4.13. *If X is a subset of a module $\mathcal{M}$, then the intersection of all submodules of $\mathcal{M}$ containing X is called the* submodule generated by X.

Suppose $X = \{\xi_1, \xi_2, \xi_3, \ldots, \xi_n\} \subseteq \mathbb{K}$; then

$$\mathcal{M} = \left\{ \sum_{i=1}^n x\xi_i : x_1, x_2, x_3, \ldots, x_n \in \mathbb{Z} \right\} \supseteq X$$

is a (free) module of $\mathbb{K}$ and $\mathcal{M}$ is generated by X. We denote this by

$$\mathcal{M} = [\xi_1, \xi_2, \ldots, \xi_n].$$

Since X is a finite set, we say that $\mathcal{M}$ if *finitely generated*; however, it is important to observe that not all modules of $\mathbb{K}$ are finitely generated. For example, $\mathbb{Q}$ is a module of $\mathbb{K}$, but $\mathbb{Q}$ cannot be generated by a finite number of rational numbers. We will be particularly concerned with modules generated by two elements of $\mathbb{K}$. For this study, we will require the following proposition.

Proposition 4.14. *Let $\mathcal{M} = [\xi_1, \xi_2]$, where $\xi_1 = a_1 + b_1\sqrt{D_0}$, $\xi_2 = a_2 + b_2\sqrt{D_0}$, and $a_1, a_2, b_1, b_2 \in \mathbb{Q}$. There exists a $\xi_3 = a_3 + b_3\sqrt{D_0}$ $(a_3, b_3 \in \mathbb{Q})$ in $\mathbb{K}$ such that $\mathcal{M} = [\xi_3]$ if and only if $a_1 b_2 - b_1 a_2 = 0$.*

Proof. If $\mathcal{M} = [\xi_3]$, then $\xi_1 = z_1\xi_3$ and $\xi_2 = z_2\xi_3$, where $z_1, z_2 \in \mathbb{Z}$. By equating rational and irrational parts, we get $a_1 = z_1 a_3$, $b_1 = z_1 b_3$, $a_2 = z_2 a_3$, and $b_2 = z_2 b_3$; hence, $a_1 b_2 - b_1 a_2 = 0$.

If $a_1 b_2 - b_1 a_2 = 0$, we may assume that $b_1 \neq 0$. For if $b_1 = 0$, then $a_1 b_2 = 0$. If $b_2 = 0$, then $\mathcal{M} = [a_1, a_2] = [g]$, where $g = (a_1, a_2)$. If $a_1 = 0$, then $\mathcal{M} = [0, \xi_2] = [\xi_2]$. Since $b_2\xi_1 - b_1\xi_2 = 0$, we see that $\mathcal{M} = [\xi_1, (b_2/b_1)\xi_1]$. Putting $b_2/b_1 = m/n$, where $m, n \in \mathbb{Z}$ and $(m, n) = 1$, we get $\mathcal{M} = (\xi_1/n)[m, n] = [\xi_1/n]$. $\qquad\square$

If $\mathcal{M} = [\xi_1, \xi_2]$, we say that $\{\xi_1, \xi_2\}$ is a $\mathbb{Z}$-*basis* of $\mathcal{M}$. Now, let $\mathrm{GL}_2(\mathbb{Z})$ be the group of invertible 2×2 matrices over $\mathbb{Z}$ and let $M \in \mathrm{GL}_2(\mathbb{Z})$. If $\mathcal{M} = [\xi_1, \xi_2]$ and

$$(\phi_1, \phi_2) = (\xi_1, \xi_2)M, \tag{4.2}$$

then $[\phi_1, \phi_2] = \mathcal{M}$. Furthermore, if $[\phi_1, \phi_2] = [\xi_1, \xi_2]$, then there must exist some $M \in \mathrm{GL}_2(\mathbb{Z})$ such that (4.2) holds.

Notice that $[1, \omega_0]$ is the module of algebraic integers in $\mathbb{K}$, but as mentioned earlier, it is also a subring of $\mathbb{K}$. This brings us to the definition of an order of $\mathbb{K}$.

Definition 4.15. *An* order *of* $\mathbb{K}$ *is a module* $\mathcal{M}$ *of* $\mathbb{K}$ *with the following properties*[6]*:*

1. $\mathcal{M}$ *is a subring of* $\mathbb{K}$ *containing* 1.
2. $\mathcal{M} = [\xi_1, \xi_2]$, *where* $\xi_1 = a_1 + b_1\sqrt{D_0}$, $\xi_2 = a_2 + b_2\sqrt{D_0}$, $a_1, b_1, a_2, b_2 \in \mathbb{Q}$, *and* $a_1 b_2 - b_1 a_2 \neq 0$.

Clearly, $[1, \omega_0]$ is an order of $\mathbb{K}$; we denote it by $\mathcal{O}_{\mathbb{K}}$.

Definition 4.16. *If* $\mathcal{O} = [\xi_1, \xi_2]$ *is any order of* $\mathbb{K}$, *we define the* discriminant Δ *of* $\mathcal{O}$, *written* $\Delta(\mathcal{O})$, *to be*

$$\Delta = \begin{vmatrix} \xi_1 & \xi_2 \\ \overline{\xi}_1 & \overline{\xi}_2 \end{vmatrix}^2 = (\xi_1 \overline{\xi}_2 - \overline{\xi}_1 \xi_2)^2 \ .$$

If $M \in \mathrm{GL}_2(\mathbb{Z})$ and (4.2) holds, then

$$\begin{pmatrix} \phi_1 & \phi_2 \\ \overline{\phi}_1 & \overline{\phi}_2 \end{pmatrix} = \begin{pmatrix} \xi_1 & \xi_2 \\ \overline{\xi}_1 & \overline{\xi}_2 \end{pmatrix} M \ ,$$

and taking determinants, we get

$$\phi_1 \overline{\phi}_2 - \overline{\phi}_1 \phi_2 = |M| \left(\xi_1 \overline{\xi}_2 - \overline{\xi}_1 \xi_2 \right) = \pm \left(\xi_1 \overline{\xi}_2 - \overline{\xi}_1 \xi_2 \right) \ .$$

Thus, Δ is an invariant of the order $\mathcal{O}$. We have $\Delta_{\mathbb{K}} = \Delta(\mathcal{O}_{\mathbb{K}}) = (\omega_0 - \overline{\omega}_0)^2 = (2/r)^2 D_0$. Note that $\Delta_{\mathbb{K}}$ is either $\equiv 1 \pmod 4$ or $\Delta_{\mathbb{K}} \equiv 8, 12 \pmod{16}$. Also, $\Delta_{\mathbb{K}}$ or $\Delta_{\mathbb{K}}/4$ is squarefree. Any value of Δ which satisfies these conditions is called a *fundamental discriminant*.

It turns out that all orders of $\mathbb{K}$ are submodules of $\mathcal{O}_{\mathbb{K}}$, which we call the *maximal order of* $\mathcal{O}_{\mathbb{K}}$.

Theorem 4.17. *If* $\mathcal{O}$ *is any order of* $\mathbb{K}$, *then* $\mathcal{O} \subseteq \mathcal{O}_{\mathbb{K}}$ *and* $\mathcal{O} = [1, f\omega_0]$ *for some* $f \in \mathbb{Z}$.

Proof. Let $\mathcal{O} = [\xi_1, \xi_2]$. Since $1 \in \mathcal{O}$, there exist $x, y \in \mathbb{Z}$ such that $x\xi_1 + y\xi_2 = 1$. Let $d = (x, y)$. Since $(x/d, y/d) = 1$, there must exist $p, q \in \mathbb{Z}$ such that

$$p \left(\frac{x}{d} \right) - q \left(\frac{y}{d} \right) = 1 \ .$$

Put

$$M = \begin{pmatrix} x/d & q \\ y/d & p \end{pmatrix} \ .$$

Clearly, $M \in \mathrm{GL}_2(\mathbb{Z})$ and $\mathcal{O} = [(x\xi_1 + y\xi_2)/d, q\xi_1 + p\xi_2] = [1/d, \gamma]$. Since $\mathcal{O}$ is a ring and $1/d \in \mathcal{O}$, we have $1/d^2 \in \mathcal{O}$, which means that $1/d^2 = s/d + t\gamma$. Since $\gamma \notin \mathbb{Q}$, this is impossible unless $d = 1$. Thus, $\mathcal{O} = [1, \gamma]$. Since $\gamma \in \mathcal{O}$, we must have $\gamma^2 \in \mathcal{O}$ and, therefore, there exist $a, b \in \mathbb{Z}$ such that $\gamma^2 = a + b\gamma$. Since γ is the zero of a monic polynomial over $\mathbb{Z}$, we must have $\gamma \in \mathcal{O}_{\mathbb{K}}$. Hence, $\gamma = c + f\omega_0$, where $c, f \in \mathbb{Z}$. Thus, $\mathcal{O} \subseteq \mathcal{O}_{\mathbb{K}}$ and $\mathcal{O} = [1, c + f\omega_0] = [1, f\omega_0]$. $\square$

With no loss of generality we may assume that $f > 0$. We call f the *conductor* of $\mathcal{O}$. We have $\Delta(\mathcal{O}) = f^2 \Delta_{\mathbb{K}}$ and we observe that $\Delta(\mathcal{O}) \equiv 0, 1 \pmod 4$. It follows from this that if we are given any $\Delta \in \mathbb{Z}$ with $\Delta \equiv 0, 1 \pmod 4$, there is only one order $\mathcal{O}_\Delta$ of discriminant Δ of $\mathbb{Q}(\sqrt{\Delta})$, and $\mathcal{O}_\Delta$ can be written as

$$\left[1, \frac{\Delta + \sqrt{\Delta}}{2} \right] = [1, f\omega_0] .$$

When Δ is assumed to be known, we will write $\mathcal{O}$ for $\mathcal{O}_\Delta$ and ω for $f\omega_0$. It is often useful to observe that

$$\Delta = T(\omega)^2 - 4N(\omega) \tag{4.3}$$

and

$$4N(b + c\omega) = (2b + cT(\omega))^2 - c^2 \Delta . \tag{4.4}$$

4.3 The Units of $\mathcal{O}$

Of particular concern to us will be the units of the order $\mathcal{O}$.

Definition 4.18. *Let $\alpha, \beta \in \mathcal{O}$. We say that α divides β in $\mathcal{O}$, denoted by $\alpha \mid \beta$, if there exists some $\gamma \in \mathcal{O}$ such that $\beta = \alpha\gamma$.*

Definition 4.19. *We say that η is a* unit *of $\mathcal{O}$ if $\eta \mid 1$ in $\mathcal{O}$.*

Notice that a unit of $\mathcal{O}$ trivially divides any element of $\mathcal{O}$. We define an *indecomposable* element of $\mathcal{O}$ to be any non-unit β of $\mathcal{O}$ for which the factorization $\beta = \alpha\gamma$ for $\alpha, \gamma \in \mathcal{O}$ is possible only when α or γ is a unit of $\mathcal{O}$. If γ is a unit of $\mathcal{O}$, we say that β and α are *associates* in $\mathcal{O}$. We next define a *prime* in $\mathcal{O}$ as a non-unit element π of $\mathcal{O}$ such that if $\pi \mid \alpha\beta$ for any $\alpha, \beta \in \mathcal{O}$, then π divides α or β in $\mathcal{O}$.

We denote the set of all units in $\mathcal{O}$ by $\mathcal{O}^*$. It is easy to see that this is a multiplicative group with identity 1. If η is a unit of $\mathcal{O}$, then

$$\eta = x + y \left(\frac{\Delta + \sqrt{\Delta}}{2} \right) = \frac{2x + y\Delta + y\sqrt{\Delta}}{2} ,$$

where $x, y \in \mathbb{Z}$ and $N(y) = \pm 1$. Hence,

$$(2x + y\Delta)^2 - y^2 \Delta = \pm 4 . \tag{4.5}$$

If $\Delta < 0$, then

$$(2x + y\Delta)^2 + y^2 |\Delta| = \pm 4 . \tag{4.6}$$

In this case we easily deduce from (4.6) that

$$\mathcal{O}^* = \begin{cases} \{1, -1, \zeta, \zeta^2, -\zeta, -\zeta^2 \ : \ \zeta^2 + \zeta + 1 = 0\} & \text{when } \Delta = -3 \\ \{1, -1, i, -i \ : \ i^2 + 1 = 0\} & \text{when } \Delta = -4 \\ \{1, -1\} & \text{when } \Delta < -4 \ . \end{cases}$$

If we put $w = |\mathcal{O}^*|$, then

$$w = \begin{cases} 6 & \text{when } \Delta = -3 \\ 4 & \text{when } \Delta = -4 \\ 2 & \text{when } \Delta < -4 \ . \end{cases}$$

If $\Delta > 0$, the problem of finding $\mathcal{O}^*$ is somewhat more complicated. If we put $X = 2x + y\Delta$ and $Y = y$ in (4.5), we get

$$X^2 - \Delta Y^2 = 4\sigma , \tag{4.7}$$

where $\sigma = \pm 1$. Since this is the same equation as (1.10) of Chapter 1 with D replaced by Δ, we know that there exists a fundamental solution (X_1, Y_1, σ_1) of (4.7). Put

$$\epsilon_\Delta = \frac{X_1 + Y_1\sqrt{\Delta}}{2} \ .$$

Since, by (4.7), $X_1 \equiv \Delta Y_1 \pmod{2}$, we see that

$$\epsilon_\Delta \in \mathcal{O} = \left[1, \frac{\Delta + \sqrt{\Delta}}{2} \right] \ .$$

Also by Theorem 1.9, we must have

$$\eta = \pm \epsilon_\Delta^n \ ;$$

hence, $\mathcal{O}^* = \langle -1, \epsilon_\Delta \rangle$. We call ϵ_Δ the *fundamental unit* of $\mathcal{O}$.

Let t, u be the fundamental solution of the Pell equation (1.7). If we put $q = ur$ and $p = t - 2uD/r$, then $t + u\sqrt{D} = p + q(\Delta + \sqrt{\Delta})/2$. Thus, $t + u\sqrt{D} \in \mathcal{O}$. By Theorem 1.9, $t + u\sqrt{D} = \epsilon_\Delta^n$ for some $n \in \mathbb{Z}^{\geq 0}$. If $r = 1$, then $\epsilon_\Delta \in [1, \sqrt{D}]$ and

$$t + u\sqrt{D} = \begin{cases} \epsilon_\Delta & \text{when } N(\epsilon_\Delta) = 1 \\ \epsilon_\Delta^2 & \text{when } N(\epsilon_\Delta) = -1 \ . \end{cases} \tag{4.8}$$

If $r = 2$ and $2 \mid Y_1$, then $2 \mid X_1$, $\epsilon_\Delta \in [1, \sqrt{D}]$ and (4.8) holds. If $r = 2$ and $2 \nmid \Delta Y_1$, then by the same reasoning as that used to produce Table 1.1, we get

$$t + u\sqrt{D} = \begin{cases} \epsilon_\Delta^3 & \text{when } N(\epsilon_\Delta) = 1 \\ \epsilon_\Delta^6 & \text{when } N(\epsilon_\Delta) = -1 \ . \end{cases}$$

Finally, if $r = 2$ and $2 \mid \Delta$, then $\epsilon_\Delta \notin [1, \sqrt{D}]$, but $\epsilon_\Delta^2 \in [1, \sqrt{D}]$; hence, $t + u\sqrt{D} = \epsilon_\Delta^2$. Thus, in summary, if $\epsilon_\Delta = (x + y\sqrt{\Delta})/2$, then

Table 4.1.

r	Δ (mod 2)	y (mod 2)	$N(\epsilon_\Delta)$	v
1	—	—	1	1
1	—	—	-1	2
2	—	0	1	1
2	—	0	-1	2
2	1	1	1	3
2	1	1	-1	6
2	0	1	—	2

$$t + u\sqrt{D} = \epsilon_\Delta^v \, ,$$

where v is given in Table 4.1.

Thus, the problem of finding the fundamental solution of the Pell equation (1.7) is equivalent to the problem of determining ϵ_Δ, the fundamental unit in the order of conductor f of $\mathbb{K}$ where $D = f^2 D_0$ and D_0 is squarefree. In view of this, we will spend much of the remainder of this work on the problem of finding ϵ_Δ.

Of course, we have already seen how this can be done in Chapter 3, but our focus will be on finding ways of doing this that are more efficient than that technique. Our main tool in this investigation will be the arithmetic of the ideals[7] of $\mathcal{O}$, a topic which we will introduce in the next section.

We conclude this section be observing that since $\epsilon_\Delta \in \mathcal{O}_\Delta \subseteq \mathcal{O}_\mathbb{K}$, we must have $\epsilon_\Delta \in \mathcal{O}_\mathbb{K}^*$ and, therefore, $\epsilon_\Delta = \epsilon_\mathbb{K}^n$, where $\epsilon_\mathbb{K}$ is the fundamental unit of the order $\mathcal{O}_\mathbb{K}$, the maximal order of $\mathbb{K}$. Since $\epsilon_\Delta, \epsilon_\mathbb{K} > 1$, n must be positive. We call n the *unit index* of ϵ_Δ. For example, when $D_0 = 7$ ($\Delta_\mathbb{K} = 28$), we have $\mathcal{O}_\mathbb{K} = [1, \sqrt{7}]$ and $\epsilon_\mathbb{K} = 8 + 3\sqrt{7}$; but if $\Delta = 4\Delta_\mathbb{K}$, then $\mathcal{O}_\Delta = [1, 2\sqrt{7}]$ and $\epsilon_\Delta = 127 + 48\sqrt{7} = \epsilon_\mathbb{K}^2$. Thus, the unit index of ϵ_Δ is 2 in this case. If $\epsilon_\mathbb{K} = (v + w\sqrt{\Delta_\mathbb{K}})/2$, then by the results in §1.4, the unit index n of ϵ_Δ, where $\Delta = f^2 \Delta_\mathbb{K}$ is given by $n = \omega(f/g)$. Here, $g = (f, w)$ and $P = x$, $Q = N(\epsilon_\mathbb{K})$. By Theorem 1.13, we know that[8] $n \leq 2f$.

4.4 The Ideals of $\mathcal{O}$

In this section we will describe the ideals of an order $\mathcal{O}$ and discuss some of their properties. In order to motivate their introduction here, we remind the reader of the Fundamental Theorem of Arithmetic, which states that any integer greater than 1 can be uniquely represented (up to order) as a product of primes in $\mathbb{Z}^{>0}$. This result, proved implicitly by Euclid,[9] and explicitly by Gauss[10] is of such importance in elementary number theory that it was often assumed by mathematicians to be true for algebraic integers. However, consider the simple example of the order $\mathcal{O}_{\mathbb{K}} = [1, \sqrt{-5}]$ of $\mathbb{K} = \mathbb{Q}(\sqrt{-5})$. It is not difficult to show by taking norms that $2, 3, 1 + \sqrt{-5}$, and $1 - \sqrt{-5}$ are all indecomposable elements of $\mathcal{O}_{\mathbb{K}}$, but

$$6 = 2 \cdot 3 = \left(1 + \sqrt{-5}\right)\left(1 - \sqrt{-5}\right) \; ;$$

that is, we have two distinct ways of expressing 6 as a product of indecomposable elements of $\mathcal{O}_{\mathbb{K}}$. In $\mathbb{Z}$, the indecomposable elements are the primes or their associates (in this case, the units are just 1 and -1), but in this $\mathcal{O}_{\mathbb{K}}$, we have that $3 \mid \left(1 + \sqrt{-5}\right)\left(1 - \sqrt{-5}\right)$ but $3 \nmid \left(1 + \sqrt{-5}\right)$ and $3 \nmid \left(1 - \sqrt{-5}\right)$. Hence, 3 is not a prime in $\mathcal{O}_{\mathbb{K}}$. Thus, there is no unique factorization of 6 into indecomposable elements of $\mathcal{O}_{\mathbb{K}}$, nor can there be any representations of 6 as a product of primes of $\mathcal{O}_{\mathbb{K}}$. Thus, the failure of unique factorization in $\mathcal{O}_{\mathbb{K}}$ is the failure of indecomposable elements of $\mathcal{O}_{\mathbb{K}}$ to be prime in general. It is this observation (in a more general setting) that has motivated much of the development of algebraic number theory.

However, unique factorization can be restored to $\mathcal{O}_{\mathbb{K}}$ by the introduction of "ideal" elements. These were introduced by Kummer[11] in 1857 as actual numbers, but it was Dedekind[12] who discovered in 1871 that unique factorization could be accomplished by the use of special modules called ideals. In order to define these objects, we begin by defining γS, where $S \subseteq \mathbb{C}$ and $\gamma \in \mathbb{C}$, to be the set $\{\gamma s : s \in S\}$. If $\gamma S_1 = \gamma S_2$, where $S_1, S_2 \subseteq \mathbb{C}$, $\gamma \in \mathbb{C}$, and $\gamma \neq 0$, then $S_1 = S_2$. We are now able to define an ideal.

Definition 4.20. *An (integral) ideal* $\mathfrak{a}$ *of an order* $\mathcal{O}$ *(an* $\mathcal{O}$-*ideal) is an additive subgroup of* $\mathcal{O}$ *such that* $\xi\mathfrak{a} \subseteq \mathfrak{a}$ *for any* $\xi \in \mathcal{O}$.

Since $\mathfrak{a}$ is an additive abelian group, it can be regarded as a submodule of $\mathcal{O}$. We remark that in order to prove that any submodule $\mathcal{M}$ of $\mathcal{O}$ is an ideal of $\mathcal{O}$, it suffices to prove that $\omega\mathcal{M} \subseteq \mathcal{M}$. We will now show that an ideal is a finitely generated submodule of $\mathcal{O}$.

Theorem 4.21. *Any ideal* $\mathfrak{a}$ *of an order* $\mathcal{O} = [1, \omega]$ *of* $\mathbb{K}$ *is a finitely generated submodule of* $\mathcal{O}$.

Proof. The result is trivially true for $\mathfrak{a} = \{0\}$. Suppose $\mathfrak{a} \neq \{0\}$, then there must exist some $\alpha \in \mathfrak{a}$, where $\alpha \neq 0$. Since $\overline{\alpha} = T(\alpha) - \alpha$, we see that $\overline{\alpha} \in \mathcal{O}$. Hence, $\alpha\overline{\alpha} \in \mathfrak{a}$. Since $N(\alpha) = \alpha\overline{\alpha} \in \mathbb{Z}$, we see that $\mathfrak{a}$ contains a non-zero rational integer $N(\alpha)$. Also, $\mathfrak{a} \supseteq N(\alpha)\mathcal{O}$. Since the set of coset representatives for $\mathcal{O}/N(\alpha)\mathcal{O}$ can be made up of precisely those elements $\beta = a + b\omega \in \mathcal{O}$ such that $0 \leq a, b < |N(\alpha)|$, we get

$$N(\alpha)^2 = |\mathcal{O}/N(\alpha)\mathcal{O}| \geq |\mathcal{O}/\mathfrak{a}| \; ;$$

we see that $|\mathcal{O}/\mathfrak{a}|$ is finite. If $\mathfrak{a} = N(\alpha)\mathcal{O}$, then $\mathfrak{a} = [N(\alpha), N(\alpha)\omega]$ is finitely generated. If $\mathfrak{a} \neq N(\alpha)\mathcal{O}$, then $|\mathcal{O}/N(\alpha)\mathcal{O}| > |\mathcal{O}/\mathfrak{a}|$ by Proposition 4.12. Let $\alpha_1 \in \mathfrak{a}$ but $\alpha_1 \notin N(\alpha)\mathcal{O}$, and consider the module $\mathcal{M}_1 = [N(\alpha), N(\alpha)\omega, \alpha_1]$. We have $\mathcal{O} \subseteq \mathcal{M}_1 \subseteq \mathfrak{a}$ and

$$|\mathcal{O}/N(\alpha)\mathcal{O}| > |\mathcal{O}/\mathcal{M}_1| \geq |\mathcal{O}/\mathfrak{a}| .$$

If $|\mathcal{O}/\mathcal{M}_1| = |\mathcal{O}/\mathfrak{a}|$, then $\mathfrak{a} = \mathcal{M}_1$ and $\mathfrak{a}$ is finitely generated. If $|\mathcal{O}/\mathcal{M}_1| > |\mathcal{O}/\mathfrak{a}|$, then there must be an element $\alpha_2 \in \mathfrak{a}$ such that $\alpha_2 \notin \mathcal{M}_1$. We continue to produce a sequence of modules in this way:

$$\mathcal{M}_1, \mathcal{M}_2, \mathcal{M}_3, \ldots$$

where $\mathcal{M}_i = [N(\alpha), N(\alpha)\omega, \alpha_1, \alpha_2, \ldots, \alpha_i]$ $(i = 1, 2, 3, \ldots)$ and $\mathcal{M}_i$ properly contains M_{i-1} and $\mathcal{M}_i \subseteq \mathfrak{a}$. This produces a strictly decreasing sequence of positive integers $|\mathcal{O}/\mathcal{M}_i|$ $(i = 1, 2, \ldots)$. Such a sequence must terminate and will when $\mathfrak{a} = \mathcal{M}_i$ for some i. Since $\mathcal{M}_i$ is finitely generated, so is $\mathfrak{a}$. $\qquad\square$

We now consider some properties of finitely generated submodules of $\mathcal{O}$.

Theorem 4.22. *If M $(\neq \{0\})$ is a finitely generated submodule of $\mathcal{O} = [1, \omega]$, we have $a, b, c \in \mathbb{Z}$, $a \geq 0$, $c \geq 0$ such that $\mathcal{M} = [a, b + c\omega]$.*

Proof. The theorem clearly holds if $\mathcal{M} = [\xi_1]$. Suppose $\mathcal{M} = [\xi_1, \xi_2]$, where $\xi_1 = a_1 + b_1\omega$ and $\xi_2 = a_2 + b_2\omega$. Let $c = (b_1, b_2)$. There must exist $p, q \in \mathbb{Z}$ such that

$$pb_1 + qb_2 = c .$$

Put

$$M = \begin{pmatrix} b_2/c & q \\ -b_1/c & p \end{pmatrix} .$$

Since $M \in \mathrm{GL}_2(\mathbb{Z})$, we have $\mathcal{M} = [(a_1b_2 - b_1a_2)/c, pa_1 + qa_2 + c\omega]$. Putting $a = |a_1b_2 - b_1a_2|/c$ and $b = pa_1 + qa_2$, we get $\mathcal{M} = [a, b + c\omega]$.

Next, suppose that $\mathcal{M} = [\xi_1, \xi_2, \xi_3]$, where $\xi_i = a_i + b_i\omega$ and $a_i, b_i \in \mathbb{Z}$ $(i = 1, 2, 3)$. Since $[\xi_1, \xi_2] = [a', b' + c'\omega]$, $a' \geq 0$, $c' \geq 0$, we see that

$$\mathcal{M} = [a', b' + c'\omega, b_3 + c_3\omega] = [a', a'', b'' + c''\omega] \quad (a'' \geq 0, c'' \geq 0) .$$

If we put $a = (a', a'')$, $b = b''$, and $c = c''$, we get $\mathcal{M} = [a, b + c\omega]$. It follows by induction that if $\mathcal{M} = [\xi_1, \xi_2, \ldots, \xi_n]$ with $\xi_i = a_i + b_i\omega$ and $a_i, b_i \in \mathbb{Z}$ $(i = 1, 2, \ldots, n)$, then $\mathcal{M} = [a, b + c\omega]$ with $a \geq 0$, $c \geq 0$, and $a, b, c \in \mathbb{Z}$. $\qquad\square$

Proposition 4.23. *If $\mathcal{M} = [a, b + c\omega]$, where $a, b, c \in \mathbb{Z}$ and $ac > 0$, then $|\mathcal{O}/\mathcal{M}| = ac$.*

Proof. It is easy to see that if $\beta \in \mathcal{O}$, then $\beta - \lambda \in \mathcal{M}$, for some $\lambda \in T = \{t_1 + t_2\omega : 0 \leq t_1 < a, 0 \leq t_2 < c; t_1, t_2 \in \mathbb{Z}\}$. Furthermore, if $\lambda_1, \lambda_2 \in T$ and $\lambda_1 \neq \lambda_2$, then $\lambda_1 - \lambda_2 \notin \mathcal{M}$. Hence, there must be exactly ac cosets of $\mathcal{O}$ modulo $\mathcal{M}$. $\qquad\square$

If $\mathfrak{a}$ is a non-zero ideal of $\mathcal{O}$, then Theorems 4.21 and 4.22 allow us to write $\mathfrak{a} = [a, b + c\omega]$, where $a, b, c \in \mathbb{Z}$ and $a, c \geq 0$. Indeed if $a = 0$, then since $\omega\mathfrak{a} \subseteq \mathfrak{a}$, we must have some integer x such that $\omega(b + c\omega) = x(b + c\omega)$, which is impossible unless $b = c = 0$. If $c = 0$, then $\mathfrak{a} = [d]$, where $d = (a, b)$. In this case, $\omega d = xd$, which is also impossible unless $d = 0$. Thus, if $\mathfrak{a}$ is a non-zero $\mathcal{O}$-ideal, then $\mathfrak{a} = [a, b + c\omega]$, where $a, b, c \in \mathbb{Z}$ and $ac > 0$. We may also assume that $0 \le b < a$. From this point on we will only deal with non-zero ideals of $\mathcal{O}$. We can now characterize all such ideals of $\mathcal{O}$.

Theorem 4.24. $\mathfrak{a}$ *is an ideal of* $\mathcal{O}$ *if and only if* $\mathfrak{a}$ *can be represented as* $[a, b + c\omega]$, *where* $a, b, c \in \mathbb{Z}$, $a > 0$, $c > 0$, $0 \le b < a$, $c \mid a$, $c \mid b$, *and* $ac \mid N(b + c\omega)$.

Proof. Suppose $\mathfrak{a} = [a, b + c\omega]$ is an ideal of $\mathcal{O}$. Since $\omega \mathfrak{a} \subseteq \mathfrak{a}$, we must have $\omega a = xa + y(b + c\omega)$ for integers x and y. By equating rational and irrational parts, we get $a = yc$ and $xa + yb = 0$. It follows that $c \mid a$ and $b = -cx$, so $c \mid b$. We next notice that if $d \in \mathfrak{a} \cap \mathbb{Z}$, then $a \mid d$; thus, a is the least positive integer in $\mathfrak{a}$. Since $b/c + \overline{\omega} \in \mathcal{O}$, we have $N(b + c\omega)/c = (b/c + \overline{\omega})(b + c\omega) \in \mathfrak{a}$. Thus, $ac \mid N(b + c\omega)$. Now, suppose that $\mathcal{M}$ is the submodule of $\mathcal{O}$ given as $[a, b + c\omega]$, where a, b, and c satisfy the conditions of the theorem. As noted earlier, $\mathcal{M}$ will be an ideal of $\mathcal{O}$ if $\omega \mathcal{M} \subseteq \mathcal{M}$. If we put $x = -N(b + c\omega)/ac \in \mathbb{Z}$ and $y = b/c + T(\omega) \in \mathbb{Z}$, then $\omega(b + c\omega) = xa + y(b + c\omega)$; if we put $p = -b/c \in \mathbb{Z}$ and $q = a/c \in \mathbb{Z}$, then $\omega a = pa + q(b + c\omega)$. Thus, $\omega \mathcal{M} \subseteq \mathcal{M}$ and $\mathcal{M}$ is an ideal of $\mathcal{O}$. $\qquad\square$

Let $\mathcal{O}$ be an order of $\mathbb{K}$ and let $\mathfrak{a}$ be any ideal of $\mathcal{O}$. We have seen that $\mathfrak{a} = [a, b + c\omega]$, where $a, b, c \in \mathbb{Z}$, $a > 0$, $c > 0$, $c \mid a$, $c \mid b$, and $ac \mid N(b + c\omega)$. If we put $S = c$, $Q = ra/c$, and $P = rb/c + f(r - 1)$, where f is the conductor of $\mathcal{O}$, then since $S^2(P^2 - D) = r^2 N(b + c\omega)$, we can represent $\mathfrak{a}$ by

$$\mathfrak{a} = S \left[\frac{Q}{r}, \frac{P + \sqrt{D}}{r} \right], \tag{4.9}$$

where $S, Q, P \in \mathbb{Z}$, $r \mid Q$, $rQ \mid D - P^2$, and $D = f^2 D_0$. Also, any such representation must be that of an ideal of $\mathcal{O}$. This representation will be of importance when we make use of continued fractions to aid us in our calculations involving ideals.

Let $\theta_1, \theta_2, \theta_3, \ldots, \theta_k \in \mathcal{O}$ and note that if we define

$$\mathfrak{a} = \left\{ \sum_{i=1}^{k} \xi_i \theta_i : \xi_i \in \mathcal{O} \text{ for } i = 1, 2, 3, \ldots, k \right\}$$
$$= \theta_1 \mathcal{O} + \theta_2 \mathcal{O} + \theta_3 \mathcal{O} + \cdots + \theta_k \mathcal{O},$$

then $\mathfrak{a}$ is an $\mathcal{O}$-ideal. We say that $\mathfrak{a}$ is the ideal *generated* by $\theta_1, \theta_2, \theta_3, \ldots, \theta_k$ and we denote this by $\mathfrak{a} = (\theta_1, \theta_2, \theta_3, \ldots, \theta_k)$. This may be somewhat confusing considering the way we have defined generators for modules; however, the following useful proposition shows that in the case of ideals, these ideas are the same.

Proposition 4.25. *If* $\mathfrak{a}$ *is the ideal given as* $[\theta_1, \theta_2, \theta_3, \ldots, \theta_k]$, *then* $\mathfrak{a} = (\theta_1, \theta_2, \theta_3, \ldots, \theta_k)$.

Proof. Clearly,

$$(\theta_1, \theta_2, \theta_3, \ldots, \theta_k) \supseteq [\theta_1, \theta_2, \theta_3, \ldots, \theta_k].$$

Now, $\theta_i \in [\theta_1, \theta_2, \theta_3, \ldots, \theta_k]$ means that

$$\theta_i \mathcal{O} \subseteq [\theta_1, \theta_2, \theta_3, \ldots, \theta_k] \quad (i = 1, 2, 3, \ldots, k).$$

Hence,

$$[\theta_1, \theta_2, \theta_3, \ldots, \theta_k] \supseteq \theta_1 \mathcal{O} + \theta_2 \mathcal{O} + \theta_3 \mathcal{O} + \cdots + \theta_k \mathcal{O}$$
$$= (\theta_1, \theta_2, \theta_3, \ldots, \theta_k).$$

$\square$

We have already seen that $\mathfrak{a} \neq [\theta]$ for any $\theta \in \mathcal{O}$, but it is possible for $\mathfrak{a} = (\theta) = \theta\mathcal{O}$. Such an ideal is said to be *principal*. If $\mathfrak{a}$ and $\mathfrak{b}$ are $\mathcal{O}$-ideals and $\mathfrak{a}$ and $\mathfrak{b}$ are both principal, then we can put $\mathfrak{a} = (\alpha)$, $\mathfrak{b} = (\beta)$ for $\alpha, \beta \in \mathcal{O}$. If $\alpha = \beta$, then clearly $\mathfrak{a} = \mathfrak{b}$, but if $\mathfrak{a} = \mathfrak{b}$, then we cannot conclude that $\alpha = \beta$. Since $(\alpha) = (\beta)$, we see that $\alpha \in \beta\mathcal{O}$ and $\beta \in \alpha\mathcal{O}$, which means that both α/β and $\beta/\alpha \in \mathcal{O}$. Thus, $\eta = \alpha/\beta$ must be a unit of $\mathcal{O}$. Also, if α/β is a unit, then $(\alpha) = (\beta\eta) = (\beta)(\eta) = (\beta)\mathcal{O} = (\beta)$. Thus, if $(\alpha) = (\beta)$, then $\alpha = \eta\beta$, where η is some unit of $\mathcal{O}$; that is, α and β are associates.

As any $\beta \in \mathcal{O}$ can have many associates when $\Delta > 0$, we distinguish a particular associate α of β which we call *primary*. We call some $\alpha \in \mathcal{O}$ primary if $\alpha > 0$ and

$$1 \leq |\alpha/\overline{\alpha}| < \epsilon_\Delta^2.$$

Theorem 4.26. *When* $\Delta > 0$, *every* $\beta \in \mathcal{O}$ *(*$\beta \neq 0$*) has precisely one primary associate.*

Proof. The most general form of an associate of β is $\gamma = \pm\epsilon_\Delta^n\beta$ ($n \in \mathbb{Z}$). We have

$$\log|\gamma| = \log|\beta| + n\log\epsilon_\Delta$$

and

$$\log|\overline{\gamma}| = \log|\overline{\beta}| + n\log|\overline{\epsilon_\Delta}|.$$

Also, since $\epsilon_\Delta|\overline{\epsilon_\Delta}| = 1$, we get $\log\epsilon_\Delta + \log|\overline{\epsilon_\Delta}| = 0$. Putting $\lambda = \log|\beta/\overline{\beta}|$, we get

$$\log|\gamma/\overline{\gamma}| = \lambda + 2n\log\epsilon_\Delta.$$

Thus, only if

$$n = \left\lfloor \frac{\lambda}{2\log\epsilon_\Delta} \right\rfloor$$

do we get

$$0 \leq \log|\gamma/\overline{\gamma}| < 2\log\epsilon_\Delta.$$

If, for this particular value of n, we select the sign such that $\gamma > 0$, we find that γ is the unique primary associate of β. $\square$

In the next section we will show how to determine a, b, and c such that a principal $\mathcal{O}$-ideal $\mathfrak{a}$ given by (θ) can be represented as $[a, b + c\omega]$, but before doing that, we turn to the definition of the product and sum of two ideals.

Definition 4.27. *If $\mathfrak{a} = (\theta_1, \theta_2, \theta_3, \ldots, \theta_k)$ and $\mathfrak{b} = (\phi_1, \phi_2, \phi_3, \ldots, \phi_m)$ are both $\mathcal{O}$-ideals, we define their* product $\mathfrak{ab}$ *to be*

$$(\theta_1\phi_1, \theta_1\phi_2, \ldots, \theta_1\phi_m, \theta_2\phi_1, \theta_2\phi_2, \ldots, \theta_n\phi_m) \, ;$$

we define their sum $\mathfrak{a} + \mathfrak{b}$ *to be*

$$\{\alpha + \beta : \alpha \in \mathfrak{a}, \beta \in \mathfrak{b}\} \, .$$

We observe that both $\mathfrak{ab}$ and $\mathfrak{a} + \mathfrak{b}$ are $\mathcal{O}$-ideals. Also, if $\mathfrak{c}$ is an $\mathcal{O}$-ideal and $\mathfrak{b} \subseteq \mathfrak{c}$, then $\mathfrak{ab} \subseteq \mathfrak{ac}$. If we let $\mathfrak{a} = [\alpha_1, \alpha_2]$, then $(\alpha)\mathfrak{a} = \alpha(\alpha_1, \alpha_2) = (\alpha\alpha_1, \alpha\alpha_2) = [\alpha\alpha_1, \alpha\alpha_2] = \alpha\mathfrak{a}$.

4.5 Equivalence and Norms

We say that two $\mathcal{O}$-ideals $\mathfrak{a}$ and $\mathfrak{b}$ are *equivalent* if there exists $\alpha, \beta \in \mathcal{O}$ such that $\alpha\beta \neq 0$ and

$$(\alpha)\mathfrak{a} = (\beta)\mathfrak{b} \, .$$

We write this as $\mathfrak{a} \sim \mathfrak{b}$. Of course, we have already shown that the product of the ideal $\mathfrak{a}$ by the principal ideal (α) is exactly the same as $\alpha\mathfrak{a}$, but we often include the parentheses for emphasis.

Proposition 4.28. *Let $\mathfrak{a}$ and $\mathfrak{b}$ be two $\mathcal{O}$-ideals. Then $\mathfrak{a} \sim \mathfrak{b}$ if and only if $\mathfrak{b} = \kappa\mathfrak{a}$, where $\kappa \in \mathbb{K}$ and $\kappa \neq 0$.*

Proof. If $\mathfrak{a} \sim \mathfrak{b}$, then there must exist $\alpha, \beta \in \mathcal{O}$ with $\alpha\beta \neq 0$ such that $\alpha\mathfrak{a} = \beta\mathfrak{b}$. Since $\beta/\alpha \in \mathbb{K}$ and $\beta/\alpha \neq 0$, we have $\mathfrak{b} = \kappa\mathfrak{a}$, where $\kappa \in \mathbb{K}$ and $\kappa \neq 0$. We next suppose that $\mathfrak{b} = \kappa\mathfrak{a}$, where $\kappa \in \mathbb{K}$ and $\kappa \neq 0$. Since $\kappa \in \mathbb{K}$ and $\kappa \neq 0$, we have $\kappa = (n_1 + n_2\omega_0)/n_3 \neq 0$, where $n_1, n_2, n_3 \in \mathbb{Z}$ and $n_3 \neq 0$. Let f be the conductor of $\mathcal{O}$; then putting $\alpha = fn_3 \neq 0$ and $\beta = fn_1 + fn_2\omega_0 \neq 0$, we have $\alpha\beta \neq 0$ and $\alpha, \beta \in \mathcal{O}$. Thus, $\alpha\mathfrak{a} = \beta\mathfrak{b}$, which means that $(\alpha)\mathfrak{a} = (\beta)\mathfrak{b}$, and therefore $\mathfrak{a} \sim \mathfrak{b}$. $\qquad\square$

The equivalence established above is a true equivlance relation on all of the ideals of $\mathcal{O}$. If $\mathfrak{a}$ is an $\mathcal{O}$-ideal, we denote by $[\mathfrak{a}]$ the set of all $\mathcal{O}$-ideals which are equivalent to $\mathfrak{a}$. We call $[\mathfrak{a}]$ an *ideal class* of $\mathcal{O}$.

Definition 4.29. *We say that an $\mathcal{O}$-ideal $\mathfrak{a}$ $(\neq (0))$ is* invertible *if there exists another $\mathcal{O}$-ideal $\mathfrak{b}$ $(\neq (0))$ such that*

$$\mathfrak{ab} \in [\mathcal{O}] \, .$$

Note that $[\mathcal{O}]$ is the set of all principal ideals of $\mathcal{O}$. Also, if $\mathfrak{a}$ is any principal $\mathcal{O}$-ideal, then $\mathfrak{a} = (\alpha)$; if we put $\mathfrak{b} = (\overline{\alpha})$, then $\mathfrak{b}$ is a principal $\mathcal{O}$-ideal and $\mathfrak{a}\mathfrak{b} = (\alpha\overline{\alpha}) = (N(\alpha))$. Thus, principal ideals are always invertible.

As an example of a non-invertible ideal, consider $\mathcal{O} = [1, \sqrt{-3}]$, the order of conductor 2 in $\mathbb{Q}(\sqrt{-3})$. We see that $\mathfrak{a} = [2, 1 + \sqrt{-3}]$ is an ideal of $\mathcal{O}$, but $\mathfrak{a}$ is not invertible in $\mathcal{O}$. For if $\mathfrak{a}$ were invertible, there would exist an $\mathcal{O}$-ideal $\mathfrak{b}$ and some $\gamma \in \mathcal{O}$ ($\gamma \neq 0$) such that $\mathfrak{a}\mathfrak{b} = (\gamma)$. If we put $\beta = (1 + \sqrt{-3})/2$, we see that $\beta \notin \mathcal{O}$, $2\beta = 1 + \sqrt{-3}$, $2\beta^2 = -1 + \sqrt{-3} = 1 + \sqrt{-3} - 2$. Hence, $(\beta)\mathfrak{a} \subseteq \mathfrak{a}$. It follows that

$$(\beta\gamma) = (\beta)\mathfrak{a}\mathfrak{b} \subseteq \mathfrak{a}\mathfrak{b} = (\gamma) .$$

However, this means that $\beta \in \mathcal{O}$, a contradiction.

We will now develop a simple criterion for determining whether or not a given $\mathcal{O}$-ideal is invertible. First, suppose that $\mathfrak{a}$ is any invertible ideal of $\mathcal{O}$. Then by definition there must exist another $\mathcal{O}$-ideal $\mathfrak{b}$ and some $\gamma \in \mathcal{O}$ such that

$$\mathfrak{a}\mathfrak{b} = (\gamma) .$$

If $\xi \in \mathbb{K}$ and $\xi\mathfrak{a} \subseteq \mathfrak{a}$, then

$$\xi\gamma\mathcal{O} = \xi\mathfrak{a}\mathfrak{b} = (\xi\mathfrak{a})\mathfrak{b} \subseteq \mathfrak{a}\mathfrak{b} = \gamma\mathcal{O} .$$

Hence, $\xi \in \mathcal{O}$; that is, the set X of all $\xi \in \mathbb{K}$ such that $\xi\mathfrak{a} \subseteq \mathfrak{a}$ is precisely $\mathcal{O}$. Ideals of $\mathcal{O}$ which possess this property are said to be *proper*. We now require two simple lemmas.

Lemma 4.30. *Let $\mathcal{O}$ be an order of $\mathbb{K}$ with conductor f and let $\mathfrak{a} = [a, b + c\omega]$ $(a, b, c \in \mathbb{Z})$ be any $\mathcal{O}$-ideal. If we define*

$$d = \left(\frac{a}{c}, \frac{T(b + c\omega)}{c}, \frac{N(b + c\omega)}{ac} \right) ,$$

then $d \mid f$.

Proof. Let $\xi = (b + c\omega)/a$, $m_1 = a/cd$, $m_2 = T(b + c\omega)/cd$, and $m_3 = N(b + c\omega)/acd$. Then since $m_1, m_2, m_3 \in \mathbb{Z}$ and by (4.1)

$$m_1\xi^2 - m_2\xi + m_3 = 0 ,$$

we must have $m_1\xi \in \mathcal{O}_\mathbb{K}$. Thus,

$$m_1\xi = (b + c\omega)/cd = x + y\omega_0$$

for $x, y \in \mathbb{Z}$. Since $\omega = f\omega_0$, we must have $f = dy$ and $d \mid f$. $\square$

Lemma 4.31. *Let $\mathfrak{a}$ and d be defined as in Lemma 4.30. If $d > 1$, then $\mathfrak{a}$ is not a proper $\mathcal{O}$-ideal.*

Proof. If $d > 1$, then $\gamma = (b + c\omega)/cd \notin \mathcal{O}$. However,

$$\gamma a = \left(\frac{a}{cd}\right)(b + c\omega) \in \mathfrak{a},$$

$$\gamma(b + c\omega) = \frac{(b + c\omega)^2}{cd} = \frac{T(b + c\omega)}{cd}(b + c\omega) + a\frac{N(b + c\omega)}{acd} \in \mathfrak{a}.$$

Thus, $\gamma\mathfrak{a} \subseteq \mathfrak{a}$, and since $\gamma \notin \mathcal{O}$, we see that $\mathfrak{a}$ is not proper. $\qquad\square$

We are now able to show that an $\mathcal{O}$-ideal is invertible if and only if it is proper.

Theorem 4.32. *Let $\mathfrak{a}$ and d be defined as in Lemma 4.30. If $d = 1$, then $\mathfrak{a}$ is invertible.*

Proof. Let $\mathfrak{a} = [a, b + c\omega]$ be an $\mathcal{O}$-ideal $(a, c > 0)$ and put $\mathfrak{b} = [a, b + c\bar{\omega}]$. Clearly, $\mathfrak{b}$ is also an $\mathcal{O}$-ideal, and by Proposition 4.25,

$$\mathfrak{a}\mathfrak{b} = [a^2, a(b + c\omega), a(b + c\bar{\omega}), N(b + c\omega)]$$

$$= ac\left[\frac{a}{c}, \frac{b}{c} + \omega, \frac{b}{c} + \bar{\omega}, \frac{N(b + c\omega)}{ac}\right].$$

Since $\bar{\omega} = T(\omega) - \omega$, we get

$$\mathfrak{a}\mathfrak{b} = ac\left[\frac{a}{c}, \frac{b}{c} + \omega, \frac{b}{c} + T(\omega) - \omega, \frac{N(b + c\omega)}{ac}\right]$$

$$= ac\left[\frac{a}{c}, \frac{T(b + c\omega)}{c}, \frac{N(b + c\omega)}{ac}, \frac{b}{c} + \omega\right].$$

Since $d = 1$, we get $[a/c, T(b + c\omega)/c, N(b + c\omega)/ac] = [1]$; hence,

$$\mathfrak{a}\mathfrak{b} = ac[1, \omega] = (ac)$$

and it follows that $\mathfrak{a}$ is invertible. $\qquad\square$

Thus, we see that $\mathfrak{a}$ is an invertible $\mathcal{O}$-ideal if and only if $d = 1$. We now need to define the norm of an $\mathcal{O}$-ideal.

Definition 4.33. *We define the* norm $N(\mathfrak{a})$ *of an $\mathcal{O}$-ideal $\mathfrak{a}$ to be the index* $|\mathcal{O}/\mathfrak{a}|$.

By Proposition 4.23 we know that $N(\mathfrak{a}) = ac$ for $\mathfrak{a} = [a, b + c\omega]$.

If $\mathfrak{a} = [\alpha, \beta]$ is an $\mathcal{O}$-ideal, we define the ideal $\bar{\mathfrak{a}}$ *conjugate* to $\mathfrak{a}$ to be $[\bar{\alpha}, \bar{\beta}]$. Notice that $\bar{\mathfrak{a}}$ is also an $\mathcal{O}$-ideal. With these definitions, we have the following corollary to Theorem 4.32.

Corollary 4.32.1. *Let $\mathfrak{a}$ be any $\mathcal{O}$-ideal. Then $\mathfrak{a}$ is proper if and only if $\mathfrak{a}$ is invertible; furthermore, if $\mathfrak{a}$ is invertible, then*

$$\mathfrak{a}\bar{\mathfrak{a}} = (N(\mathfrak{a})).$$

We also notice that if $\mathfrak{a}$ is any $\mathcal{O}$-ideal and $(N(\mathfrak{a}), f) = 1$, then $d = 1$ and therefore $\mathfrak{a}$ is invertible.

If $\mathfrak{a}$ is an $\mathcal{O}$-ideal, but $\mathfrak{a}$ is not a proper $\mathcal{O}$-ideal, then there must exist some γ, where $\gamma \in \mathbb{K}$ but $\gamma \notin \mathcal{O}$ such that $\gamma\mathfrak{a} \subseteq \mathfrak{a}$. If $\mathfrak{b}$ is any $\mathcal{O}$-ideal equivalent to $\mathfrak{a}$, then $\mathfrak{b} = \kappa\mathfrak{a}$ for some $\kappa \in \mathbb{K}$ ($\kappa \neq 0$) and

$$\gamma\mathfrak{b} = \kappa\gamma\mathfrak{a} \subseteq \kappa\mathfrak{a} = \mathfrak{b} .$$

Thus, $\mathfrak{b}$ is also not a proper ideal of $\mathcal{O}$. If, on the other hand, $\mathfrak{a}$ is a proper $\mathcal{O}$-ideal and $\mathfrak{b}$ is an $\mathcal{O}$-ideal equivalent to $\mathfrak{a}$, then $\mathfrak{b}$ must also be a proper $\mathcal{O}$-ideal. We will next show that if we are given any integer M and a proper $\mathcal{O}$-ideal $\mathfrak{a}$, there exists an $\mathcal{O}$-ideal $\mathfrak{b}$ such that $\mathfrak{b} \sim \mathfrak{a}$ and $(N(\mathfrak{b}), M) = 1$. We first require the following simple result.

Proposition 4.34. *If $f(x, y) = ax^2 + bxy + cy^2$, where $a, b, c \in \mathbb{Z}$ and $(a, b, c) = 1$, then for any given integer M there exists a pair of values $x', y' \in \mathbb{Z}$ such that $(f(x', y'), M) = 1$.*

Proof. Let p be any prime divisor of M. Since $(a, b, c) = 1$, at least one of $f(1, 0) = a$, $f(0, 1) = c$, and $f(1, 1) = a + b + c$ must be relatively prime to p. Thus, we can use the Chinese remainder theorem to compute values x' and y' such that $(f(x', y'), M) = 1$. $\qquad\square$

Now, let $\mathfrak{a} = [a, b + c\omega]$ be any invertible $\mathcal{O}$-ideal of $\mathcal{O} = [1, \omega]$ and let $\lambda = xa + y(b + c\omega) \in \mathfrak{a}$ for $x, y \in \mathbb{Z}$. If we put $\mathfrak{b} = \kappa\mathfrak{a}$, where $\kappa = \overline{\lambda}/N(\mathfrak{a})$, then

$$\mathfrak{b} = \left[x\frac{a}{c} + y\left(\frac{b}{c} + \overline{\omega}\right), x\left(\frac{b}{c} + \omega\right) + y\frac{N(b + c\omega)}{ac} \right]$$

is certainly a submodule of $\mathcal{O}$. Also, $\omega\mathfrak{b} = \kappa\omega\mathfrak{a} \subseteq \kappa\mathfrak{a} = \mathfrak{b}$; hence, $\mathfrak{b}$ is an $\mathcal{O}$-ideal and

$$(N(\mathfrak{a}))\mathfrak{b} = (\overline{\lambda})\mathfrak{a} . \tag{4.10}$$

We now need some important results concerning the norm of an ideal.

Theorem 4.35. *Let $\mathcal{O} = [1, \omega]$ be an order of $\mathbb{K}$, $\mathfrak{a}$ be an $\mathcal{O}$-ideal, and $\beta \in \mathcal{O}$. Then*

$$N((\beta)\mathfrak{a}) = |N(\beta)|N(\mathfrak{a}) .$$

Proof. Since $\mathfrak{a}$ is an $\mathcal{O}$-ideal, we have $\mathfrak{a} = [a, \alpha]$, where $\alpha = b + c\omega$ for $a, b, c \in \mathbb{Z}$ and $c \mid a$, $c \mid b$ and $a, c > 0$. By Proposition 4.23 we have $N(\mathfrak{a}) = ac$. Since $\beta \in \mathcal{O}$ and $\alpha\beta \in \mathcal{O}$, we have $m, n, s, t \in \mathbb{Z}$ such that $\beta = m + n\omega$, $\alpha\beta = s + t\omega$, where $c \mid s$ and $c \mid t$. It follows that $t\beta - n\alpha\beta = tm - ns$. Since $\beta - m = n\omega$, we get

$$\beta(t - n\alpha) = \beta(t - nb - nc\omega) = \beta(t - nb - c\beta + cm) = tm - ns .$$

Hence, $\beta^2 - \beta(t - nb + cm)/c + (tm - ns)/c = 0$ and therefore $N(\beta) = (tm - ns)/c$. Now, by Theorem 4.22, $(\beta)\mathfrak{a} = [(m + n\omega)a, s + t\omega] = [a(tm - ns)/g, h + g\omega]$ for some $h \in \mathbb{Z}$, where $g = (na, t)$. Thus, by Proposition 4.23 we get $N((\beta)\mathfrak{a}) = a|tm - ns| = |N(\beta)|N(\mathfrak{a})$. $\qquad\square$

Notice that if $\mathfrak{a} = \mathcal{O}$, then $N(\mathfrak{b}) = |N(\beta)|$, where $\mathfrak{b} = (\beta)$. Thus, the proof of this result shows us how to find a $\mathbb{Z}$-basis for a principal $\mathcal{O}$-ideal $\mathfrak{b}$ when we know a generator β of $\mathfrak{b}$. We show later that the problem of finding a generator for a principal $\mathcal{O}$-ideal, given its $\mathbb{Z}$-basis, is much more difficult.

Another important result concerning the norm is given in the following theorem.

Theorem 4.36. *If $\mathfrak{a}$ and $\mathfrak{b}$ are both invertible $\mathcal{O}$-ideals, then $\mathfrak{ab}$ is invertible and $N(\mathfrak{ab}) = N(\mathfrak{a})N(\mathfrak{b})$.*

Proof. Since $\mathfrak{a}$ and $\mathfrak{b}$ are both invertible, we know from Corollary 4.32.1 that $\mathfrak{a}\bar{\mathfrak{a}} = (N(\mathfrak{a}))$ and $\mathfrak{b}\bar{\mathfrak{b}} = (N(\mathfrak{b}))$. Thus,

$$\mathfrak{ab}\bar{\mathfrak{a}}\bar{\mathfrak{b}} = (N(\mathfrak{a})N(\mathfrak{b}))$$

and therefore $\mathfrak{ab}$ is invertible. It follows that

$$(N(\mathfrak{ab})) = \mathfrak{ab}\overline{\mathfrak{ab}} = \mathfrak{ab}\bar{\mathfrak{a}}\bar{\mathfrak{b}} = (N(\mathfrak{a})N(\mathfrak{b})) \ .$$

Since $N(\mathfrak{ab}), N(\mathfrak{a}), N(\mathfrak{b}) \in \mathbb{Z}^{>0}$, we must have $N(\mathfrak{ab}) = N(\mathfrak{a})N(\mathfrak{b})$. $\square$

We now return to (4.10). By using the results above, we get

$$N(\mathfrak{a})^2 N(\mathfrak{b}) = N(\bar{\lambda})N(\mathfrak{a}) \ .$$

Hence, $N(\mathfrak{b}) = N(\lambda)/N(\mathfrak{a})$. Also,

$$\frac{N(\lambda)}{N(\mathfrak{a})} = \frac{a}{c}x^2 + \frac{T(b + c\omega)}{c}xy + \frac{N(b + c\omega)}{ac}y^2 \ .$$

Since $\mathfrak{a}$ is invertible, we have $d = 1$ and we can use Proposition 4.34 to prove the following theorem.

Theorem 4.37. *If $\mathfrak{a}$ is an given invertible $\mathcal{O}$-ideal and M is any given integer, there is always some $\mathcal{O}$-ideal $\mathfrak{b}$ such that $\mathfrak{b} \sim \mathfrak{a}$ and $(N(\mathfrak{b}), M) = 1$.*

We conclude this long section by proving the cancellation law.

Proposition 4.38. *Let $\mathfrak{a}, \mathfrak{b}, \mathfrak{c}$ be $\mathcal{O}$-ideals such that*

$$\mathfrak{ab} = \mathfrak{ac} \ .$$

If $\mathfrak{a}$ is invertible, then $\mathfrak{b} = \mathfrak{c}$.

Proof. We have

$$\mathfrak{a}\bar{\mathfrak{a}}\mathfrak{b} = \mathfrak{a}\bar{\mathfrak{a}}\mathfrak{c} \ ;$$

hence, by Corollary 4.32.1,

$$N(\mathfrak{a})\mathfrak{b} = N(\mathfrak{a})\mathfrak{c}$$

and $\mathfrak{b} = \mathfrak{c}$. $\square$

4.6 Divisibility and Prime Ideals

If $\mathfrak{a}$ and $\mathfrak{b}$ are non-zero $\mathcal{O}$-ideals, we say that $\mathfrak{a}$ *divides* $\mathfrak{b}$ if there exists an $\mathcal{O}$-ideal $\mathfrak{c}$ such that $\mathfrak{b} = \mathfrak{ac}$. We denote this by $\mathfrak{a} \mid \mathfrak{b}$. We next show that to divide is to contain.

Theorem 4.39. *If $\mathfrak{a}$ and $\mathfrak{c}$ are non-zero $\mathcal{O}$-ideals and $\mathfrak{a}$ is invertible, then $\mathfrak{a} \mid \mathfrak{c}$ if and only if $\mathfrak{a} \supseteq \mathfrak{c}$.*

Proof. If $\mathfrak{a} \mid \mathfrak{c}$, then there exists an $\mathcal{O}$-ideal $\mathfrak{b}$ such that $\mathfrak{c} = \mathfrak{ab}$. If $\gamma \in \mathfrak{c}$, then $\gamma \in \mathfrak{ab}$. Let $\mathfrak{a} = [\alpha_1, \alpha_2]$ and $\mathfrak{b} = [\beta_1, \beta_2]$; then by Proposition 4.25, $\gamma \in [\alpha_1\beta_1, \alpha_1\beta_2, \alpha_2\beta_1, \alpha_2\beta_2]$. Since $\alpha_i\beta_j \in \beta_j\mathfrak{a} \subseteq \mathfrak{a}$, we have $\gamma \in \mathfrak{a}$ and $\mathfrak{a} \supseteq \mathfrak{c}$.

If $\mathfrak{a} \supseteq \mathfrak{c}$, then $\mathfrak{a}\bar{\mathfrak{a}} \supseteq \mathfrak{c}\bar{\mathfrak{a}}$ and $(N(\mathfrak{a})) \supseteq \mathfrak{c}\bar{\mathfrak{a}}$. Thus, every element of $\mathfrak{c}\bar{\mathfrak{a}}$ is divisible by $N(\mathfrak{a})$ and $\mathfrak{c}\bar{\mathfrak{a}} = (N(\mathfrak{a}))\mathfrak{b}$, where $\mathfrak{b}$ is an $\mathcal{O}$-ideal. It follows that

$$\bar{\mathfrak{a}}\mathfrak{a} = (N(\mathfrak{a}))\mathfrak{ab} ;$$

hence, $N(\mathfrak{a})\mathfrak{c} = N(\mathfrak{a})\mathfrak{ab}$ and $\mathfrak{c} = \mathfrak{ab}$. □

Now, suppose that $\mathfrak{c}$ is an invertible $\mathcal{O}$-ideal and an $\mathcal{O}$-ideal $\mathfrak{a}$ divides $\mathfrak{c}$. Then $\mathfrak{c} = \mathfrak{ab}$ for some $\mathcal{O}$-ideal $\mathfrak{b}$. If $\kappa\mathfrak{a} \subseteq \mathfrak{a}$ for some $\kappa \in \mathbb{K}$, then $\kappa\mathfrak{c} = \kappa\mathfrak{ab} \subseteq \mathfrak{ab} = \mathfrak{c}$. Since $\mathfrak{c}$ is invertible, we must have $\kappa \in \mathcal{O}$; hence, $\mathfrak{a}$ is proper and therefore invertible.

From this result, we see from Proposition 4.12 that there can only be a finite number of $\mathcal{O}$-ideals which divide a given invertible $\mathcal{O}$-ideal.

Since we now have the concept of divisibility of ideals, it seems appropriate to consider the concept of primality of an ideal.

Definition 4.40. *A* prime *ideal of $\mathcal{O}$ is an invertible $\mathcal{O}$-ideal $\mathfrak{p} \neq \mathcal{O}$ with the property that if $\mathfrak{p} \mid \mathfrak{ab}$, where $\mathfrak{a}$ and $\mathfrak{b}$ are any two $\mathcal{O}$-ideals, then $\mathfrak{p} \mid \mathfrak{a}$ or $\mathfrak{p} \mid \mathfrak{b}$.*

We will now attempt to characterize all the prime $\mathcal{O}$-ideals. We first discuss some properties of the sum[13] of two $\mathcal{O}$-ideals.

If $\mathfrak{a}_1$ and $\mathfrak{a}_2$ are two $\mathcal{O}$-ideals and $\mathfrak{a}_1 = [a_1, b_1 + c_1\omega]$, $\mathfrak{a}_2 = [a_2, b_2 + c_2\omega]$, then by definition of $\mathfrak{a}_1 + \mathfrak{a}_2$ we have

$$\mathfrak{a}_1 + \mathfrak{a}_2 = \{\alpha_1 + \alpha_2 : \alpha_1 \in \mathfrak{a}_1, \alpha_2 \in \mathfrak{a}_2\} .$$

However this is the same as

$$\mathfrak{a}_1 + \mathfrak{a}_2 = [a_1, b_1 + c_1\omega, a_2, b_2 + c_2\omega]$$
$$= [a_3, b_3 + c_3\omega] ,$$

where $a_3 = (a_1, a_2)$, $c_3 = (c_1, c_2)$, and $b_3 \in \mathbb{Z}$.

We next prove the following theorem.

Theorem 4.41. *If $\mathfrak{p}$ is any prime $\mathcal{O}$-ideal, there exists a unique rational prime p such that $\mathfrak{p} \mid (p)$.*

Proof. Since $\mathfrak{p}$ is an $\mathcal{O}$-ideal, it must contain some rational integer a. Since $\mathfrak{p} \supseteq a\mathcal{O}$, we see that $\mathfrak{p} \mid (a)$. Since a can be written as a product of rational primes, we see by Definition 4.40 that we must have $\mathfrak{p} \mid (p)$ for some prime p such that $p \mid a$. If $\mathfrak{p} \mid (q)$ for some other prime $q \neq p$, then $\mathfrak{p} \mid (p) + (q)$. However, since $(p, q) = 1$, we can only have $(p) + (q) = \mathcal{O}$, which means that $\mathfrak{p} \mid \mathcal{O}$ or $\mathfrak{p} = \mathcal{O}$, a contradiction.   □

At this point, it is useful to introduce the concept of an indecomposable $\mathcal{O}$-ideal.

Definition 4.42. *An* indecomposable *ideal of $\mathcal{O}$ is an invertible $\mathcal{O}$-ideal $\mathfrak{r}$ ($\neq \mathcal{O}$) which has no divisors other than $\mathcal{O}$ and $\mathfrak{r}$.*

Now, let p be any rational prime and let $\mathfrak{p} \mid (p)$, where $\mathfrak{p}$ is a prime ideal. Since $\mathfrak{p}$ is an $\mathcal{O}$-ideal, we have $\mathfrak{p} = [a, b + c\omega]$, where $a, b, c \in \mathbb{Z}$, $a > 0$, and $c > 0$. Since $\mathfrak{p} \supseteq (p)$, we have $p \in \mathfrak{p}$ and $p = ax + y(b + c\omega)$ for some $x, y \in \mathbb{Z}$. It follows that $p = ax$; hence, $p \mid a$ or $p \nmid a$ and $p \mid x$. In the latter case, we must have $a = 1$ and $\mathfrak{p} = \mathcal{O}$, a contradiction; hence, $a = p$. Since $c \mid a$, we can only have $c = 1$ or p. If $c = p$, then $\mathfrak{p} = (p)$. If $c = 1$, then $\mathfrak{p} = [p, b + \omega]$. Since $\mathfrak{p}$ is an ideal, we must also have $p \mid N(b + \omega)$, which is equivalent to

$$(2b + T(\omega))^2 \equiv \Delta \pmod{4p} .$$

If the Kronecker symbol $(\Delta/p) = -1$, this is impossible. Thus, if $(\Delta/p) = -1$, then $\mathfrak{p}$ can only be (p). In this case we say that p is *inert* in $\mathcal{O}$. If $\mathfrak{r}$ is some ideal that divides (p), by Theorem 4.36 we must have $N(\mathfrak{r}) \mid p^2$. If $N(\mathfrak{r}) = 1$, then $\mathfrak{r} = \mathcal{O}$. If $N(\mathfrak{r}) = p$, then $\mathfrak{r} = [p, b' + c'\omega]$, where $b', c' \in \mathbb{Z}$ and $c' \mid p$. If $c' = p$, then $(p) \mid \mathfrak{r}$ and therefore $\mathfrak{r} = (p)$. If $c' = 1$, then $\mathfrak{r} = [p, b' + \omega]$ and $p \mid N(b' + \omega)$, which is impossible by (4.4) as $(\Delta/p) = -1$. Thus, (p) is an indecomposable ideal of $\mathcal{O}$.

If $(\Delta/p) \neq -1$, then there must exist some $t \in \mathbb{Z}$ such that

$$t^2 \equiv \Delta \pmod{4p} .$$

By (4.3), $t \equiv T(\omega) \pmod 2$. We put

$$\mathfrak{q} = \left[p, \frac{t + \sqrt{\Delta}}{2} \right] = [p, t' + \omega] ,$$

where $t' = (t - T(\omega))/2$. Then $\mathfrak{q}$ is an $\mathcal{O}$-ideal and we cannot have $(p) \mid \mathfrak{q}$.

Now, $N(\mathfrak{q}) = p$ and therefore $\mathfrak{q}\bar{\mathfrak{q}} = (p)$. Since $\mathfrak{p} \mid (p)$, we get $\mathfrak{p} \mid \mathfrak{q}$ or $\mathfrak{p} \mid \bar{\mathfrak{q}}$. If $\mathfrak{p} \mid \mathfrak{q}$, then $c = 1$ and $\mathfrak{p} = [p, b + \omega] \supseteq \mathfrak{q} = [p, t' + \omega]$. It follows that $b \equiv t' \pmod p$ and $\mathfrak{p} = \mathfrak{q}$. Similarly, if $\mathfrak{p} \mid \bar{\mathfrak{q}}$, then $\mathfrak{p} = \bar{\mathfrak{q}}$. If $\mathfrak{r}$ is any $\mathcal{O}$-ideal such that $\mathfrak{r} \mid \mathfrak{q}$, then $N(\mathfrak{r}) \mid N(\mathfrak{q})$ and therefore $N(\mathfrak{r}) = 1$ or p. If $N(\mathfrak{r}) = 1$, then $\mathfrak{r} = \mathcal{O}$; if $N(\mathfrak{r}) = p$, then $\mathfrak{r} = [p, b' + \omega] \supseteq [p, t' + \omega]$ and $b' \equiv t' \pmod p$. Thus, $\mathfrak{r} = \mathfrak{q}$. Also, if $\mathfrak{r} \mid \bar{\mathfrak{q}}$, then it is easy to show that $\mathfrak{r} = \bar{\mathfrak{q}}$. Thus, we have shown that any prime ideal of $\mathcal{O}$ is indecomposable.

If $\mathfrak{q} = \bar{\mathfrak{q}}$, then $p \mid t$ and $(\Delta/p) = 0$. Thus, if $(\Delta/p) = 1$, then $(p) = \mathfrak{q}\bar{\mathfrak{q}}$. In this case we say that p *splits* in $\mathcal{O}$. Finally, if $(\Delta/p) = 0$, then $p \mid t$ and $\mathfrak{q} = \bar{\mathfrak{q}}$. In this case, $(p) = \mathfrak{q}^2$, and we say that p *ramifies* in $\mathcal{O}$. Notice that in this latter case we must have $(p, f) = 1$; otherwise $\mathfrak{q}$ is not invertible.

We have therefore found that if $\mathfrak{p}$ is a prime ideal of $\mathcal{O}$, then $\mathfrak{p} = (p)$ or $\mathfrak{p} = \mathfrak{q}$ or $\mathfrak{p} = \bar{\mathfrak{q}}$. Also, $(p, f) = 1$.

It remains to show that these are indeed prime ideals. For this, we now require the following theorem.

Theorem 4.43. *If $\mathfrak{r}$ is an indecomposable $\mathcal{O}$-ideal such that $(N(\mathfrak{r}), f) = 1$, then $\mathfrak{r}$ is a prime ideal of $\mathcal{O}$.*

Proof. Suppose $\mathfrak{r} \mid \mathfrak{ab}$, where $\mathfrak{a}$ and $\mathfrak{b}$ are any two $\mathcal{O}$-ideals, and put $\mathfrak{d} = \mathfrak{a} + \mathfrak{b}$. We observe that $\mathfrak{d} \supseteq \mathfrak{r}$ and $\mathfrak{d} \supseteq \mathfrak{a}$. Also, by the remarks following Definition 4.40, we have $(N(\mathfrak{d}), f) = 1$. Since $\mathfrak{d}$ is invertible, we have $\mathfrak{d} \mid \mathfrak{r}$ and $\mathfrak{d} \mid \mathfrak{a}$. Since $\mathfrak{r}$ is indecomposable, we can only have $\mathfrak{d} = \mathcal{O}$ or $\mathfrak{d} = \mathfrak{r}$. If $\mathfrak{d} = \mathfrak{r}$, then $\mathfrak{r} \mid \mathfrak{a}$; if $\mathfrak{d} = \mathcal{O}$, then since $1 \in \mathcal{O}$, there must exist some $\rho \in \mathfrak{r}$ and some $\alpha \in \mathfrak{a}$ such that $\rho + \alpha = 1$. Let β be any element of $\mathfrak{b}$. We have

$$\beta = \beta\rho + \beta\alpha .$$

Now, $\mathfrak{r} \supseteq \mathfrak{ab}$ and $\alpha\beta \in \mathfrak{ab}$; hence, $\alpha\beta \in \mathfrak{r}$. Since $\rho \in \mathfrak{r}$ and $\mathfrak{r}$ is an $\mathcal{O}$-ideal, we must have $\beta\mathfrak{r} \subseteq \mathfrak{r}$, which means that $\beta\rho \in \mathfrak{r}$. It follows that $\beta(= \beta\rho + \beta\alpha) \in \mathfrak{r}$ and this is true for all $\beta \in \mathfrak{b}$. Hence, $\mathfrak{b} \subseteq \mathfrak{r}$ and $\mathfrak{r} \mid \mathfrak{b}$ by Theorem 4.39. $\square$

It might seem that all indecomposable $\mathcal{O}$-ideals should be prime ideals, but consider the ideal $\mathfrak{r} = [p^2, p\omega_0]$ in the order $\mathcal{O} = [1, \omega]$, where $\omega = p\omega_0$. If we select p to be some rational prime such that $p \nmid N(\omega_0)$, then $\mathfrak{r}$ is invertible. Also, it is not difficult to show that $\mathfrak{r}$ is indecomposable in $\mathcal{O}$, but it is certainly not a prime ideal of $\mathcal{O}$ because $\mathfrak{r} \mid (p)[p, \omega]$, but $\mathfrak{r} \nmid (p)$ and $\mathfrak{r} \nmid [p, \omega]$.

If we produce any $\mathcal{O}$-ideal $\mathfrak{a}$ as a product of prime ideals of $\mathcal{O}$, we see by Theorem 4.36 that $(N(\mathfrak{a}), f) = 1$. We will now show that if $(N(\mathfrak{a}), f) = 1$ for any $\mathcal{O}$-ideal $\mathfrak{a}$, then $\mathfrak{a}$ can be represented as a unique (up to order) product of prime ideals of $\mathcal{O}$.

Theorem 4.44. *If $\mathfrak{a}$ is any $\mathcal{O}$-ideal such that $(N(\mathfrak{a}), f) = 1$, then $\mathfrak{a}$ can be written uniquely (up to order) as a product of prime ideals of $\mathcal{O}$.*

Proof. Clearly, $\mathfrak{a}$ is invertible, and since $\mathfrak{a}$ has only a finite number of divisors, it can be written as a product of indecomposable ideals of $\mathfrak{a}$. Furthermore, by Theorem 4.36, if $\mathfrak{r}$ is any one of these indecomposable ideals, we have $(N(\mathfrak{r}), f) = 1$, which, by Theorem 4.43, means that $\mathfrak{r}$ is a prime ideal of $\mathcal{O}$. That this product is unique follows in the same way as the proof for the unique factorization of any rational integer. $\square$

Thus, if $\mathcal{O} = \mathcal{O}_{\mathbb{K}}$, then $f = 1$ and any ideal of $\mathcal{O}_{\mathbb{K}}$ can be written uniquely (up to order) as a product of prime ideals of $\mathcal{O}_{\mathbb{K}}$. Note, however, that this is not the case when $f > 1$ unless the ideal $\mathfrak{a}$ has norm relatively prime to f.

Notes and References

[1] See, for example, [NZM91].

[2] This was proved in a different form in art. 42 of [Gau86]. For an interesting discussion of this result and its generalization, see [MM05].

[3] For an elementary introduction to this subject, see [AW04], [ST87], or [PD98]. For more advanced texts, see, for example, [Mar77a], [Lan91], [IR82], and the encyclopaedic [Nar04].

[4] We confine our discussions to quadratic fields. For further information on these structures, see [Coh62] and [Cox89].

[5] For a more general discussion of modules, consult any standard text in abstract algebra such as [DF04] or [Hun74].

[6] An order of $\mathbb{K}$ can be defined more generally as a subset $\mathcal{O}$ of $\mathbb{K}$ such that $\mathcal{O}$ is a subring of $\mathbb{K}$ containing 1, $\mathcal{O}$ is a finitely generated module of $\mathbb{K}$, and $\mathcal{O}$ contains a basis for $\mathbb{K}$ over $\mathbb{Q}$. However, as mentioned in [Cox89], p. 133, this is equivalent to the definition provided here.

[7] It has become customary to use binary quadratic forms instead of ideals for these kinds of investigations, but we will use ideals because they are computationally more convenient. For more information on the use of binary quadratic forms in this context, the reader is referred to [BV07], [Bue89], [Coh93], or [Hua82].

[8] We are grateful to John P. Robertson for bringing this result to our attention.

[9] See Prop. VII.30 in [Hea56].

[10] See art. 16 in [Gau86].

[11] A good introduction to Kummer's work can be found in [Smi65].

[12] This appeared in Supplement XI (second edition, 1871) appended to Dirichlet's *Vorlesungen über Zahlentheorie*. For an English version, see [Ded96].

[13] For a very comprehensive treatment of the problem of adding two ideals, the reader is referred to [Wei06].

Ideals and Continued Fractions

5.1 Reduced Ideals of $\mathcal{O}$

Throughout this chapter we will let $\mathcal{O} = [1, \omega]$ be the order of discriminant Δ in the quadratic field $\mathbb{K} = \mathbb{Q}(\sqrt{D})$. If $\mathfrak{a}$ is any ideal of $\mathcal{O}$, it is evident that its corresponding ideal class, $[\mathfrak{a}]$, contains an infinitude of ideals. In order to deal with this difficulty in managing $[\mathfrak{a}]$, we will restrict our attention to a finite subset of particular ideals of $[\mathfrak{a}]$. To this end we provide the following definitions.[1]

Definition 5.1. *We say that an ideal $\mathfrak{a}$ of $\mathcal{O}$ is* primitive *if it cannot be written as*

$$\mathfrak{a} = m\mathfrak{b} \, ,$$

where $\mathfrak{b}$ is an ideal of $\mathcal{O}$ and $m \in \mathbb{Z}$, $|m| > 1$.

By Theorem 4.24, we know that any $\mathcal{O}$-ideal $\mathfrak{a}$ can be represented as

$$\mathfrak{a} = [a, b + c\omega] \, ,$$

where $a, b, c \in \mathbb{Z}$; $c \mid b$, $c \mid a$; $a, c > 0$, $ac \mid N(b + c\omega)$. Since $\mathfrak{a} = c[a/c, b/c + \omega] = c[-a/c, b/c + \omega]$, it follows that we can represent any primitive ideal $\mathfrak{b}$ by

$$\mathfrak{b} = [s, t + \omega] \, ,$$

where $s, t \in \mathbb{Z}$, $N(\mathfrak{b}) = |s|$, and $N(\mathfrak{b}) \mid N(t + \omega)$. That $N(\mathfrak{b}) = |s|$ must, of course, follow from Proposition 4.23. Clearly, we see that $[\mathfrak{b}]$ contains primitive ideals. We will now derive some properties of such ideals. We use $\lfloor x \rceil$ to denote $\lfloor x + 1/2 \rfloor$, the nearest integer to $x \in \mathbb{R}$.

Proposition 5.2. *If $\mathfrak{a} = [a, b + \omega]$ is any primitive ideal of $\mathcal{O}$, there exists some $\alpha \in \mathfrak{a}$ such that $\mathfrak{a} = [a, \alpha]$ and $|T(\alpha)| \le |a|$.*

Proof. Put $q = \lfloor T(b + \omega)/2a \rceil$ and $\alpha = b + \omega - qa$. Certainly, $\mathfrak{a} = [a, \alpha]$ and $|T(\alpha)| = |T(b + \omega) - 2qa| = 2|a||T(b + \omega)/2a - q| \le |a|$. $\qquad\square$

We will now show that for a given $\mathfrak{a}$ the value of $|T(\alpha)|$ in Proposition 5.2 is unique.

Proposition 5.3. *Let $\mathfrak{a}$ be a primitive $\mathcal{O}$-ideal and suppose that $\mathfrak{a} = [a, \alpha]$ and $\mathfrak{a} = [a, \beta]$. If $|T(\alpha)| \leq |a|$ and $|T(\beta)| \leq |a|$, then $|T(\alpha)| = |T(\beta)|$.*

Proof. Suppose that $|T(\alpha)| \leq |a|$ and $|T(\beta)| \leq |a|$. Since $\beta \in [a, \alpha]$ and $\alpha \in [a, \beta]$, we must have $x, y, s, t \in \mathbb{Z}$ such that

$$\beta = x|a| + y\alpha, \quad \alpha = s|a| + t\beta .$$

Thus, $\beta = x|a| + y(s|a| + t\beta)$ and, therefore, $ty = 1$, which means that $|y| = 1$. If $x = 0$, then $|T(\alpha)| = |T(\beta)|$. Suppose $x \neq 0$. Since $T(\beta) = 2x|a| + yT(\alpha)$, we have

$$|T(\beta)| \geq |2x|a| - (-y)T(\alpha)| \geq 2x|a| - |T(\alpha)| \geq 2|a| - T(\alpha) .$$

If $|a| > |T(\alpha)|$, then $T(\beta) > |a|$, which is not possible. If $|a| = |T(\alpha)|$, then $|T(\beta)| \geq |a|$. Since $T(\beta) \leq |a|$, we must have $|T(\beta)| = |a| = |T(\alpha)|$. □

When we perform arithmetic in $\mathbb{Z}$, it is often very useful to simplify our calculations by reducing the results modulo some integer m. This allows us to reduce all our intermediate results to numbers whose values are bounded by m, thereby guaranteeing that our computations will not produce numbers that are unmanageably large. It turns out that we can also perform a reduction operation on ideals. As we shall see, this operation is extremely useful when we need to perform arithmetic on the ideals of $\mathcal{O}$. Indeed, ideal reduction is the most important of all the basic arithmetic operations that we will perform on ideals. It is for this reason that we discuss it in considerable detail in this chapter.

Definition 5.4. *If $\mathfrak{a}$ is an $\mathcal{O}$-ideal, then $\mathfrak{a}$ is said to be a reduced ideal of $\mathcal{O}$ if $\mathfrak{a}$ is primitive and there does not exist a non-zero $\alpha \in \mathfrak{a}$ such that both $|\alpha| < N(\mathfrak{a})$ and $|\overline{\alpha}| < N(\mathfrak{a})$ hold.*

Notice that if $\mathfrak{a}$ is a reduced $\mathcal{O}$-ideal, then so is $\overline{\mathfrak{a}}$. Note further that when $\Delta < 0$, we have $|\alpha| = |\overline{\alpha}|$; thus, in this case we could shorten the definition of a reduced ideal $\mathfrak{a}$ in $\mathcal{O}$ by simply demanding that there be no non-zero $\alpha \in \mathfrak{a}$ such that $|\alpha| < N(\mathfrak{a})$.

We will now distinguish between the two cases of $\Delta < 0$ and $\Delta > 0$. We first consider the case of $\Delta < 0$.

Theorem 5.5. *If $\mathfrak{a}$ is a primitive ideal of $\mathcal{O}$, $\Delta < 0$ and $\mathfrak{a} = [a, \alpha]$, with $|T(\alpha)| \leq |a|$, then $\mathfrak{a}$ is a reduced ideal of $\mathcal{O}$ if and only if $|\alpha| \geq |a|$.*

Proof. If $\mathfrak{a}$ is reduced, then $|\alpha| \geq |a| = N(\mathfrak{a})$. (We will often simply use the shorter term "reduced" to mean a reduced $\mathcal{O}$-ideal when there is no doubt about the underlying order $\mathcal{O}$.) Now assume that $|\alpha| \geq |a|$. By (4.4), we also have

$$4|\alpha|^2 = 4\alpha\overline{\alpha} = T(\alpha)^2 + |\Delta| .$$

Hence,

$$|\Delta| = 4|\alpha|^2 - T(\alpha)^2 \geq 4a^2 - a^2 = 3a^2 .$$

If $\mathfrak{a}$ is not a reduced $\mathcal{O}$-ideal, there must exist some $\beta \in \mathfrak{a}$ such that $|\beta| < |a|$ and $\beta \neq 0$. Thus, there must exist $x, y \in \mathbb{Z}$ such that

$$\beta = x|a| + y\alpha \quad (y \neq 0)$$

and, by (4.4),

$$4|\beta|^2 = 4\beta\overline{\beta} = T(\beta)^2 + y^2|\Delta| . \tag{5.1}$$

Since $|\Delta| \geq 3a^2$, we get $4|\beta|^2 \geq 3y^2a^2 > 4a^2$ when $|y| > 1$. Thus, since $|\beta| < |\alpha|$, we must have $|y| = 1$ and therefore $\mathfrak{a} = [a, \beta]$. If $|T(\beta)| > |a|$, then by (5.1), $4|\beta|^2 > 4a^2$ and $|\beta| > |a|$, which is impossible. Thus, $|T(\beta)| \leq |a|$. By Proposition 5.3, we must have $|T(\beta)| = |T(\alpha)|$, which by (5.1) and the fact that $|y| = 1$ means that $|\beta| = |\alpha| \geq |a|$, a contradiction. $\qquad\square$

Corollary 5.5.1. *If $\mathfrak{a}$ is a reduced $\mathcal{O}$-ideal and $\Delta < 0$, then $N(\mathfrak{a}) \leq \sqrt{|\Delta|/3}$.*

We also have a companion result to this corollary below.

Theorem 5.6. *If $\mathfrak{a} = [a, b + \omega]$ is a primitive $\mathcal{O}$-ideal, $\Delta < 0$, and $N(\mathfrak{a}) < \sqrt{|\Delta|}/2$, then $\mathfrak{a}$ is reduced.*

Proof. Let $\beta \in \mathfrak{a}$, $\beta \neq 0$. By (5.1) we have

$$4|\beta|^2 \geq |\Delta| > 4a^2$$

and $|\beta| > |a|$. Thus, $\mathfrak{a}$ must be a reduced $\mathcal{O}$-ideal by definition. $\qquad\square$

We next point out that if $\mathfrak{a} = [a, \alpha]$ is a primitive $\mathcal{O}$-ideal ($\alpha = b + \omega$), then so is $\mathfrak{b} = [-N(\alpha)/a, -\overline{\alpha}]$. Also, it is easy to verify that

$$(-\overline{\alpha})\mathfrak{a} = (a)\mathfrak{b} \tag{5.2}$$

and therefore $\mathfrak{a} \sim \mathfrak{b}$. This result is independent of the sign of Δ. We now put

$$q = \begin{cases} \lfloor T(\alpha)/2a \rceil & \text{when } \Delta < 0 \\ \lfloor \alpha/a \rfloor & \text{when } \Delta > 0 \end{cases} \tag{5.3}$$

and $\beta = \alpha - qa$. We define the operation[2] ρ acting on $\mathfrak{a} = [a, b + \omega]$ as that action that produces the ideal $\mathfrak{a}' = [-N(\beta)/a, -\overline{\beta}]$. We denote this by $\mathfrak{a}' = \rho(\mathfrak{a})$, where $\mathfrak{a}' = [a', b' + \omega]$ with $b' = qa - b - T(\omega)$ and $a' = -N(b' + \omega)/a$, and by (5.2), since $\mathfrak{a} = [a, \alpha] = [a, \beta]$, we get

$$\rho(\mathfrak{a}) = \gamma\mathfrak{a} , \tag{5.4}$$

where $\gamma = (b' + \omega)/a$. Notice also that since $b' + \overline{\omega} = qa - b - \omega$, we have $\mathfrak{a} = [a, b' + \overline{\omega}]$ and

$$\frac{b' + \omega}{a'} = \frac{-a}{b' + \overline{\omega}} = \frac{-a}{qa - b - \omega} = \frac{1}{\frac{b+\omega}{a} - q} . \tag{5.5}$$

It is important to realize that although the ideal $\rho(\mathfrak{a})$ is independent of the value of b used in the representation of $\mathfrak{a} = [a, b + \omega]$, it does depend on the sign of a. For example, if $D = 67$, $\omega = \sqrt{67}$, $a = -3$, and $b = 1$, then $\rho(\mathfrak{a}) = [18, 11 + \omega]$, but if we change a to 3, then $\rho(\mathfrak{a}) = [1, 8 + \omega]$. Thus, when we use the shorthand notation $\rho(\mathfrak{a})$, we are assuming that we have been given a and b in the representation of $\mathfrak{a}$ as $[a, b + \omega]$ and then $\rho(\mathfrak{a}) = \mathfrak{a}'$ as defined above.

We now have a simple criterion for determining when the primitive $\mathcal{O}$-ideal $\mathfrak{a}$ when $\Delta < 0$ is reduced.

Theorem 5.7. *If $\mathfrak{a} = [a, b + \omega]$ is a primitive $\mathcal{O}$-ideal and $\Delta < 0$, then $\mathfrak{a}$ is reduced if and only if $N(\rho(\mathfrak{a})) \geq N(\mathfrak{a})$.*

Proof. We define β as above and observe that $|T(\beta)| \leq |a| = N(\mathfrak{a})$. Thus, by Theorem 5.5, $\mathfrak{a} = [a, \beta]$ is reduced if and only if $|\beta| \geq |a| = N(\mathfrak{a})$. Also, $|\beta|^2 = N(\mathfrak{a})N(\mathfrak{a}')$, where $\mathfrak{a}' = \rho(\mathfrak{a})$. If $N(\mathfrak{a}') \geq N(\mathfrak{a})$, then $|\beta|^2 \geq N(\mathfrak{a})^2$. It follows that $|\beta| \geq |a|$ and $\mathfrak{a}$ is reduced. If, on the other hand, $\mathfrak{a}$ is reduced, then $|\beta| \geq |a|$ and, therefore, $N(\mathfrak{a}') = |\beta|^2/N(\mathfrak{a}) \geq N(\mathfrak{a})$. $\qquad\square$

We remark here that since $|T(\beta)| \leq |a|$ and $N(\rho(\mathfrak{a})) = |\beta|^2/|a|$, we get $N(\rho(\mathfrak{a})) \leq (1/4)(T(\beta)^2 + |\Delta|)/|a|$. Hence,

$$N(\rho(\mathfrak{a})) \leq \frac{|a| + |\Delta|/|a|}{4} \quad (\Delta < 0) . \tag{5.6}$$

Notice that we have proved that every ideal class of $\mathcal{O}$, when $\Delta < 0$, must contain a reduced ideal. For if we start with the primitive $\mathcal{O}$-ideal $\mathfrak{a} = [a, b+\omega]$ and define $\rho^n(\mathfrak{a})$ recursively by $\rho^n(\mathfrak{a}) = \rho(\rho^{n-1}(\mathfrak{a}))$, then the sequence of positive integers

$$N(\mathfrak{a}), N(\rho(\mathfrak{a})), N(\rho^2(\mathfrak{a})), \ldots$$

cannot be strictly decreasing indefinitely. Thus, at some point, we must have

$$N(\rho^i(\mathfrak{a})) \geq N(\rho^{i+1}(\mathfrak{a}))$$

for some $i \geq 0$. It follows that $\rho^i(\mathfrak{a})$ is a reduced ideal of $\mathcal{O}$. Also, since $\rho(\mathfrak{a}) \sim \mathfrak{a}$, we have $\rho^i(\mathfrak{a}) \sim \mathfrak{a}$ by induction.

We now turn our attention to the case of $\Delta > 0$. We first need to develop a result like Theorem 5.5 for this case. This is provided in the next theorem.

Theorem 5.8. *If $\mathfrak{a}$ is a primitive ideal of $\mathcal{O}$, $\Delta > 0$, and $\mathfrak{a} = [a, b + \omega]$, then $\mathfrak{a}$ is a reduced ideal of $\mathcal{O}$ if and only if there exists some $\beta \in \mathfrak{a}$ such that $\mathfrak{a} = [a, \beta]$, $\beta > |a|$, and $-|a| < \overline{\beta} < 0$.*

Proof. Suppose $\mathfrak{a}$ is a reduced $\mathcal{O}$-ideal. There certainly exists an infinitude of pairs $(x, y) \in \mathbb{Z}^2$ such that

$$|xa + y\overline{\alpha}| < |a| ,$$

for $\alpha = b + \omega$. (For example, $x = \lfloor -y\overline{\alpha}/a \rfloor$, $y = 1, 2, 3, \ldots$.) Let (s, t) be one such pair and put $\gamma = sa + t\alpha$. We note that there can only be a finite number of elements $\lambda \in \mathfrak{a}$ such that $|\lambda| < |\gamma|$ and $|\overline{\lambda}| < |a|$. If there are no such elements, we put $\beta = |\gamma| \in \mathfrak{a}$. If there are such elements, we put $\beta = |\lambda|$, where $|\lambda|$ is minimal. Hence, we may properly define β to be the least positive element of $\mathfrak{a}$ such that $|\overline{\beta}| < |a|$. Since $\mathfrak{a}$ is a reduced ideal of $\mathcal{O}$, we must have $\beta \geq |a|$, but since $|\overline{\beta}| < |a|$, we cannot have $\beta = |a|$; hence, $\beta > |a|$ and $0 < \beta - |a| < \beta$. It follows that since β is the least positive element of $\mathfrak{a}$ such that $|\overline{\beta}| < a$, $\beta - |a| \in \mathfrak{a}$, and $0 < \beta - |a| < \beta$, we must have $|\overline{\beta} - |a|| \geq |a| > |\overline{\beta}|$ and therefore $-|a| < \overline{\beta} < 0$.

Since $\beta \in \mathfrak{a}$, we must have $(p, q) \in \mathbb{Z}^2$ such that $\beta = pa + q\alpha$. Suppose $|q| > 1$ and let $u \equiv p \pmod{q}$, where $|u| \leq |q/2|$; then $\mu = |(\beta - ua)/q| \in \mathfrak{a}$. Hence,

$$|\overline{\mu}| \leq |\overline{\beta}/q| + |ua/q| \leq |\overline{\beta}|/2 + |a|/2 < |a| .$$

Also, $\mu > 0$ and

$$\mu \leq |\beta/q| + |ua/q| < \beta/2 + |a|/2 < \beta ;$$

however, such a $\mu \in \mathfrak{a}$ cannot exist by selection of β. Hence, we must have $|q| \leq 1$. Since $q \neq 0$, we can only have $q = \pm 1$ and $\mathfrak{a} = [a, \beta]$.

Next, suppose that $\mathfrak{a} = [a, b + \omega]$, where $\beta = b + \omega > |a|$ and $-|a| < b + \overline{\omega} < 0$. If $\mathfrak{a}$ is not a reduced ideal of $\mathcal{O}$, there must exist some $\lambda \in \mathfrak{a}$ such that $\lambda \neq 0$, $|\lambda| < |a|$, and $|\overline{\lambda}| < |a|$. Since $\lambda = x|a| + y\alpha$ for some $x, y \in \mathbb{Z}$, we have

$$|x|a| + y\beta| < |a|, \quad |x|a| + y\overline{\beta}| < |a| .$$

We see that if $x = 0$, then $y = 0$, and if $y = 0$, then $x = 0$; hence, $xy \neq 0$. Also, $xy < 0$ by the first inequality and $xy > 0$ by the second. Hence, no such λ can exist in $\mathfrak{a}$ and, therefore, $\mathfrak{a}$ must be reduced. $\square$

Corollary 5.8.1. *If $\mathfrak{a}$ is a reduced $\mathcal{O}$-ideal and $\Delta > 0$, then $N(\mathfrak{a}) < \sqrt{\Delta}$.*

Proof. By the theorem $\mathfrak{a} = [a, \beta]$, where $\beta > |a|$ and $-|a| < \overline{\beta} < 0$. Thus, $N(\mathfrak{a}) = |a| < \beta - \overline{\beta} = \omega - \overline{\omega} = \sqrt{\Delta}$. $\square$

We also have a result similar to Theorem 5.6.

Theorem 5.9. *If $\Delta > 0$, $\mathfrak{a}$ is a primitive ideal of $\mathcal{O}$, and $N(\mathfrak{a}) < \sqrt{\Delta}/2$, then $\mathfrak{a}$ is reduced.*

Proof. Let $\mathfrak{a} = [a, \alpha]$ and put $\beta = \alpha + \lfloor -\overline{\alpha}/|a| \rfloor |a|$. Then $\mathfrak{a} = [a, \beta]$, where $-|a| < \overline{\beta} < 0$. Since $\beta - \overline{\beta} = \omega - \overline{\omega}$ ($\mathfrak{a}$ is primitive) and $\overline{\beta} > -|a|$, we get $\beta > \omega - \overline{\omega} - |a| = \sqrt{\Delta} - |a| > |a|$. Hence, $\mathfrak{a}$ is reduced by Theorem 5.8. $\square$

We also have a simple criterion for determining when the primitive $\mathcal{O}$-ideal $\mathfrak{a} = [a, b + \omega]$ is reduced.

Theorem 5.10. *Let* $\mathfrak{a} = [a, b + \omega]$ *be any primitive* $\mathcal{O}$-*ideal, where* $\Delta > 0$. *Put* $\beta = k|a| + b + \omega$, *where* $k = \lfloor -(b + \overline{\omega})/|a| \rfloor$. $\mathfrak{a}$ *is reduced if and only if* $\beta > |a|$.

Proof. We have $-|a| < \overline{\beta} < 0$. If $\beta > |a|$, then $\mathfrak{a}$ is reduced by Theorem 5.8. If $|a| > \beta$, then $\beta \neq 0$ and

$$\beta = \beta - \overline{\beta} + \overline{\beta} > \beta - \overline{\beta} - |a| = \omega - \overline{\omega} - |a| > -|a| .$$

Hence, $|\beta| < |a|$, and since $|\overline{\beta}| < |a|$, we see that $\mathfrak{a}$ cannot be a reduced $\mathcal{O}$-ideal. $\qquad\square$

We now denote any primitive $\mathcal{O}$-ideal $\mathfrak{a} = [a, b + \omega]$ by the equivalent representation given by (4.9):

$$\mathfrak{a} = \left[\frac{Q}{r}, \frac{P + \sqrt{D}}{r} \right] , \qquad (5.7)$$

where $a = Q/r$ and $b + \omega = (P + \sqrt{D})/r$. If we put $Q_0 = Q$, $P_0 = P$, and $q_0 = q$, where q is defined by (5.3), then

$$\rho(\mathfrak{a}) = \left[\frac{Q_1}{r}, \frac{P_1 + \sqrt{D}}{r} \right] ,$$

where

$$\frac{P_1 + \sqrt{D}}{Q_1} = \frac{1}{(P_0 + \sqrt{D})/Q_0 - q_0}$$

by (5.5). In computing the sequence of ideals

$$\mathfrak{a}, \rho(\mathfrak{a}), \rho^2(\mathfrak{a}), \dots, \rho^n(\mathfrak{a}), \dots$$

we produce a sequence of integers representing the value of q at each stage in the process. If these are represented as

$$q_0, q_1, q_2, \dots, q_n, \dots ,$$

then we can consider the continued fraction given by

$$\frac{P + \sqrt{D}}{Q} = \left\langle q_0, q_1, q_2, \dots, q_{n-1}, \frac{P_n + \sqrt{D}}{Q_n} \right\rangle . \qquad (5.8)$$

In this case we find by (5.5) and our results in §3.1 of Chapter 3 that

$$\rho^j(\mathfrak{a}) = \left[\frac{Q_j}{r}, \frac{P_j + \sqrt{D}}{r} \right] \qquad (j = 0, 1, 2, \dots, n) .$$

Also,
$$\rho^j(\mathfrak{a}) = \psi_j \rho^{j-1}(\mathfrak{a})$$
by (5.4). If we define $\mathfrak{a}_{j+1} = \rho^j(\mathfrak{a})$ $(j = 0, 1, 2, \ldots, n)$, we get

$$\mathfrak{a}_j = \psi_{j-1}\mathfrak{a}_{j-1} \quad (j = 2, 3, \ldots, n+1) \tag{5.9}$$

and, by (3.17),
$$\mathfrak{a}_j = \theta_j \mathfrak{a}_1 \quad (j = 1, 2, 3, \ldots, n+1) . \tag{5.10}$$

We also mention that

$$\psi_j = \frac{P_j + \sqrt{D}}{Q_{j-1}} = q_{j-1} + \frac{\sqrt{D} - P_{j-1}}{Q_{j-1}} = q_{j-1} - \overline{\phi}_{j-1} ,$$

and since $\mathfrak{a}_j = [Q_{j-1}/r, (Q_{j-1}/r)\phi_{j-1}]$, we get $\mathfrak{a}_j = [Q_{j-1}/r, (Q_{j-1}/r)\overline{\psi}_j]$. Hence, by (5.10), (3.18), and (3.16), we get

$$(Q_{j-1})\mathfrak{a}_1 = (Q_0\overline{\theta}_j)\mathfrak{a}_j = \left[\frac{Q_0 Q_{j-1}\overline{\theta}_j}{r}, \frac{Q_0 Q_{j-1}\overline{\theta}_j \overline{\psi}_j}{r}\right]$$

$$= (Q_{j-1})\left[\frac{Q_0\overline{\theta}_j}{r}, \frac{Q_0\overline{\theta}_{j+1}}{r}\right] .$$

Thus,

$$\mathfrak{a}_1 = \left[\frac{Q_0\overline{\theta}_j}{r}, \frac{Q_0\overline{\theta}_{j+1}}{r}\right] . \tag{5.11}$$

In the case of $\Delta > 0$, the definition of ρ implies that the continued fraction expansion given by (5.8) is simple. Thus, by Theorem 3.5, we must find some $j \geq 0$ such that

$$\rho^j(\mathfrak{a}) = \left[\frac{Q_j}{r}, \frac{P_j + \sqrt{D}}{r}\right]$$

has $\overline{\phi}_j = (P_j - \sqrt{D})/Q_j < 0$. By Proposition 3.4 we get

$$\rho^{j+1}(\mathfrak{a}) = \left[\frac{Q_{j+1}}{r}, \frac{P_{j+1} + \sqrt{D}}{r}\right] ,$$

where

$$\frac{P_{j+1} + \sqrt{D}}{Q_{j+1}} > 0 \text{ and } -1 < \frac{P_{j+1} - \sqrt{D}}{Q_{j+1}} < 0 .$$

By Theorem 5.8, $\rho^{j+1}(\mathfrak{a})$ is reduced. Thus, we have shown that when $\Delta > 0$, there is always a reduced ideal in any ideal class of $\mathcal{O}$.

Indeed, from the discussion in §3.3 we know that there must be some minimal $m \in \mathbb{Z}^{>0}$ such that

$$0 < Q_{m-1} < \sqrt{D} .$$

Since $\mathfrak{a}_m = \rho^{m-1}(\mathfrak{a}) = \theta_m \mathfrak{a}$ and $N(\mathfrak{a}_m) = Q_{m-1}/r < \sqrt{\Delta}/2$, we see that $\mathfrak{a}_m$ is a reduced ideal equivalent to $\mathfrak{a}$, and if $Q_0 > 0$, we also know from Theorem 3.7 that $|\theta_m| < 2$. Since, by (3.17),

$$|\theta_m| = \frac{|Q_{m-1}|}{Q_0} \prod_{k=1}^{m-1} \phi_k > \frac{|Q_{m-1}|}{Q_0} \geq \frac{r}{Q_0} \, ,$$

we have

$$\frac{r}{Q_0} < |\theta_m| < 2 \, . \tag{5.12}$$

5.2 Reduction Algorithms

We have seen in the previous section that if we are given any $\mathcal{O}$-ideal $\mathfrak{a}$, we can find a reduced $\mathcal{O}$-ideal $\mathfrak{b}$ such that $\mathfrak{b} \sim \mathfrak{a}$. We call this the process of *reducing* $\mathfrak{a}$, and we have provided algorithms for doing this when $\Delta < 0$ and when $\Delta > 0$. Ideal reduction (or the equivalent process involving quadratic forms) has been studied for many years, starting with Lagrange[3] in the 18th century, but not a lot of progress has been made in significantly reducing the computational complexity of the various techniques developed for performing this process. For the most part, these procedures are of bit complexity[4] $O(nM(n))$, where n is the bit length of $N(\mathfrak{a})$ ($\lceil \log_2 N(\mathfrak{a}) \rceil$) and $M(n)$ is the number of bit operations required to multiply two n-bit integers. However, this can be improved[5] to $O(n^2)$ by using a more sophisticated analysis of the runtime similar to that used to show that the bit complexity of the Euclidean algorithm is $O((\log N)^2)$ (see §3.2).

It is not our intention here to review all of the algorithms[6] that have been proposed for ideal reduction. We should, however, point out that currently the best algorithm, from the point of view of asymptotic complexity, is that of Schönhage.[7] This technique is of asymptotic bit complexity $O(\log n M(n))$; however, tests[8] indicate that the benefit of using this algorithm is not realized in any practical computational setting until the value of $n = \log \Delta$ exceeds 10^5. Since Schönage's algorithm is somewhat intricate and impractical for numbers the size that we will be considering, we offer here a simple reduction technique that performs well in practice[9] whether Δ is positive or negative.

Let

$$\mathfrak{a} = \left[\frac{Q}{r}, \frac{P + \sqrt{D}}{r} \right]$$

be a primitive $\mathcal{O}$-ideal. If $|Q| < \sqrt{|D|}$, we already know that $\mathfrak{a}$ is reduced by Theorems 5.6 and 5.9. Thus, we will consider the case of $|Q| > \sqrt{|D|}$. We put $K = P$ and $L = |Q|$ and expand K/L into a simple continued fraction by using the Euclidean algorithm. We then compute R_j and C_j ($j = 0, 1, 2, \ldots, n$) by (3.20), (3.21), and (3.24). Since $R_{-1} = L > \sqrt{|D|}$ and $R_n = 0$, there must exist some integer i such that $0 \leq i \leq n$ and

$$R_i < \sqrt{|Q|}|D|^{1/4} < R_{i-1} \, .$$

If $(P + \sqrt{D})/Q = \langle q_0, q_1, q_2, \ldots, q_i, (P_{i+1} + \sqrt{D})/Q_{i+1} \rangle$, then by (3.13) and the definition of C_i,

$$(-1)^{i+1} Q_{i+1} = \frac{G_i{}^2 - DC_i{}^2}{|Q|} \, .$$

Now,

$$G_i = |Q|A_i - PB_i = (-1)^{i+1} R_i$$

by (3.22); hence, $G_i{}^2 = R_i{}^2 < |Q|\sqrt{|D|}$. Also, by (3.23),

$$|Q| = B_i R_{i-1} + B_{i-1} R_i \, ,$$

and it follows that $B_i \leq |Q|/R_{i-1} < \sqrt{|Q|}|D|^{-1/4}$. Thus, $|D|B_i{}^2 = |D|C_i{}^2 < |Q|\sqrt{|D|}$ and

$$|Q_{i+1}| \leq \begin{cases} 2\sqrt{|D|} & \text{when } D < 0 \\ \sqrt{D} & \text{when } D > 0 \, . \end{cases}$$

In the latter case we have

$$N(\mathfrak{a}_{i+2}) < \sqrt{\Delta}/2 \, ;$$

thus,

$$\mathfrak{a}_{i+2} = \left[\frac{|Q_{i+1}|}{r}, \frac{P_{i+1} + \sqrt{D}}{r} \right] \, ,$$

where

$$Q_{i+1} = (-1)^{i+1}(R_i{}^2 - DC_i{}^2)/|Q|$$

and (by (3.12))

$$P_{i+1} = \left((-1)^{i+1} R_i - Q_{i+1} B_{i-1} \right)/B_i$$
$$= (R_i + Q_{i+1} C_{i-1})/C_i$$

is a reduced $\mathcal{O}$-ideal by Theorem 5.9 when $D > 0$. In the case of $D < 0$, we can only show that $N(\mathfrak{a}_{i+2}) < \sqrt{|\Delta|}$; however, this case is easily handled by making use of the following result.

Theorem 5.11. Let $\mathfrak{a} = [a, b + \omega]$ be a primitive ideal of $\mathcal{O}$ where $\Delta < 0$. If $N(\mathfrak{a}) < \sqrt{|\Delta|}$, then either $\mathfrak{a}$ or $\rho(\mathfrak{a})$ is reduced.

Proof. Put $\mathfrak{a}' = [a', b' + \omega] = \rho(\mathfrak{a})$. If $\mathfrak{a}'$ is not a reduced ideal of $\mathcal{O}$, there must exist some non-zero $\lambda \in \mathfrak{a}'$ such that $|\lambda| < N(\mathfrak{a}')$. Since $|a| = N(\mathfrak{a})$ and by (5.4),

$$\mathfrak{a}\mathfrak{a}' = -\bar{\beta}\mathfrak{a} \, ,$$

where $-\overline{\beta} = b' + \omega$, there must exist some $\mu \in \mathfrak{a}$ such that $\mu\overline{\beta} = \lambda a$. Hence,

$$|\mu| = \frac{|\lambda||a|}{|\beta|} < \frac{|a'||a|}{|\beta|} = |\beta| \ .$$

Also, since $\mu \in \mathfrak{a}$ and $\mathfrak{a} = [a, \beta]$, we must have $x, y \in \mathbb{Z}$ such that $\mu = xa + y\beta$ and

$$4|\mu|^2 = T(\mu)^2 + |\Delta|y^2 \ . \tag{5.13}$$

By our remarks in the proof of Theorem 5.7 we have $|T(\beta)| \le |a|$; thus, if $\mathfrak{a}$ is not a reduced $\mathcal{O}$-ideal, then $|\beta| < |a|$ by Theorem 5.5. Hence,

$$|a| > |\beta| > |\mu| \ge \left(\frac{\sqrt{|\Delta|}}{2} \right) |y| > \left(\frac{N(\mathfrak{a})}{2} \right) |y|$$

and therefore $|y| \le 1$. If $y = 0$, then $|\mu| = |xa|$. Since $\mu \ne 0$, we have $|\mu| \ge |a|$, which is impossible. If $|y| = 1$, then $\mathfrak{a} = [a, \mu]$. Thus, since $\mathfrak{a}$ is primitive, we have

$$4|\beta|^2 = T(\beta)^2 + |\Delta| \ ,$$

with $|\beta| > |\mu|$. It follows by (5.13) with $|y| = 1$ that $|T(\mu)| < |T(\beta)| \le |a|$. However, by Proposition 5.3, this must mean that $|T(\mu)| = |T(\beta)|$, an impossibility. Thus, if $\mathfrak{a}$ is not a reduced ideal of $\mathcal{O}$, then $\rho(\mathfrak{a})$ must be. $\quad\square$

From this result we see that if $\mathfrak{a}_{i+2}$ is not a reduced $\mathcal{O}$-ideal, then $\rho(\mathfrak{a}_{i+2})$ is a reduced $\mathcal{O}$-ideal when $\Delta < 0$. It is easy to determine whether or not $\mathfrak{a}_{i+2}$ is reduced by comparing $|Q_{i+2}|$ and $N(\rho(\mathfrak{a}_{i+2}))$. The advantage in using this reduction technique is that all we need to do is use the Euclidean algorithm on P and $|Q|$. The operations are simple and we can easily perform the reduction process in essentially the same time complexity as that of the Euclidean algorithm. Tests carried out on this reduction algorithm have shown that it is very efficient in practice.[10]

It will be helpful in what follows to derive some results analogous to Theorem 5.11 in the case of $\Delta > 0$. We begin by observing that in this case we have

$$b' + \overline{\omega} = -\eta a \ , \quad b' + \omega = \sqrt{\Delta} - \eta a \ , \tag{5.14}$$

where $\eta = (b + \omega)/a - q$. Hence, $0 < \eta < 1$. We also have

$$a' = \eta(b' + \omega) = \eta(\sqrt{\Delta} - \eta a) \ . \tag{5.15}$$

Here, the symbols a, b, a', b', and q have the meanings assigned to them in §5.1 when we write $[a', b' + \omega] = \rho(\mathfrak{a})$ and $\mathfrak{a} = [a, b + \omega]$.

Theorem 5.12. *Let $\Delta > 0$. If $a > 0$ and $\mathfrak{a} = [a, b + \omega]$ is a reduced $\mathcal{O}$-ideal, then $\rho(\mathfrak{a})$ is also a reduced $\mathcal{O}$-ideal.*

Proof. Let $\mathfrak{a}' = \rho(\mathfrak{a}) = [a', b' + \omega]$ and recall that $\mathfrak{a} = [a, b + \omega] = [a, b' + \overline{\omega}]$. Since $\mathfrak{a}$ is reduced, we know by Corollary 5.8.1 that $a < \sqrt{\Delta}$. This means by (5.14) that $b' + \omega > 0$ and by (5.15) that $a' > 0$, $b' + \omega > a'$. Also, by (5.14), we have $-a < b' + \overline{\omega} < 0$. Thus, since $\mathfrak{a}$ is a reduced ideal, we must have $b' + \omega > a$ because $b' + \omega \in \overline{\mathfrak{a}}$ and $\overline{\mathfrak{a}}$ is reduced.

Since

$$\frac{|b' + \overline{\omega}|}{a'} = \frac{a}{|b' + \omega|},$$

we get $|b' + \overline{\omega}|/a' < 1$. Hence, $-a' < b' + \overline{\omega} < 0$ and $b' + \omega > a'$. It follows that $\rho(\mathfrak{a}) = \mathfrak{a}'$ is reduced by Theorem 5.8. □

Notice that if $\mathfrak{a} = [a, b + \omega]$ is reduced and $a < 0$, then $\rho(\mathfrak{a}) = [a', b' + \omega]$ is not necessarily reduced. For example, consider $D = 67$, $\omega = \sqrt{67}$, $a = -3$, and $b = -14$. Certainly, $[-3, -14 + \omega]$ is a primitive ideal of $\mathcal{O} = [1, \omega]$ and $\mathfrak{a}$ is reduced because $N(\mathfrak{a}) = 3$ and $3 < \sqrt{D} = \sqrt{\Delta}/2$. However, $q = [-14 + \omega/(-3)] = 1$, $b' = qa - b - T(\omega) = 11$, and $a' = -N(11 + \sqrt{67})/(-3) = 18 > 2\sqrt{D} = \sqrt{\Delta}$. Hence, $\mathfrak{a}'$ is not reduced by Corollary 5.8.1.

We are now able to present a theorem analogous to Theorem 5.9 for $\Delta > 0$.

Theorem 5.13. *Let $\Delta > 0$. If $0 < a < 3\sqrt{\Delta}/2$ and $\mathfrak{a} = [a, b + \omega]$ is a primitive $\mathcal{O}$-ideal, then $\rho(\mathfrak{a})$ is a reduced $\mathcal{O}$-ideal. Furthermore, if $\mathfrak{a}$ is not a reduced $\mathcal{O}$-ideal, then $\rho(\mathfrak{a}) = \psi\mathfrak{a}$, where $|\psi|, |\overline{\psi}| < 1$.*

Proof. As before, we put $\mathfrak{a}' = \rho(\mathfrak{a}) = [a', b' + \omega]$. We have $\rho(\mathfrak{a}) = \psi\mathfrak{a}$, where $\psi = (b' + \omega)/a$ and $|\overline{\psi}| < 1$ by (5.14). If $\mathfrak{a}$ is a reduced ideal, we have our result by Theorem 5.12. Suppose that $\mathfrak{a}$ is not reduced. If $\sqrt{\Delta} > \eta a$, then $b' + \omega > a' > 0$ by (5.15). Since $\overline{\mathfrak{a}}$ is not reduced, $b' + \omega \in \overline{\mathfrak{a}}$, and $-a < b' + \overline{\omega} < 0$ by (5.14), we must, by Theorem 5.8, have $b' + \omega < a$. Thus, $|b' + \overline{\omega}|/a' = a/(b' + \omega) > 1$ and we have $-a' < b' + \overline{\omega} < 0$, $b' + \omega > a'$. It follows from Theorem 5.8 that $\mathfrak{a}'$ is reduced. Also, $0 < \psi < 1$. In the case of $\sqrt{\Delta} < \eta a$, we get $b' + \omega < 0$ and $-b' - \omega = \eta a - \sqrt{\Delta} < \sqrt{\Delta}/2$. Thus, $-\sqrt{\Delta}/2 < b' + \omega < 0$ and we find that $|a'| = \eta|b' + \omega| < \sqrt{\Delta}/2$. Hence, $\rho(\mathfrak{a})$ is reduced by Theorem 5.9. Also, $|\psi| = |b' + \omega|/a < \sqrt{\Delta}/2a < 1$ because $\mathfrak{a}$ is not a reduced $\mathcal{O}$-ideal. □

Remark 5.14. Notice that if $\mathfrak{a}$ is a reduced ideal and $N(\mathfrak{a}) > \sqrt{\Delta}/2$, then $N(\rho(\mathfrak{a})) < \sqrt{\Delta}/2$. This is simply because both $\mathfrak{a}$ and $\mathfrak{a}' = \rho(\mathfrak{a})$ are reduced and, therefore, by Corollary 5.8.1, $N(\mathfrak{a})N(\mathfrak{a}') < \Delta$. In the appendix (Theorem A.1), we extend Theorem 5.13 somewhat further.

We will next concern ourselves with the problem of finding all of the reduced $\mathcal{O}$-ideals in a given ideal class of $\mathcal{O}$. We first give the following simple proposition.

Proposition 5.15. *If $\mathfrak{a}$ and $\mathfrak{b}$ are $\mathcal{O}$-ideals, $\mathfrak{b}$ is primitive, and $\mathfrak{a} \sim \mathfrak{b}$, then there exists some non-zero $\lambda \in \mathfrak{a}$ such that $(\lambda)\mathfrak{b} = (N(\mathfrak{b}))\mathfrak{a}$.*

Proof. Since $\mathfrak{a} \sim \mathfrak{b}$, there must exist $\alpha, \beta \in \mathcal{O}$ such that $\alpha\beta \neq 0$ and $(\alpha)\mathfrak{a} = (\beta)\mathfrak{b}$. Since $\mathfrak{b}$ is primitive, we have $N(\mathfrak{b}) \in \mathfrak{b}$, and therefore there must exist some $\lambda \in \mathfrak{a}$ such that $\alpha\lambda = \beta N(\mathfrak{b})$. Consider the $\mathcal{O}$-ideal $\mathfrak{c} = (\lambda)\mathfrak{b}$. We have

$$(\alpha)\mathfrak{c} = (\alpha\lambda)\mathfrak{b} = (\beta N(\mathfrak{b}))\mathfrak{b} = (N(\mathfrak{b})\alpha)\mathfrak{a} .$$

Since $\alpha \neq 0$, we get $\mathfrak{c} = (N(\mathfrak{b}))\mathfrak{a}$ by Proposition 4.38 and the result follows immediately. $\qquad\square$

We next require a simple lemma.

Lemma 5.16. *Let* $\Delta < 0$ *and* $\mathfrak{a} = [a, \alpha]$ *be a primitive* $\mathcal{O}$-*ideal with* $|T(\alpha)| \leq |a|$. *If* $\mathfrak{a}$ *is a reduced* $\mathcal{O}$-*ideal and* $\lambda \in \mathfrak{a}$ *such that* $|\lambda| = |a|$, *then either* $\lambda = \pm a$ *or* $\mathfrak{a} = [a, \lambda]$ *and* $|T(\lambda)| = |T(\alpha)|$, $|\lambda| = |\alpha|$.

Proof. Since $\lambda \in \mathfrak{a}$, there must exist $x, y \in \mathbb{Z}$ such that $\lambda = xa + y\alpha$ and, by (4.4),
$$4a^2 = 4|\lambda|^2 = T(\lambda)^2 + y^2|\Delta| .$$
Since $\mathfrak{a}$ is a reduced ideal, we must have $|\Delta| \geq 3a^2$, which means that $|y| \leq 1$. If $y = 0$, we get $\lambda = xa$, and since $|\lambda| = |a|$, we can only have $\lambda = \pm a$. If $|y| = 1$, then $T(\lambda)^2 \leq 4|\lambda|^2 - |\Delta| \leq 4|\lambda|^2 - 3a^2 = a^2$ and therefore $|T(\lambda)| \leq |a|$. Since $\mathfrak{a} = [a, \lambda]$, we must have $|T(\lambda)| = |T(\alpha)|$ by Proposition 5.3. Since $4|\alpha|^2 = T(\alpha)^2 + |\Delta|$, we get $|\alpha| = |\lambda|$. $\qquad\square$

We are now able to show that in the case of $\Delta < 0$, there can never be more than two reduced $\mathcal{O}$-ideals in any ideal class of $\mathcal{O}$.

Theorem 5.17. *Let* $\Delta < 0$. *If* $\mathfrak{a}$ *and* $\mathfrak{b}$ *are reduced* $\mathcal{O}$-*ideals such that* $\mathfrak{a} \sim \mathfrak{b}$, *then either* $\mathfrak{a} = \mathfrak{b}$ *or* $\mathfrak{a} = \overline{\mathfrak{b}}$.

Proof. By Proposition 5.2, we may assume with no loss of generality that $\mathfrak{a} = [N(\mathfrak{a}), \alpha]$, $\mathfrak{b} = [N(\mathfrak{b}), \beta]$, where $|T(\alpha)| \leq N(\mathfrak{a})$ and $|T(\beta)| \leq N(\mathfrak{b})$. By Proposition 5.15, we must have $\mu \in \mathfrak{a}$ and $\nu \in \mathfrak{b}$ such that $\mu\nu \neq 0$ and

$$(\mu)\mathfrak{b} = (N(\mathfrak{b}))\mathfrak{a} , \quad (\nu)\mathfrak{a} = (N(\mathfrak{a}))\mathfrak{b} . \tag{5.16}$$

If we take norms of both sides of these two equations, we get

$$|\mu|^2 N(\mathfrak{b}) = N(\mathfrak{b})^2 N(\mathfrak{a}), \quad |\nu|^2 N(\mathfrak{a}) = N(\mathfrak{a})^2 N(\mathfrak{b})$$

by Theorem 4.35. Hence, $|\mu|^2 = |\nu|^2 = N(\mathfrak{a})N(\mathfrak{b})$. If $N(\mathfrak{b}) < N(\mathfrak{a})$, then $|\mu|^2 < N(\mathfrak{a})^2$ and $|\mu| < N(\mathfrak{a})$ which is impossible because $\mathfrak{a}$ is reduced. Similarly, if $N(\mathfrak{a}) < N(\mathfrak{b})$, we find that $|\nu| < N(\mathfrak{b})$, which is also impossible. Thus, we must have $|\nu| = |\mu| = N(\mathfrak{a}) = N(\mathfrak{b})$. If $\mu = \pm N(\mathfrak{a})$, then $\mathfrak{b} = \mathfrak{a}$ by the first equation in (5.16). If $\nu = \pm N(\mathfrak{b})$, then $\mathfrak{b} = \mathfrak{a}$ by the second equation in (5.16). Suppose that $\mu \neq \pm N(\mathfrak{a})$ and $\nu \neq \pm N(\mathfrak{b})$. By Lemma 5.16 we must have $\mathfrak{a} = [N(\mathfrak{a}), \mu]$, $\mathfrak{b} = [N(\mathfrak{b}), \nu]$, $|T(\mu)| = |T(\alpha)|$, $|T(\nu)| = |T(\beta)|$, and

$|\alpha| = |\mu| = |\nu| = |\beta|$. Since $4|\mu|^2 = T(\mu)^2 + |\Delta|$ and $4|\nu|^2 = |T(\nu)|^2 + |\Delta|$, we get $|T(\mu)| = |T(\nu)|$ and therefore $|T(\alpha)| = |T(\beta)|$. Since

$$0 = \alpha^2 - (\alpha + \overline{\alpha})\alpha + \alpha\overline{\alpha} = \alpha^2 \mp (\beta + \overline{\beta})\alpha + \beta\overline{\beta} = (\alpha \mp \beta)(\alpha \mp \overline{\beta}) \,,$$

we see that $\alpha = \pm\beta$ or $\alpha = \pm\overline{\beta}$. If $\alpha = \pm\beta$, then $\mathfrak{a} = \mathfrak{b}$; if $\alpha = \pm\overline{\beta}$, then $\mathfrak{a} = \overline{\mathfrak{b}}$.

$\square$

It is possible[11] for $\mathfrak{a}$ and $\mathfrak{b}$ to be reduced, $\mathfrak{a} \sim \mathfrak{b}$, $\mathfrak{a} = \overline{\mathfrak{b}}$, and $\mathfrak{a} \neq \mathfrak{b}$. We have already seen in the proof of Theorem 5.17 that if $\mathfrak{a}$ and $\mathfrak{b}$ are reduced and equivalent and $\mathfrak{a} \neq \mathfrak{b}$, then $N(\mathfrak{a}) = |\alpha|$, where $\mathfrak{a} = [N(\mathfrak{a}), \alpha]$ and $\alpha = b + \omega$. Now, suppose that $\mathfrak{a} = [N(\mathfrak{a}), \alpha]$ is any primitive $\mathcal{O}$-ideal such that $N(\mathfrak{a}) = |\alpha|$. We have

$$N(\mathfrak{a})^2 = |\alpha|^2 = T(\alpha)^2 + \Delta > T(\alpha)^2$$

and therefore $|T(\alpha)| < N(\mathfrak{a})$. Since $N(\mathfrak{a}) = |\alpha|$, we know that $\mathfrak{a}$ is reduced by Theorem 5.5. Also, since

$$N(\mathfrak{a})[N(\mathfrak{a}), \alpha] = \alpha[N(\mathfrak{a}), \overline{\alpha}] \,,$$

we have $\mathfrak{a} \sim \overline{\mathfrak{a}}$. However, if $\mathfrak{a} = \overline{\mathfrak{a}}$, we must have $\overline{\alpha} \in \mathfrak{a}$. This means that there exist $x, y \in \mathbb{Z}$ such that $\overline{\alpha} = xN(\mathfrak{a}) + y\alpha$. Since $\alpha = b + \omega$, we get

$$b + \overline{\omega} = xN(\mathfrak{a}) + y(b + T(\omega) - \overline{\omega}) \,.$$

It follows that $y = -1$ and $T(\alpha) = 2b + T(\omega) = xN(\mathfrak{a})$. Since $|T(\alpha)| < N(\mathfrak{a})$, this is impossible.

The set of reduced ideals in any equivalence class of $\mathcal{O}$ is more complicated when $\Delta > 0$. In this case it is possible to have many more than two reduced ideals in a given class. Since, by Proposition 5.2, the norm of any reduced ideal in this case is bounded by $\sqrt{\Delta}$, there can only be a finite number of reduced ideals in any ideal class of $\mathcal{O}$. We will devote the next section to a detailed discussion of this case.

5.3 Reduced Ideals When $\Delta > 0$

We have seen that if $\mathfrak{a} = [a, b + \omega]$ ($a > 0$) is any reduced $\mathcal{O}$-ideal, then $\rho(\mathfrak{a})$ is a reduced $\mathcal{O}$-ideal. Also, $\rho(\mathfrak{a}) = \mathfrak{a}' = [a', b' + \omega]$, and since $a < \sqrt{\Delta}$, we have $a' > 0$ by (5.15). Since $\mathfrak{a} \sim \rho(\mathfrak{a})$, we see that each ideal in the sequence

$$\mathfrak{a}, \rho(\mathfrak{a}), \rho^2(\mathfrak{a}), \ldots, \rho^n(\mathfrak{a}), \ldots \qquad (5.17)$$

is reduced and equivalent to $\mathfrak{a}$. We will now show that the sequence (5.17) contains all of the reduced ideals equivalent to $\mathfrak{a}$. We initially recall that if $\mathfrak{a} \sim \mathfrak{b}$, then $\gamma\mathfrak{b} = \mathfrak{a}$ for some $\gamma \in \mathbb{K}$. We may certainly assume that $\gamma > 0$, and if $\gamma > N(\mathfrak{a})/N(\mathfrak{b})$, we can multiply γ by that power (ϵ_Δ^k) of ϵ_Δ (> 1) such that $\kappa = \epsilon_\Delta^k \gamma \leq N(\mathfrak{a})/N(\mathfrak{b})$. Since $(\epsilon_\Delta)\mathfrak{b} = \mathfrak{b}$, we get

$$\kappa \mathfrak{b} = \mathfrak{a} ,$$

where $0 < \kappa < N(\mathfrak{a})/N(\mathfrak{b})$. We now require the following result.

Theorem 5.18. *Let* $\mathfrak{a} = [a, b + \omega]$ *($a > 0$) and* $\mathfrak{b}$ *be reduced* $\mathcal{O}$-*ideals. If* $\mathfrak{a} = \kappa \mathfrak{b}$, *where* $\kappa \in \mathbb{Q}(\sqrt{\Delta})$ *and* $0 < \kappa < N(\mathfrak{a})/N(\mathfrak{b})$, *then* $\kappa = \theta_m^{-1}$, *where* $\rho^{m-1}(\mathfrak{a}) = \theta_m \mathfrak{a}$ *($m \geq 1$).*

Proof. We have $\theta_1 = 1$, and by (3.17) and (3.18),

$$|\overline{\theta}_j| = (-1)^{j-1}\overline{\theta}_j = \prod_{k=1}^{j-1} \phi_k^{-1} \quad (j = 2, 3, \dots) . \tag{5.18}$$

In the case of $\Delta > 0$, the continued fraction (5.8) is a simple continued fraction; hence, $\phi_k > 1$. Also, because $\phi_{k+1} = 1/(\phi_k - \lfloor \phi_k \rfloor)$, we get $\phi_{k+1}\phi_k = \lfloor \phi_k \rfloor \phi_{k+1} + 1 > 2$ for $k \geq 1$.

By (5.18), there must, therefore, exist some integer $m \geq 1$ such that

$$|\overline{\theta}_{m+1}| < \frac{N(\mathfrak{b})\kappa}{N(\mathfrak{a})} \leq |\overline{\theta}_m| . \tag{5.19}$$

Suppose $N(\mathfrak{b})\kappa/N(\mathfrak{a}) \neq |\overline{\theta}_m|$. Since $\kappa N(\mathfrak{b}) \in \mathfrak{a}$, we must have

$$\kappa N(\mathfrak{b}) = N(\mathfrak{a})(x\overline{\theta}_m + y\overline{\theta}_{m+1}) \tag{5.20}$$

by (5.11), and by (5.19), we have

$$|\overline{\theta}_{m+1}| < |x\overline{\theta}_m + y\overline{\theta}_{m+1}| < |\overline{\theta}_m| . \tag{5.21}$$

Also, since $N(\mathfrak{a})\overline{\theta}_{m+1} \in \mathfrak{a}$, there must exist some $\lambda \in \mathfrak{b}$ such that

$$N(\mathfrak{a})\overline{\theta}_{m+1} = \kappa\lambda ;$$

hence, $|\lambda| < N(\mathfrak{b})$ by (5.19). Since $\mathfrak{b}$ is a reduced $\mathcal{O}$-ideal, we must have $|\overline{\lambda}| \geq N(\mathfrak{b})$. By (5.20) this means that

$$|x\theta_m + y\theta_{m+1}| \leq \theta_{m+1} . \tag{5.22}$$

However, θ_m and θ_{m+1} are both positive and therefore (5.22) cannot hold unless $xy \leq 0$. However, if $xy \leq 0$, then (5.21) cannot hold because $\overline{\theta}_m$ and $\overline{\theta}_{m+1}$ have opposite signs; hence, we must have $N(\mathfrak{b})\kappa/N(\mathfrak{a}) = |\overline{\theta}_m|$. It follows that $(N(\mathfrak{a})\overline{\theta}_m)\mathfrak{b} = (N(\mathfrak{b}))\mathfrak{a}$. By taking norms of both sides of this equality we get $|N(\overline{\theta}_m)| = \theta_m|\overline{\theta}_m| = N(\mathfrak{b})/N(\mathfrak{a})$ and, therefore, $\kappa = \theta_m^{-1}$. □

By our earlier remarks and Theorem 5.18, we see that if $\mathfrak{b}$ is any reduced $\mathcal{O}$-ideal equivalent to $\mathfrak{a}$, then $\mathfrak{b} = \theta_m\mathfrak{a} = \mathfrak{a}_m$; hence, $\mathfrak{b}$ is in the sequence (5.17). Since the number of reduced $\mathcal{O}$-ideals in any equivalence class is finite, there must be some i and j ($j > i \geq 0$) such that

$$\rho^i(\mathfrak{a}) = \rho^j(\mathfrak{a}) \ . \tag{5.23}$$

We will now show that $\rho^{j-i}(\mathfrak{a}) = \mathfrak{a}$.

Let $\mathfrak{a} = [a, b + \omega]$ be any primitive $\mathcal{O}$-ideal. If we put $a^* = -N(b+\omega)/a$, $q^* = \lfloor (b+\omega)/a^* \rfloor$, $b^* = q^*a^* - b - T(\omega)$, then

$$\mathfrak{a}^* = [a^*, b^* + \omega]$$

is a primitive ideal of $\mathcal{O}$, and by (5.2),

$$(a)\mathfrak{a}^* = (b + \overline{\omega})\mathfrak{a} \ . \tag{5.24}$$

Suppose that $(b + \omega)/a > 1$. In this case, since

$$\frac{b^* + \omega}{a^*} = q^* - \frac{b + \overline{\omega}}{a^*} = q^* + \frac{a}{b + \omega} \ , \tag{5.25}$$

we see that

$$\left\lfloor \frac{b^* + \omega}{a^*} \right\rfloor = q^* \ .$$

If we compute $\mathfrak{a}' = [a', b' + \omega] = \rho(\mathfrak{a}^*)$, we find that

$$b' = \lfloor (b^* + \omega)/a^* \rfloor a^* - b^* - T(\omega) = q^*a^* - b^* - T(\omega) = b \ ,$$
$$a' = -N(b' + \omega)/a^* = -N(b + \omega)/a^* = a \ .$$

Hence, $\rho([a^*, b^* + \omega]) = [a, b + \omega] = \mathfrak{a}$. Thus, if $\mathfrak{a} = [a, b + \omega]$ and $(b+\omega)/a > 1$, we can define the operation ρ^{-1} whose action on $\mathfrak{a}$ is to produce $\mathfrak{a}^* = [a^*, b^* + \omega]$ and $\rho(\rho^{-1}(\mathfrak{a})) = \mathfrak{a}$. If, in addition, we suppose that $-1 < (b + \overline{\omega})/a < 0$ and $\mathfrak{a}' - \rho(\mathfrak{a}) - [a', b' + \omega]$, we have $(b' + \omega)/a' > 1$ by (5.5) and, therefore, we can apply ρ^{-1} to $\mathfrak{a}'$ to obtain $\tilde{\mathfrak{a}} = [\tilde{a}, \tilde{b} + \omega]$, where

$$\tilde{a} = \frac{-N(b' + \omega)}{a'} = a \ ,$$
$$\tilde{q} = \left\lfloor \frac{b' + \omega}{\tilde{a}} \right\rfloor = q + \left\lfloor \frac{-(b + \overline{\omega})}{a} \right\rfloor = q \ ,$$
$$\tilde{b} = \tilde{q}\tilde{a} - b' - T(\omega) = qa - b' - T(\omega) = b \ .$$

Hence, $\tilde{\mathfrak{a}} = \mathfrak{a}$ and, consequently, $\rho^{-1}(\rho(\mathfrak{a})) = \mathfrak{a}$. We also point out that if $\mathfrak{a} = [a, b + \omega]$ with $(b + \omega)/a > 1$, $-1 < (b + \overline{\omega})/a < 0$, then $\rho^{-1}(\mathfrak{a}) = \mathfrak{a}^* = [a^*, b^* + \omega]$, where $q^* = \lfloor -a/(b + \overline{\omega}) \rfloor \geq 1$ and $-1 < (b^* + \overline{\omega})/a^* < 0$, $(b^* + \omega)/a^* > 1$ by (5.25).

If we represent $\mathfrak{a}$ by $[Q/r, (P + \sqrt{D})/r]$ as in (4.9) and we have $(P + \sqrt{D})/Q > 1$, $-1 < (P - \sqrt{D})/Q < 0$, we can repeatedly apply ρ^{-1} to $\mathfrak{a}$. We put $Q_0^* = Q$, $P_0^* = P$ and define

$$Q_i^* = \frac{D - P_{i-1}^{*2}}{Q_{i-1}^*}, \quad q_i^* = \left\lfloor \frac{P_{i-1}^* + \lfloor \sqrt{D} \rfloor}{Q_i^*} \right\rfloor \geq 1 \ ,$$

$$P_i{}^* = q_i{}^* Q_i{}^* - P_{i-1}{}^* \quad (i = 1, 2, 3, \dots) \, .$$

Thus, if we define $\rho^{-n}(\mathfrak{a})$ recursively by $\rho^{-n}(\mathfrak{a}) = \rho^{-1}(\rho^{-n+1}(\mathfrak{a}))$, we get

$$\rho^{-j}(\mathfrak{a}) = \left[\frac{Q_j{}^*}{r}, \frac{P_j{}^* + \sqrt{D}}{r} \right] \tag{5.26}$$

and

$$-1 < \frac{P_{i-1}{}^* - \sqrt{D}}{Q_{i-1}{}^*} < 0 \, .$$

Also, by (5.24) we get $\rho^{-j-1}(\mathfrak{a}) = \xi_{j+1} \rho^{-1}(\mathfrak{a})$, where

$$1 > \xi_i = \frac{\sqrt{D} - P_{i-1}{}^*}{Q_{i-1}{}^*} > 0 \, .$$

It follows that

$$\rho^{-j}(\mathfrak{a}) = \chi_j \mathfrak{a} \, , \tag{5.27}$$

where $\chi_0 = 1$ and $\chi_n = \prod_{i=1}^{n} \xi_i$ $(n \geq 1)$. Tenner's variant for this is given by

$$Q_i{}^* = Q_{i-2}{}^* - q_{i-1}{}^*(P_{i-1}{}^* - P_{i-2}{}^*) \quad (i \geq 2) \, ,$$
$$P_i{}^* = \lfloor \sqrt{D} \rfloor - R_i{}^* \quad (i \geq 1) \, ,$$

where $R_i{}^*$ is the remainder on dividing $P_{i-1}{}^* + \lfloor \sqrt{D} \rfloor$ by $Q_i{}^*$. In this case, we use $Q_1{}^* = (D - P_0{}^{*2})/Q_0{}^*$.

It is easy to verify that $\xi_i \bar{\xi}_i = (P_{i-1}{}^{*2} - D)/Q_{i-1}^{*} = -Q_i{}^*/Q_{i-1}{}^*$; hence,

$$N(\chi_j) = \chi_j \bar{\chi}_j = (-1)^j Q_j{}^*/Q_0{}^* \, . \tag{5.28}$$

Since

$$\xi_{i+1} = -q_i{}^* + \frac{\sqrt{D} + P_{i-1}{}^*}{Q_i{}^*} \, ,$$

we get

$$\xi_{i+1} \xi_i = -q_i{}^* \xi_i + 1 \, .$$

From this it is easy to deduce that

$$\chi_{i+1} = \xi_{i+1} \xi_i \chi_{i-1} = -q_i{}^* \chi_i + \chi_{i-1} \, , \tag{5.29}$$

and this can be used to establish that

$$\chi_j = (-1)^j (B_{j-2}{}^* + A_{j-2}{}^* \bar{\phi}_0) = (-1)^j (G_{j-2}{}^* - \sqrt{D} A_{j-2}{}^*)/Q_0{}^* \, , \tag{5.30}$$

where we define $G_i{}^*$ by $G_i{}^* = Q_0{}^* B_i{}^* + P_0{}^* A_i{}^*$ $(i = -2, -1, 0, \dots)$. Here, $A_{-2}{}^* = 0$, $A_{-1}{}^* = 1$, $B_{-2}{}^* = 1$, $B_{-1}{}^* = 0$, and

$$A_{i+1}{}^* = q_{i+2}{}^* A_i{}^* + A_{i-1}{}^*, \quad B_{i+1}{}^* = q_{i+2}{}^* B_i{}^* + B_{i-1}{}^* \quad (i = -1, 0, 1, \dots) \, .$$

Also, by (5.28) and (5.30), we get

$$\frac{1}{\chi_n} = \frac{G_{n-2}{}^* + \sqrt{D}A_{n-2}{}^*}{Q_n{}^*} \tag{5.31}$$

Notice that if $\mathfrak{a}$ is a reduced $\mathcal{O}$-ideal and $\mathfrak{b} = \rho^j(\mathfrak{a})$, where $a > 0, j > 1$, then $\mathfrak{b} = [c, d + \omega]$, where $(d + \omega)/c > 1$ and $-1 < (d + \overline{\omega})/c < 0$ by Theorem 5.12. Thus, we can apply ρ^{-1} repeatedly on both sides of (5.23) until we get

$$\mathfrak{a} = \rho^{j-i}(\mathfrak{a}) .$$

Thus, there must be a minimal integer $p \geq 1$ for which $\mathfrak{a} = \rho^p(\mathfrak{a})$. Also, the p ideals in the set

$$\left\{\mathfrak{a}, \rho(\mathfrak{a}), \rho^2(\mathfrak{a}), \ldots, \rho^{p-1}(\mathfrak{a})\right\} \tag{5.32}$$

must all be distinct, for, otherwise, there would be a positive integer n such that $\mathfrak{a} = \rho^n(\mathfrak{a})$ and $n < p$, which is impossible by the definition of p. Also, (5.32) must include all of the reduced ideals equivalent to $\mathfrak{a}$. We call the set (5.32) the *cycle of reduced ideals* in the ideal class of $\mathfrak{a}$. Thus, we now have shown that if $\mathfrak{a}$ $(= \mathfrak{a}_1)$ is any primitive $\mathcal{O}$-ideal, to which the simple continued fraction of a corresponding ϕ $(= \phi_0)$ is applied, we must ultimately produce a reduced ideal $\mathfrak{a}_k \sim \mathfrak{a}_1$, and once this has occurred, the subsequent ideals determined by this process will be only the reduced ideals of $\mathcal{O}$ which are equivalent to $\mathfrak{a}$. This, of course, means, as we already showed in Chapter 3, that the simple continued fraction expansion of ϕ must ultimately become periodic. We have also shown that the preperiod of the simple continued fraction expansion of ϕ corresponds to the process of finding a reduced ideal equivalent to $\mathfrak{a}$.

Since $(\epsilon_\Delta^{-1})\mathfrak{a} = \mathfrak{a}$ and $0 < \epsilon_\Delta^{-1} < 1$, we see by Theorem 5.18 with $\kappa = \epsilon_\Delta^{-1}$ and $\mathfrak{a} = \mathfrak{b}$ that $\epsilon_\Delta = \theta_m$ for some $m > 1$. Also, since $\rho^{m-1}(\mathfrak{a}) = \theta_m \mathfrak{a} = \epsilon_\Delta \mathfrak{a} = \mathfrak{a}$, we must have $m - 1 \geq p$. Now, $\mathfrak{a} = \rho^p(\mathfrak{a}) = \theta_{p+1}\mathfrak{a}$ means that θ_{p+1} is a unit of $\mathcal{O}$. Thus, $\theta_{p+1} = \epsilon_\Delta^k$ $(k \geq 1)$. However, since $m - 1 \geq p$, we also must have $\epsilon_\Delta = \theta_m \geq \theta_{p+1} = \epsilon_\Delta^k$ and $k = 1$. Hence,

$$\epsilon_\Delta = \theta_{p+1} . \tag{5.33}$$

Also, since $Q_{p+1} = Q_0$, we get

$$\epsilon_\Delta = \prod_{i=1}^{p} \phi_i = \prod_{i=1}^{p} \psi_i . \tag{5.34}$$

From this and (3.41), we would expect that the number of reduced ideals in the cycle (5.32) is $p \approx \log \epsilon_\Delta / \lambda$, where λ is the Khinchine-Lévy constant. Certainly we have

$$p < 2.08 \log \epsilon_\Delta + 1$$

by (3.40). The invariant R_Δ or $R(\Delta) = \log \epsilon_\Delta$ of $\mathcal{O}$ will be of great importance in much of our subsequent work. We call R_Δ the *regulator*[12] of $\mathcal{O}$. If $\mathcal{O} = \mathcal{O}_\mathbb{K}$, then $R = R(\mathbb{K})$ or $R_\mathbb{K} = \log \epsilon_\mathbb{K}$ is called the regulator of $\mathbb{K}$. Notice that

$$\epsilon_\Delta < (\sqrt{\Delta})^p$$

by (3.28) and (5.34). Thus,

$$R_\Delta < p \log \sqrt{\Delta} . \qquad (5.35)$$

We have seen that if $\mathfrak{a} = \mathfrak{a}_1 = [Q_0/r, (P_0 + \sqrt{D})/r]$ is a reduced ideal and $Q_0 > 0$, then by expanding $\phi_0 = (P_0 + \sqrt{D})/Q_0$ into a simple continued fraction, we find that $\rho^{j-1}(\mathfrak{a}) = \mathfrak{a}_j = [Q_j/r, (P_j + \sqrt{D})/r]$ and the sequence of ideals (5.17) is periodic with period p. Since $(P_1 + \sqrt{D})/Q_1 > 1$ and $-1 < (P_1 - \sqrt{D})/Q_1 < 0$, the continued fraction expansion of ϕ_0 is either purely periodic or has a preperiod consisting of only one element. We are now able to extend some of the results in §3.4. If the $\mathcal{O}$-ideal $\mathfrak{a}$ satisfies $\mathfrak{a} = \bar{\mathfrak{a}}$, we say that $\mathfrak{a}$ is *ambiguous* in $\mathcal{O}$. For example, if $\mathfrak{a} = \mathcal{O}$, then since $\mathcal{O} = \bar{\mathcal{O}}$, we know that $\mathfrak{a}$ is ambiguous. If, however, $\mathfrak{a} = [Q/r, (P + \sqrt{D})/r]$ and $\mathfrak{a}$ is ambiguous, it is easy to show that since $(P - \sqrt{D})/r \in \mathfrak{a}$, then $Q \mid 2P$. Thus, by Theorem 3.12 we have the following result.

Theorem 5.19. *If $\mathfrak{a}$ is a reduced ambiguous $\mathcal{O}$-ideal and $Q_0 > 0$, then*

$$Q_{p-i} = Q_i, \quad P_{p-i} = P_{i+1} \quad (0 \le i \le p)$$

and

$$q_{p-i} = q_i \quad (0 < i < p) .$$

Corollary 5.19.1. *If $\mathfrak{a}$ is a reduced ambiguous ideal, then*

$$\mathfrak{a}_{i+1} = \bar{\mathfrak{a}}_{p+1-i} \quad (i = 0, 1, 2, \ldots, p) .$$

The next result allows us to extend some results in §3.3 and §3.4.

Theorem 5.20. *Let $\mathfrak{a}$ be a reduced ambiguous $\mathcal{O}$-ideal with $Q_0 > 0$. If $p = 2s$ ($s \ge 1$), then*

$$\epsilon_\Delta = Q_0 \theta_{s+1}^2/Q_s$$

and

$$G_{p-1} = G_s B_{s-1} + G_{s-1} B_{s-2}, \quad B_{p-1} = B_{s-1}(B_j + B_{j-2}) ;$$

if $p = 2s + 1$ ($s \ge 1$), then

$$\epsilon_\Delta = Q_0 \theta_{s+1} \theta_{s+2}/Q_s$$

and

$$G_{p-1} = G_s B_s + G_{s-1} B_{s-1} , \quad B_{p-1} = B_s^2 + B_{s-1}^2 .$$

Proof. If $p = 2s$, then by Corollary 5.19.1 we have $\mathfrak{a}_{s+1} = \bar{\mathfrak{a}}_{s+1}$. Since $\mathfrak{a}_{s+1} = \theta_{s+1} \mathfrak{a}_1$, it follows that $\theta_{s+1} \mathfrak{a}_1 = \bar{\theta}_{s+1} \bar{\mathfrak{a}}_1$. Thus, $\theta_{s+1}/|\bar{\theta}_{s+1}|$ must be a unit of $\mathcal{O}$. Also, since $s = p/2 < p$, we have $\epsilon_\Delta = \theta_{p+1} > \theta_{s+1} > 1$ and $|\bar{\theta}_{p+1}| < |\bar{\theta}_{s+1}| < 1$; hence, $1 < \theta_{s+1}/|\bar{\theta}_{s+1}| < \theta_{p+1}/|\bar{\theta}_{p+1}| = \theta_{p+1}^2 = \epsilon_\Delta^2$. Thus, we must have $\epsilon_\Delta = \theta_{s+1}/|\bar{\theta}_{s+1}| = Q_0 \theta_{s+1}^2/Q_s$ by (3.19). In the case of $p = 2s+1$, we get $\mathfrak{a}_{s+1} = \bar{\mathfrak{a}}_{s+2}$ and $\epsilon_\Delta = \theta_{s+1}/|\bar{\theta}_{s+2}| = Q_0 \theta_{s+1} \theta_{s+2}/Q_s$. The remaining results follow as in the proof of Theorems 3.13 and 3.15. □

We can also extend Theorem 3.15 by using the same techniques.

Corollary 5.20.1. *If in the simple continued fraction expansion of ϕ_0 we get $P_k = P_{k+1}$ for some minimal $k \geq 1$, then $p = 2k$; if $Q_k = Q_{k+1}$ for some minimal $k \geq 1$, then $p = 2k + 1$.*

Corollary 5.20.2. *Let $\mathfrak{a}$ $(= \mathfrak{a}_1)$ be any reduced ambiguous ideal of an order $\mathcal{O}$. If we have k reduced ideals $\mathfrak{b}_1, \mathfrak{b}_2, \ldots, \mathfrak{b}_k$ in $[\mathfrak{a}]$ such that*

$$N(\mathfrak{b}_i) \neq N(\mathfrak{b}_j) \quad (i \neq j)$$

and

$$N(\mathfrak{b}_i) \nmid 2D/r \quad (i = 1, 2, \ldots, k),$$

then there are at least $2k + 1$ distinct reduced ideals in $[\mathfrak{a}]$.

Proof. As above, we let $\mathfrak{a} = [Q_0/r, (P_0 + \sqrt{D})/r]$. Since each $\mathfrak{b}_i$ is equivalent to $\mathfrak{a}$ and is reduced, we must have $\mathfrak{b}_m = \mathfrak{a}_{i_m}$ for some $i_m \leq p$. Also, $Q_{i_m+1}/r = N(\mathfrak{b}_m)$ are all distinct for $m = 1, 2, \ldots, k$. If $p = 2s+1$, then since $Q_s = Q_{s+1}$, we must have $s \geq k$. If $p = 2s$, then since $P_{s+1} = P_s$, we get $2P_s = q_s Q_s$ and $Q_s \mid 2P_s$. Since $Q_s \mid D - P_s^2$, we get $Q_s \mid 2D$. Since $Q_{i_m+1} \nmid 2D$ $(m = 1, 2, \ldots, k)$, we must have $k > s$. $\square$

Suppose $p > 1$. By Theorem 5.20, (5.18) and (3.19), we see that if $P_s = P_{s+1}$ when $s \geq 1$ is minimal, then

$$\epsilon_\Delta = \frac{Q_s}{Q_0} \prod_{i=1}^{s} \phi_i^2 \, ;$$

if $Q_s = Q_{s+1}$ when $s \geq 1$ is minimal, then

$$\epsilon_\Delta = \frac{P_{s+1} + \sqrt{D}}{Q_0} \prod_{i=1}^{s} \phi_i^2 \, .$$

In the first case we get

$$R_\Delta = \log\left(\frac{Q_s}{Q_0}\right) + 2\sum_{i=1}^{s} \log \phi_i$$

and in the second case we get

$$R_\Delta = \log\left(\frac{P_{s+1} + \sqrt{D}}{Q_0}\right) + 2\sum_{i=1}^{s} \log \phi_i \, .$$

These are very simple formulas for computing R_Δ when p is not large.[13] We can get similar formulas from the nearest integer continued fraction expansion (NICF) of $\phi = \sqrt{D}$. Such results were used[14] to compute the values of $R_\mathbb{K}$ for all real quadratic fields $\mathbb{Q}(\sqrt{D})$ with $D < 10^6$. This approach is about 25% faster than that utilizing the simple continued fraction (SCF) expansion of ϕ. Adams[15] derived a relation concerning the speed of convergence of the nearest integer and simple continued fractions for almost all irrational numbers. Indeed, if we let $l(D)$ be the length of the period of the SCF expansion of $\sqrt{D}$ and let $n(D)$ be that for the NICF, he conjectured, based on his findings, that for almost all squarefree D,

$$\lim_{D \to \infty} \frac{n(D)}{l(D)} = \frac{\log\left(\frac{\sqrt{5}+1}{2}\right)}{\log 2} \approx 0.6942419 .$$

Thus, we would expect that by using the NICF instead of the SCF, we might get almost a 30% speedup. Unfortunately, the mid-period criteria[16] for the NICF are more complicated than those for the SCF, which explains why the speedup is only 25%.

Before closing this section we mention another useful feature of the reduced ideals in any ideal class of $\mathcal{O}$. If $\mathfrak{b}$ is any given ideal of $\mathcal{O}$, there always exists a reduced ideal $\mathfrak{c}$ of $\mathcal{O}$ such that $\mathfrak{c} \sim \mathfrak{b}$ and $\mathfrak{c} = \gamma\mathfrak{b}$, where $1/\sqrt{\Delta} < \gamma < 1$; that is, there is always a reduced $\mathcal{O}$-ideal which is equivalent to any given ideal $\mathfrak{b}$ and it is not very far away from $\mathfrak{b}$. To show this, we observe that there must exist some reduced $\mathcal{O}$-ideal $\mathfrak{a}$ such that $\mathfrak{a} \sim \mathfrak{b}$. Hence, there exists some $\kappa \in \mathbb{K}$ ($\kappa > 0$) such that $\mathfrak{a} = \kappa\mathfrak{b}$. Put $\lambda = \kappa\epsilon_\Delta^k$ such that

$$\epsilon_\Delta^{-1} < \lambda \leq 1 .$$

If we define θ_i by

$$\mathfrak{a}_i = \rho^{i-1}(\mathfrak{a}) = \theta_i\mathfrak{a}$$

as in (5.10), then because $1 \leq \lambda^{-1} < \epsilon_\Delta$, we must have $\theta_i \leq \lambda^{-1} < \theta_{i+1}$ for some i such that $1 \leq i \leq p$ ($\theta_{p+1} = \epsilon_\Delta$). It follows that

$$\psi_i^{-1} = \frac{\theta_i}{\theta_{i+1}} < \lambda\theta_i \leq 1 .$$

Since $\mathfrak{a}_i = \theta_i\mathfrak{a} = \theta_i\lambda\mathfrak{b}$, we get $\mathfrak{c} = \gamma\mathfrak{b}$, where $\mathfrak{c} = \mathfrak{a}_i$ is reduced and $\gamma = \lambda\theta_i < 1$ and $\gamma > \psi_i^{-1} > 1/\sqrt{\Delta}$ by (3.33). Indeed, since $\lfloor \psi_i \rfloor = q_i$, we expect by our remarks at the end of §3.4 that, in most cases, $\gamma > 1/10$.

5.4 Ideal Products and NUCOMP

In subsequent work it will be necessary to find a representation for the product of two ideals. This is an old problem, called *composition*, which was originally dealt with in the case of quadratic forms by Legendre and more extensively by Gauss,[17] but it is still rather difficult to find a simple description of how to do this in the literature. We will discuss, based on the observations of Shanks,[18] how to do this here. For simplicity, we will assume that

$$\mathfrak{a}' = \left[\frac{Q'}{r}, \frac{P' + \sqrt{D}}{r} \right] \quad \text{and} \quad \mathfrak{a}'' = \left[\frac{Q''}{r}, \frac{P'' + \sqrt{D}}{r} \right]$$

are two primitive invertible $\mathcal{O}$-ideals. Since the product of two primitive $\mathcal{O}$-ideals need not be primitive, we will write

$$\mathfrak{a}'\mathfrak{a}'' = S \left[\frac{Q}{r}, \frac{P + \sqrt{D}}{r} \right] .$$

We need to show how to compute $S, Q,$ and P. We know by Proposition 4.25 that

$$\mathfrak{a}'\mathfrak{a}'' = \left[\frac{Q'Q''}{r^2}, \frac{Q'(P'' + \sqrt{D})}{r^2}, \frac{Q''(P' + \sqrt{D})}{r^2}, \frac{P'P'' + D + (P' + P'')\sqrt{D}}{r^2} \right] .$$

Also,

$$N(\mathfrak{a}'\mathfrak{a}'') = \frac{S^2 Q}{r} = N(\mathfrak{a}')N(\mathfrak{a}'') = \frac{Q'Q''}{r^2} .$$

Thus,

$$Q = \frac{Q'Q''}{S^2 r} . \tag{5.36}$$

Since $Q'(P'' + \sqrt{D})/r^2 \in \mathfrak{a}'\mathfrak{a}''$, we must have $Q'(P'' + \sqrt{D})/r^2 = xQS + SPy + S\sqrt{D}y$ $(x, y \in \mathbb{Z})$; it follows that $S \mid Q'/r$ on equating irrational parts. Similarly, $S \mid Q''/r$ and $S \mid (P' + P'')/r$.

Also, $S(P + \sqrt{D})/r \in \mathfrak{a}'\mathfrak{a}''$ means that there must exist $X, V, W, Y \in \mathbb{Z}$ such that

$$\frac{S(P + \sqrt{D})}{r} = X\frac{Q'Q''}{r^2} + V\frac{Q'P''}{r^2} + W\frac{Q''P'}{r^2} + Y\frac{P'P'' + D}{r^2}$$
$$+ \left(V\frac{Q}{r^2} + W\frac{Q''}{r^2} + Y\left(\frac{P' + P''}{r^2} \right) \right) \sqrt{D} .$$

Hence,

$$S = V\left(\frac{Q'}{r} \right) + W\left(\frac{Q''}{r} \right) + Y\left(\frac{P' + P''}{r} \right) .$$

Since $S \mid \gcd(Q'/r, Q''/r, (P' + P'')/r)$, we see that $S = \gcd(Q'/r, Q''/r, (P' + P'')/r)$. It remains to determine P, but we need only compute P modulo Q. We have

$$SP = X\frac{Q'Q''}{r} + V\frac{Q'P''}{r} + W\frac{Q''P'}{r} + Y\frac{P'P'' + D}{r} ;$$

thus,

$$P \equiv V\frac{Q'P''}{rS} + W\frac{Q''P'}{rS} + Y\frac{P'P'' + D}{rS} \pmod{Q} .$$

Since

$$
V \frac{Q'}{rS} + W \frac{Q''}{rS} + Y \left(\frac{P' + P''}{rS} \right) = 1 \,,
$$

we get

$$
\begin{aligned}
P &\equiv \left(1 - W \frac{Q''}{rS} - Y \left(\frac{P' + P''}{rS} \right) \right) P'' \\
&\quad + W \frac{Q'' P'}{rS} + Y \left(\frac{P' P'' + D}{rS} \right) \pmod{Q} \\
&= P'' + \frac{Q''}{rS} (W(P' - P') + Y R'') \pmod{Q} \,,
\end{aligned}
$$

where $R'' = (D - P''^2)/Q''$.

Thus, to find P, we use the extended Euclidean algorithm to find V, W, and Y such that

$$
V \frac{Q'}{r} + W \frac{Q''}{r} + Y \left(\frac{P' + P''}{r} \right) = S
$$

and put

$$
U \equiv W(P' - P'') + Y R'' \pmod{Q'/S} \,;
$$

then

$$
P \equiv P'' + U \frac{Q''}{rS} \pmod{Q} \,. \tag{5.37}
$$

If we assume that $\mathfrak{a}'$ and $\mathfrak{a}''$ are reduced and put $\mathfrak{a} = [Q/r, (P + \sqrt{D})/r]$, it is very likely that $\mathfrak{a}$ is not reduced; however, we often need a reduced ideal which is equivalent to $\mathfrak{a}$. We have already mentioned that there are several ways of finding a reduced $\mathcal{O}$-ideal equivalent to $\mathfrak{a}$, and we produced such an algorithm in §5.2 which is very effective for values of D that are not very large. To use this algorithm we must first calculate Q and P and these numbers could be as large as $O(D)$, whereas Q', Q'', P', and P'' are $O(\sqrt{D})$.

For example, if we were to reduce $\mathfrak{a}$ by finding the least $m \in \mathbb{Z}^{\geq 0}$ for which

$$
0 < Q_{m-1} < \sqrt{D} \,,
$$

then since

$$
\frac{Q_0}{r} = \frac{Q'Q''}{S^2 r^2} < \frac{\Delta}{S^2} \,,
$$

we see that $\mathfrak{a}_m = \theta_m \mathfrak{a}$ and

$$
\frac{S^2}{\Delta} < |\theta_m| < 2 \tag{5.38}
$$

by (5.12).

Shanks[19] discovered that there is a more efficient technique for finding a reduced ideal equivalent to $\mathfrak{a}'\mathfrak{a}''$ than first multiplying $\mathfrak{a}'$ by $\mathfrak{a}''$ and then using a reduction algorithm on $\mathfrak{a}$. He was guided in searching for such an algorithm by his need to keep the numbers involved in the calculations as small as possible.[20] Since Q could be as large as about the size of D, and he wanted to keep all the values computed by his algorithm to be of size roughly $\sqrt{D}$, the technique of first multiplying $\mathfrak{a}'$ and $\mathfrak{a}''$ and then carrying out the reduction phase was not acceptable. Instead, he developed a new technique which he called NUCOMP, standing for "New COMPosition". We will not discuss Shanks' version of this algorithm or its later improvements by Atkin, van der Poorten, and Jacobson[21] here. Instead we will provide a version[22] of the algorithm that is congenial to our previously developed continued fraction theme.[23]

When we used the technique in §5.2 to effect the reduction of $\mathfrak{a} = [Q/r, (P + \sqrt{D})/r]$, we simply developed the simple continued fraction expansion of P/Q. If we were to do that here, the values of P and Q might be as large as D. However, we note that

$$\frac{P}{Q} = \frac{P'' + UQ''/rS}{(Q'/S)(Q''/rS)} \approx \frac{SU}{Q'} \ .$$

Thus, instead of finding the simple continued fraction expansion of P/Q, we instead will look at that of SU/Q'. We will need the following result.

Theorem 5.21. *Suppose* Q, D, P, N, L, K, P', *and* P'' *are integers such that* $D > 0$, $\sqrt{D} \notin \mathbb{Q}$, $Q \mid D - P^2$, *and*

$$P = P'' + NK, \quad Q = NL, \quad P \equiv P' \pmod{L} \ .$$

If $K/L = [q_0, q_1, q_2, \ldots, q_n]$ *and we put*

$$\frac{P + \sqrt{D}}{Q} = \langle q_0, q_1, \ldots, q_i, \phi_{i+1} \rangle \quad (i < n) \ ,$$

then

$$Q_{i+1} = (-1)^{i-1}(R_i M_1 - C_i M_2) \ ,$$
$$M_1 = (NR_i + (P' - P'')C_i)/L \in \mathbb{Z} \ ,$$
$$M_2 = (R_i(P' + P'') + TC_i)/L \in \mathbb{Z} \ ,$$
$$P_{i+1} = (NR_i + Q_{i+1}C_{i-1})/C_i - P'' \ ,$$

where $T = (D - P''^2)/N$.

Proof. From (3.13) we know that

$$Q_{i+1} = (-1)^{i+1}(G_i^2 - DB_i^2)/Q$$
$$= (-1)^{i+1}\left[G_i\left(\frac{G_i + PB_i}{Q} \right) - B_i\left(\frac{DB_i + G_iP}{Q} \right) \right] \ . \tag{5.39}$$

Also, by (3.22)

$$(-1)^{i+1}R_i = LA_i - KB_i = LA_i - (-1)^{i+1}KC_i .$$

Since

$$G_i = A_iQ - PB_i ,$$

we get

$$G_i = NLA_i - PB_i = (-1)^{i+1}(NR_i - P''C_i) \qquad (5.40)$$

and

$$\frac{G_i + PB_i}{Q} = \frac{LA_i}{L} = \frac{(-1)^{i+1}(R_i + KC_i)}{L} , \qquad R_i \equiv -KC_i \ (\text{mod } L) . \quad (5.41)$$

Furthermore,

$$\begin{aligned}
(DB_i + G_iP)/Q &= ((D - P''P)B_i + NPR_i(-1)^{i+1})/Q \\
&= \left((D - P''^2)B_i - NKP''B_i + NPR_i(-1)^{i+1}\right)/Q \quad (5.42) \\
&= (-1)^{i+1}(TC_i - KP''C_i + PR_i)/L .
\end{aligned}$$

Since $DB_i + G_iP \equiv DB_i - P^2B_i \equiv 0 \ (\text{mod } Q)$, we must have

$$TC_i \equiv -(P'' + P)R_i \equiv -(P'' + P')R_i \ (\text{mod } L) .$$

If we substitute (5.40), (5.41), and (5.42) into (5.39), we get the value for Q_{i+1} in the theorem. The value for P_{i+1} follows from (3.13) and (5.4). Since $R_i \equiv -KC_i \ (\text{mod } L)$ and $TC_i \equiv -(P''+P')R_i \ (\text{mod } L)$, we see that $M_1, M_2 \in \mathbb{Z}$.
□

With this result we can now provide a simple description of our version of NUCOMP. We will begin by assuming that both $\mathfrak{a}'$ and $\mathfrak{a}''$ are reduced and $N(\mathfrak{a}') \geq N(\mathfrak{a}'')$. Thus, $2\sqrt{D} > Q' \geq Q'' > 0$. We will also assume that $Q > \sqrt{|D|}$. For if $Q < \sqrt{|D|}$, then $N(\mathfrak{a}) < \sqrt{|\Delta|}/2$, and by Theorems 5.6 and 5.9, $\mathfrak{a}$ is already reduced. Since $\mathfrak{a}''$ is also a reduced ideal, there is no loss in generality by assuming that $0 < P'' < \sqrt{|D|}$ or $|2P''/r| < N(\mathfrak{a}) = Q''/r$ when $D < 0$. We now put $K = U$, $N = Q''/rS$, and $L = Q'/S$. It is easy to see that we must have $P \equiv P' \ (\text{mod } Q'/S)$ and we can now invoke Theorem 5.21. We select i such that $0 \leq i \leq n$ and

$$R_i < \sqrt{rQ'/Q''}|D|^{1/4} < R_{i-1} . \qquad (5.43)$$

Such an i must exist because $Q > \sqrt{|D|}$ means that $R_{-1} = Q'/S > \sqrt{rQ'/Q''}|D|^{1/4}$. We now find Q_{i+1} and P_{i+1} by computing R_j and C_j $(j = 0, 1, 2, \ldots, i)$ and using the formulas

$$\begin{aligned}
Q_{i+1} &= (-1)^{i+1}(R_iM_1 - C_iM_2) , \\
P_{i+1} &= ((Q''/rS)R_i + Q_{i+1}C_{i-1})/C_i - P'' ,
\end{aligned} \qquad (5.44)$$

where

$$M_1 = ((Q''/rS)R_i + (P' - P'')C_i)/(Q'/S) \, ,$$
$$M_2 = (R_i(P' + P'') + rSR''C_i)/(Q'/S) \, ,$$
$$R'' = (D - P''^2)/Q'' \, .$$

We now have

$$G_i = QA_i - PB_i = (-1)^{i+1}((Q''/rS)R_i - P''C_i)$$

and

$$|Q_{i+1}| = |G_i{}^2 - DC_i{}^2|/Q$$
$$= |(Q''/rS)R_i{}^2 + 2P''(-1)^i R_i C_i - rSR''C_i{}^2|/(Q'/S) \, .$$

Put $x = (Q''/rS)R_i{}^2/(Q'/S)$. Since $R_i < \sqrt{rQ'/Q''}|D|^{1/4}$, we get $0 < x < \sqrt{|D|}$. If we put $y = (-1)^i 2P''R_iC_i/(Q'/S)$, then since $Q'/S \geq B_iR_{i-1}$ by (3.23), we get $|C_i| \leq (Q'/S)/R_{i-1}$ and $|y| < 2|P''|$. We now consider $z = rSR''C_i{}^2/(Q'/S)$. We have

$$C_i{}^2 \leq (Q'/S)^2/R_{i-1}{}^2 < (Q'/S)^2/[r(Q'/Q'')\sqrt{|D|}] = Q/\sqrt{|D|} \, .$$

Hence,

$$rSC_i{}^2/(Q'/S) < Q''/\sqrt{|D|} \, .$$

If $D > 0$, then $R'' = (D - P''^2)/Q'' > 0$ and

$$0 < z < (D - P''^2)/\sqrt{D} \, .$$

In this case,

$$-2|P''| + P''^2/\sqrt{D} - \sqrt{D} < x + y - z < \sqrt{D} + 2|P''|$$

and

$$|Q_{i+1}| < \sqrt{D} + 2|P''| < 3\sqrt{D} \, .$$

If $D < 0$, then $R'' = (D - P''^2)/Q'' = -(|D| + P''^2)/Q''$. Here, we have

$$-2|P''| < x + y - z < 2\sqrt{|D|} + 2|P''| + P''^2/\sqrt{|D|} \, .$$

Since $\mathfrak{a}''$ is a reduced $\mathcal{O}$-ideal, we must have $N(\mathfrak{a}'') < \sqrt{|\Delta|/3}$. It follows that $Q'' < 2\sqrt{|D|/3}$, $P'' < \sqrt{|D|/3}$ and therefore

$$|Q_{i+1}| < 3.5\sqrt{|D|}$$

in this case.

Thus, once U has been computed we need use little more than the Euclidean algorithm to find a primitive $\mathcal{O}$-ideal

$$\mathfrak{a}_{i+2} = \left[\frac{|Q_{i+1}|}{r}, \frac{P_{i+1} + \sqrt{D}}{r} \right] \sim \mathfrak{a} = \left[\frac{Q}{r}, \frac{P + \sqrt{D}}{r} \right]$$

such that

$$|Q_{i+1}| < \begin{cases} 3\sqrt{D} & \text{if } D > 0 \\ 3.5\sqrt{|D|} & \text{if } D < 0. \end{cases}$$

Also, as we usually have $Q' \approx Q''$, we would expect that $R_i \approx |D|^{1/4}$, and we know that $|C_i| < 2|D|^{1/4}$; thus, the numbers involved in this step of NUCOMP do not get much larger than $|D|^{1/4}$ until we compute Q_{i+1} and P_{i+1}. However, even in the worst case, the numerators of M_1 and M_2 are about $|D|^{3/4}$ and the denominator about $|D|^{1/2}$.

Since $N(\mathfrak{a}_{i+2}) < 3\sqrt{\Delta}/2$ when $\Delta > 0$, we see by Theorem 5.13 that if $\mathfrak{a}_{i+2}$ is not already reduced, then $\rho(\mathfrak{a}_{i+2})$ is reduced. Computations have revealed that $\mathfrak{a}_{i+2}$ is reduced about 98% of the time; thus, we only infrequently have to evaluate $\rho(\mathfrak{a}_{i+2})$. When $\Delta < 0$, we have $N(\mathfrak{a}_{i+2}) < 2\sqrt{|\Delta|}$. If $N(\mathfrak{a}_{i+2}) < \sqrt{|\Delta|}$, then by Theorem 5.11 if $\mathfrak{a}_{i+2}$ is not reduced, then $\rho(\mathfrak{a}_{i+2})$ is. If $N(\mathfrak{a}_{i+2}) > \sqrt{|\Delta|}$, then $\mathfrak{a}_{i+2}$ is not reduced and, by (5.6), $N(\rho(\mathfrak{a}_{i+2})) < \sqrt{|\Delta|}$. Hence, either $\rho(\mathfrak{a}_{i+2})$ or $\rho^2(\mathfrak{a}_{i+2})$ must be reduced. Thus, we can always find a reduced ideal equivalent to $\mathfrak{a}$ in no more than the application of two ρ-steps to $\mathfrak{a}_{i+2}$. We do not have to evaluate Q or use numbers near in value to $|D|$ in order to find this ideal by NUCOMP.

There might be some concern for the need to evaluate $\sqrt{rQ'/Q''}|D^{1/4}|$, particularly if NUCOMP has to be performed many times in $\mathcal{O}$. In the case of $D > 0$ (the case of most concern to us), we can avoid this by simply evaluating $\sqrt{2r}D^{1/4}$. This value does not change with the ideals being multiplied. With this in mind, we are now able to present our version of NUCOMP for $D > 0$. In order to improve the flow of the material which we will present in this and some further chapters, we have elected to record our most useful algorithms by providing only the inputs and outputs. The interested reader can find the complete pseudocode for these procedures, together with proofs of their correctness, in the Appendix.

Algorithm 5.1: NUCOMP

Input: $\mathfrak{a}' = [Q'/r, (P'+\sqrt{D})/r]$, $\mathfrak{a}'' = [Q''/r, (P''+\sqrt{D})/r]$ reduced invertible $\mathcal{O}$-ideals with $Q' \geq Q'' > 0$.

Output: A reduced $\mathcal{O}$-ideal $\mathfrak{b} = [Q/r, (P + \sqrt{D})/r]$ such that $\mathfrak{b} \sim \mathfrak{a}'\mathfrak{a}''$. (Optional output: μ where $\mathfrak{a}'\mathfrak{a}'' = \mu\mathfrak{b}$.)

At the conclusion of Algorithm 5.1 we will have a reduced ideal $\mathfrak{b}$ such that

$$\mu\mathfrak{b} = \mathfrak{a}'\mathfrak{a}''.$$

We show in the Appendix (§A.1) that

$$1 \leq \mu < 2\Delta^{3/4} \qquad (5.45)$$

and

$$\frac{1}{q'+3} < \frac{\mu}{\Delta^{1/4}} < \frac{2(\tilde{q}+2)}{q'} , \qquad (5.46)$$

where both $\tilde{q}$ and q' are partial quotients in the simple continued fraction expansion of quadratic irrationals. Thus, we would expect that they would not normally be very large. Since, by (5.46), we have

$$|\log\mu - \log\Delta^{1/4}| < \max\left\{\log(q'+3), \log\left(\frac{2(\tilde{q}+2)}{q'}\right)\right\} , \qquad (5.47)$$

we observe that $\log\mu$ would normally not differ very much from $(\log\Delta)/4$, and this is what occurs when we compute values of $\log\mu$, particularly when D is fairly large ($> 10^{10}$).

Notes and References

[1]Much of the information in this section is based on earlier work of Berwick [Ber28] and, particularly Ince [Inc34]. Further work relevant to this chapter can be found in [WW87], [BW88a], and [SW88b].

[2]This operation was introduced in [JSW01].

[3]An account of this can be found in the third volume of [Dic19], p. 5ff. An interesting historical account of the work of Lagrange is presented in [Wei84].

[4]By bit complexity, we mean the number of bit operations (add, subtract, multiply, shift, etc.) that need to be performed in the process of executing the algorithm. Lagarias [Lag80b] seems to have been the first to examine the problem of determining the bit complexity of ideal reduction.

[5][BB97].

[6]Several of these are reviewed in some detail in [JSW06a].

[7][Sch91].

[8]See the remarks in [BB97] and the results of testing in [JSW06a].

[9]This is a slightly modified version of the new algorithm presented in [JSW06a].

[10]See the discussion in [JSW06a].

[11]This is not the case in the theory of positive definite quadratic forms, where there can only be one reduced form equivalent to a given one. The reason for the difference between the situations for ideals and forms is a slight difference in the definition of reduction for forms as opposed to the definition given here for a reduced ideal.

[12]The term "regulator" was first used (in a much wider context than this) by Dedekind [Dir93], p. 597, in 1893. However, as pointed out in a footnote, he borrowed the term from an 1844 paper of Eisenstein [Eis44a], p. 313, who used it in a somewhat different context.

[13]These formulas were used in [WB76] to compute the values of $R_{\mathbb{K}}$ for all positive $D_0 < 1.5 \times 10^5$.

[14][WB79].

[15][Ada79].

[16][Wil80b] and [Wil85b].

[17]An account of the history of this early work on composition is given in [Wei84], Ch. IV, §VI. See also [Edw07].

[18][Sha71].

[19][Sha89].

[20]At the time, Shanks was performing all of his computations by making use of a pocket calculator which had a fixed word size of 10 decimal digits. As he was often working with quadratic fields of discriminants near 20 decimal digits, it was important for him to keep the size of the numbers small in order to simplify the programming on the calculator.

[21]This is described in [vdP03] and [JvdP02]. Atkin's work, while providing some important insights with respect to NUCOMP, was never published.

[22]An earlier version of this was first discussed in [JSW06b].

[23]The version that we give here will work regardless of the sign of Δ, but a bit more work may need to be done if $\Delta < 0$. Shanks' version of NUCOMP was intended to be used only when $\Delta < 0$. It was van der Poorten [vdP03] who discovered that the basic idea of NUCOMP could be applied to the case of $\Delta > 0$.

6

Some Special Pell Equations

6.1 Introduction

Let D be a positive non-square integer and let (t, u) denote the fundamental solution of the Pell equation

$$T^2 - DU^2 = 1 .\qquad(6.1)$$

In this chapter we will be concerned with Pell equations for which the values of t and u tend to be small. As we shall see later in Chapter 9, this appears to be a very unusual circumstance; for this reason, then, if no other, it is of some interest to investigate this phenomenon. For example, consider the case of $D = M^2 - 1$ for $M \in \mathbb{Z}^{>0}$. Clearly, $(M, 1)$ is a solution of the Pell equation (6.1). Furthermore, it is easy to verify that this is the fundamental solution. Of course, we might regard such Pell equations as being easy to solve, and in a sense this is the case; however, there are a few problems that do develop.

We have seen in Chapter 3 that we can solve the Pell equation by expanding $\sqrt{D}$ into a simple continued fraction. In this case we have

$$\sqrt{D} = [q_0, \overline{q_1, q_2, \ldots, q_p}] ,$$

$Q_p = Q_0 = 1$, $P_p = P_1 = q_0$, $q_p = 2q_0$, and $q_i = q_{p-i}$ $(i = 1, 2, \ldots, p-1)$. Also, by (3.13)

$$A_{p-1}^2 - DB_{p-1}^2 = (-1)^p .$$

Thus, if $2 \mid p$, then $t = A_{p-1}$, $u = B_{p-1}$; if $2 \nmid p$, then

$$t + u\sqrt{D} = (A_{p-1} + \sqrt{D}B_{p-1})^2$$

and

$$t = A_{p-1}^2 + DB_{p-1}^2 = 2A_{p-1}^2 + 1 , \quad u = 2A_{p-1}B_{p-1} .$$

Euler[1] showed how to solve (6.1) for certain values of D for which p is small. For example, if $p = 1$, then $q_1 = q_p = 2q_0$, $A_{p-1} = A_0 = q_0$, and

$B_{p-1} = 1$. Hence, $q_0^2 - D = -1$ or $D = q_0^2 + 1$. Thus, the fundamental solution of (6.1) for $D = M^2 + 1$ ($M \in \mathbb{Z}^{>0}$) is given by

$$t = 2M^2 + 1 , \quad u = 2M .$$

However, when $p = 2$, the problem becomes somewhat more difficult. In this case,

$$\sqrt{D} = [q_0, \overline{q_1, 2q_0}] ,$$

$A_{p-1} = A_1 = q_0 q_1 + 1$, $B_{p-1} = q_1$, and

$$(q_0 q_1 + 1)^2 - D q_1^2 = 1 .$$

From this we see that we must have

$$D q_1 = q_0^2 q_1 + 2q_0$$

and, therefore, $q_1 \mid 2q_0$. In this case, then,

$$D = q_0^2 + 2q_0/q_1 .$$

If we put $m = 2q_0/q_1$, then $q_0 = q_1 m/2$ and $D = (q_1 m/2)^2 + m$, where $2 \mid q_1 m$. If $m = 2M$ and $q_1 = N$, then

$$D = M^2 N^2 + 2M \quad (M > 0) \tag{6.2}$$

and $t = MN^2 + 1$ and $u = N$; if $2 \nmid m$ and $m = M$, then $q_1 = 2N$ ($N \in \mathbb{Z}^{>0}$),

$$D = M^2 N^2 + M \quad (M > 0) , \tag{6.3}$$

$t = 2MN^2 + 1$, and $u = 2N$.

If we next examine the forms

$$D = N^4 M^2 + 2MN \text{ with } x = N^3 M + 1 , \ y = N , \ N \in \mathbb{Z}^{>0} ,$$

or

$$D = N^4 M^2 + MN \text{ with } x = 2N^3 M + 1 , \ y = 2N , \ N \in \mathbb{Z}^{>0} ,$$

we see that in each case we get

$$x^2 - Dy^2 = 1 \text{ and } y \mid D .$$

We can, of course, prove that $t = x$ and $u = y$ by making use of the continued fraction approach, but there is another method which makes use of a pretty observation of Störmer.[2] In order to establish Störmer's theorem (Theorem 6.3), we need two simple results concerning Lucas functions. We employ the notation introduced in §1.4 of Chapter 1.

Proposition 6.1. *If p is a prime such that $p > 3$, $p \nmid Q$, and $p \mid d$, then $\omega(p^n) = p^n$.*

Proof. From

$$\frac{v_{pn} + u_{pn}\sqrt{d}}{2} = \left(\frac{v_n + u_n\sqrt{d}}{2}\right)^p$$

we get that

$$2^{p-1}u_{pn} = \binom{p}{1}v_n^{p-1}u_n + \binom{p}{3}v_n^{p-3}u_n^3 d + \cdots + u_n^p d^{(p-1)/2} .$$

Thus, if $p > 3$, then $p \mid \binom{p}{3}$ and

$$2^{p-1}u_p = pv_1^{p-1} \pmod{p^2} .$$

Since $p \mid d$ and $d = P^2 - 4Q$, we cannot have $p \mid P$; hence, $p \nmid v_1$ and $p \parallel u_p$.

By the Law of Repetition, we must have $p^k \parallel u_{p^k}$. Thus, $\omega(p^k) = p^\nu$ ($\nu \le n$). Since $p^\nu \parallel u_{p^\nu}$, we cannot have $p^n \mid u_{p^\nu}$ if $\nu < n$; hence, $\omega(p^n) = p^n$. $\square$

Lemma 6.2. *Suppose $d, Q > 0$. If p is a prime such that $p > 3$ and $p \mid (d, u_n)$, then there exists a prime q such that $q \mid u_n$ and $q \nmid d$.*

Proof. We first note that since $(P, Q) = 1$, we must have $(d, Q) = 1$. Let $p^\alpha \parallel u_n$. By Proposition 6.1 we know that $\omega(p^\alpha) = p^\alpha$ and $p^\alpha \parallel n$ by the Law of Repetition. Put $T = |u_{p^\alpha}/(pu_{p^{\alpha-1}})|$; we must have $T \in \mathbb{Z}$ and $p \nmid T$. Also, $T \mid u_n$. If we put $m = p^{\alpha-1}$ and refer to the proof of Proposition 6.1, we see that

$$2^{p-1}u_{pm} = \binom{p}{1}v_m^{p-1}u_m + \binom{p}{3}v_m^{p-3}u_m^3 d + \cdots + u_m^p d^{(p-1)/2} .$$

Since

$$v_m^2 - du_m^2 = 4Q^m$$

and $d, Q > 0$, we see that $|v_m| > \sqrt{p}$ and $d > p$; hence, $2^{p-1}|u_{pm}/pu_m| > 1$ and $T > 1$. Since $p > 3$, we have

$$2^{p-1}T \equiv v_m^{p-1} \pmod{d} .$$

Suppose $2 \nmid d$. There must be a prime divisor q of T and if $q \mid d$, then $q \mid v_m$, which means that $q \mid Q$, an impossibility. If $2 \mid d$, then $2 \mid P$. Since $(P, Q) = 1$, we have $2 \nmid Q$, and $u_{2i+1} \equiv 1 \pmod 2$. It follows that u_{pm} and T are odd. As before, any prime divisor q of T cannot divide d. $\square$

Theorem 6.3. *Suppose $x^2 - Dy^2 = 1$ ($x, y \ge 1$) and each prime that divides y also divides D. Then $x = t$ and $y = u$; that is, (x, y) is the fundamental solution of the Pell equation.*

Proof. Certainly, we must have

$$x + y\sqrt{D} = (t + u\sqrt{D})^n$$

for some $n \geq 1$. Hence, $y = uu_n(P, Q)$, where $P = 2t$, $Q = 1$. If $y \neq u$, then there must exist a prime p such that $p \mid u_n$ and $p \mid y$. Since $d = 4u^2 D$ and $p \mid D$, we have $p \mid d$. If $2 \mid n$, then $u_2 \mid u_n$ ($u_2 = P = 2t$) and $t \mid u_n$. Since $t > 2$, there must be a prime q such that $q \mid t$ and $q \mid D$, but this is impossible because $t^2 - Du^2 = 1$. If $3 \mid n$, then $u_3 \mid u_n$, and since $u_3 = 4t^2 - 1 = (2t - 1)(2t + 1)$ and $t > 2$, there must be a prime p such that $p \mid u_n$, $p \mid d$, and $p \neq 2, 3$. Suppose $(n, 6) = 1$. Since $p \mid d$, we must have $p \mid u_2$ if $p = 2$ or $p \mid u_3$ if $p = 3$; however, since $(n, 6) = 1$ and $p \mid u_n$, this is not possible. Thus, $p > 3$. By Lemma 6.2, there must exist a prime q such that $q \mid u_n$ and $q \nmid d$, a contradiction. □

6.2 Continued Fractions

In view of our earlier remarks, we might expect that as the period length p becomes larger, the problem of finding suitable values of D would become much more complicated.[3] However, we do have a general result concerning this problem.[4] We now assume that we are given some $p \, (> 2)$, $q_1, q_2, q_3, \ldots, q_{p-1} \in \mathbb{Z}^{>0}$ ($q_i = q_{p-i}$; $i = 1, 2, \ldots, p - i$), and we want[*]

$$\phi_0 = [q_0, \overline{q_1, q_2, \ldots, q_{p-1}, 2q_0}] \tag{6.4}$$

to be the square root of an integer D. This problem, then, reduces to that of finding those values of q_0 for which ${\phi_0}^2 = D \in \mathbb{Z}$. Define $q_i' = q_{i+1}$ ($i = 0, 1, 2, \ldots, p - 2$) and use the formulas (3.4) to compute the convergents

$$A'_{p-2}/B'_{p-2} = [q_0', q_1', \ldots, q'_{p-2}] = [q_1, q_2, \ldots, q_{p-1}] \,,$$
$$A'_{p-3}/B'_{p-3} = [q_0', q_1', \ldots, q'_{p-3}] = [q_1, q_2, \ldots, q_{p-2}] \,.$$

Now, $[q_1, q_2, \ldots, q_{p-1}] = [q_{p-1}, q_{p-2}, \ldots, q_1] = B_{p-1}/B_{p-2}$ by (3.5). Thus,

$$A'_{p-2} = B_{p-1} \,, \quad B'_{p-2} = B_{p-2} \,. \tag{6.5}$$

Notice that the values of A'_{p-2}, B'_{p-2} A'_{p-3}, B'_{p-3} are all independent of the choice of q_0. Also, by the third formula of (3.5) and the symmetry of the partial quotients (Theorem 3.12),

$$\frac{A_{p-1}}{B_{p-1}} = [q_0, q_{p-1}, \ldots, q_1] = q_0 + \frac{B_{p-2}}{B_{p-1}} \,;$$

thus,

---

[*] We remind the reader that $[a, b, c, \ldots]$ denotes a simple continued fraction with partial quotients $a, b, c, \ldots$.

$$A_{p-1} = q_0 B_{p-1} + B_{p-2} . \qquad (6.6)$$

If we consider

$$\phi_0 = [q_0, q_1, \ldots, q_{p-1}, \phi_p] ,$$

by (3.6) we must have

$$\phi_0 = \frac{A_{p-1}\phi_p + A_{p-2}}{B_{p-1}\phi_p + B_{p-2}} .$$

Also, by (6.4), we have $\phi_p = q_0 + \phi_0$, and we get

$$B_{p-1}\phi_0{}^2 + (q_0 B_{p-1} + B_{p-2})\phi_0 = A_{p-1}\phi_0 + q_0 A_{p-1} + A_{p-2} .$$

By (6.6), this reduces to

$$B_{p-1}\phi_0{}^2 = q_0 A_{p-1} + A_{p-2} .$$

Putting $\phi_0{}^2 = D$, we get

$$B_{p-1}D = q_0 A_{p-1} + A_{p-2} . \qquad (6.7)$$

Computing both A_{p-1} and A_{p-2} involves using the value of q_0, so we now need to rewrite (6.7) in such a way that the value of q_0 is independent of all the coefficients. By (6.5) and (6.6), this is not difficult for A_{p-1}. It remains to consider A_{p-2}. Since

$$A_{p-2}/B_{p-2} = [q_0, q_1, \ldots, q_{p-2}] ,$$

we get $A_{p-2}/B_{p-2} = q_0 + B'_{p-3}/A'_{p-3}$; hence,

$$A_{p-2} = q_0 A'_{p-3} + B'_{p-3} , \qquad (6.8)$$
$$B'_{p-2} = B_{p-2} = A'_{p-3} \qquad (6.9)$$

and (6.7) becomes

$$A'_{p-2}D = q_0(q_0 A'_{p-2} + B'_{p-2}) + q_0 B'_{p-2} + B'_{p-3}$$

or

$$A'_{p-2}(D - q_0{}^2) - 2q_0 B'_{p-2} = B'_{p-3} . \qquad (6.10)$$

Now, (6.10) is a linear Diophantine equation in $D - q_0{}^2$ and $2q_0$ and

$$A'_{p-2}B'_{p-3} - B'_{p-2}A'_{p-3} = (-1)^{p-1} \qquad (6.11)$$

by (3.4). It follows that from our observations in §3.2 that we must have

$$2q_0 = (-1)^{p-1}A'_{p-3}B'_{p-3} + mA'_{p-2} ,$$
$$D - q_0{}^2 = (-1)^{p-1}B'_{p-3}{}^2 + mB_{p-2} ,$$

for some integer parameter m. Thus, if

$$\sqrt{D} = [q_0, \overline{q_1, q_2, \ldots, q_{p-1}, 2q_0}] \;,$$

it is necessary that

$$q_0 = \frac{(-1)^{p-1} A'_{p-3} B'_{p-3} + m A'_{p-2}}{2} \tag{6.12}$$

and

$$D = q_0{}^2 + m A'_{p-3} + (-1)^{p-1} B'_{p-3}{}^2 \;, \tag{6.13}$$

where

$$(-1)^{p-1} A'_{p-3} B'_{p-3} + m A'_{p-2}$$

is a positive even integer.

We next suppose that q_0 and D are given by (6.12) and (6.13). By the results in §3.1 we know that

$$\sqrt{D} = \langle q_0, q_1, q_2, \ldots, q_{p-1} \tilde{\phi}_p \rangle \;,$$

where $\tilde{\phi}_p = (\tilde{P} + \sqrt{D})/\tilde{Q}$ and $\tilde{P}, \tilde{Q} \in \mathbb{Z}$. Also, by (3.13),

$$(-1)^p \tilde{Q} = A_{p-1}^2 - D B_{p-1}^2 \;, \tag{6.14}$$

$$(-1)^{p-1} \tilde{P} = A_{p-1} A_{p-2} - D B_{p-1} B_{p-2} \;. \tag{6.15}$$

By (6.5), (6.6), and (6.8), we have

$$A_{p-1} = q_0 A'_{p-2} + B'_{p-2} \;, \qquad A_{p-2} = q_0 A'_{p-3} + B'_{p-3} \;.$$

Furthermore, since q_0 and D must satisfy (6.10), we get

$$D B_{p-1} = D A'_{p-2} = q_0{}^2 A'_{p-2} + 2 q_0 B'_{p-2} + B'_{p-3} \;.$$

By substituting this into (6.14) and (6.15) and using (6.11), we get $\tilde{Q} = 1$ and $\tilde{P} = q_0$. Hence,

$$\sqrt{D} = \langle q_0, q_1, q_2, \ldots q_{p-1}, q_0 + \sqrt{D} \rangle \;,$$

and therefore $\sqrt{D} = [q_0, \overline{q_1, q_2, \ldots, q_p}]$, where $q_p = 2q_0$. We have proved the following theorem.

Theorem 6.4. *Suppose we are given an integer $p > 2$ and a finite sequence of $p - 1$ positive integers $q_1, q_2, \ldots, q_{p-1}$ such that $q_i = q_{p-i}$ $(i = 1, 2, \ldots, p-1)$. There exists an integer D such that*

$$\sqrt{D} = [q_0, \overline{q_1, q_2, \ldots q_{p-1}, 2q_0}] \tag{6.16}$$

if and only if D is given by (6.13) and q_0 by (6.12), where $m \in \mathbb{Z}$ and

$$(-1)^{p-1} A'_{p-3} B'_{p-3} + m A'_{p-2}$$

is a positive even integer. Furthermore,

$$A_{p-1} = q_0 A'_{p-2} + A'_{p-3} \text{ and } B_{p-1} = A'_{p-2} \;. \tag{6.17}$$

Example 6.5. Suppose $p = 4$ and we are given values for q_1, q_2, and q_3 with $q_3 = q_1$. We get

$$B'_{p-2} = A'_{p-3} = q_1 q_2 + 1, \quad B'_{p-3} = q_2, \quad A'_{p-2} = q_1{}^2 q_2 + 2q_1 .$$

Thus, if

$$q_0 = \frac{m(q_1{}^2 q_2 + 2q_1) - q_2(q_1 q_2 + 1)}{2}$$

and

$$D = q_0{}^2 + m(q_1 q_2 + 1) - q_2{}^2 ,$$

then $\sqrt{D} = [q_0, \overline{q_1, q_2, q_1, 2q_0}]$ when $m, q_0 \in \mathbb{Z}$ and $q_0 > 0$. Note that if $2 \mid q_2$, then $q_0 \in \mathbb{Z}$. If $2 \nmid q_2$, then $q_0 \in \mathbb{Z}$ if and only if $2 \nmid q_1$ and $2 \mid m$. If we consider the special case of $q_1 = 1$ and $q_2 = n$, then we must have $m(n+2) - n(n+1) \equiv 0 \pmod 2$ or $2 \mid mn$. If $2 \nmid n$, then $2 \mid m - n + 1$ and we put $2M = m - n + 1$ and $N = n + 2 > 0$. Since $2q_0 = m(n+2) - n(n+1) \geq 2$, we must have

$$m \geq \frac{n^2 + n + 2}{n + 2} > n - 1$$

and $M \geq 1$. We get $q_0 = NM - 1$ and

$$D = N^2 M^2 - 2M \quad (M > 0, N > 2 \text{ and } 2 \nmid N). \tag{6.18}$$

Here, $A_3 = N^2 M - 1$ and $B_3 = N$; thus, $t = N^2 M - 1$ and $u = N$ in this case.

The forms (6.2), (6.3), and (6.18) are all instances of a more general form referred to as a *Richaud-Degert type*.[5] Such forms are given by

$$D = Q^2 + R ,$$

where $R \mid 4Q$, $Q \geq 1$, and $R \neq 0$. There is no loss in generality in making the futher assumption that

$$-2Q < R \leq 2Q . \tag{6.19}$$

For if $R > 2Q$, then since $R \mid 4Q$, we must have $R = 4Q$, $D = Q^2 + 4Q = (Q+2)^2 - 4$ and $-4 > -2(Q+2)$. If $R \leq -2Q$, then $R = -4Q$ or $R = -2Q$. In the first case, $D = Q^2 - 4Q = (Q-2)^2 - 4$, and since $Q > 4$, $-4 > -2(Q-2)$; in the second case, $D = Q^2 - 2Q = (Q-1)^2 - 1$ and $-1 > -2(Q-1)$. Note that (6.19) also guarantees that D is not a perfect square and that $\lfloor \sqrt{D} \rfloor = Q - 1$ or Q. For such values of D it is possible to develop the simple continued fraction (SCF) expansion[6] of $\sqrt{D}$, and for each possible value, we have $p \leq 12$. From this, of course, it is possible to determine t and u. However, it is instructive to do this in another way.[7]

Suppose we put $\gamma = Q + \sqrt{D}$; then,

$$N(\gamma) = -R .$$

If $|R| = 1$, it is easy to see that if ϵ is defined as in §1.3, then $\epsilon = \gamma$. If $R = -4$, then $\epsilon = \gamma/2$ and $N(\epsilon) = 1$. Thus,

$$t + u\sqrt{D} = \begin{cases} \epsilon^2 & \text{when } 2 \mid Q \\ \epsilon^3 & \text{otherwise} . \end{cases}$$

Suppose for the remainder of our discussions that $R \neq \pm 1, -4$, and consider

$$\lambda = \frac{\gamma^2}{|R|} = \frac{\frac{4Q^2}{|R|} + \frac{2R}{|R|} + \frac{4Q}{|R|}\sqrt{D}}{2} .$$

We now define $\eta = \epsilon$ when $N(\epsilon) = 1$ and $\eta = \epsilon^2$ when $N(\epsilon) = -1$. Since we have $N(\lambda) = 1$, $4Q^2/|R| + 2R/|R| > 1$, and $4Q/|R| \geq 1$, we must have

$$\lambda = \eta^n \quad (n \in \mathbb{Z}^{>0}) .$$

Also,

$$\frac{4Q^2}{|R|} + \frac{2R}{|R|} = \frac{4D - 2R}{|R|} < 4D + 2 .$$

Let

$$\eta = \frac{w_1 + z_1\sqrt{D}}{2} \quad (w_1, z_1 \in \mathbb{Z}^{>0}) .$$

Since $N(\eta) = 1$, we have $w_1{}^2 = z_1{}^2 D + 4$ and, as a consequence, $w_1 > 2$. If we define w_j and z_j by

$$\frac{w_j + z_j\sqrt{D}}{2} = \eta^j ,$$

then

$$w_{j+1} = \frac{w_j w_1 + z_j z_1 D}{2} > w_j .$$

If $n > 1$, then

$$w_n \geq w_2 = \frac{w_1{}^2 + z_1{}^2 D}{2} = 2 + z_1{}^2 D \geq 4D + 2$$

when $z_1 \geq 2$. If $z_1 = 1$, then $D = w_1{}^2 - 4$. Since, in this case, $\lfloor \sqrt{D} \rfloor = w_1 - 1 \in \{Q, Q-1\}$, we can only have $R = -4$ or $R = 2Q - 3$. Now, $2Q - 3 \mid 4Q$ only if $Q = 1, 2$; thus, since $R \neq \pm 1, -4$, we cannot have $z_1 = 1$. For the remaining cases, then,

$$w_n \neq \frac{4Q^2 + 2R}{|R|} < 4D + 2 \quad (|R| > 1)$$

for $n \geq 2$ and, therefore, $n = 1$ and $\eta = \lambda$. Now, suppose $\eta = \epsilon^2$. In this case we have $N(\epsilon) = -1$, $\epsilon = (x_1 + y_1\sqrt{D})/2$ $(x_1, y_1 \in \mathbb{Z}^{>0})$, and $Dy_1{}^2 = x_1{}^2 + 4$. Also,

$$\frac{4Q^2 + 2R}{|R|} = \frac{x_1{}^2 + Dy_1{}^2}{2} = x_1{}^2 + 2 ,$$

$$\frac{4Q}{|R|} = x_1 y_1 .$$

It follows that

$$x_1{}^2 = 4Q^2 + 2R - 2|R| .$$

If $R = -|R|$, then $x_1{}^2 = 4Q^2 + 4R = 4D$, which is impossible because D is not square. If $R = |R|$, then $x_1 = 2Q$, $y_1 = 2/R$, and $R \in \{1,2\}$. If $R = 2$, then $y_1 = 1$ and

$$4Q^2 + 4 = D = Q^2 + R ,$$

which is impossible. If $R = 1$, then $y_1 = 2$ and $D = Q^2 + 1$. Thus, $\eta = \epsilon^2$ only when $D = Q^2 + 1$; in this case, $\epsilon = Q + \sqrt{D}$. Thus, if $R \notin \{-1, 1, -4\}$, we get $\lambda = \epsilon$ and, by Table 1.1,

$$t + u\sqrt{D} = \begin{cases} \lambda & \text{if } R \mid 2Q \\ \lambda^2 & \text{if } R \nmid 2Q \text{ and } R \mid 2Q^2 \\ \lambda^3 & \text{otherwise} . \end{cases}$$

On returning to Theorem 6.4, we observe that if we put $R = A'_{p-3}$, $S = B'_{p-3}$, and $T = A'_{p-2}$, then by (6.9) and (6.11) we get

$$TS - R^2 = (-1)^{p-1} . \tag{6.20}$$

From this it is clear that if $2 \nmid RS$, then $2 \mid T$ and $2 \nmid (-1)^{p-1}RS + mT$ for any integer m. Thus, there can be no positive, integral values of q_0 and D such that $\sqrt{D}$ satisfies (6.16). We now assume that $2 \mid RS$. When $2 \mid m$, then putting $X = m/2$, we get $q_0 = q_0(X) = XT + (-1)^{p-1}RS/2$ and

$$D = D_1(X) = a_1{}^2 X^2 + b_1 X + c_1 ,$$

where $a_1 = T$, $b_1 = (-1)^{p-1}RST + 2R$, and $c_1 = R^2 S^2/4 + (-1)^{p-1}S^2$. When $2 \nmid m$, we must have $2 \mid T$. If we put $X = (m-1)/2$, then $q_0 = q_0(X) = XT + (T + (-1)^{p-1}RS)/2$,

$$D = D_2(X) = a_2{}^2 X^2 + b_2 X + c_2 ,$$

where

$$a_2 = T,$$
$$b_2 = (-1)^{p-1}RST + T^2 + 2R ,$$
$$c_2 = \left[\frac{(-1)^{p-1}RS + T}{2}\right]^2 + R + (-1)^{p-1}S^2 .$$

Thus, if we completely specify

$$p, q_1, q_2 \ldots, q_{p-1} \text{ with } q_i = q_{p-i} \quad (i = 1, 2, \ldots, p-1) ,$$

then there can be at most two single parameter families $\{D_1(X)\}$, $\{D_2(X)\}$ such that $\sqrt{D_i(X)} = [q_0(X), q_1, q_2, \ldots, q_{p-1}, 2q_0(X)]$. Furthermore, because of (6.20), we also have

$$b_i{}^2 - 4a_i{}^2 c_i = 4(-1)^p \quad (i = 1 \text{ or } 2) . \tag{6.21}$$

6.3 Schinzel's Families

If ϕ is any quadratic irrational, we will denote by $l(\phi)$ the length of the period in the SCF expansion of ϕ. We would expect by (5.34) that the solution of (6.1) will tend to be small when $l(\sqrt{D})$ is. As mentioned in the previous section, the search for values for which this is the case goes back to at least the time of Euler. Since Euler's work, many results[8] concerning this problem have been produced. For example, as early as 1834, Stern[9] noted 42 different forms of D for which $l(\sqrt{D})$ is small. These include

$$D = m^2 n^2 + 2m , \quad D = (6n \pm 1)^2 + (8n \pm 1)^2 , \quad \text{and } D = (16n + 7)(25n + 11) .$$

By far the greater portion of these examples have the general form of $D = aX^2 + bX + c$, where a, b, and c are fixed integers and X is allowed to vary. However, it was not until 1961 that Schinzel[10] examined in a systematic manner the problem of finding univariate polynomials $D(X)$ in $\mathbb{Z}[X]$ for which $l(\sqrt{D(X)})$ is small. He first proved the following important result.

Theorem 6.6. *Let* $f(X) = a_0 X^k + a_1 X^{k-1} + \cdots + a_k, \in \mathbb{Z}[x]$ *with* $a_0 > 0$. *If*

1. $2 \nmid k$

or

2. $2 \mid k$ *and* a_0 *is not a perfect integral square,*

then for $n \in \mathbb{Z}$

$$\varlimsup_{n \to \infty} l\left(\sqrt{f(n)}\right) = \infty .$$

Notice that, unlike the case of $D(X) = D_1(X)$ or $D_2(X)$ in the previous section, we will have an infinitude of values of n such that $l(\sqrt{f(n)})$ is arbitrarily large when (1) or (2) is true. As the proof of this result is quite lengthy and somewhat outside the scope of this work, we suggest that the interested reader consult Schinzel's original paper.

Schinzel went on to prove another important result for the case of $k = 2$.

Theorem 6.7. *If* $f(X) = a^2 X^2 + bX + c$, *where* $a, b, c \in \mathbb{Z}$, *then for* $n \in \mathbb{Z}$,

$$\varlimsup_{n \to \infty} l\left(\sqrt{f(n)}\right) < \infty$$

if and only if

$$d \mid 4(2a^2, b)^2 , \tag{6.22}$$

where $d = b^2 - 4a^2 c$.

These two remarkable results completely solved the problem of when $l(\sqrt{f(n)})$ could be bounded independently of n when $f(X)$ is a quadratic polynomial in $\mathbb{Z}[X]$. In a second paper[11] he also provided a result which generalized Theorem 6.7 to polynomials in $\mathbb{Z}[X]$ of degree greater than 2. In this section we will confine our attention to the case of quadratic polynomials. Since Schinzel's seminal work on this case, further work has been done which provides more detail than Schinzel's techniques were able to produce. While his methods were very powerful, they were not constructive and therefore could not be used to exhibit explicit bounds on $l(\sqrt{f(n)})$. We will briefly describe here some more recent constructive approaches to Theorem 6.3.

Put $f(X) = a^2 X^2 + bX + c$, $d = b^2 - 4a^2 c$, and $\delta = d/(2a, b)^2$. We can write

$$f(X) = \frac{(2a^2 X + b)^2 - d}{(2a)^2} . \tag{6.23}$$

If we put $e = (2a^2, b)$, then Theorem 6.3 asserts that

$$\varlimsup_{n \to \infty} l\left(\sqrt{f(n)}\right) = \infty$$

if $d \mid 4e^2$. We will prove this here by making use of a clever idea of Louboutin.[12]

For any prime p we define $\nu_p(x)$ $(x \in \mathbb{Z})$ to be that value of m such that $p^m \parallel x$. Note that

$$\nu_p(xy) = \nu_p(x) + \nu_p(y) \quad (x, y \in \mathbb{Z}) .$$

We now observe that if $\bar{a} = 2a^2/e$ and $\bar{b} = b/e$, then

$$4(2a^2 X + b)^2 = 4e^2(\bar{a}X + \bar{b})^2 .$$

Also, if $d \nmid 4e^2$, there must exist a prime p such that $\nu_p(d) > \nu_p(4e^2)$. If $p \mid \bar{a}$, then $\nu_p(\bar{a}X + \bar{b}) = 0$ for all $X \in \mathbb{Z}$; if $p \nmid \bar{a}$, then there must exist an infinitude of values of $X \in \mathbb{Z}$ such that $(\bar{a}X + \bar{b}, p) = 1$ [i.e., $\nu_p(\bar{a}X + \bar{b}) = 0$]. In either case, then, there must exist infinitely many values of $X \in \mathbb{Z}$ such that

$$\nu_p(d) > \nu_p\left(4(2a^2 X + b^2)^2\right) = 2\nu_p\left(2(2a^2 X + b)\right) .$$

Let x be any one of these values of X and put $m = 2a^2 x + b$; then $\nu_p(d) > 2\nu_p(2m) \geq 2\nu_p(x)$, and by definition of δ, we have $\nu_p(\delta) > 2\nu_p(2a')$, where $a' = m/(2a, b)$. If we put $b' = 2a/(2a, b)$, by (6.23) we get

$$f(x) = \frac{a'^2 - \delta}{b'^2} ;$$

hence, we have

$$0 \leq \nu_p(f(x)) = 2\left(\nu_p(a') - \nu_p(b')\right) ,$$

which means that $\nu_p(a') \geq \nu_p(b')$. Since $(a', b') = 1$, we find that $\nu_p(b') = 0$ and $\nu_p(f(x)) = 2\nu_p(a')$.

We next put $P = 2a'$, $Q = \delta$, and $d' = P^2 - 4Q = 4b'^2 f(x)$. We have $\nu_p(Q) > 2\nu_p(P)$. We now require a simple lemma involving the Lucas functions introduced in §1.4.

Lemma 6.8. *For the Lucas functions $v_n(P,Q)$ and $u_n(P,Q)$, if $\nu_p(Q) > 2\nu_p(P)$, then*

$$\nu_p(v_n) = n\nu_p(P), \quad \nu_p(u_n) = (n-1)\nu_p(P) .$$

Proof. We see that since $v_1 = P$ and $u_1 = 1$, the result holds for $n = 1$. Also, since $v_2 = P^2 - 2Q$ and $u_2 = P$, we have $\nu_p(v_2) = 2\nu_p(P)$, $\nu_p(u_2) = \nu_p(P)$ and the result also holds for $n = 2$. It is easy to show that

$$v_{n+1} = Pv_n - Qv_{n-1}, \quad u_{n+1} = Pu_n - Qu_{n-1} ;$$

thus, we can prove the lemma by using induction on n. □

We are now able to prove Louboutin's version of Schinzel's result above.

Theorem 6.9. *If $d \nmid 4m^2$, then*

$$l\left(\sqrt{f(x)}\right) \geq 1 + 2\left\lfloor \frac{\log\sqrt{f(x)}}{\log|\delta|} \right\rfloor .$$

Proof. Since

$$v_k{}^2 - d'u_k{}^2 = 4Q^k ,$$

we have

$$v_k{}^2 - 4f(x)b'^2u_k{}^2 = 4\delta^k .$$

Put $N = \lfloor \log\sqrt{f(x)}/\log|\delta| \rfloor$ and observe that $\delta^N < \sqrt{f(x)}$. By Theorem 3.3, we see that $(v_k/2)/(b'u_k)$ must be some convergent C_{i_k} in the SCF expansion of $\sqrt{f(x)}$ whenever $1 \leq k \leq N$. If we put $g_k = (v_k/2, b'u_k)$, then $A_{i_k} = v_k/2g_k$ and $B_{i_k} = b'u_k/g_k$, and since

$$A_{i_k}^2 - f(x)B_{i_k}^2 = (-1)^{i_k+1}Q_{i_k+1} ,$$

we get $Q_{i_k+1} = |\delta|^k/g_k{}^2$. By definition of g_k and Lemma 6.8, we have

$$\nu_p(g_k) = \min\left\{\nu_p\left(\frac{v_k}{2}\right),\ \nu_p\left(b'u_k\right)\right\}$$
$$= \min\{k\nu_p(2a') - \nu_p(2),\ (k-1)\nu_p(2a')\}$$
$$= (k-1)\nu_p(2a') ;$$

hence,

$$\nu_p(Q_{i_k+1}) = k\left(\nu_p(\delta) - 2\nu_p(2a')\right) + 2\nu_p(2a') .$$

Since $\nu_p(\delta) > 2\nu_p(2a')$, it follows that all the values of Q_{i_k+1} are distinct for $k = 1, 2, \ldots, N$. Also, since $\nu_p(Q_{i_k+1}) > 2\nu_p(a') \geq \nu_p(2f(x))$, we have $Q_{i_k+1} \nmid 2f(x)$. Thus, the theorem now follows from Corollary 5.20.2. □

Since $f(x)$ becomes arbitrarily large as x does, Louboutin's result proves that $l(\sqrt{f(x)})$ is unbounded as x goes to infinity whenever $d \nmid 4e^2$. We now turn our attention to the case where $d \mid 4e^2$. Notice that by (6.21), this includes the cases discussed in §6.1. Also, it is easy to deduce that if D is of Richard-Degert type, then

$$D = a^2X^2 \pm a, \ a^2X^2 \pm 2a, \ \text{or} \ a^2X^2 \pm 4a$$

and such forms are, of course, covered by Schinzel's results.

It has been found convenient to deal with those $f(X)$ such that

$$f(X) = a^2X^2 + bX + c, \tag{6.24}$$

where $2 \mid a$ and $2 \mid b$. This is not really a restiction because we can divide the possible values of X into even $(X = 2W)$ and odd $(X = 2W + 1)$ integers and write

$$f(X) = g(W) = a'^2W^2 + b'W + c',$$

where if $X = 2W$,

$$a' = 2a, \quad b' = 2b, \quad c' = c,$$

or if $X = 2W + 1$,

$$a' = 2a, \quad b' = 4a^2 + 2b, \quad c' = a^2 + b + c.$$

In either case we get $d' = b'^2 - 4a'^2c' = 4d$ and $2e \mid (2a'^2, b')$; hence, $d' \mid 4(2a', b')^2$ whenever $d \mid 4e^2$. As we may always assume that $2 \mid b$, we will replace b by $2B$ and rewrite the polynomial in (6.24) as

$$D(X) = A^2X^2 + 2BX + C. \tag{6.25}$$

In this case, Schinzel's condition $d \mid 4e^2$ becomes $\Delta_0 \mid 4(A^2, B)^2$, where $\Delta_0 = B^2 - A^2C$.

If, in (6.25), $(A^2, 2B, C)$ is squarefree, then $D(X)$ is of Richard-Degert type when $C \leq 0$ or C is a perfect square.[13] If $C > 0$ and not a perfect square, then with no real loss of generality we may assume that $B > 0$ and $C > RT^2$, where $G = (A, B)$ and $\Delta_0/G^2 = RT^2$, with R squarefree. In this case it was shown[14] that $l(\sqrt{C}) \mid l(\sqrt{D(X)})$ and $l(\sqrt{D(X)})/l(\sqrt{C})$ is independent of the value of X. Later,[15] the condition that $(A^2, 2B, C)$ be squarefree was removed and the complete continued fraction expansion of $\sqrt{D(X)}$ was determined whenever $\Delta_0 \mid 4(A^2, B)^2$.

We put $|\Delta_0| = \Delta_1\Delta_2{}^2\Delta_4{}^4$, where Δ_1 and Δ_2 are squarefree; we set $A' = A/G$, $\Delta' = \Delta_2\Delta_4{}^2/G$, $\sigma = \Delta/|\Delta| = \text{sign}(\Delta)$, and

$$\eta = \begin{cases} 1 & \text{if } A \mid B \text{ and } \sigma = 1 \\ 0 & \text{otherwise}. \end{cases}$$

For integers $a \geq r \geq 0$, define an ordered set

$$S(a, r) = \begin{cases} \varnothing & \text{if } r = 0 \\ \{s_0, s_1, \ldots, s_{m-1}\} & \text{otherwise ,} \\ & \text{where } a/r = [s_0, s_1, \ldots, s_{m-1}] \\ & \text{with } (-1)^{m-1} = \sigma \, . \end{cases}$$

Finally, we define $\delta_i \in \{0, 1\}$ and $\delta_i \equiv i \pmod 2$. Note that $\delta_{i+1} = 1 - \delta_i$. Our subsequent results will be conditional on X being sufficiently large. Here, this means that

$$X > \frac{|\Delta|}{AG^2} + \frac{1}{2A} - \frac{B}{A^2} \text{ and } X > 1 + \frac{1}{A} - \frac{B}{A^2} \, . \tag{6.26}$$

We next assume that $X \equiv K \pmod{\Delta'}$, where $0 \leq K < \Delta'$, and write $B = Aq + r$, where $0 \leq r < A$.

The following complicated theorem completely characterizes the SCF expansion of $\sqrt{D(X)}$.

Theorem 6.10. *Suppose that $D(X) = A^2 X^2 + 2BX + C$ satisfies the Schinzel condition $\Delta_0 \mid 4(A^2, B)^2$ where X satisfies (6.26). Write $X = W\Delta' + K$ for some $W \geq 0$. Put $d_0 = \Delta'$ and $r_0 = (r + A\eta)/G$ and inductively define the following.*

For $i \geq 0$, define

$$S_i = S(A'\Delta'/d_i, r_i) \, , \quad d_{i+1} = (\Delta'/d_i, r_i) \, ,$$

where the parity of $|S_i|$ is even if $\sigma = -1$ and odd if $\sigma = 1$. Put $g_{i+1} = 0$ or $g_{i+1} = d_i$ according to whether $r_i = 0$ or $r_i = A'\Delta'/d_i$. If $r_i \not\equiv 0 \pmod{A'\Delta'/d_i}$, then choose $g_{i+1} \in \mathbb{Z}$ so that

$$\frac{g_{i+1}}{d_i} \frac{r_i}{d_{i+1}} \equiv \sigma \left(\bmod \, \frac{A'\Delta'}{d_i d_{i+1}} \right) \text{ and } 0 < g_{i+1} < \frac{A'\Delta'}{d_{i+1}} \, .$$

Also, set

$$q_{i+1}(W) = \frac{2AW d_{i+1}^2}{\Delta_1^{\delta_{i+1}} \Delta'} \left| \frac{2A^2 K + 2B - \Delta_1^{\delta_{i+1}} g_{i+1} \left(\frac{A}{A'} \right) \left(\frac{\Delta'}{d_{i+1}} \right) - A\eta - r}{A\Delta_1^{\delta_{i+1}} \left(\frac{\Delta'}{d_{i+1}} \right)^2} \right| \, .$$

Compute r_{i+1} such that

$$r_{i+1} \equiv \frac{d_{i+1}(2A^2 K + 2B)}{\Delta_1^{\delta_{i+1}} \Delta_2 \Delta_4^2} - g_{i+1} \left(\bmod \, \frac{A'\Delta'}{d_{i+1}} \right) \, ,$$

where $0 \leq r_{i+1} < A'\Delta'/d_{i+1}$ when $\sigma = -1$ and $0 < r_{i+1} \leq A'\delta'/d_{i+1}$ when $\sigma = 1$.

Then the simple continued fraction expansion of $\sqrt{D(X)}$ is given by

$$[AX + q - \eta, \overline{S_0, q_1(W), S_1, q_2(W), \ldots, S_{\kappa-1}, q_\kappa(W)}] \, ,$$

where κ is the least natural number such that

$$d_\kappa = \Delta' \text{ and } \Delta_1^{\delta_\kappa} = 1 \, .$$

The proof of this result and those in the next section is both lengthy and intricate; thus, for the sake of brevity we will not provide them in this chapter. The interested reader can find the proofs in the references cited in the notes.

We notice that S_0 is independent of the residue classes of X modulo Δ', but S_i with $i \geq 1$ depends on the residue classes. In particular, if $r = 0$ (i.e., $A \mid B$), then $S_0 = \varnothing$ or $\{1\}$ according to whether $\sigma = -1$ or 1. Also, when $r_i \equiv 0 \pmod{A'\Delta'/d_i}$ for $i \geq 1$, $S_i = \varnothing$ or $\{1\}$ according to whether $\sigma = -1$ or 1. Furthermore, if $r > 0$, then $r_i > 0$ for $i \geq 0$.

Example 6.11. [16] Consider $D(X) = 119^2 X^2 + 2(2205)X + 343$. We first look at $X \equiv 1 \pmod 7$, where $X > 1$ and list $q_i(0)$, S_i, g_i, d_i, and r_i:

$$q_0(0) = 137, \quad S_0 = \{1,1,8\}, \qquad\qquad d_0 = \Delta' = 7, \qquad\qquad r_0 = 9,$$
$$q_1(0) = 2, \quad S_1 = \{1,2,4,1,1,1,2\}, \quad d_1 = 1, \qquad g_1 = 14, \quad r_1 = 82,$$
$$q_2(0) = 5, \quad S_2 = \{4,3,1\}, \qquad\qquad d_2 = 1, \qquad g_2 = 45, \quad r_2 = 28,$$
$$q_3(0) = 136, \quad S_3 = \{1,3,4\}, \qquad\qquad d_3 = 7, \qquad g_3 = 13, \quad r_3 = 13,$$
$$q_4(0) = 5, \quad S_4 = \{2,1,1,1,4,2,1\}, \quad d_4 = 1, \qquad g_4 = 28, \quad r_4 = 45,$$
$$q_5(0) = 2, \quad S_5 = \{8,1,1\}, \qquad\qquad d_5 = 1, \qquad g_5 = 82, \quad r_5 = 14,$$
$$q_6(0) = 247, \quad S_6 = \{1,1,8\}, \qquad\qquad d_6 = \Delta' = 7, \qquad\qquad\qquad r_6 = 9.$$

Since $\delta_6 = 0$ by definition of δ_i, we have $\Delta_1{}^{\delta_6} = 1$. Also, since $d_6 = \Delta'$, the computation of S_i of the continued fraction expansion of $\sqrt{D(1)}$ is complete and $\kappa = 6$. Hence, when $X = 7W + 1$ for $W \geq 0$, we have

$$\sqrt{D(7W+1)} = [833W + 137, \overline{1,1,8,q_1(W),1,2,4,1,1,1,2,q_2(W),}$$
$$\overline{4,3,1,q_3(W),1,3,4,q_4(W),2,1,1,1,4,2,1,q_5(W),}$$
$$\overline{8,1,1,2(833W+137)}],$$

where $q_1(W) = q_5(W) = 17W + 2$, $q_2(W) = q_4(W) = 34W + 5$ and $q_3(W) = 833W + 136$.

Similarly, when $X = 7W + 2$ and $W \geq 0$,

$$\sqrt{D(7W+2)} = [833W + 256, \overline{1,1,8,17W+5,8,1,1,2(833W+256)}].$$

From Theorem 6.10 it is also possible to deduce[17] that when X satisfies (6.26),

$$\ell\left(\sqrt{D(X)}\right) \leq \frac{3\Delta' \log(\sqrt{5}A'\Delta')}{\log \tau},$$

where $\tau = (1 + \sqrt{5})/2$. Finally, if $\eta_D(X)$ is the fundamental unit of the order $[1, D(X)]$, then the SCF expansion of $\sqrt{D(X)}$ can be used to prove[18] that

$$\eta_D(X) = \left(\frac{A^2 X + B + A\sqrt{D(X)}}{\sqrt{|\Delta_0|}}\right)^{\kappa} \tag{6.27}$$

and $N(\eta_D(X)) = \sigma^\kappa$.

We now require some further notation. Let $f(x)$ and $g(x)$ be two functions such that $f : \mathbb{R}^{\geq 0} \to \mathbb{R}^{\geq 0}$ and $g : \mathbb{R}^{\geq 0} \to \mathbb{R}^{\geq 0}$. We write $f(x) \gg g(x)$ if there exists some fixed constant c, independent of x, such that

$$f(x) > cg(x)$$

for all sufficiently large x.

For example, if Δ (> 0) is the discriminant of an order $\mathcal{O}$, then $\epsilon_\Delta = (x + y\sqrt{\Delta})/2$ and

$$x^2 - \Delta y^2 = \pm 4 .$$

Thus, since $x^2 = \Delta y^2 \pm 4 \geq \Delta - 4$ and $y \geq 1$, we get

$$\epsilon_\Delta \geq \frac{\sqrt{\Delta - 4} + \sqrt{\Delta}}{2} > \frac{\sqrt{\Delta}}{2} . \qquad (6.28)$$

Hence, $R_\Delta \gg \log \Delta$.

We write $f(x) = \Theta(g(x))$ when $f(x) = O(g(x))$ and $f(x) \gg g(x)$. Now, the value of κ above depends only on the values of A, B, and C; thus, if $R(D(X))$ is the regulator of $[1, D(X)]$, then by (6.27), we have

$$R(D(X)) = \Theta(\log D(X)) .$$

We also see that we can easily solve the Pell equation for $D = D(X)$ as

$$t + u\sqrt{D} = \begin{cases} \eta & \text{when } \sigma^k = 1 \\ \eta^2 & \text{when } \sigma^k = -1 \end{cases}$$

for any X satisfying (6.26).

6.4 Creepers and Kreepers

We have seen how to obtain the fundamental unit of $[1, D(X)]$ when $D(X)$ is any of the Richard-Degert types: $D(X) = X^2 - 1$, $X^2 - 2$, and $X^2 - 4$. In 1969 Shanks[19] considered the case of $D(X) = X^2 - 8$, a form that does not obey the Schinzel condition. He observed that the fundamental unit of $[1, D(X)]$ tended to be large as the values of X increased, but for $D(X) = 4481 = 67^2 - 8$, it was uncharacteristically small. This caused him to investigate numbers of the form

$$S_n = (2^n + 3)^2 - 8 = (2^n + 1)^2 + 2^{n+2} .$$

He also stated[20] that if $\eta_n = t_n + u_n\sqrt{S_n}$, where (t_n, u_n) is the fundamental solution of (6.1) with $D = S_n$, then $\log \eta_n = 2n^2 \log 2 + O(n2^{-n})$. Later, Yamamoto[21] pointed out that in the SCF expansion of $(\sqrt{S_n} + 1)/2$, we get $P_0 = 1$, $Q_0 = 2$, $P_{2i-1} = 2^n + 1$, $Q_{2i-1} = 2^{n+2-i}$, $q_{2i-1} = 2^{i+1}$, $P_{2i} = 2^n - 1$, $Q_{2i} = 2^{i+1}$, and $q_{2i} = 2^{n-i}$. Thus,

$$\epsilon_n = \frac{\alpha\gamma^n}{2^n} ,$$

where $\alpha = (2^n + 1 + \sqrt{S_n})/2$ and $\gamma = (2^n + 3 + \sqrt{S_n})/2$. In this case, $l((\sqrt{S_n} + 1)/2) = 2n + 1$. This seems to be the first example ever found of a parametric family $\{S_n\}$ for which the fundamental unit of the corresponding order $\mathcal{O}_n$ can be easily predicted, even though (unlike the cases of our previous examples $\{D(X)\}$) the period length of the associated continued fraction becomes arbitrarily large.

Since 1969, a number of generalizations of Shanks' sequence have been described.[22] Indeed, it turns out that the first such sequences to be mentioned in the literature are those of Nyberg[23] in a little noticed work published in 1949. These sequences are given by

$$N_n = (x^n \pm (x - 1)/2)^2 + x .$$

Here, $l(\sqrt{N_n}) = 6n - 2$ for the negative sign and $l(\sqrt{N_n}) = 6n$ for the positive sign.

Before proceeding any further, it will be useful to introduce some more notation. If D (> 0) is the discriminant of some order $\mathcal{O}$, by our results in §4.2 we know that $D \equiv i \pmod 4$, where $i \in \{0, 1\}$. If we put

$$\omega = \begin{cases} \sqrt{D}/2 & \text{when } i = 0 \\ (\sqrt{D} + 1)/2 & \text{when } i = 1 , \end{cases}$$

then $\mathcal{O} = [1, \omega]$. Thus, if we are given a family $\{D_n\}$ of positive discriminants of the orders $\mathcal{O}_n$, then $\mathcal{O}_n = [1, \omega_n]$, where $D_n \equiv i_n \pmod 4$ and

$$\omega_n = \begin{cases} \sqrt{D_n} & \text{when } i_n = 0 \\ (1 + \sqrt{D_n})/2 & \text{when } i_n = 1 . \end{cases}$$

We now consider Hendy's[24] generalization of Shanks' sequence:

$$D_n = (qx^n + (x - 1)/q)^2 + 4x^n ,$$

where $x \equiv 1 \pmod q$ and $2 \nmid (qx^n + (x - 1)/q)$. In this case,

$$\omega_n = [(qx^n + (x - 1)/q + 1)/2, \overline{q_1, q_2, \ldots, q_p}] ,$$

where $p = l(\omega_n) = 2n + 1$, $q_{2j+1} = qx^j$, $q_{j+2} = qx^{n-j-1}$ $(j = 0, 1, 2, \ldots, n-1)$, and $q_p = 2q_0 - 1$. For the family $\{D_k\}$ given by

$$D_n = (qx^n - (x + 1)/4q)^2 + x^n ,$$

when $x \equiv 1 \pmod{4q}$, we get[25]

$$\sqrt{D_n} = [qx^n - (x + 1)/4q, \overline{q_1, q_2, \ldots, q_p}] ,$$

where

$$q_{3j+1} = 2qx^j - 1 , \quad q_{3j+2} = 1 , \quad q_{3j+3} = 2qx^{n-j-1} - 1$$

$(j = 0, 1, \ldots, n-1)$ and $p = l(\sqrt{D_n}) = 3n + 1$. Bernstein[26] regarded subsequences like $\{qx^j, qx^{n-j-1}\}$ in the period of ω_n above and $\{2qx^j - 1, 1, 2qx^{n-j-1} - 1\}$ in the period of $\sqrt{D_n}$ to be *cycles* within the respective continued fraction periods. By the length of these cycles he meant the number of items in the relevant subsequences. It seems that he was the first individual to examine the cycle structure of periodic continued fractions to any great extent. As a result of his investigations into the continued fraction expansion of $\sqrt{D_n}$ for certain parametric families $\{D_n\}$, he was able to produce a rather complicated definition of a cycle. He also found examples of such families for which there were cycles of length $2, 4, 5, 6, 8, 10, 11,$ and 12. Indeed, he expressed some surprise[27] that a cycle length as large as 12 could even exist. In fact, as was shown later,[28] it is possible to find a family $\{D_n\}$ for which the cycle length of the SCF expansion of ω_n can be any preselected integer. Such examples have $l(\omega_n) = an + b$ for some fixed $a, b \in \mathbb{Z}$. For example, if we put $q = 1450042921$,

$$D_n = (qx^{6n+1} + (x^{3n} - 1)/q) + 4x^{6n+1}$$

and $x \equiv 84498480 \pmod{q}$, then $l(\sqrt{D_n}) = 59n + 2$ and the SCF expansion of $\sqrt{D_n}$ has a cycle structure with cycle length 59.

However, it is possible to find families $\{D_n\}$ such that $l(\omega_n) = an + b$, where $a, b \in \mathbb{Q}$. In these examples, of course, the values of n must be selected from certain residue classes. For example, consider the rather peculiar family[29] given by

$$D_n = 1018^2 \cdot 1319011^{2n} + 132932744752 \cdot 1319077^n + 65290957^2 . \tag{6.29}$$

Here,

$$l(\omega_n) = \begin{cases} \dfrac{284}{5}n - \dfrac{94}{5} & \text{when } n \equiv 1 \pmod{60} \\[2ex] \dfrac{604}{15}n - \dfrac{8}{15} & \text{when } n \equiv 2 \pmod{60} \\[2ex] \dfrac{203}{5}n - \dfrac{29}{5} & \text{when } n \equiv 3 \pmod{60} \\[2ex] \dfrac{392}{3}n - 48 & \text{when } n \equiv 21 \pmod{60} \\[2ex] \dfrac{53}{3}n + \dfrac{13}{3} & \text{when } n \equiv 37 \pmod{60} . \end{cases}$$

In fact, there is a value for a and b for n in each residue class modulo 60. Kaplansky[30] named families like the $\{D_n\}$ given above *creepers*.

"Creepers"? Along with "sleepers" these are my silly nicknames. A sleeper is a family of continued fractions bounded in length; Schinzel pretty well wrapped these up. In a creeper the lengths go to infinity, but gently, forming one or more arithmetic progressions when sorted out into residue classes; there may be a waiting period before the arithmetic progressions begin.

We have already dealt with several types of sleepers in the previous section. In the remainder of this section we will discuss what (little) is known about creepers.

By selecting discriminants from families of sleepers, it is possible to form a sequence of discriminants with linear period length. These are now referred to as *beepers*.[31] As an example, we mention one discovered by Buck and Williams[32]:

$$D_n = (2F_{6n+1} + 1)^2 + 8F_{6n} + 4 \,,$$

where F_n is the nth Fibonacci number. Here, we get

$$\omega_n = [F_{6n+1} + 1, \overline{1, 1, 1, \ldots, 1, 2F_{6n+1} + 1}]$$

and $l(\omega_n) = 6n + 1$. It follows that since all the partial quotients are bounded except for the obvious ones, we must have $R(D_n) = O(\log D_n)$ by (5.34).

Madden[33] constructed sequences $\{D_n\}$ for which the SCF expansion of $\sqrt{D_n}$ possesses a slowly growing period length. These examples are distinct from our earlier examples of creepers because they are not polynomially parameterized in terms of x^n. However, they can be viewed as a selection of specific discriminants from the various families that we will discuss in sequel, just as beepers are specifically selected sleepers.

Let $\{f(X, n)\}$ be a set of polynomials in $\mathbb{Q}[X]$ parameterized by $n \in \mathbb{Z}$. In his discussion of a creeper, Kaplansky was thinking of an infinite family of discriminants $\{D_n\}$ such that for a fixed $x \in \mathbb{Z}$, we have $D_n = f(x, n)$ satisfying

$$l(\omega_n) = an + b \quad (a, b \in \mathbb{Q})$$

and

$$R(D_n) = \Theta((\log D_n)^2) \,.$$

He made several conjectures about creepers that are quadratic in x^n and suggested that each such creeper could be written as $\{D_n\}$, where

$$D_n = A^2 x^{2n} + Bx^n + C^2$$

and $A, B, C \in \mathbb{Q}$. Each of the corresponding orders $\mathcal{O}_n$ on which his conjectures were based contains a principal ideal with norm x^g, where g ($\in \mathbb{Z}^{>0}$) is independent of n.

Given this information, we define a Kaplansky creeper or *kreeper* as an infinite family of discriminants D_n such that the following hold:

1. $D_n = A^2 x^{2n} + Bx^n + C^2$, where $A, B, C \in \mathbb{Q}$, $x \in \mathbb{Z}^{>0}$.
2. $l(\omega_n) = an + b$, where $a, b \in \mathbb{Q}$ are independent of n.

3. In the principal ideal cycle of $\mathcal{O}_n$, there exists an ideal whose norm is x^g for some fixed g ($\in \mathbb{Z}^{>0}$) independent of n.

We point out that condition (3) implies that $R(D_n) = \Theta((\log D_n)^2)$; that is, every kreeper is a creeper.

Recently, the following two results,[34] which completely characterize kreepers, have been demonstrated.

Theorem 6.12. *Any kreeper D_n can be written as*

$$d^2 D_n = c^2 \left(\left(qrx^n + (mz^2 x^k - ly^2)/q \right)^2 + 4ly^2 rx^n \right) , \qquad (6.30)$$

where each term in the above equation is an element of $\mathbb{Z}$; the terms r, l, and m are squarefree, r and x are positive, and the following conditions hold:

$$\begin{array}{c} (qrx, mlzy) = 1 , \quad (qr, x) = 1 , \quad (mz, ly) = 1 , \\ q \mid mz^2 x^k - ly^2 , \quad c^2 rly^2 mz^2 \mid d^2 D_n . \end{array} \qquad (6.31)$$

Theorem 6.13. *Any sequence of discriminants given by (6.32) and satisfying the conditions (6.33) as above must in fact be a kreeper.*

For example, if we return to one of Nyberg's forms

$$N_n = (x^n + (x-1)/2)^2 + x ,$$

we see that

$$4N_n = (2x^n + x + 1)^2 - 8x^n .$$

Putting $D_n = 4N_n$, we have a family of discriminants $\{D_n\}$ which satisfy the conditions for a kreeper with $r = 2$, $k = d = c = z = y = m = 1$, and $l = -1$. Thus, from Theorem 6.13, we can deduce that $l(\sqrt{D_n}/2) = l(\sqrt{N_n}) = an + b$ $(a, b \in \mathbb{Q})$. Of course, we know in this case that $a = 6$ and $b = 0$. If we put $d = c = k = 1$, $q = 2$, $r = 509$, $m = 11$, $z = 3$, $l = 7$, $y = 5$, and $x = 1319011$, we get the D_n in (6.29).

For any kreeper $\{D_n\}$, it is also possible to determine ϵ_n, the fundamental unit of $\mathcal{O}_n$. If, for a fixed n, we put

$$\alpha = \frac{S_1 + d\sqrt{D_n}}{2d} , \qquad \beta = \frac{S_2 + d\sqrt{D_n}}{2d} ,$$

where

$$S_1 = cqrx^n + \frac{c(z^2 mx^k - y^2 l)}{q} , \qquad S_2 = cqrx^n - \frac{c(z^2 mx^k - y^2 l)}{q} ,$$

then

$$\epsilon_n = \lambda^w ,$$

where

$$\lambda = \frac{(\alpha\beta)^n \alpha^k}{\sqrt{|N(\alpha\beta)^n N(\alpha)^k|}} \ .$$

Here, the value of w depends only on a number which divides $2cyz$, and as it is a rank of appartition of this number in a certain Lucas function $u_n(P,Q)$, we know by Theorem 1.13, that it must be bounded above by $4cyz$. Also,

$$N(\alpha) = \frac{-c^2 r l y^2 x^n}{d^2} \ , \quad N(\beta) = \frac{-c^2 r m z^2 x^{n+k}}{d^2} \ ,$$

and

$$\alpha\beta = cqrx^n \gamma / d \ ,$$

where $\gamma = (S_3 + d\sqrt{D_n})/2d$ and $S_3 = cqrx^n + c(z^2 m x^k + y^2 l)$. Thus, we can write

$$\epsilon_n = \left(\frac{\gamma^n \alpha^k d^{n+k} q^n}{c^{n+k} y^{n+k} z^n \sqrt{r^k l^{n+k} m^n}} \right)^w \ ,$$

a generalization of the results of Shanks and Yamamoto when $D_n = S_n$. Of course, we can then find (t,u) for $D = D_n$ by using this result and Table 4.1. Notice that $R(D_n) = \Theta((\log D_n)^2)$. Very little is known about creepers that are not kreepers; they do, however, exist. For example,

$$D_n = (x^{2n+2} + x^{n+2} + x^n - 1)^2 + 4x^n$$

is a creeper, but not a kreeper. Indeed, the theory of creepers does not enjoy the same state of development as that of sleepers. We conclude this section with some unsolved problems concerning these objects.

Problem 1. Show that if $d^2 D_n = AX^{2n} + BX^n + C$, then $\{D_n\}$ is a creeper implies that A and C must be squares in $\mathbb{Q}$.

Problem 2. Show that if $d^2 D_n = A^2 X^2 + BX^n + C^2$ and $\{D_n\}$ is a creeper, then $\{D_n\}$ is a kreeper.

Problem 3. Show that there are no *jeepers*.[35] These are polynomially parameterized families $\{D_n\}$ which are not creepers such that $l(w_n)$ can be explicitly given and $R(D_n) = O((\log D_n)^k)$ for $k \geq 3$.

6.5 Yamamoto's Results

Let

$$\mathcal{O}_1, \mathcal{O}_2, \ldots, \mathcal{O}_k, \ldots \tag{6.32}$$

be a sequence of real quadratic orders with discriminants $\Delta_i = \Delta(\mathcal{O}_i)$ and $0 < \Delta_i < \Delta_{i+1}$, $i = 1, 2, \ldots, k, \ldots$. In the last two sections we have shown that such sequences of orders exist with $R(\Delta_i) = \Theta(\log \Delta_i)$ in the case of sleepers and $R(\Delta_i) = \Theta((\log \Delta_i)^2)$ in the case of kreepers. We also mentioned that no jeepers are known, but Yamamoto[36] has shown that we do have a sequence of orders such that $R(\Delta_i) \gg (\log \Delta_i)^3$. Notice that by (5.35), in this case we have

$$l(\omega_i) \gg (\log \Delta_i)^2 \, ,$$

where $\mathcal{O}_i = [1, \omega_i]$.

In this section we will present Yamamoto's proof of this result. We begin by assuming that for the sequence (6.32) there exists a set of n rational primes $\{p_1, p_2, \ldots, p_n\}$ such that in each $\mathcal{O}_i$, each of the primes p_j $(j = 1, 2, \ldots, n)$ splits into the product of two principal ideals $\mathfrak{p}_j$ and $\overline{\mathfrak{p}}_j$ (i.e., $(\Delta_i/p_j) = 1$). Let $\mathcal{O} = [1, \omega]$ by any one of these orders and put $\Delta = \Delta(\mathcal{O})$.

We now consider the set

$$S = \left\{ \prod_{j=1}^{n} \mathfrak{p}_j^{e_j} \overline{\mathfrak{p}}_j^{f_j} : e_j, f_j \geq 0; \ e_j f_j = 0; \ \prod_{i=1}^{n} p_j^{e_j + f_j} < \sqrt{\Delta}/2 \right\} .$$

We observe that all the ideals in S are distinct, principal, and primitive and by Theorem 5.9 they are also reduced. Thus, if $\mathcal{C}$ denotes the cycle of the reduced principal ideals of $\mathcal{O}$, then $\mathcal{C} \supseteq S$.

Let $\mathfrak{a} \in S$. Since $\mathfrak{a}$ is a reduced ideal of $\mathcal{O}$, we may assume from the results developed in §5.3 that $\mathfrak{a} = [N(\mathfrak{a}), b + \omega]$, where $(b + \omega)/N(\mathfrak{a}) > 1$ and $-1 < (b + \overline{\omega})/N(\mathfrak{a}) < 0$. Hence, $b + \omega > \omega - \overline{\omega} = \sqrt{\Delta}/2$ and

$$\frac{b + \omega}{N(\mathfrak{a})} > \frac{\sqrt{\Delta}/2}{N(\mathfrak{a})} \, .$$

Also, for some $m \in \mathbb{Z}^{>0}$ and $m \leq l(\omega)$, $(b + \omega)/N(\mathfrak{a}) = \phi_m$ in the SCF expansion of ω. Thus, if $\mathfrak{a}_k = [N(\mathfrak{a}_k), b_k + \omega] \in S$, by (5.34) we must have

$$\epsilon_\Delta \geq \prod_{k=1}^{|S|} \frac{b_k + \omega}{N(\mathfrak{a}_k)}$$

$$> {\prod}' \frac{\sqrt{\Delta}/2}{p_1^{e_1 + f_1} p_2^{e_2 + f_2} \cdots p_n^{e_n + f_n}} \, ,$$

where the product is over all the pairs $(e_k, f_k) \in \mathbb{Z}^2$ satisfying the following:

1. $e_k \geq 0$, $f_k \geq 0$, $e_k f_k = 0$ $(k = 1, 2, \ldots, n)$
2. $\sum_{k=1}^{n} (e_k + f_k) \log p_k < \log(\sqrt{\Delta}/2)$.

Let $\alpha_1, \alpha_2, \ldots, \alpha_n$, and B be positive reals. In order to estimate a lower bound on $R(\Delta) = \log \epsilon_\Delta$, we need some results concerning lattice points in the tetrahedron in

$$\mathbb{R}^n = \{(y_1, y_2, \ldots, y_n) : y_i \in \mathbb{R} \ (i = 1, 2, \ldots, n)\}$$

defined by $y_i \geq 0$ $(i = 1, 2, \ldots, n)$ and $\sum_{i=1}^{n} \alpha_i y_i = B$. Let $(x_1, x_2, \ldots, x_n) \in (\mathbb{Z}^{\geq 0})^n$ and let

$$T(n, B) = (x_1, x_2, \ldots, x_n)$$

denote the number of distinct lattice points such that

$$\sum_{i=1}^{n} \alpha_i x_i \le B .$$ (6.33)

Then

$$T(n, B) = \frac{B^n}{n!A} + O(B^{n-1}) ,$$ (6.34)

where $A = \prod_{i=1}^{n} \alpha_i$. Also, if we take the sum $S(n, B)$ of all the values of $\sum_{i=1}^{n} \alpha_i x_i$ satisfying (6.33), we have

$$S(n, B) = \frac{nB^{n+1}}{(n+1)!A} + O(B^n) .$$ (6.35)

Both of these results can be established analytically[37] or by induction[38] on n. For example, if we wish to prove (6.35), we first observe that if $n = 1$, it is clear that $S(1, B) = \sum_{x_1=0}^{\lfloor B/\alpha_1 \rfloor} \alpha_1 x_1 = B^2/2A + O(B)$ (here $A = \alpha_1$). We put $B(x_1) = B - \alpha_1 x_1$ and assume (6.35) holds for $n - 1$. Then

$$S(n, B) = \prod_{x_1=0}^{\lfloor B/\alpha_1 \rfloor} \sideset{}{'}\sum (\alpha_1 x_1 + \alpha_2 x_2 + \cdots + \alpha_n x_n) ,$$

where $\sum'$ denotes the sum over all $x_2, x_3, \ldots, x_n \in \mathbb{Z}^{\ge 0}$ such that

$$\alpha_2 x_2 + \alpha_3 x_3 + \cdots + \alpha_n x_n \le B(x_1) .$$

It follows that

$$S(n, B) = \prod_{x_1=0}^{\lfloor B/\alpha_1 \rfloor} [\alpha_1 x_1 T(n-1, B(x_1)) + S(n-1, B(x_1))] .$$

We can now use (6.34) and (6.35) with n replaced by $n - 1$ and B replaced by $B(x_1)$, the well-known result

$$\sum_{i=1}^{m} i^k = \frac{m^{k+1}}{k+1} + O(m^k) ,$$

and the combinatorial identity

$$\frac{n}{(n+1)!} = \frac{1}{(n-1)!} \sum_{i=0}^{n-1} \binom{n-1}{i} (-1)^i \frac{1}{i+2} + \frac{n-1}{n!} \sum_{i=0}^{n} \binom{n}{i} (-1)^i \frac{1}{i+1}$$ (6.36)

to prove (6.35). The identity (6.36) is easily derived ($p = 1$ and $p = 2$) from the identity[39]

$$\binom{n+p}{n}^{-1} = \sum_{i=0}^{n}(-1)^i\binom{n}{i}\frac{p}{i+p} .$$

If we now put $\alpha_i = \log p_i$ and $B = \log\sqrt{\Delta}/2$, we see that

$$\log\epsilon_\Delta = \log\prod{}' \frac{\sqrt{\Delta}/2}{p_1^{e_1+f_1}p_2^{e_2+f_2}\cdots p_n^{e_n+f_n}}$$
$$= 2^n BT(n,B) - 2^n S(n,B)$$
$$= \frac{2^n}{(n+1)!}\frac{B^{n+1}}{A} + O(B^n) .$$

Hence, for a fixed value of n,

$$R(\Delta) \gg (\log\Delta)^{n+1} .$$

Of course, in order to use this result we have to find the sequence (6.32) and the primes $\{p_1, p_2, \ldots, p_n\}$. Yamamoto considered

$$D_k = (p^k q + p + 1)^2 - 4p ,$$

where p and q are distinct primes and either $p = 2$ or p and q are both odd. We note that

$$D_k \equiv 1 \pmod 4 ,$$
$$D_k \equiv 1 \pmod p ,$$
$$D_k \equiv (p-1)^2 \pmod q .$$

If we put $\omega_k = (\sqrt{D_k}+1)/2$ and $\mathcal{O}_k = [1, \omega_k]$, then $D_k = \Delta(\mathcal{O}_k)$ and $(D_k/p) = (D_k/q) = 1$. Since

$$(p^k q + p + 1)^2 - D_k = 4p ,$$

in $\mathcal{O}_{\mathbb{K}}$, we have

$$(p) = \mathfrak{p}\bar{\mathfrak{p}}, \text{ where } \mathfrak{p} = \left(\frac{p^k q + p + 1 + \sqrt{D_k}}{2}\right) .$$

Also, $(q) = \mathfrak{q}\bar{\mathfrak{q}}$, and from

$$(p^k q + p - 1)^2 - D_k = -4p^k q ,$$

we see that either $\mathfrak{p}^k\mathfrak{q}$ or $\bar{\mathfrak{p}}^k\mathfrak{q}$ must be a principal ideal of $\mathcal{O}_{\mathbb{K}}$. However, since $\mathfrak{p}$ and $\bar{\mathfrak{p}}$ are both principal, this means that $\mathfrak{q}$ and, therefore, $\bar{\mathfrak{q}}$ are both principal. Thus, the conditions on our sequence of ideals is fulfilled for $n = 2$ and $\{p_1, p_2\} = \{p, q\}$. Consequently, for the family given by $\{D_k\}$, we have

$$R(D_k) \gg (\log D_k)^3 . \tag{6.37}$$

Table 6.1. Values for $l(\omega_n)$

n	$l(\omega_n)$		n	$l(\omega_n)$		n	$l(\omega_n)$
2	29		6	1801		10	20968
3	81		7	2216		11	61748
4	217		8	22206		12	566474
5	652		9	44776			

Thus, $\{D_k\}$ cannot be a sleeper or a creeper; furthermore, it seems to be very difficult to predict the value of $l(\omega_k)$. For example, when $p = 3$ and $q = 5$, we get the values in Table 6.1.

If D_k is squarefree, then $\mathcal{O}_\mathbb{K}$ is the maximal order of $\mathbb{Q}(\sqrt{D_k})$. If $2 \mid k$, we have

$$D_k = D_{2j} = (p^{2j}q + p + 1)^2 - 4p$$
$$= q^2 p^{4j} + 2q(p+1)p^{2j} + (p-1)^2 .$$

Suppose $f^2 \mid D_{2j}$ and we put $D_{2j} = f^2 D$; then

$$f^2 D = q^2 x^4 + 2q(p+1)x^2 + (p-1)^2 ,$$

where $x = p^j$. For a fixed D, this Diophantine equation can only have a finite number of solutions (f, x) by Siegel's Theorem.[40] Thus, $\mathbb{Q}(\sqrt{D_{2j}})$ represents an infinitude of distinct, real quadratic fields as $j = 1, 2, 3, \ldots$. It follows that there exists an infinite sequence of distinct, real quadratic fields

$$\mathbb{K}_1, \mathbb{K}_2, \ldots, \mathbb{K}_n, \ldots, \tag{6.38}$$

where $\mathbb{K}_i = \mathbb{Q}(\sqrt{D_i})$, such that

$$R_{\mathbb{K}_i} \gg (\log D_i)^3 . \tag{6.39}$$

This is the best result of this type currently known.[41] Indeed, in Chapter 9 we will provide reasons that suggest the existence of some sequence (6.38) for which

$$R_{\mathbb{K}_i} \gg D_i^{1/2 - \epsilon} ,$$

but the best rigorously proved result is still (6.39).

However, if we are allowed to deal with orders with non-unit conductors, we can do much better than (6.39). To show this we will make use of an idea of Lagarias.[42] We put $\mathcal{O} = [1, \omega]$, where $\Delta = \Delta(\mathcal{O}) = (\omega - \overline{\omega})^2$ is fixed, and $(\Delta, 3z) = 1$, where $\epsilon_\Delta = w + z\omega$ $(w, z \in \mathbb{Z})$. Clearly, such orders must exist, as $\epsilon_\Delta = M + \sqrt{\Delta} = M - 1 + 2\omega$ when $\Delta = M^2 + 1$ and $2 \mid M$. Select some odd f $(\in \mathbb{Z})$ such that each prime divisor of f must divide Δ. Define $\mathcal{O}_k = [1, f^k\omega]$.

Put $P = \epsilon_\Delta + \bar{\epsilon}_\Delta$, $Q = \epsilon_\Delta\bar{\epsilon}_\Delta = \pm 1$, and $d = (\epsilon_\Delta - \bar{\epsilon}_\Delta)^2 = z^2\Delta$ and let $u_n(P, Q)$ be the Lucas function in §1.4. If ϵ_k (> 1) is the fundamental unit of $\mathcal{O}_k$, then we know that

$$\epsilon_k = \epsilon_\Delta^m$$

and $m = \omega(f^k/g)$, where $g = (f^k, z) = 1$. From Proposition 6.1, it follows that if $p^n \parallel f$, then $p \mid \Delta$ and $p \mid d$. Thus, $\omega(p^{nk}) = p^{nk}$; therefore,

$$m = \omega(f^k) = f^k$$

and

$$R(f^{2k}\Delta) = f^k R_\Delta \;.$$

If we put $\Delta_k = f^{2k}\Delta = \Delta(\mathcal{O}_k)$, then $f^k = \sqrt{\Delta_k}/\sqrt{\Delta}$ and $R(\Delta_k) = \sqrt{\Delta_k}(R_\Delta/\sqrt{\Delta})$. Since $R_\Delta/\sqrt{\Delta}$ is fixed, we get

$$R(\Delta_k) \gg \sqrt{\Delta_k}$$

in this case.

Notes and References

[1] See [Dic19], Vol. II, pp. 354–355.

[2] This result was used in [Stö97] to address the following problem: Let $P = \{p_1, p_2, \ldots, p_t\}$ be a set of distinct primes and let Q be the set $Q = \{\prod_{i=1}^{t} p_i^{\alpha_i} : \alpha_i \geq 0\}$; find s such that s and $s + 1 \in Q$. This is another instance of a Diophantine problem in which the Pell equation arises somewhat unexpectedly. Several years later, Lehmer [Leh64] simplified Störmer's procedure and extended his results.

[3] See, for example, [Kra26].

[4] [Per57], Vol. I, Satz 3.17. There is also a version of this result for $(\sqrt{D} + 1)/2$ in Satz 3.34 of Vol. I. See also [Fri88], [HK91], and [Che03], pp. 67–72.

[5] These were named by Hasse in [Has65], p. 51 in honour of the earlier work done on them by Richaud [Ric66] and Degert [Deg58]. Much information on these forms can be found in [Mol95].

[6] [HK90], §3. These results can also be found as Theorem 3.2.1 in [Mol95].

[7] We use the ideas in [Deg58] for this.

[8] See Chapter XII of [Dic19] for a summary of many of these results.

[9] [Ste34].

[10] [Sch61].

[11] [Sch62].

[12] [Lou89]. Later, this work was extended by Farhane [Far94]. Dubois and Paysant-Le Roux [DPLR91] applied Louboutin's idea to the case of polynomials of degree greater than 2.

[13] See Theorem 2.2 of [vdPW99].

[14] [vdPW99]. See also [Che03], pp. 60–66, where a minor error in [vdPW99] is corrected.

[15] [CW05]; [Che03], Ch. 4.

[16] Additional examples can be found in [Che03], Appendix C.

[17] [Sha69] and [Sha71].

[18] This is an extension of a result of Stender [Ste79].

[19] [Sha69].

[20] [Sha71].

[21] [Yam70].

[22] [Hen74], [Ber76a], [Ber76b], [Azu84], [Azu87], [LR86], [Wil85a], [Lev88], [deM88], [HK89a], [MW92b], [MW92c], [Wil95a], [vdP94], and [MZ].

[23] [Nyb49].

[24] [Hen74].

[25] [Wil85a], p. 203.

[26] [Ber76a] and [Ber76b].

[27] [Ber76a], p. 446.

[28] [Wil00].

[29] [Pat03], pp. 97–100.

[30] [Kap98].

[31] These families are called beepers in honour of the beer won by van der Poorten. See [vdP99a]. Much further work on beepers has been done by Mollin and Cheng. See [MC02a], [MC02b], [MC04], and [MCG02].

[32] [BW94].

[33] [Mad01].

[34] [PvdPW07] and [Pat03].

[35]The reader will perhaps be relieved to learn that so far there are no families named peepers.

[36][Yam70].

[37]See [Yam70] or [HK89b].

[38]In the case of (6.34), see, for example, the work of Lehmer [Leh40]. Stronger estimates were found later by Granville [Gra91].

[39]See [Rio68], p. 47.

[40]See [Mor69], pp. 264, 268.

[41]Halter-Koch [HK89b] thought he had found a family $\{D_i\}$ for which $R_{\mathbb{K}_i} \gg (\log D_i)^4$, but, unfortunately, an error in his main theorem invalidates this result. For a brief discussion of this error, see [Pat03], p. 66. Nevertheless, [HK89b] contains some valuable insights that were very useful in establishing Theorem 6.12.

[42][Lag80a], Appendix A.

7

The Ideal Class Group

7.1 Introduction

As we have seen in §4.3 of Chapter 4, solutions of the Pell equation are closely related to the fundamental unit of a real quadratic field. In particular, the regulator, defined in §5.3 to be the logarithm of the fundamental unit, is the number of bits of a fundamental solution up to a small constant factor. In Chapter 6 we have seen that the size of the regulator can vary greatly; indeed, there are many special families of discriminants that yield very small regulators.

In Chapter 8 we will show that the size of the regulator is closely linked to another invariant of a quadratic field, namely the ideal class number. This is the number of elements in the ideal class group, a finite abelian group consisting of equivalence classes of integral ideals.

In this chapter we introduce the ideal class group and class number of a quadratic field. A number of properties concerning the size and structure of the class group will be presented, including classical results on its 2-Sylow subgroup and heuristics on its odd part due to Cohen and Lenstra. Finally, we will introduce Shanks' idea of the infrastructure of an ideal equivalence class and show how it can be used to improve dramatically the speed of computing the regulator. In order to do this, we will discuss the notion of the distance of a reduced ideal in a given cycle and develop some useful results concerning it.

We begin by defining the ideal class group. Recall from §4.5 that ideal equivalence, as defined in Proposition 4.28, partitions the set of invertible $\mathcal{O}_\Delta$-ideals into equivalence classes.

Definition 7.1. *The set of equivalence classes of invertible $\mathcal{O}_\Delta$-ideals of a quadratic order is a finite abelian group called the* ideal class group, *denoted by Cl_Δ.*

Definition 7.2. *The* class number *of $\mathcal{O}_\Delta$ is defined as the order of Cl_Δ and is denoted by h_Δ.*

When we wish to emphasize that $\mathcal{O}_\Delta$ is maximal, we write $Cl_\mathbb{K}$ and $h_\mathbb{K}$ to denote the class group and class number of the quadratic field $\mathbb{K} = \mathbb{Q}(\sqrt{\Delta})$, respectively.

The group operation, written multiplicatively, is given by $[\mathfrak{a}][\mathfrak{b}] = [\mathfrak{ab}]$; that is, an ideal representative of the product of two equivalence classes is computed by selecting ideal representatives of each equivalence class and multiplying them. It is fairly straightforward to verify the group axioms and commutativity from the definition of ideal multiplication (Definition 4.27) and the discussion on invertibility in §4.5. In particular, closure, associativity, and commutativity follow easily from Definition 4.27, the identity element of the class group is the principal class $[\mathcal{O}_\Delta]$, and the inverse of $[\mathfrak{a}]$ is $[\overline{\mathfrak{a}}]$. Finiteness[1] follows from the facts that each ideal equivalence class contains at least one reduced ideal and that the coefficients of a reduced ideal are bounded (see §5.1).

The ideal class group can also be defined using fractional ideals.

Definition 7.3. *A fractional ideal of $\mathcal{O}_\Delta$ is a subset $\mathfrak{a}$ of $\mathbb{Q}(\sqrt{\Delta})$ such that $d\mathfrak{a}$ is an integral ideal of $\mathcal{O}_\Delta$ for some $d \in \mathbb{Z}^{>0}$. The minimal such d is called the denominator of $\mathfrak{a}$, denoted by $d(\mathfrak{a})$.*

Recall that $\mathcal{O}_\Delta$ acts as the identity for multiplication of integral ideals. However, most integral ideals do not have integral ideal inverses in the sense that, given an integral ideal $\mathfrak{b}$, there is in general no integral ideal $\mathfrak{b}^{-1}$ such that $\mathfrak{b}\mathfrak{b}^{-1} = \mathcal{O}_\Delta$. As a result, the set of integral ideals is not a group under ideal multiplication, Fractional ideals, on the other hand, do have inverses in this sense. To show this, we first generalize the notion of ideal norm to fractional ideals.

Definition 7.4. *The norm $N(\mathfrak{a})$ of a fractional ideal $\mathfrak{a}$ of $\mathcal{O}_\Delta$ is defined to be the unique non-negative rational generator of the $\mathbb{Z}$-module containing the norms of all the elements in $\mathfrak{a}$. We have $N(\mathfrak{a}) = N(\mathfrak{b})/d(\mathfrak{a})^2$, where $\mathfrak{a} = \mathfrak{b}/d(\mathfrak{a})$ for an integral $\mathcal{O}_\Delta$-ideal $\mathfrak{b}$.*

By Corollary 4.32.1, $\mathfrak{b}\overline{\mathfrak{b}} = (N(\mathfrak{b}))$ for any integral ideal $\mathfrak{b}$. It follows that for any fractional ideal $\mathfrak{a}$ we have

$$\mathfrak{a}\left(\frac{\overline{\mathfrak{a}}}{N(\mathfrak{a})}\right) = \mathcal{O}_\Delta \, ,$$

and $\mathfrak{a}^{-1} = \overline{\mathfrak{a}}/N(\mathfrak{a})$. Thus, it is easy to see that $\mathcal{I}_\Delta$, the set of all invertible fractional ideals of $\mathcal{O}_\Delta$, is an abelian group under ideal multiplication with identity $\mathcal{O}_\Delta$. Furthermore, the set of principal ideals $\mathcal{P}_\Delta$ is a subgroup of $\mathcal{I}_\Delta$, so $\mathcal{I}_\Delta/\mathcal{P}_\Delta$ is a finite abelian group. Clearly, this is another description of the ideal class group as defined in Definition 7.1. This formulation allows us to represent equivalence classes using fractional ideals as well as integral ideals, and it allows us to write $[\mathfrak{a}]^{-1} = [\mathfrak{a}^{-1}]$ when $\mathfrak{a}$ is fractional. These generalizations will be useful when describing the algorithms for computing class groups and regulators in Chapter 13.

In practice, reduced ideals are most often used as representatives of ideal classes, allowing NUCOMP, as presented in §5.4, to be used as the group operation in the class group. As shown in §5.1, the coefficients of reduced ideals are bounded by $\sqrt{\Delta}$, so using reduced ideals provides the additional computational advantage that the operands used are relatively small in terms of Δ. Furthermore, in imaginary quadratic fields, the results in §5.2 show that using reduced ideal representatives results in a unique representative of each non-ambiguous ideal equivalence class and two reduced representatives for ambiguous classes.[2] Thus, testing whether two ideal classes are equal can be done efficiently by comparing their reduced representatives. Reduced ideal representatives are also used in the real case, but, as shown in §5.3, the number of reduced equivalent ideals can be approximately the same size as the regulator. Thus, testing equality of ideal equivalence classes can be difficult if the regulator is large, a fact exploited by a number of public-key cryptosystems (see Chapter 14).

Adding the additional restriction that $N(\kappa) > 0$ in Proposition 4.28 yields the notion of *narrow* or *proper* equivalence. The class group and class number can also be defined with respect to narrow equivalence, resulting in the narrow class group, denoted by Cl_Δ^+, and the narrow class number, denoted by h_Δ^+. The usual class group and class number, as defined above, are sometimes referred to as the wide class group and wide class number. As all elements of an imaginary quadratic field have positive norm, narrow and wide equivalence are the same in this case, so $h_\Delta = h_\Delta^+$. In real quadratic fields, $h_\Delta^+ = h_\Delta$ if and only if $N(\epsilon_\Delta) = -1$, because two equivalent ideals $\mathfrak{a}$ and $\mathfrak{b}$ are also properly equivalent in this case; if $\mathfrak{a} = \kappa \mathfrak{b}$ and $N(\kappa) < 0$, then $\mathfrak{a} = \epsilon_\Delta \kappa \mathfrak{b}$ and $N(\epsilon_\Delta \kappa) > 0$. If $N(\epsilon_\Delta) = 1$, then improperly equivalent ideals form two distinct proper equivalence classes, so we have $h_\Delta^+ = 2h_\Delta$.

As mentioned earlier, the class number and regulator are closely linked, so in order to understand the behaviour of the regulator and, hence, fundamental solutions of Pell's equation,[3] it is necessary to understand how large the class number can be as a function of the discriminant Δ. Results in this direction will be presented in detail in Chapter 9; we summarize a few highlights here.

For negative discriminants, solutions of Pell's equation are trivial, so bounds on the class number in terms of Δ alone can be obtained. For example, Cohen[4] stated that

$$h_\Delta < \frac{1}{\pi}\sqrt{|\Delta|}\log|\Delta| \quad \text{if } \Delta < -4.$$

Siegel[5] proved asymptotic lower and upper bounds on h_Δ. His lower bound, that h_Δ tends to infinity as fast as $|\Delta|^{1/2-\epsilon}$ for every $\epsilon > 0$, immediately suggests that there should be only finitely many imaginary quadratic fields with a given class number. Gauss[6] was the first to conjecture that there are only nine imaginary quadratic fields with class number 1 and that their discriminants are -1, -2, -3, -7, -11, -19, -43, -67, and -163. Baker and Stark,[7] using methods anticipated by Heegner,[8] proved that this list is in fact complete. The most recent work in this area is by Watkins,[9] in which he listed all imaginary quadratic fields with class number $h_\Delta \leq 100$, a dramatic improvement over the previous bound of class numbers less than or equal to 7 and odd values up to 23 due to Arno, Robinson, Wheeler, and Wagner.[10] Although significant progress has been made in terms of solving Gauss' class number problems for imaginary quadratic fields, as we will see below, there is still much that is unknown about this fundamental invariant.

The situation is much different in real quadratic fields, due to the fact that the class number and regulator are so closely related. Indeed, analogues of the bounds for the imaginary case typically bound the product $h_\Delta R_\Delta$ as opposed to h_Δ. For example, Hua[11] showed that

$$h_\Delta R_\Delta < (1 + (1/2)\log\Delta)\sqrt{\Delta} \quad \text{if } \Delta > 0 , \tag{7.1}$$

and Siegel's bound implies that $h_\Delta R_\Delta$ tends to infinity as fast as $\Delta^{1/2-\epsilon}$ for every $\epsilon > 0$. In Chapters 8 and 9 we will explore the connection between h_Δ and R_Δ more closely in the context of the analytic class number formula and present even tighter bounds whose correctness depends on the extended Riemann hypothesis[12].

For families of fields with small regulators, one would expect exhaustive lists of fields with a given class group similar to the imaginary case. Mollin and Williams[13] have produced a series of papers on this subject, in which, for example, methods similar to those used in the imaginary case were used to enumerate all real quadratic fields with continued fraction period length of ω less than 25 that have class number 1 or 2. However, in general, the regulator is usually large, meaning that the class number can be as small as 1 no matter what size the discriminant is. Assuming that the regulator is large for most real quadratic fields, one might conjecture that the class number is 1 infinitely often, and, in fact, Gauss posed the following:

Conjecture 7.5 (Gauss Conjecture[14]). There are infinitely many real quadratic fields with class number 1.

There is a great deal of numerical evidence supporting this conjecture,[15] but to date a proof remains elusive.

In addition to the size of the class group, one can ask various questions concerning divisibility properties of the class number and the structure of the class group. By the structure of the class group, we mean its canonical decomposition as a direct product of cyclic subgroups; that is,

$$Cl_\Delta \cong C(m_1) \times \cdots \times C(m_s) ,$$

where the positive integers $m_1, \ldots, m_s$ satisfy $m_1 \geq 1$, $m_{j+1} \mid m_j$ for $1 \leq j < s$, and $C(x)$ denotes the cyclic group of order x. We call the m_i's the *elementary divisors* of Cl_Δ. The *p-rank* of the class group is the number of elementary divisors that are divisible by p—in other words, it is the number of cyclic factors of the p-Sylow subgroup.

In §7.2, we will discuss the odd part of the class group, the subgroup of ideal classes whose orders are odd. Although very little can be proved in this case, we have a good heuristic understanding thanks to the work of Cohen and Lenstra, which we will describe. The situation with the even part of the class group is much better understood. There are many proved results, several of which are classical and were known to Gauss. We will give an overview of these results in §7.3.

7.2 The Cohen-Lenstra Heuristics

Let Cl_Δ be the class group of $\mathbb{Q}(\sqrt{\Delta})$ and let Cl_Δ^* be the odd part of Cl_Δ, the subgroup of ideal classes with odd order. Cohen and Lenstra[16] presented some heuristics on the distribution and structure of various Cl_Δ^* and divisibility properties of $h_\Delta^* = |Cl_\Delta^*|$. For example, the probability that $h_\Delta^* = 1$ is conjectured to be approximately 0.75446 if $\Delta > 0$, a figure supported by extensive computations.[17] As another example, Cl_Δ^* is conjectured to be cyclic over 97% of the time when $\Delta < 0$, and this is also supported by extensive numerical evidence.[18]

7.2.1 Imaginary Quadratic Fields

The fundamental heuristic assumption used by Cohen and Lenstra came from the observation that tables of class groups of imaginary quadratic fields available at the time did indeed indicate that Cl_Δ^* was cyclic much more frequently than non-cyclic. As the automorphism group of a cyclic group is smaller than that of any other abelian group of the same size, Cohen and Lenstra hypothesized that when computing probabilities of occurrences of particular abelian groups G, each isomorphism class of G should have a weight associated with it equal to $1/|\text{Aut}(G)|$, where $|\text{Aut}(G)|$ is the order of the automorphism group of G. The Cohen-Lenstra heuristics for imaginary quadratic fields are derived from this heuristic assumption.

Define

$$w(n) = \sum_{\substack{G \\ |G|=n}} \frac{1}{|\text{Aut}(G)|} \,,$$

where the sum is taken over all abelian groups of order n up to isomorphism. The main idea for deriving heuristics on class groups is that $w(n)$ is the sum of weights of all groups of order n. The probability of occurrence of a particular odd order group or set of groups is computed by dividing the sum of weights for the target groups by the sum of the weights of all finite abelian groups of odd order, $\sum_{d\,\text{odd}}^{\infty} w(d)$.

As an example, we now derive the probability that Cl_Δ^* is cyclic when $\Delta < 0$ using the above heuristic assumption. The number of automorphisms of a cyclic group of order n is equal to the number of generators, $\phi(n)$. Thus, we need to compute

$$\Pr(Cl_\Delta^* \text{ is cyclic}) = \frac{\sum_{\substack{d \geq 1 \\ d \text{ odd}}} 1/\phi(d)}{\sum_{\substack{d \geq 1 \\ d \text{ odd}}} w(d)} . \tag{7.2}$$

We begin with the numerator of (7.2). Define

$$\Phi(x) = \sum_{\substack{n \leq x \\ n \text{ odd}}} \frac{1}{\phi(n)} .$$

From Landau[19] we have that

$$\Phi(x) = E_1' \log x + E_2' + O\left(\frac{\log x}{x}\right) , \tag{7.3}$$

where $E_1' = 315\,\zeta(3)/6\pi^4$ and E_2' are explicit constants and $\zeta(s)$ denotes the Riemann zeta function, which we will discuss in more detail in §9.3.

We now derive an analogous result for the denominator of (7.2). If we define

$$W(x) = \sum_{\substack{n \leq x \\ n \text{ odd}}} w(n) , \tag{7.4}$$

then we have the following theorem.

Theorem 7.6. *There exist constants E_1 and E_2 such that*

$$W(x) = E_1 \log x + E_2 + O\left(\frac{\log x}{x}\right) ,$$

where $E_1 = 1/2C = \eta_\infty(2)C_\infty$,

$$\eta_k(p) = \prod_{i=1}^{k} (1 - 1/p^i) ,$$

and

$$C_\infty = \prod_{i=2}^{\infty} \zeta(i) \approx 2.294856589 .$$

Before proving this theorem, we first describe how to approximate $\eta_\infty(2)$ efficiently. An identity of Euler [20] allows us to express functions of the form

$$\frac{1}{(1 - ax)(1 - ax^2) \cdots (1 - ax^i) \cdots}$$

as the more rapidly converging

$$1 + \frac{ax}{1-x} + \frac{a^2 x^2}{(1-x)(1-x^2)} + \frac{a^3 x^3}{(1-x)(1-x^2)(1-x^3)} + \cdots . \qquad (7.5)$$

At $a = 1$ and $x = 1/2$ this gives us

$$\frac{1}{\eta_\infty(2)} = 1 + \frac{1}{2(1-1/2)} + \frac{1}{4(1-1/2)(1-1/4)}$$
$$+ \frac{1}{8(1-1/2)(1-1/4)(1-1/8)} + \cdots$$

and we can compute

$$\eta_\infty(2) = 0.288788095\ldots,$$

an approximation correct to nine digits. This gives us

$$C = \frac{1}{2\eta_\infty(2)C_\infty} \approx 0.754458173$$

in the statement of the theorem.

Proof (of Theorem 7.6). We first need some additional results[21] on $w(n)$. We have

$$w(n) = \prod_{p^\alpha || n} \left(p^\alpha \left(1 - \frac{1}{p}\right)\left(1 - \frac{1}{p^2}\right) \cdots \left(1 - \frac{1}{p^\alpha}\right) \right)^{-1}, \qquad (7.6)$$

$$\sum_{d|n} w(d) = nw(n), \qquad (7.7)$$

and

$$\frac{A}{\phi(n)} < w(n) < \frac{B}{\phi(n)}, \qquad (7.8)$$

where A and B are constants such that $0 < A < B$. It is also known[22] that

$$\sum_{d>x} \frac{1}{d\,\phi(d)} = O\left(\frac{1}{x}\right), \qquad (7.9)$$

$$\sum_{d>x} \frac{\log d}{d\,\phi(d)} = O\left(\frac{\log x}{x}\right), \qquad (7.10)$$

$$\sum_{\substack{d\leq x \\ (d,l)=1}} \frac{1}{\phi(d)} = O(\log x), \qquad (7.11)$$

and

$$\sum_{\substack{d\leq x \\ (d,l)=1}} \frac{1}{d} = \frac{\phi(l)}{l} \log x + E_0(l) + O\left(\frac{1}{x}\right), \qquad (7.12)$$

where $E_0(l)$ is a constant which only depends on l; for example, $E_0(1) = \gamma$. From (7.8), it follows that

$$\sum_{d>x} \frac{w(d)}{d} = O\left(\frac{1}{x}\right) , \tag{7.13}$$

$$\sum_{d>x} \frac{w(d)\log d}{d} = O\left(\frac{\log x}{x}\right) , \tag{7.14}$$

and

$$\sum_{\substack{d\leq x \\ (d,l)=1}} w(d) = O(\log x) . \tag{7.15}$$

Now, let

$$\Omega(x,l) = \sum_{\substack{n\leq x \\ (n,l)=1}} w(n) .$$

We apply standard analytic methods similar to those employed by Landau. From (7.7) we have

$$\Omega(x,l) = \sum_{\substack{n\leq x \\ (n,l)=1}} \left(\frac{1}{n}\sum_{d\mid n} w(d)\right)$$

$$= \sum_{\substack{d\leq x \\ (d,l)=1}} \left(w(d)\sum_{\substack{n\leq x, d\mid n \\ (n,l)=1}} \frac{1}{n}\right)$$

$$= \sum_{\substack{d\leq x \\ (d,l)=1}} \left(\frac{w(d)}{d}\sum_{\substack{m\leq x/d \\ (m,l)=1}} \frac{1}{m}\right) ,$$

and from (7.9) we have

$$\Omega(x,l) = \sum_{\substack{d\leq x \\ (d,l)=1}} \frac{w(d)}{d}\left(\frac{\phi(l)}{l}\log\frac{x}{d} + E_0(l) + O\left(\frac{d}{x}\right)\right)$$

$$= \sum_{\substack{d\leq x \\ (d,l)=1}} \frac{w(d)}{d}\frac{\phi(l)}{l}\log\frac{x}{d} + E_0(l)\sum_{\substack{d\leq x \\ (d,l)=1}} \frac{w(d)}{d} + O\left(\frac{1}{x}\sum_{\substack{d\leq x \\ (d,l)=1}} w(d)\right) .$$

By (7.15) we can set

$$O\left(\frac{1}{x}\sum_{\substack{d\leq x\\(d,l)=1}}w(d)\right)=O\left(\frac{\log x}{x}\right),$$

so we now have

$$\Omega(x,l)=\frac{\phi(l)}{l}\log x\sum_{\substack{d\leq x\\(d,l)=1}}\frac{w(d)}{d}-\frac{\phi(l)}{l}\sum_{\substack{d\leq x\\(d,l)=1}}\frac{w(d)\log d}{d}$$

$$+E_0(l)\left(\sum_{\substack{d\geq 1\\(d,l)=1}}\frac{w(d)}{d}-\sum_{\substack{d>x\\(d,l)=1}}\frac{w(d)}{d}\right)+O\left(\frac{\log x}{x}\right)$$

$$=\frac{\phi(l)}{l}\log x\left(\sum_{\substack{d\geq 1\\(d,l)=1}}\frac{w(d)}{d}-\sum_{\substack{d>x\\(d,l)=1}}\frac{w(d)}{d}\right)$$

$$-\frac{\phi(l)}{l}\left(\sum_{\substack{d\geq 1\\(d,l)=1}}\frac{w(d)\log d}{d}-\sum_{\substack{d>x\\(d,l)=1}}\frac{w(d)\log d}{d}\right)$$

$$+E_0(l)\left(\sum_{\substack{d\geq 1\\(d,l)=1}}\frac{w(d)}{d}-\sum_{\substack{d>x\\(d,l)=1}}\frac{w(d)}{d}\right)+O\left(\frac{\log x}{x}\right).$$

Since $\phi(l)/l$ and $E_0(l)$ are constants depending only on l, we can use (7.13) and (7.14) to obtain

$$\Omega(x,l)=\frac{\phi(l)}{l}\log x\sum_{\substack{d\geq 1\\(d,l)=1}}\frac{w(d)}{d}$$

$$-\frac{\phi(l)}{l}\sum_{\substack{d\geq 1\\(d,l)=1}}\frac{w(d)\log d}{d}+E_0(l)\sum_{\substack{d\geq 1\\(d,l)=1}}\frac{w(d)}{d}+O\left(\frac{\log x}{x}\right).$$

Put

$$E_2(l)=E_0(l)\sum_{\substack{d\geq 1\\(d,l)=1}}\frac{w(d)}{d}-\frac{\phi(l)}{l}\sum_{\substack{d\geq 1\\(d,l)=1}}\frac{w(d)\log d}{d}$$

and

$$E_1(l)=\frac{\phi(l)}{l}\sum_{\substack{d\geq 1\\(d,l)=1}}\frac{w(d)}{d}.$$

Then we have

$$\Omega(x, l) = E_1(l) \log x + E_2(l) + O\left(\frac{\log x}{x}\right).$$

Now, $W(x) = \Omega(x, 2)$, so if we set $E_1 = E_1(2)$ and $E_2 = E_2(2)$, then

$$W(x) = E_1 \log x + E_2 + O\left(\frac{\log x}{x}\right).$$

Finally, we show that $E_1 = 1/2C$. We have

$$E_1 = E_1(2) = \frac{\phi(2)}{2} \sum_{\substack{d \geq 1 \\ d \text{ odd}}} \frac{w(d)}{d}$$

and need to evaluate the sum in this expression. Since $w(d)/d$ is multiplicative and $\sum w(d)/d$ converges[23], we can apply the Euler product formula, yielding

$$\sum_{d=1}^{\infty} \frac{w(d)}{d} = \prod_p \left(\sum_{i=0}^{\infty} \frac{w(p^i)}{p^i}\right) = \left(\sum_{i=0}^{\infty} \frac{w(2^i)}{2^i}\right)\left(\sum_{\substack{d=1 \\ d \text{ odd}}}^{\infty} \frac{w(d)}{d}\right).$$

We have[24]

$$\sum_{n \geq 1} \frac{w(n)}{n^s} = \zeta(s+1)\zeta(s+2)\cdots$$

for $s > 0$, so

$$\sum_{d=1}^{\infty} \frac{w(d)}{d} = \zeta(2)\zeta(3) \cdots = C_\infty.$$

Also, from (7.6) we have

$$\sum_{i=0}^{\infty} \frac{w(2^i)}{2^i} = \sum_{i=0}^{\infty} \frac{1}{2^{2i}(1 - 1/2)(1 - 1/2^2) \cdots (1 - 1/2^i)}.$$

Applying Euler's identity (7.5) with $a = 1/2$ and $x = 1/2$ gives us

$$\sum_{i=0}^{\infty} \frac{w(2^i)}{2^i} = \frac{1}{(1 - 1/2^2)(1 - 1/2^3)(1 - 1/2^4) \cdots} = \frac{1}{2\eta_\infty(2)}.$$

Hence,

$$\sum_{\substack{d \geq 1 \\ d \text{ odd}}} \frac{w(d)}{d} = 2\eta_\infty(2)C_\infty \tag{7.16}$$

and we obtain

$$E_1 = \frac{\phi(2)}{2} \sum_{\substack{d \geq 1 \\ d \text{ odd}}} \frac{w(d)}{d} = \frac{1}{2}\left(2\eta_\infty(2)C_\infty\right) = \frac{1}{2C}$$

as required. $\qquad\qquad\qquad\qquad\qquad\qquad\qquad\qquad\qquad\qquad\qquad\qquad\quad$ $\square$

In order to estimate $\Pr(Cl^*_\Delta \text{ is cyclic})$, we write (7.2) as

$$\Pr(Cl^*_\Delta \text{ is cyclic}) = \lim_{x \to \infty} \frac{\Phi(x)}{W(x)} .$$

Applying (7.3) and Theorem 7.6 yields

$$
\begin{aligned}
\Pr(Cl^*_\Delta \text{ is cyclic}) &= \lim_{x \to \infty} \frac{E_1' \log x + E_2' + O\left(\frac{\log x}{x}\right)}{E_1 \log x + E_2 + O\left(\frac{\log x}{x}\right)} \\
&= \lim_{x \to \infty} \frac{E_1' + E_2'/\log x + O(1/x)}{E_1 + E_2/\log x + O(1/x)} \\
&= \frac{E_1'}{E_1} \\
&= \frac{315\,\zeta(3)}{6\pi^4 \eta_\infty(2) C_\infty} .
\end{aligned}
$$

Approximations of $\eta_\infty(2)$ and C_∞ are given above, and the function $\zeta(3)$ is easy to approximate efficiently as it converges rapidly using its Euler product representation. Hence, we obtain

$$\Pr(Cl^*_\Delta \text{ is cyclic}) \approx 0.977575 .$$

We now summarize some of the main heuristics obtained using this method.

Conjecture 7.7 (Cohen-Lenstra Heuristics for Imaginary Quadratic Fields[25]). Let Cl^*_Δ be the odd part of the class group of an imaginary quadratic field.

1. The probability that Cl^*_Δ is cyclic is

$$\Pr(Cl^*_\Delta \text{ is cyclic}) = \frac{315\,\zeta(3)}{6\pi^4 \eta_\infty(2) C_\infty} \approx 0.977575 .$$

2. The probability that an odd prime $p \mid h_\Delta$ is

$$\Pr(p \mid h_\Delta) = 1 - \eta_\infty(p) \approx 1/p + 1/p^2 .$$

For example, $\Pr(3 \mid h_\Delta) \approx 0.43987$, $\Pr(5 \mid h_\Delta) \approx 0.23967$, and $\Pr(7 \mid h_\Delta) \approx 0.16320$, and these are significantly higher than the expected value of $1/p$ for divisibility of a random integer by p.

3. The probability that the p-rank of Cl_Δ is equal to r for an odd prime p is

$$\Pr(p\text{-rank} = r) = \frac{\eta_\infty(p)}{p^{r^2} \eta_r(p)^2} .$$

All the numerical evidence produced to date supports the validity of Conjecture 7.7. For example, Jacobson, Ramachandran, and Williams[26] have computed the class group for each imaginary quadratic field with absolute value of the discriminant less than 10^{11} unconditionally (i.e., without having to assume the extended Riemann hypothesis). The tabulation was extended to $2 \cdot 10^{11}$ in Ramachandran's M.Sc. thesis,[27] and in both cases, the results completely support the Cohen-Lenstra heuristics.[28] Some of the computational techniques used to produce these results will be discussed in Chapter 10.

7.2.2 Real Quadratic Fields

For real quadratic fields, we assign the weight $w(n)/n$ to those Cl_Δ^* with $|Cl_\Delta^*| = n$. Dividing by n can be justified by the fact that the ideal classes of real quadratic fields partition themselves into h_Δ distinct cycles of reduced ideals and that each of these cycles exhibits a group-like structure called the infrastructure (see §7.4). Now, $w(n)$ is the sum of the weights of all groups G of order n up to isomorphism, so since we are considering real quadratic field class groups, we divide $w(n)$ by n because we do not want to count the n "groups" corresponding to the infrastructures of the n ideal classes.[29]

As an example, we show how to derive the heuristic result on the probability that $h^* = l$. If the weight $w(n)/n$ is assigned to those Cl_Δ^* with $h^* = n$, then we would expect that

$$\Pr(h^* = l) = \frac{w(l)/l}{\sum_{\substack{d \geq 1 \\ d \text{ odd}}} w(d)/d} \ .$$

This is simply the weight assigned to groups with $|G^*| = l$ divided by the sum of the weights of groups of all odd orders. Applying (7.16) yields

$$\Pr(h_\Delta^* = l) = C \frac{w(l)}{l} \ .$$

This gives us $\Pr(h_\Delta^* = 1) = 0.754458173\ldots$, $\Pr(h_\Delta^* = 3) = 0.125743028\ldots$, and $\Pr(h_\Delta^* = 5) = 0.037722908\ldots$ for the first few values of l.

We also derive the probability that Cl_Δ^* is cyclic for real quadratic fields, because, to the best of our knowledge, this result does not appear elsewhere in the literature. In this case, we need to compute

$$\Pr(Cl_\Delta^* \text{ is cyclic}) = \frac{\sum_{\substack{d \geq 1 \\ d \text{ odd}}} 1/\phi(d)d}{\sum_{\substack{d \geq 1 \\ d \text{ odd}}} w(d)/d} \ . \tag{7.17}$$

The denominator of (7.17) is evaluated in (7.16), so we only need to evaluate the numerator. Because $\phi(d)d$ is multiplicative and $\sum 1/(\phi(d)d)$ converges,[30] we can write

$$\sum_{d=1}^{\infty} \frac{1}{\phi(d)d} = \prod_p \left(\sum_{i=0}^{\infty} \frac{1}{\phi(p^i)p^i} \right) = \left(\sum_{i=0}^{\infty} \frac{1}{\phi(2^i)2^i} \right) \left(\sum_{\substack{d \geq 1 \\ d \text{ odd}}} \frac{1}{\phi(d)d} \right) . \quad (7.18)$$

Using the fact that $\phi(p^i) = p^i(1 - 1/p)$ when $i \geq 1$, we obtain

$$\sum_{i=0}^{\infty} \frac{1}{\phi(2^i)2^i} = 1 + \sum_{i=1}^{\infty} \frac{1}{4^i(1 - 1/2)} = 1 + 2 \sum_{i=1}^{\infty} \frac{1}{4^i} = \frac{5}{3} \quad (7.19)$$

and

$$\sum_{d=1}^{\infty} \frac{1}{\phi(d)d} = \prod_p \left(1 + \sum_{i=1}^{\infty} \frac{1}{p^{2i}(1 - 1/p)} \right)$$

$$= \prod_p \left(1 + (1 - 1/p)^{-1} \sum_{i=1}^{\infty} \frac{1}{p^{2i}} \right) \quad (7.20)$$

$$= \prod_p \left(\frac{p^3 - p^2 + 1}{(p - 1)(p^2 - 1)} \right) .$$

Thus, from (7.16), (7.19), and (7.20) we obtain

$$\Pr(Cl_\Delta^* \text{ is cyclic}) = \frac{3}{10\eta_\infty(2)C_\infty} \prod_p \left(\frac{p^3 - p^2 + 1}{(p - 1)(p^2 - 1)} \right) \approx 0.997631 .$$

We now summarize some of the main heuristics obtained using this method.

Conjecture 7.8 (Cohen-Lenstra Heuristics for Real Quadratic Fields[31]). Let Cl_Δ^* be the odd part of the class group of a real quadratic field.

1. The probability that Cl_Δ^* is cyclic is

$$\Pr(Cl_\Delta^* \text{ is cyclic}) = \frac{3}{10\eta_\infty(2)C_\infty} \prod_p \left(\frac{p^3 - p^2 + 1}{(p - 1)(p^2 - 1)} \right) \approx 0.997631 .$$

2. The probability that an odd prime $p \mid h_\Delta$ is

$$\Pr(p \mid h_\Delta) = 1 - \frac{\eta_\infty(p)}{1 - 1/p} .$$

For example, $\Pr(3 \mid h_\Delta) \approx 0.15981$, $\Pr(5 \mid h_\Delta) \approx 0.049584$, and $\Pr(7 \mid h_\Delta) \approx 0.023739$.

3. The probability that the p-rank of Cl_Δ is equal to r for an odd prime p is

$$\Pr(p\text{-rank} = r) = \frac{\eta_\infty(p)}{p^{r(r+1)}\eta_r(p)\eta_{r+1}(p)} .$$

4. The probability that $h_\Delta^* = l$ for an odd integer l is

$$\Pr(h_\Delta^* = l) = C\frac{w(l)}{l}.$$

For example, $\Pr(h_\Delta^* = 1) \approx 0.754458173$, $\Pr(h_\Delta^* = 3) \approx 0.125743028$, and $\Pr(h_\Delta^* = 5) \approx 0.037722908$.

5. (Hooley's Conjecture[32]) Let h_p be the class number of the field $\mathbb{Q}(\sqrt{p})$, where p is a prime congruent to 1 (mod 4). Then
   a) $\Pr(h_p > x) = 1/2x$ as $x \to \infty$,
   b) $\sum_{p \le x} h_p \sim x/8$.

Using the same heuristic assumptions, Jacobson, Lukes, and Williams[33] were able to show that part (a) of the last conjecture holds for all h_Δ^*, thereby providing even further heuristic evidence in support of the hypothesis that h_Δ^* is almost always small.

Conjecture 7.9. The probability that $h_\Delta^* > x$ is given by

$$\Pr(h_\Delta^* > x) = \frac{1}{2x} + O\left(\frac{\log x}{x^2}\right).$$

This result is derived as follows. Assuming the conjecture on $\Pr(h_\Delta^* = l)$, we would expect $\Pr(h_\Delta^* > x)$ to be given by

$$\Pr(h_\Delta^* > x) = C \sum_{\substack{j > x \\ j \text{ odd}}} \frac{w(j)}{j}. \tag{7.21}$$

Thus, we need to estimate the sum in this expression.

Theorem 7.10. *For n and x both odd,*

$$\sum_{n > x} \frac{w(n)}{n} = \frac{E_1}{x} + O\left(\frac{\log x}{x^2}\right).$$

Proof. Consider the sum

$$\sum_{\substack{n > 2r+1 \\ n \text{ odd}}} \frac{w(n)}{n}. \tag{7.22}$$

Using the fact

$$w(2j+1) = W(2j+1) - W(2j-1),$$

we apply partial summation[34] to (7.22) and obtain

$$\sum_{\substack{n>2r+1 \\ n \text{ odd}}} \frac{w(n)}{n} = \sum_{\substack{n>2r+1 \\ n \text{ odd}}} \frac{1}{n} \left(W(n) - W(n-2)\right)$$

$$= -\frac{W(2r+1)}{2r+3} + \sum_{\substack{n>2r+1 \\ n \text{ odd}}} W(n) \left(\frac{1}{n} - \frac{1}{n+2}\right)$$

$$= -\frac{W(2r+1)}{2r+3} + 2 \sum_{\substack{n>2r+1 \\ n \text{ odd}}} \frac{W(n)}{n(n+2)} .$$

Now, consider

$$2 \sum_{\substack{n>2r+1 \\ n \text{ odd}}} \frac{W(n)}{n(n+2)} .$$

By Theorem 7.6 we have

$$\sum_{\substack{n>2r+1 \\ n \text{ odd}}} \frac{W(n)}{n(n+2)} = E_1 \sum_{\substack{n>2r+1 \\ n \text{ odd}}} \frac{\log n}{n(n+2)} + E_2 \sum_{\substack{n>2r+1 \\ n \text{ odd}}} \frac{1}{n(n+2)} \tag{7.23}$$

$$+ O\left(\sum_{n>x} \frac{\log n}{n^2(n+2)}\right) .$$

We know that

$$\sum_{n>x} \frac{\log n}{n^3} = O\left(\int_x^\infty \frac{\log t}{t^3} \, dt\right) = O\left(\frac{\log x}{x^2}\right) ,$$

so

$$\sum_{n>x} \frac{\log n}{n^2(n+2)} = O\left(\frac{\log x}{x^2}\right) . \tag{7.24}$$

Since

$$\sum_{\substack{n>2r+1 \\ n \text{ odd}}} \frac{1}{n(n+2)} = \frac{1}{2(2r+3)} ,$$

we can write

$$\sum_{\substack{n>x \\ n,\, x \text{ odd}}} \frac{1}{n(n+2)} = \frac{1}{2x} + O\left(\frac{1}{x^2}\right) . \tag{7.25}$$

From

$$\frac{\log n}{n^2} - \frac{2\log n}{n(n+2)} = \frac{2\log n}{n^2(n+2)}$$

we have by (7.24)

$$\sum_{\substack{n>x \\ n,\, x \text{ odd}}} \frac{\log n}{n^2} - 2 \sum_{\substack{n>x \\ n,\, x \text{ odd}}} \frac{\log n}{n(n+2)} = \sum_{\substack{n>x \\ n,\, x \text{ odd}}} \frac{\log n}{n^2(n+2)} = O\left(\frac{\log x}{x^2}\right) .$$

Thus,

$$2 \sum_{\substack{n>x \\ n,\, x \text{ odd}}} \frac{\log n}{n(n+2)} = \sum_{\substack{n>x \\ n,\, x \text{ odd}}} \frac{\log n}{n^2} + O\left(\frac{\log x}{x^2}\right)$$

$$= \int_{\frac{x+1}{2}}^{\infty} \frac{\log(2t+1)}{(2t+1)^2}\, dt + O\left(\frac{\log x}{x^2}\right),$$

and by evaluating the integral, we obtain

$$2 \sum_{\substack{n>x \\ n,\, x \text{ odd}}} \frac{\log n}{n(n+2)} = \frac{1}{2}\left(\frac{\log x+2}{x+2} + \frac{1}{x+2}\right) + O\left(\frac{\log x}{x^2}\right). \qquad (7.26)$$

Substituting (7.24), (7.25), and (7.26) into (7.23) yields

$$2 \sum_{\substack{n>2r+1 \\ n \text{ odd}}} \frac{W(n)}{n(n+2)} = E_1 \frac{\log x + 2}{x+2} + \frac{E_1}{x+2} + \frac{E_2}{x} + O\left(\frac{\log x}{x^2}\right). \qquad (7.27)$$

We now apply Theorem 7.6 to

$$-\frac{W(2r+1)}{2r+3} = -\frac{W(x)}{x+2}$$

and combine the result with (7.27), giving us

$$\sum_{\substack{n>x \\ n,\, x \text{ odd}}} \frac{w(n)}{n} = \frac{-E_1 \log x}{x+2} + \frac{E_1 \log x + 2}{x+2} + \frac{E_1}{x+2} - \frac{E_2}{x+2}$$

$$+ \frac{E_2}{x} + O\left(\frac{\log x}{x^2}\right)$$

$$= \frac{E_1(\log(x+2) - \log x + 1)}{x+2} + O\left(\frac{\log x}{x^2}\right)$$

$$= \frac{E_1}{x+2} + \frac{E_1 \log(1+2/x)}{x+2} + O\left(\frac{\log x}{x^2}\right).$$

Finally, since

$$\log\left(1 + \frac{2}{x}\right) = O\left(\frac{1}{x}\right),$$

we have

$$\sum_{\substack{n>x \\ n,\, x \text{ odd}}} \frac{w(n)}{n} = \frac{E_1}{x} + O\left(\frac{\log x}{x^2}\right)$$

and the theorem is proved.   □

Proof (of Conjecture 7.9). Combining (7.21) and Theorem 7.10 gives us

$$C \sum_{\substack{j>x \\ j \text{ odd}}} \frac{w(j)}{j} = C \left(\frac{E_1}{x} + O\left(\frac{\log x}{x^2} \right) \right)$$

$$= C \frac{1}{2Cx} + O\left(\frac{\log x}{x^2} \right)$$

$$= \frac{1}{2x} + O\left(\frac{\log x}{x^2} \right),$$

as required. $\qquad\qquad\qquad\qquad\qquad\qquad\qquad\qquad\qquad\qquad\qquad\qquad$ □

As in the imaginary case, there is a fair amount of numerical evidence supporting the truth of these conjectures. The most extensive tabulations to date[35] consist of statistics for class groups of real quadratic fields with $\Delta < 10^9$ and class numbers for all primes $p \equiv 1 \pmod 4$ and $p < 2 \cdot 10^{11}$ (computed primarily to provide evidence in support of Hooley's conjecture). In both cases, the numerical data completely support the conjectures. However, the correctness of the class groups computed in these cases is conditional on the Extended Riemann Hypothesis. As we will see in Chapter 10, it is still an open problem to construct an unconditionally correct tabulation algorithm whose efficiency is close to that used in the imaginary case.

7.3 The 2-Sylow Subgroup

In contrast to the odd part of the class group, there is a great deal that can be proved about the even part. For example, given only the prime factorization of the discriminant, it is easy to determine whether the class number is even or odd and to determine the 2-rank of the class group.

Much of what is known about the 2-Sylow subgroup begins with the connection between divisors of the discriminant and ambiguous ideals. Recall that an ambiguous $\mathcal{O}_\Delta$-ideal $\mathfrak{a}$ satisfies $\mathfrak{a} = \bar{\mathfrak{a}}$. We refer to an ideal equivalence class $[\mathfrak{a}]$ satisfying $[\mathfrak{a}]^2 = [\mathcal{O}_\Delta]$ as an ambiguous class and, in the real case, the cycle of reduced ideals in an ambiguous class as an ambiguous cycle. As an ambiguous class is equal to its own inverse in Cl_Δ, it must contain both $\mathfrak{a}$ and $\bar{\mathfrak{a}}$. Thus, if an equivalence class contains an ambiguous ideal, then that class is ambiguous. The converse is not true; if $\Delta > 0$ and $N(\epsilon_\Delta) > 0$, then there may exist at most one ambiguous class without an ambiguous ideal.[36]

Suppose that $\mathfrak{a} = [Q/r, (P + \sqrt{D})/r]$ is a primitive ambiguous ideal. We show that every ambiguous ideal corresponds to a divisor of the discriminant Δ.

Theorem 7.11. *If $\mathfrak{a} = [Q/r, (P + \sqrt{D})/r]$ with $N(\mathfrak{a}) = Q/r$ is a primitive ambiguous ideal, then $N(\mathfrak{a}) \mid \Delta$.*

Proof. As $\mathfrak{a}$ is assumed to be ambiguous, by definition $\mathfrak{a} = \bar{\mathfrak{a}}$, so both $(P + \sqrt{D})/r$ and $(P - \sqrt{D})/r \in \mathfrak{a}$. Thus, $(P + \sqrt{D})/r + (P - \sqrt{D})/r = 2P/r \in \mathfrak{a}$, and because Q/r is the least positive integer in $\mathfrak{a}$, we have that $Q \mid 2P$. Since $Q \mid D - P^2$, we also have $Q \mid 2D - 2P^2$, which, together with the fact that $Q \mid 2P$, implies $Q \mid 2D$ and $Q/r \mid 2D/r$. If $r = 1$, we have $Q/r \mid 2D$, and thus, $Q/r \mid 4D = \Delta$. If $r = 2$, $Q/r \mid D = \Delta$. In either case we have $Q/r \mid \Delta$.   □

In fact, if Δ is fundamental and has k distinct prime divisors, then there are precisely 2^k ambiguous ideals in $\mathcal{O}_\Delta$. This follows from the fact that, by Theorem 4.44, every ambiguous ideal has a unique factorization into prime ideals. As an ambiguous ideal is equal to its conjugate, every prime ideal divisor of an ambiguous ideal must be ambiguous itself and thus ramified. The ramified prime ideals are those whose norms divide Δ, so if Δ has k distinct prime factors, then there are exactly 2^k ambiguous ideals.

To determine the 2-rank of the class group, it suffices to determine how many independent ambiguous ideal classes are formed from the ambiguous ideals and the one possible additional ambiguous class. In the imaginary case, two ambiguous ideals, one reduced and one non-reduced, exist in each ambiguous class, yielding 2^{k-1} independent ambiguous classes,[37] where k is the number of distinct prime divisors of Δ. This implies that $k - 1$ is the 2-rank of the class group, that 2^{k-1} divides the class number, and the class number of an imaginary quadratic field is odd if and only if $|\Delta|$ is a prime congruent to 3 (mod 4).

In the real case, it is possible to have multiple reduced ambiguous ideals in the same equivalence class. Barrucand and Cohn[38] stated that the 2-rank of the class group is $k - 1$ if all odd prime divisors of Δ are congruent to 1 (mod 4) and $k - 2$ otherwise, indicating that either 2^{k-1} or 2^{k-2} divides the class number. As a result,[39] we see that the class number of a real quadratic field $\mathbb{Q}(\sqrt{D})$ is odd if and only if $D = 2$, p, $2q_1$, or q_1q_2, where p is any odd prime and q_1 and q_2 are primes congruent to 3 (mod 4).

The connection between ambiguous ideals and divisors of the discriminant has not been overlooked by those studying the integer factorization problem. Clearly, being able to produce an ambiguous ideal different from $\mathcal{O}_\Delta$ yields a non-trivial divisor of Δ, allowing one partially to factor Δ. A variety of methods have been proposed.[40] We briefly mention a few noteworthy examples here.

Using imaginary quadratic fields, one approach to factor a positive integer N is to compute the class number of the imaginary quadratic field of discriminant $-N$ or $-4N$, after which raising a random element in the class group to the power of $h_\Delta/2$ will likely yield a non-trivial ambiguous ideal. Shanks' CLASNO algorithm[41] uses this approach and was part of his motivation for developing the baby-step giant-step method to speed the computation of class numbers.[42] Schoof[43] presented an improved version of this approach and proved that it factors N in time $O(N^{1/5+\epsilon})$ under the extended Riemann hypothesis (ERH). The algorithm Schoof used to compute h_Δ will be described in Chapter 10. Schnorr and Lenstra presented a Monte Carlo algorithm for factoring N, also known as SPAR,[44] which attempts to find an ambiguous ideal by raising a random element in the class group to a highly-composite exponent consisting of a large number of small odd primes raised to large powers. If the random element has smooth order (i.e., the largest prime divisor of the order is small), then the result will be an element whose order is a power of 2, from which an ambiguous ideal can be found quickly by repeated squaring. Assuming a certain smoothness conjecture on the class number, this algorithm has subexponential complexity in $\log N$. Finally, the index-calculus method was applied to construct ambiguous ideals by Seysen[45] and Lenstra and Pomerance.[46] The latter algorithm is noteworthy in that it is the first factoring algorithm for which subexponential running time, namely $O(\exp((1+o(1))(\log N \log \log N)^{1/2}))$, could be rigorously proved without any assumptions, including the ERH.

Computing the regulator often allows one to factor the discriminant as well. If the norm of the fundamental unit is 1, then there is a single ambiguous ideal not equal to $\mathcal{O}_\Delta$ halfway through the cycle of reduced principal ideals.[47] Given the regulator, this ideal can be found efficiently using Shanks' infrastructure techniques (see §7.4). Schoof[48] described how this idea can be used to factor N deterministically in time $O(N^{1/5+\epsilon})$ under the ERH. Shanks[49] also developed a factoring algorithm based on finding ambiguous ideals in real quadratic fields called SQUFOF, which stands for "SQUare FOrm Factorization." The idea is to search for a reduced principal ideal whose norm is a square. Then the square root can be computed and will lie in an ambiguous class. Using Shanks' infrastructure techniques, an ambiguous ideal in that class can be located rapidly. Gower and Wagstaff[50] have recently provided a detailed description and analysis of SQUFOF. Although it has exponential complexity, $O(N^{1/4+\epsilon})$, it is a very simple, elegant algorithm that performs extremely well for integers between 10 and 18 decimal digits.

The fact that integer factorization reduces to computing the class number of a quadratic field has not been overlooked by cryptographers, as this can be interpreted as evidence that the discrete logarithm problem (computing the order of an element is a special case of the discrete logarithm problem) in the class group is at least as hard as factoring. One can draw an analogous conclusion based on the fact that, due to Schoof's algorithm, integer factorization also reduces to computing the regulator. We will revisit these issues in Chapter 14, in which we discuss cryptosystems based on quadratic fields.

Although the subgroup of order 2 elements can be determined completely as described above, the picture is not yet complete for higher powers of 2. Genus theory[51] yields some results; for example, it was known to Gauss[52] that the 4-rank is equal to the 2-rank if the ambiguous ideals are all squares and that there are simple quadratic residuosity conditions on the prime divisors of the discriminant that indicate when this is the case. Other than that, it is usually conjectured that the 2-Sylow subgroup of Cl_Δ^2, the subgroup of squares, behaves like the odd part of the class group and that, in particular, the same heuristic assumptions and results from the Cohen-Lenstra heuristics should hold. Numerical evidence[53] does indeed support his hypothesis. Recently, Fouvry and Klüners[54] were able to prove that the predicted probability for the 2-rank of Cl_Δ^2 (i.e., the 4-rank of Cl_Δ) to be equal to r is exactly as predicted by the Cohen-Lenstra heuristics, one of the few rigorously proved results in this area.

7.4 Infrastructure

Recall from Chapter 5 that each ideal equivalence class of a real quadratic field contains a finite set of reduced ideals. Shanks[55] noticed that the set of reduced ideals of any ideal class has certain additional group-like structural properties, which he called the *infrastructure*. This discovery allowed Shanks to adapt his baby-step giant-step method to this setting, resulting in significantly improved algorithms for computing the regulator, testing ideal equivalence classes for equality, and, therefore, for computing class groups. Most modern methods for computing the regulator and class number rely explicitly on Shanks' discovery. In this section, we describe the infrastructure, how to compute in it, and how it can be used to compute the regulator.

We will confine our discussion of infrastructure here to the cycle of reduced ideals in the principal class

$$\mathcal{C} = \left\{ \mathfrak{a}_1 = (1), \mathfrak{a}_2 = \rho(1), \mathfrak{a}_3 = \rho^2(1), \ldots, \mathfrak{a}_p = \rho^{p-1}(1) \right\} ,$$

although the principles presented here also apply to other ideal classes. We will refer to $\mathcal{C}$ as the *principal cycle* in what follows.

Notice that, by (5.10), we have $\mathfrak{a}_i = (\theta_i)$ for $i \geq 1$. Shanks' first observation was that the size of the generators θ_i increases with i, thereby providing a measure of how far along $\mathfrak{a}_i$ is on the principal cycle. Shanks dubbed this quantity the *distance* of an ideal.

Definition 7.12. *The* distance[56] *of the reduced principal ideal* $\mathfrak{a}_m$ *is*

$$\delta(\mathfrak{a}_m) = \log \theta_m .$$

The following notion of the relative distance between two equivalent ideals is also sometimes useful.

Definition 7.13. *Let* $\mathfrak{a}_m$ *and* $\mathfrak{a}$ *be two reduced, equivalent ideals such that* $\mathfrak{a}_m = \rho^{m-1}(\mathfrak{a}) = (\theta_m)\mathfrak{a}$. *Then the* relative distance *from* $\mathfrak{a}$ *to* $\mathfrak{a}_m$ *is defined as*

$$\delta(\mathfrak{a}_m, \mathfrak{a}) = \log \theta_m .$$

Notice that the first notion of distance is equivalent to the relative distance of a reduced principal ideal from (1), because

$$\delta(\mathfrak{a}_m, (1)) = \log \theta_m = \delta(\mathfrak{a}_m) .$$

It can be shown that $\delta(\mathfrak{a}_m)$ is a strictly increasing function of m and, recalling (5.33), that $\delta(\mathfrak{a}_{p+1}) = \log \theta_{p+1} = R_\Delta$. Hence, as $\mathfrak{a}_1 = \mathfrak{a}_{p+1}$, the regulator can be viewed as the distance around the entire cycle of reduced principal ideals. Furthermore, if $k \in \mathbb{Z}$ and $\delta(\mathfrak{a}_m) = kR + \delta(\mathfrak{a}_s)$ for ideals $\mathfrak{a}_m$ and $\mathfrak{a}_s$, then $\mathfrak{a}_m = \mathfrak{a}_s$; in other words, we may work with distances reduced modulo R_Δ.

Example 7.14. Consider the real quadratic field $\mathbb{Q}(\sqrt{193})$. Computing the continued fraction expansion of $\omega = (1 + \sqrt{193})/2$ yields 15 distinct reduced principal ideals $\mathfrak{a}_1 = (1)$ through $\mathfrak{a}_{15} = \rho^{14}(1)$, listed in Table 7.1 along with their distances. Notice that $\mathfrak{a}_{16} = \mathfrak{a}_1$, so $\delta(\mathfrak{a}_{16}) \approx 15.07631652$ is the regulator

Table 7.1. Principal Cycle for $\Delta = 193$

j	(Q_j, P_j), for $\mathfrak{a}_j = [Q_j/r, (P_j + \sqrt{D})/r]$	$\delta(\mathfrak{a}_j)$
1	$(2, 1)$	0
2	$(12, 13)$	2.59869817
3	$(6, 11)$	3.32835583
4	$(4, 13)$	4.82844171
5	$(18, 11)$	6.65671165
6	$(8, 7)$	6.80572746
7	$(14, 9)$	7.85709282
8	$(12, 5)$	8.15679754
9	$(12, 7)$	8.71127845
10	$(14, 5)$	9.16513386
11	$(8, 9)$	9.65688343
12	$(18, 7)$	10.61682945
13	$(4, 11)$	10.94102199
14	$(6, 13)$	12.84657298
15	$(12, 11)$	14.26937782
16	$(2, 13)$	15.07631652

of $\mathbb{Q}(\sqrt{193})$. Notice also that $\delta(\mathfrak{a}_j) \approx j$ in all cases, as one would expect from the Khintchine-Lévy Theorem (Theorem 3.17) and its consequence (3.41).

As mentioned above, Shanks' goal in developing the infrastructure was to be able to apply his baby-step giant-step method for computing the order of an element in a finite abelian group to the problem of computing the regulator. Consider first the cyclic group generated by an element g. Computing consecutive powers of g eventually produces all elements in the group because $g^{\mathrm{ord}(g)} = 1$, so the elements in $\langle g \rangle$ form a cycle ordered by their base-g

discrete logarithms. This discrete logarithm measures how far around the cycle a particular element is, and, clearly, ord(g) measures the "circumference" of the cycle. In addition, one can compute ord(g) by iteratively computing $g^{i-1}g = g^i$ for $i = 1, 2, \ldots$ until $g^i = 1$, in which case $i = $ ord(g). These "baby steps," consisting of multiplication by g, walk through the entire cycle, increasing the distance traversed by 1 each time.

The infrastructure as described so far has properties similar to those of $\langle g \rangle$. The elements of $\mathcal{C}$ form a cycle and are ordered in terms of distance, as described above. The regulator, the distance around the entire cycle, is analogous to ord(g). The ρ operation acts as a baby step, in that iteratively applying ρ walks through the cycle one step at a time. By the Khintchine-Lévy Theorem (Theorem 3.17), the distance traversed through each step is close to 1. Walking through the entire cycle using baby steps yields the regulator.

The baby-step giant-step algorithm improves on this method by utilizing so-called giant steps that allow one to move t steps through the cycle in a single operation. In the case of $\langle g \rangle$, multiplication by g^t acts as such a giant step, as $g^s g^t = g^{s+t}$, the element precisely t steps further along in the cycle from g^s. Shanks[57] described how to combine baby steps and giant steps in $\langle g \rangle$ to compute ord(g) in time $O(\sqrt{H})$ for an upper bound H of ord(g) as opposed to $O(\text{ord}(g))$. In order to realize the same improvement for computing the regulator, Shanks required an analogue of the giant step. In particular, he needed a giant-step operation that would efficiently compute a reduced ideal $\mathfrak{a}_k$ equivalent to $\mathfrak{a}_s \mathfrak{a}_t$ with $\delta(\mathfrak{a}_k) \approx \delta(\mathfrak{a}_s) + \delta(\mathfrak{a}_t)$. The second key observation in developing the infrastructure was a solution to this problem.

Let $\mathfrak{a}_s$ and $\mathfrak{a}_t$ be reduced principal ideals in $\mathcal{C}$ and consider the product $\mathfrak{a}_s \mathfrak{a}_t$. As $\mathfrak{a}_s$ and $\mathfrak{a}_t$ are both principal, their product must also be principal, but it may not be reduced. Let $\mathfrak{c}$ be defined via the ideal product of $\mathfrak{a}_s$ and $\mathfrak{a}_t$ as described in §5.4 such that

$$(S)\mathfrak{c} = \mathfrak{a}_s \mathfrak{a}_t \ .$$

If $\mathfrak{c} = [Q_0'/r, (P_0' + \sqrt{D})/r]$, then by Theorem 5.9 we can compute a reduced ideal $\mathfrak{c}' = [Q_m'/r, (P_m' + \sqrt{D})/r]$ equivalent to $\mathfrak{c}$ by expanding the continued fraction corresponding to $\mathfrak{c}$ as in §5.2 until we have $0 < Q_{m-1}' < \sqrt{D}/2$. Since $\mathfrak{c}'$ is reduced and in the principal class, we must have $\mathfrak{c}' = \mathfrak{a}_k$ for some $k \geq 1$ and, because by (5.10) $\mathfrak{c}' = (\theta_m')\mathfrak{c}$, we have

$$\mathfrak{a}_k = \mathfrak{a}_s \mathfrak{a}_t (\theta_m'/S) \ .$$

Thus, we have

$$\theta_k = \theta_s \theta_t \frac{|\theta_m'|}{S} \ ,$$

so if we set $\kappa = \log |\theta_m'/S|$, we have

$$\delta(\mathfrak{a}_k) = \delta(\mathfrak{a}_s) + \delta(\mathfrak{a}_t) + \kappa \ . \tag{7.28}$$

By (5.38) we know that $-\log \Delta < \kappa < \log 2$, so $\delta(\mathfrak{a}_k) \approx \delta(\mathfrak{a}_s) + \delta(\mathfrak{a}_t)$, as required. NUCOMP, as described in §5.4, yields a similar result, the difference being that in this case, $|\theta'_m|/S = 1/\mu$ and by (5.45) we have $(1/2)\Delta^{-3/4} < 1/\mu \leq 1$, so $(-3/4)\log \Delta - \log 2 < \kappa \leq 0$. We will use the symbol "$*$" to denote such a giant-step operation, namely the reduced product of two input ideals computed using either ideal multiplication followed by reduction or NUCOMP.

The infrastructure of the principal cycle $\mathcal{C}$ is depicted in Figure 7.1 as a circle. The first ideal in the cycle of reduced principal ideals, $\mathfrak{a}_1$, is at the

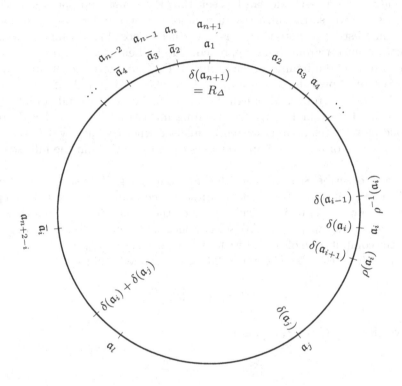

Fig. 7.1. The infrastructure of the principal class

top of the circle, and the first few baby steps follow it clockwise. The figure emphasizes that the distances between consecutive ideals are not equal, but that in most cases, they are all roughly the same, approximately 1 by the Khintchine-Lévy Theorem. The distance of each ideal measures how far around the circle it is placed, and the distance around the entire circle, corresponding to $\mathfrak{a}_{n+1} = \mathfrak{a}_1$, is equal to the regulator R_Δ. The symmetry of $\mathcal{C}$ is also illustrated, in that $\mathfrak{a}_{n+1-i} = \overline{\mathfrak{a}_{i+1}}$ for $i \geq 1$. Finally, the giant step

$\mathfrak{a}_l = \mathfrak{a}_i \mathfrak{a}_j$ is also illustrated in the figure. Notice that, as described above, $\delta(\mathfrak{a}_l)$ is slightly less than $\delta(\mathfrak{a}_i) + \delta(\mathfrak{a}_j)$ but that $\mathfrak{a}_l$ is nevertheless roughly distance $\delta(\mathfrak{a}_j)$ from $\mathfrak{a}_i$ around the circle, thereby traversing a greater distance than a baby step.

Despite the similarities of the infrastructure to a cyclic group, it must be emphasized that the infrastructure as presented here is *not* a group. For example, in the cyclic group generated by g, the $(i + j)$th element in the cycle is always equal to the product of the ith and the jth, as $g^{i+j} = g^i g^j$. This need not be the case in the infrastructure, as $\delta(\mathfrak{a}_i * \mathfrak{a}_j)$ is only close to $\delta(\mathfrak{a}_i) + \delta(\mathfrak{a}_j)$. As a result, it can be seen that the giant step operation need not be associative, so the infrastructure is not a group under this operation.

It is interesting to note that it is nevertheless possible to embed the infrastructure into a group. One approach, due to Lenstra,[58] embeds the infrastructure into the infinite cyclic group $\mathbb{R}/R_\Delta\mathbb{Z}$. A more recent approach, due to Schoof,[59] provides a description of the class group and infrastructure via Arakelov class groups. Although these approaches are certainly of interest theoretically, for example, in understanding infrastructure in higher-degree algebraic number fields, it is currently unclear whether they will have any impact in terms of practical improvements to computations in the infrastructure.

We now describe Shanks' algorithm for computing R_Δ using baby steps and giant steps in the infrastructure. First, we compute a list of baby steps starting with $\mathfrak{a}_1 = (1)$ by iteratively applying the ρ operator, yielding the list $\mathcal{L} = \{\mathfrak{a}_1, \mathfrak{a}_2, \ldots, \mathfrak{a}_t, \mathfrak{a}_{t+1}, \mathfrak{a}_{t+2}\}$. We select t such that $\delta(\mathfrak{a}_t) > \sqrt[4]{\Delta} > \delta(\mathfrak{a}_{t-1})$. If during the computation of the list we find an ideal $\mathfrak{a}_{n+1} = [Q_n/r, (P_n + \sqrt{D})/r]$ with $P_n = P_{n+1}$, then by Theorem 5.20 we can immediately set

$$R_\Delta = 2\delta(\mathfrak{a}_{n+1}) + \log \frac{Q_0}{Q_n} .$$

Similarly, if $Q_n = Q_{n+1}$, then we set

$$R_\Delta = 2\delta(\mathfrak{a}_{n+1}) + \log \frac{Q_0 \psi_{n+1}}{Q_n} .$$

In either case, we terminate the algorithm.

The list $\mathcal{L}$ must be maintained in such a way that creating it and searching in it can be done efficiently. Lexicographical ordering on the ideal coefficients, yielding worst-case search and insert times in $O(\log n)$ for lists of size n, is sufficient for the purposes of complexity analysis. In practice, a hash table using some portion of the Q coefficients as the key is usually used, as one can obtain average-case times of $O(1)$ for these operations.

Once the list $\mathcal{L}$ has been computed, we put $\mathfrak{b}_1 = \mathfrak{a}_t$ and compute giant steps $\mathfrak{b}_2, \mathfrak{b}_3, \ldots$ using $\mathfrak{b}_{i+1} = \mathfrak{b}_i * \mathfrak{b}_1$. Notice that by (7.28),

$$\delta(\mathfrak{b}_{i+1}) - \delta(\mathfrak{b}_i) < \delta(\mathfrak{b}_1) + \log 2 = \delta(\mathfrak{a}_t) + \log 2 ,$$

and Proposition 3.16, taking $m = t$ and $i = 2$, yields

$$\theta_{t+2} > F_3 \, \theta_t = 2\theta_t \ .$$

Thus, $\log \theta_{t+2} > \log 2 + \log \theta_t$ and we have

$$\delta(\mathfrak{b}_{i+1}) - \delta(\mathfrak{b}_i) < \delta(\mathfrak{a}_{t+2}) \ .$$

This implies that the distance traversed by a single giant step with distance $\delta(\mathfrak{a}_t)$ will always be less than $\delta(\mathfrak{a}_{t+2})$, the distance encompassed by the baby-step list, so eventually we will find some giant step $\mathfrak{b}_i \in \mathcal{L}$ with $\mathfrak{b}_i = \mathfrak{a}_j$, and can compute

$$R_\Delta = \delta(\mathfrak{b}_i) - \delta(\mathfrak{a}_j) \ .$$

The following algorithm summarizes this method.

Algorithm 7.1: Regulator of a Real Quadratic Order (Shanks)

Input: discriminant $\Delta > 0$ of a real quadratic order
Output: R_Δ
  /* *Compute baby steps*/
1: Set $\mathfrak{a}_1 = (1)$ and compute $\mathcal{L} = \{\mathfrak{a}_1, \mathfrak{a}_2, \ldots, \mathfrak{a}_t, \mathfrak{a}_{t+1}, \mathfrak{a}_{t+2}\}$ where $\mathfrak{a}_i = \rho(\mathfrak{a}_{i-1})$ and $\delta(\mathfrak{a}_t) > \sqrt[4]{\Delta} > \delta(\mathfrak{a}_{t-1})$.
2: **if** $\mathfrak{a}_{n+1} = [Q_n/r, (P_n + \sqrt{D})/r] \in \mathcal{L}$ such that $P_n = P_{n+1}$ **then**
3:     Set $R_\Delta = 2\delta(\mathfrak{a}_{n+1}) + \log(Q_0/Q_n)$.
4: **else if** $\mathfrak{a}_{n+1} = [Q_n/r, (P_n + \sqrt{D})/r] \in \mathcal{L}$ such that $Q_n = Q_{n+1}$ **then**
5:     Set $R_\Delta = 2\delta(\mathfrak{a}_{n+1}) + \log(Q_0\psi_{n+1}/Q_n)$.
6: **else**
      /* *Compute giant steps*/
7:     Set $\mathfrak{b}_1 = \mathfrak{a}_t$, $\mathfrak{b}_2 = \mathfrak{b}_1 * \mathfrak{a}_t$, and $i = 2$.
8:     **while** $\mathfrak{b}_i \notin \mathcal{L}$ **do**
9:         Set $i = i + 1$ and compute $\mathfrak{b}_i = \mathfrak{b}_{i-1} * \mathfrak{a}_t$.
10:    **end while**
11:    Find $\mathfrak{a}_j \in \mathcal{L}$ such that $\mathfrak{b}_i = \mathfrak{a}_j$.
12:    Set $R_\Delta = \delta(\mathfrak{b}_i) - \delta(\mathfrak{a}_j)$.
13: **end if**

The idea behind this algorithm is depicted in Figure 7.2. As before, $\mathcal{C}$, the principal cycle, is represented by a circle. The first ideal in the cycle, $\mathfrak{a}_1$, is shown at the top of the circle, and the baby steps in $\mathcal{L}$ are listed on the circle clockwise from $\mathfrak{a}_1$. The first few giant steps, $\mathfrak{b}_1, \mathfrak{b}_2, \mathfrak{b}_3$, and $\mathfrak{b}_4$, are also illustrated. The first giant step $\mathfrak{b}_1$ is equal to $\mathfrak{a}_t$. Notice that the baby steps are all fairly close together, whereas the giant steps each move distance approximately $\delta(\mathfrak{a}_t)$ around the circle. Finally, the last giant step $\mathfrak{b}_i$ shown in the figure ends up in the list of baby steps and is equal to $\mathfrak{a}_j$. This ideal has made one complete traversal of the circle before landing in $\mathcal{L}$. Thus, the total distance around the circle traversed by $\mathfrak{b}_i$ is $R_\Delta + \delta(\mathfrak{a}_j)$, and it can be seen that $R_\Delta = \delta(\mathfrak{b}_i) - \delta(\mathfrak{a}_j)$.

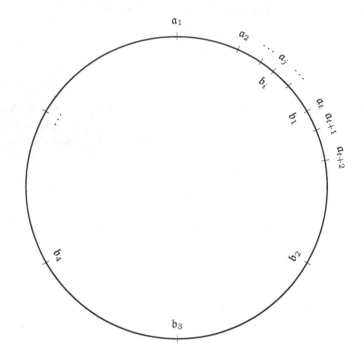

Fig. 7.2. Baby steps and giant steps in the infrastructure of the principal class

Example 7.15. Consider once again the example with $\Delta = 193$. We will now compute R_Δ using the baby-step giant-step method. We have $\sqrt[4]{\Delta} \approx 3.7$, so we take $t = 4$. The list of baby steps consists of

$$\mathcal{L} = \{\mathfrak{a}_1, \mathfrak{a}_2, \mathfrak{a}_3, \mathfrak{a}_4, \mathfrak{a}_5, \mathfrak{a}_6\} \,,$$

where throughout this example we use (Q, P) to denote the ideal $[Q/r, (P + \sqrt{D})/r]$, with the ideals taken from Table 7.1. We use $\mathfrak{a}_4 = (4, 13)$ with distance $\delta(\mathfrak{a}_4) \approx 4.82844171$ for the giant steps. The sequence of giant steps is

$$\begin{aligned}
\mathfrak{b}_2 &= \mathfrak{a}_4 * \mathfrak{a}_4 = (8, 1) \,, & \delta(\mathfrak{b}_2) &\approx 9.65688343 \,, \\
\mathfrak{b}_3 &= \mathfrak{b}_2 * \mathfrak{a}_4 = (12, 11) \,, & \delta(\mathfrak{b}_3) &\approx 14.26937782 \,, \\
\mathfrak{b}_4 &= \mathfrak{b}_3 * \mathfrak{a}_4 = (6, 5) \,, & \delta(\mathfrak{b}_4) &\approx 18.40467235 \,.
\end{aligned}$$

We find that $\mathfrak{b}_4 \in \mathcal{L}$ with $\mathfrak{b}_4 = \mathfrak{a}_3$, so

$$R_\Delta = \delta(\mathfrak{b}_4) - \delta(\mathfrak{a}_3) \approx 18.40467235 - 3.32835583 = 15.07631652 \,.$$

Shanks' baby-step giant-step algorithm is a significant improvement for computing R_Δ over simply computing the continued fraction expansion of ω as described in §3.3, a process which clearly has complexity $O(R_\Delta^{1+\epsilon})$. We know from (7.1) that $\Delta^{1/2+\epsilon}$ is an upper bound on R_Δ, so we would expect that, as in the order computation algorithm, the complexity would be the square root of this upper bound, namely $O(\Delta^{1/4+\epsilon})$, a considerable savings over the continued-fraction-based method, especially when Δ is large. We will now prove that this is in fact the case.

Theorem 7.16. *Shanks' baby-step giant-step algorithm as described above computes the regulator R_Δ of $\mathcal{O}_\Delta$ in time $O(\Delta^{1/4+\epsilon})$.*

Proof. First, we know that $R_\Delta > \delta(\mathfrak{b}_{i-1})$ because $\mathfrak{b}_i$ is the first giant step to have traversed the entire cycle and to have distance greater than R_Δ. Second, by (7.28) and because $\mathfrak{b}_{i-1} = \mathfrak{a}_t^{i-1}$ where "*" is the operation used in $\mathfrak{a}_t^{i-1}$, we know that

$$\delta(\mathfrak{b}_{i-1}) > (i-1)\delta(\mathfrak{a}_t) - (i-2)\log\Delta \; .$$

Thus, since t is chosen such that $\delta(\mathfrak{a}_t) > \sqrt[4]{\Delta} > \delta(\mathfrak{a}_{t-1})$, we have

$$\Delta^{1/2+\epsilon} > (i-1)\delta(\mathfrak{a}_t) - (i-2)\log\Delta > (i-1)\sqrt[4]{\Delta} - (i-2)\log\Delta \; ,$$

so i, the number of giant steps, is $O(\Delta^{1/4+\epsilon})$. Also, by the Khintchine-Lévy Theorem (Theorem 3.17), we have that t, the number of baby steps, is also $O(\Delta^{1/4+\epsilon})$. It follows that the complexity of determining R_Δ by Shanks' method is $O(\Delta^{1/4+\epsilon})$ as claimed. $\qquad\qquad\square$

One shortcoming of this method is that its efficiency depends on the accuracy of the upper bound on R_Δ used. Ideally, one would like to have roughly the same number of baby steps and giant steps,[60] but a coarse bound such as $\Delta^{1/2}$ will almost certainly not achieve an optimal balance, especially in cases where $h_\Delta > 1$. With a tighter bound on R_Δ, it should be possible to obtain a runtime of $O(R_\Delta^{1/2+\epsilon})$ as opposed to $O(\Delta^{1/4+\epsilon})$.

Two modifications of the basic baby-step giant-step algorithm presented above have been proposed that circumvent this problem, both of which have complexity $O(R_\Delta^{1/2}\Delta^\epsilon)$. The first of these, which we will sketch briefly here, is due to Buchmann and Williams.[61] The basic idea is to select a small value of v and use the baby-step giant-step method to test whether $0 < R_\Delta < v^2$ using baby steps with distance bounded by v and a giant step of distance v, requiring roughly v baby steps and v giant steps. If the regulator is not found in this interval (i.e., $R_\Delta > v^2$), we double v and test whether $v^2 < R_\Delta < (2v)^2$ by extending the baby-step list to include baby steps with distance bounded by $2v$ and using a giant step of distance $2v$. This doubling process is repeated until the regulator is found in some interval $(2^{k-1}v)^2 < R_\Delta < (2^k v)^2$. In the worst case, this algorithm requires $2^k v > R_\Delta^{1/2}$ baby steps and giant steps, so the overall complexity is $O(R_\Delta^{1/2}\Delta^\epsilon)$.

Buchmann and Vollmer[62] presented an improved version of this approach based on an algorithm due to Terr[63] for computing the order of an element g in a finite abelian group G. Terr's algorithm relies on the following result[64].

Lemma 7.17. *Let* $g \in G$. *Then there exist* $e \in \mathbb{N}$ *and* $f \in \{0, \ldots, e-1\}$ *with* $g^{e(e+1)/2} = g^f$. *If* e *is chosen minimal with this property, then* $e(e-1)/2 < \mathrm{ord}(g) \le e(e+1)/2$ *and* $\mathrm{ord}(g) = e(e+1)/2 - f$.

The algorithm consists of finding the minimal value of e for which $g^{e(e+1)/2} = g^f$ with $f \in \{0, \ldots, e-1\}$ by maintaining a set of baby steps $\{(g^f, f) \mid 0 \le f < e\}$ and a current giant step $g^{e(e+1)/2}$, beginning with $e = 1$. In the eth iteration we add a new baby step g^e to the list, compute the next baby step $gg^e = g^{e+1}$, and compute a new giant step $g^{e+1}g^{e(e+1)/2} = g^{(e+1)(e+2)/2}$. As soon as the giant step is equal to one of the elements g^f in the baby-step list, Lemma 7.17 tells us that $\mathrm{ord}(g) = e(e+1)/2 - f$. Using the bounds on $\mathrm{ord}(g)$ from Lemma 7.17, Terr proved that this algorithm computes $\mathrm{ord}(g)$ using $2\lceil \sqrt{2(\mathrm{ord}(g) - 1)} \rceil - 2$ multiplications in G and $\lceil \sqrt{2(\mathrm{ord}(g) - 1)} \rceil - 1$ searches in the baby-step list. Given the similarities between the cyclic group $\langle g \rangle$ and the infrastructure we have discussed previously, it is clear that a similar strategy can be employed to compute R_Δ. Buchmann and Vollmer described the details of such an algorithm and proved that it computes R_Δ in time $O((\log \Delta + \sqrt{R_\Delta})(\log \Delta)^3)$.

In addition to computing the regulator, the baby-step giant-step method can be used in the infrastructure to solve the principal ideal problem, namely computing the distance of a given reduced principal ideal. This is the infrastructure analogue of the discrete logarithm problem in $\langle g \rangle$, for which, given $g, h \in \langle g \rangle$, one finds the smallest integer n such that $g^n = h$, if it exists. Once again, the discrete logarithm n can be viewed as the distance the element h is around the cycle of elements in $\langle g \rangle$. Thus, using reasoning similar to that above, it is fairly straightforward to adapt the baby-step giant-step algorithm for solving the discrete logarithm problem in $\langle g \rangle$ to an algorithm that solves the principal ideal problem with complexity $O(\Delta^{1/4+\epsilon})$, assuming $\Delta^{1/2+\epsilon}$ is an upper bound on the unknown distance.[65] More generally, this idea can be used to decide whether two reduced ideals are equivalent, also in time $O(\Delta^{1/4+\epsilon})$. It is also possible to generalize the improved baby-step giant-step algorithms mentioned above to decide equivalence or principality in time $O(\delta^{1/2}\Delta^\epsilon)$, where δ is the unknown distance (for principality testing) or relative distance (for equivalence testing).[66]

Another application of infrastructure is the ability to efficiently compute a reduced principal ideal with distance close to a given quantity. Once again, we draw on the analogy to $\langle g \rangle$, in which an element with distance n around the cycle (i.e., g^n) can be computed efficiently using binary exponentiation. A similar method can be used in the infrastructure. In particular, we can find an ideal with distance close to n by performing about $n/\delta(\mathfrak{a}_s)$ multiply-reduction steps using an ideal $\mathfrak{a}_s$ with distance $\delta(\mathfrak{a}_s)$, as opposed to n/γ continued fraction steps. A more precise formulation of such an algorithm will be given in

Chapter 11, along with a description of a method called (f, p)-representations for approximating distances in such a way that we can ensure that sufficient precision is used to guarantee that our results are numerically accurate.

Finally, it is also possible, and sometimes useful, to perform a "backward" giant step. Notice that if $\mathfrak{a}_l$ is the first reduced ideal obtained by expanding the continued fraction of $\mathfrak{a}_s \bar{\mathfrak{a}}_t$, then, because $\bar{\mathfrak{a}}_t = (\bar{\theta}_m) = (N(\theta_t)/\theta_t) = (N(\mathfrak{a})/\theta_t)$, we have

$$\delta(\mathfrak{a}_l) = \delta(\mathfrak{a}_s) - \delta(\mathfrak{a}_t) + \log N(\mathfrak{a}_t) + \kappa \pmod{R_\Delta} .$$

Thus, multiplying a reduced principal ideal by $\bar{\mathfrak{a}}_t$ and reducing results in a reduced principal ideal whose distance is roughly $\delta(\mathfrak{a}_t)$ less than the original distance, effectively a giant step backwards in the principal cycle.

In Chapters 10, 13, and 15, we will discuss additional algorithms for computing R_Δ, including state-of-the-art methods for computing it conditionally and unconditionally, some of which are refinements of the basic baby-step giant-step method presented above. A number of these require accurate approximations of the product $h_\Delta R_\Delta$. In the next two chapters, we will present analytic results related to this quantity and describe how to approximate it efficiently.

Notes and References

[1] The finiteness of the class group is more generally proved using the geometry of numbers, in particular Minkowski's convex body theorem. For details, see [AW04].

[2] Using a different representation of ideals, more closely aligned with positive definite quadratic forms, it is possible to define reduced ideals in such a way that *every* equivalence class has a unique reduced representative. See [Bue89] or [BV07] for details.

[3] Our interest in the ideal class number for the purposes of this work is mainly its connection to solutions of Pell's equation. However, early interest was due to the fact that class number 1 implies that the maximal order of the quadratic field is a principal ideal domain and, hence, is also a unique factorization domain. Class numbers larger than 1 provide a measure of sorts as to how far away $\mathcal{O}_\Delta$ is from having unique factorization. See [Mas79] for a survey and examples illustrating this phenomenon.

[4] [Coh93], p. 295.

[5] [Sie35].

[6] Gauss' study of class groups was in the context of equivalence classes of binary quadratic forms, which are closely related to ideal equivalence classes discussed here. See [Bue89], [Coh93], and [BV07] for more details on the correspondence between these two models of the class group.

[7] [Bak66] and [Sta67].

[8] [Hee52].

[9] [Wat04].

[10] [Arn92], [Wag96], and [ARW98].

[11] See [Hua82], Ch. 12.

[12] The extended Riemann hypothesis (ERH) is a hypothesis on the locations of non-trivial zeros of Dirichlet L-functions. Dirichlet L-functions are defined in Chapter 8, and the ERH is discussed in §9.5.

[13] See, for example, [MW91b] and [MW91c].

[14] Again, Gauss' original conjecture was posed with respect to the related class number of indefinite binary quadratic forms.

[15] See [SW88a], [Jac98], and [tRW03] for some examples of computations providing numerical evidence in support of the Gauss conjecture.

[16] The Cohen-Lenstra heuristics, mostly for quadratic fields, are presented in [CL83] and [CL84]. Analogous heuristic results were derived for number fields of degree ≥ 2 by Cohen and Martinet [CM87].

[17] See, for example, [SW88a] and [Jac98].

[18] [Bue99], [JRW06], and [Ram06].

[19] [Lan36], Eq. (3).

[20] See, for example, [Hua82], p. 194.

[21] These results are taken from [CL84].

[22] [Lan36].

[23] See [CL84].

[24] [CL84].

[25] These are conjectures C1, C2, and C5 from [CL84].

[26] [JRW06].

[27] [Ram06].

[28] The tabulations presented in [JRW06] and [Ram06] are the latest in numerous tabulations of class groups of imaginary quadratic fields, most notably by Buell [Bue76], [Bue87], and [Bue99].

[29] Further justification of this is provided in [CL83].

[30] [Lan36], Eq. (6).

[31] The first of these conjectures is not listed in [CL84] and is presented here for the first time. The others are conjectures C7, C9, C11, and C12 from [CL84].

[32] This was independently conjectured by Hooley [Hoo84].

[33] [JLW95].

[34] This process is described in some detail in §8.3

[35] [Jac98] and [tRW03]. .

[36] This result is sketched in [Coh62], pp. 189–190.

[37] [Coh62], p. 225, Table 1.

[38] [BC70], Theorem 3.1.

[39] See [BC70], Corollary 3.1.1.

[40] See [Bue89], Ch. 10, for a partial survey.

[41] [Sha71].

[42] It was also numbers produced from Lehmer's number sieve the DLS-127, an earlier version of the DLS-157 (see, for example, [Wil98], pp. 194–195), which inspired Shanks' discovery of the baby-step giant-step technique. In 1968, Lehmer had produced three values of D for which he wanted the value of h_D, the class number of $\mathbb{Q}(\sqrt{-D})$. These values were selected in order to minimize the value of $h_D/\sqrt{D}$ (see [TW99]), but they were too large for computing h_D by the means currently available. The first time the baby-step giant-step idea was used was in the determination of $h_D = 29351$ for $D = -229565917267$ in August 1968.

[43] [Sch83].

[44] This algorithm, published in [SL84] by Schnorr and Lenstra, was independently discovered by Atkin and Rickert, who called it SPAR after Shanks, Pollard, Atkin, and Rickert.

[45] [Sey87].

[46] [LP92].

[47] [Sch83].

[48] [Sch83].

[49] [Bue89] and [GW08].

[50] [GW08].

[51] Genus theory partitions the class group into genera according to which congruence classes of integers belong to the ideals in a particular collection of equivalence classes (a genus). Once again, this theory was developed by Gauss and, as such, is usually described in the language of binary quadratic forms and when integers can be represented by forms. See [Coh62], Ch. 13, Sec. 3, and [Bue89], Ch. 4, for brief introductions to this area.

[52] See [Bue89], Ch. 9.

[53] For example, [Bue99] and [JRW06].

[54] [FK06].

[55] [Sha72].

[56] We are using Shanks' original formulation of distance from [Sha72]. Lenstra [Len82] used the alternative formulation

$$\delta(\mathfrak{a}_m) = \frac{1}{2} \log \left| \frac{\theta_m}{\overline{\theta}_m} \right| ,$$

which has some computational advantages when ideal conjugation is involved. We will revisit Lenstra's distances in Chapter 13 when discussing subexponential index-calculus algorithms.

[57] [Sha71].

[58] [Len82]

[59] [Sch08]

[60] In order to further optimize baby-step giant-step methods in the infrastructure, it is necessary to account for the fact that baby steps are much faster than giant steps. This idea is applied to the unconditional verification algorithm for R_Δ presented in Chapter 15.

[61] [BW88b].

[62] [BV06] and [BV07], Section 10.2.

[63] [Ter00].

[64] [BV07], Lemma 9.7.7.

[65] This procedure is described in much greater detail in §16.1 of Chapter 16. There we assume that we are given a rational approximation to R_Δ.

[66] [BW88b] and [BV07].

8

The Analytic Class Number Formula

8.1 Dirichlet Characters

In much of what will follow we will be concerned with problems which require knowledge of bounds on h_Δ and R_Δ. To this end, we need to develop the analytic class number formula,[1] a remarkable result which relates h_Δ, R_Δ, Δ, and a particular value of a function called the Dirichlet L-function.[2] In order to derive this formula we first need to discuss some results concerning *characters*.

Definition 8.1. *Let G be any group. A complex-valued function f defined on G is called a* character *of G if*

$$f(ab) = f(a)f(b)$$

for all $a, b \in G$ and $f(c) \neq 0$ for some $c \in G$.

If f is any character of a group G, it is easy to see by Definition 8.1 that f must also be a character of any subgroup of G. From this definition it is possible to prove the following results.

Theorem 8.2. *If f is a character of a finite group G with identity element e, then $f(e) = 1$, and e is the only element of G such that $f(e) = 1$ for all possible characters of G. Furthermore, each functional value $f(a)$ is a root of unity. In fact, if $a^n = e$, then $f(a)^n = 1$.*

Theorem 8.3. *A finite abelian group of order n has exactly n distinct characters.*

If $f(a) = 1$ for all $a \in G$, we call f the *principal* character of G. Every group has this character. There are a number of other results that can be derived concerning these general characters, but we will now confine our attention to characters defined on the group of reduced residue classes modulo m, where $m \in \mathbb{Z}^{\geq 0}$.

Definition 8.4. *Let G be the group of reduced residue classes modulo m. Corresponding to each character f of G we define an arithmetical function $\chi = \chi_f$ as follows:*

$$\chi(n) = \begin{cases} f([n]) & \text{if } (n,m) = 1 \\ 0 & \text{if } (n,m) > 1 . \end{cases}$$

The function χ is called a Dirichlet *character modulo m.*

Notice that if $\chi(n) \neq 0$, then $\chi(n)$ is a root of unity, and we must have $|\chi(n)| = 1$; also, $\chi(1) = 1$ and $\chi(-1) = \pm 1$. If $\chi(-1) = 1$, χ is said to be an *even* character; if $\chi(-1) = -1$, χ is said to be *odd*.

The principal character χ_0 has the property that

$$\chi_0(n) = \begin{cases} 1 & \text{if } (n,m) = 1 \\ 0 & \text{if } (n,m) > 1 . \end{cases}$$

The least positive integer k such that $\chi^k(n) = \chi_0(n)$ for all n is called the *order* of χ. The following result is an easy consequence of the definition of χ.

Theorem 8.5. *If χ is a Dirichlet character modulo m, then we have the following:*

1. $\chi(rs) = \chi(r)\chi(s)$ *for all $r, s \in \mathbb{Z}$*
2. $\chi(r + m) = \chi(r)$.

Conversely, if χ obeys (1) and (2) and if $\chi(n) = 0$ for $(n,m) > 1$, then χ is one of the Dirichlet characters modulo m.

We will require the following results in the sequel.

Theorem 8.6. *If χ is a Dirichlet character modulo m, then*

$$\sum_n \chi(n) = \begin{cases} \phi(m) & \text{if } \chi = \chi_0 \\ 0 & \text{if } \chi \neq \chi_0 , \end{cases}$$

where the sum is taken over a complete set of residues modulo m.

Proof. From the definition of χ_0, we see that the theorem clearly holds for $\chi = \chi_0$. If $\chi \neq \chi_0$, there must exist some a such that $\chi(a) \neq 1$ and $(a,m) = 1$. Now,

$$\chi(a) \sum_n \chi(n) = \sum_n \chi(an) = \sum_n \chi(n) .$$

Thus,

$$(\chi(a) - 1) \sum_n \chi(n) = 0$$

and the result follows. $\square$

Corollary 8.6.1. *If χ is a non-principal Dirichlet character modulo m and $b \geq a \geq 0$, then*

$$\sum_{n=a}^{b} \chi(n) = \sum_{n=0}^{r} \chi(n+a) ,$$

where $b - a \equiv r \pmod{m}$ and $r \geq 0$.

Proof. Let $b - a = qm + r$. Then by the theorem we have

$$\sum_{n=a}^{b} \chi(n) = \sum_{n=qm+a}^{qm+r+a} \chi(n) .$$

The corollary follows from (2) of Theorem 8.5.   □

Theorem 8.7. *If χ is a non-principal Dirichlet character modulo m and $\chi(-1) = 1$, then we have the following:*

1. $\displaystyle\sum_{n=1}^{\lfloor m/2 \rfloor} \chi(n) = 0$

2. $\displaystyle\sum_{n=1}^{m} n\chi(n) = 0.$

Proof. Since $\chi(-1) = 1$, we get $\chi(m-n) = \chi(n)$. Also, if $2 \mid m$, then $\chi(m/2) = 0$. Thus, by Theorem 8.6, we can easily deduce part 1. Also,

$$\sum_{n=1}^{m} n\chi(n) = \sum_{n=1}^{\lfloor m/2 \rfloor} n\chi(n) + \sum_{n=1}^{m-\lfloor m/2 \rfloor - 1} (m-n)\chi(m-n)$$

$$= \sum_{n=1}^{\lfloor m/2 \rfloor} n\chi(n) + m\sum_{n=1}^{\lfloor m/2 \rfloor} \chi(n) - \sum_{n=1}^{\lfloor m/2 \rfloor} n\chi(n)$$

$$= 0 ,$$

by part 1.   □

We use these results to produce a bound on $|\sum_{n=1}^{N} \chi(n)|$.

Theorem 8.8. *If χ is any non-principal character modulo m, then*

$$\left| \sum_{n=1}^{N} \chi(n) \right| \leq \frac{\phi(m)}{2} .$$

Proof. By Corollary 8.6.1,

$$\left| \sum_{n=1}^{N} \chi(n) \right| = \left| \sum_{n=0}^{N} \chi(n) \right| = \left| \sum_{n=0}^{r} \chi(n) \right| \leq \sum_{n=0}^{r} |\chi(n)| ,$$

where $r \equiv N \pmod{m}$. Suppose $\chi(-1) = 1$. By part 1 of Theorem 8.7, we get

$$\sum_{n=0}^{r} \chi(n) = \begin{cases} \sum_{n=0}^{r} \chi(n) & \text{if } r \leq \lfloor m/2 \rfloor \\ \sum_{n=\lfloor m/2 \rfloor+1}^{r} \chi(n) & \text{if } r > \lfloor m/2 \rfloor . \end{cases}$$

Since $(n, m) = 1$ if and only if $(m - n, m) = 1$, we must have

$$\sum_{n=0}^{r} |\chi(n)| \leq \frac{\phi(m)}{2} .$$

If $\chi(-1) = -1$, then $\chi(m - n) = -\chi(n)$ and, therefore,

$$\chi\left(\frac{m}{2} - j\right) = -\chi\left(\frac{m}{2} + j\right)$$

when $2 \mid m$ or

$$\chi\left(\frac{m-1}{2} - j\right) = -\chi\left(\frac{m+1}{2} + j\right)$$

when $2 \nmid m$. It follows that $\sum_{n=0}^{r} |\chi(n)| \leq \phi(m)/2$.   □

If a Dirichlet character can assume only real values, we call such a character a *real* character. For example, if $d \in \mathbb{Z}$, $d \equiv 1 \pmod{4}$, or $d \equiv 8, 12 \pmod{16}$ and d is not a perfect square, then it is easy to verify that the Kronecker symbol satisfies[3]

$$\left(\frac{d}{k}\right) = \left(\frac{d}{|d| + k}\right)$$

and $(d/k) = \chi(k)$ is a real character modulo $|d|$ with $\chi(-1) = 1$ when $d > 0$ and $\chi(-1) = -1$ when $d < 0$. Furthermore, we have the following result.

Theorem 8.9. *If d is as defined above, then (d/k) is a non-principal character modulo $|d|$.*

Proof. Suppose $2^n \parallel d$ and n is odd; we must have $n \geq 3$. Put $s \equiv 1 \pmod{d/2^n}$ and $s \equiv 5 \pmod{2^n}$. We get

$$\left(\frac{d}{s}\right) = \left(\frac{2}{s}\right)^n = -1 .$$

If n is even, there must exist an odd prime q and an odd m such that $q^m \parallel d$. Let r be a quadratic nonresidue of q. Find s such that

$$s \equiv 1 \left(\operatorname{mod} \frac{d}{q^m}\right) \text{ and } s \equiv r \pmod{q^m} .$$

We have

$$\left(\frac{d}{s}\right) = \left(\frac{s}{d}\right) = \left(\frac{r}{q}\right)^m = -1 .$$

□

We now obtain some results concerning products of Dirichlet characters.

Theorem 8.10. *If $m_1, m_2 \in \mathbb{Z}^{\geq 0}$, $(m_1, m_2) = 1$ and χ_i $(i = 1, 2)$ are Dirichlet characters modulo m_i $(i = 1, 2)$, then $\chi(n) = \chi_1(n)\chi_2(n)$ is a Dirichlet character modulo $m_1 m_2$.*

Proof. Clearly part 1 of Theorem 8.5 is satisfied for $m = m_1 m_2$. Also,

$$\chi(r + m) = \chi_1(r + m)\chi_2(r + m) = \chi_1(r)\chi_2(r) = \chi(r) .$$

Let $(n, m) = g$. If $g > 1$, there exists $g_1 > 1$, where $g_1 \mid (m_1, n)$ or $g_1 \mid (m_2, n)$, and $\chi(n) = \chi_1(n)\chi_2(n) = 0$. $\qquad \square$

In fact, we can produce a sort of converse of Theorem 8.10.

Theorem 8.11. *Let χ be a Dirichlet character modulo m, where $m = m_1 m_2$ and $(m_1, m_2) = 1$. Then there exist unique characters χ_1 and χ_2 modulo m_1 and m_2, respectively, such that*

$$\chi(n) = \chi_1(n)\chi_2(n) .$$

Furthermore, $\chi_i(n) = \chi(n_i)$, where we define n_i by

$$n_i \equiv n \pmod{m_i}, n_i \equiv 1 \pmod{m_j} \quad (i, j = 1, 2; j \neq i) .$$

Proof. Define n_1 and n_2 as above. Then $n \equiv n_1 n_2 \pmod{m_1}$ and $n \equiv n_1 n_2 \pmod{m_2}$; hence, $n \equiv n_1 n_2 \pmod{m}$. Now, define

$$\chi_i(n) = \chi(n_i) \quad (i = 1, 2) .$$

Then χ_i is a Dirichlet character modulo m_i $(i = 1, 2)$. We see immediately that if $a \equiv b \pmod{m}$, then $a \equiv b \pmod{m_i}$ and $\chi_i(a) = \chi_i(b)$. Also, if $r \equiv s \pmod{m_i}$, then $r_i \equiv s_i \pmod{m}$ and

$$\chi_i(r) = \chi(r_i) = \chi(s_i) = \chi_i(s) .$$

Since

$$\chi(n) = \chi(n_1 n_2) = \chi(n_1)\chi(n_2) = \chi_1(n)\chi_2(n) ,$$

we have expressed χ as a product of χ_1 and χ_2.

To show uniqueness, suppose that $\chi = \chi_1' \chi_2'$ is another decomposition of χ into characters modulo m_1' and m_2', respectively. Then

$$\chi_i(n) = \chi(n_i) = \chi_1'(n_i)\chi_2'(n_i) = \chi_i'(n_i) = \chi_i'(n) .$$

$\qquad \square$

At this point it is convenient to introduce a particular exponential sum, called a Gauss sum. We will need some properties of this object in the next chapter.

Definition 8.12. *Let χ be a Dirichlet character modulo m and define the* Gauss sum

$$S(n, \chi) = \sum_{j=1}^{m} \chi(j) e^{2\pi i j n/m} \, ,$$

where $i^2 + 1 = 0$.

Theorem 8.13. *Let $\chi, \chi_1, \chi_2, m, m_1$, and m_2 be defined as in the previous theorem. We have*

$$S(n, \chi) = \chi_1(m_2)\chi_2(m_1)S(n, \chi_1)S(n, \chi_2) \, .$$

Proof. By multiplying we get

$$S(n, \chi_1)S(n, \chi_2) = \sum_{j=1}^{m_1} \sum_{k=1}^{m_2} \chi_1(j)\chi_2(k) e^{2\pi i n(m_2 j + m_1 k)/m} \, .$$

Now, $m_2 j + m_1 k$ runs through a complete set of residues modulo $m_1 m_2$ as $j = 1, 2, \ldots, m_1$ and $k = 1, 2, \ldots, m_2$. Also,

$$\chi_1(m_2 j + m_1 k) = \chi_1(m_2 j) = \chi_1(m_2)\chi_1(j) \, ,$$
$$\chi_2(m_2 j + m_1 k) = \chi_2(m_2 k) = \chi_2(m_1)\chi_2(k) \, ;$$

thus,

$$\chi_1(m_2)\chi_2(m_1)S(n, \chi_1)S(n, \chi_2) = S(n, \chi) \, .$$

$\square$

We next show an important factorization property of this Gauss sum. Here, we will use the symbol $\bar{z}$ to denote the complex conjugate of $z \in \mathbb{C}$. We also use $\overline{\chi}(n)$ to denote $\overline{\chi(n)}$.

Theorem 8.14. *If χ is any Dirichlet character modulo m, then*

$$S(n, \chi) = \overline{\chi}(n)S(1, \chi) \tag{8.1}$$

whenever $(n, m) = 1$.

Proof. Note that $\chi(n)\overline{\chi}(n) = 1$; hence,

$$\chi(nk) = \chi(n)\chi(k) = \chi(k)/\overline{\chi}(n)$$

or $\chi(k) = \overline{\chi}(n)\chi(nk)$. It follows that

$$S(n, \chi) = \sum_{k=1}^{m} \chi(k) e^{2\pi i k n/m} = \overline{\chi}(n) \sum_{k=1}^{m} \chi(nk) e^{2\pi i k n/m} \, .$$

Since $(n, m) = 1$, nk will run through a complete set of residues modulo m as k does; thus,

$$S(n, \chi) = \overline{\chi}(n)S(1, \chi) \, .$$

$\square$

8.2 Primitive Characters

We next investigate under what conditions (8.1) can hold when $(n, m) > 1$. We note that if $(n, m) > 1$, then $\overline{\chi}(n) = 0$; hence, we can only have (8.1) when $S(n, \chi) = 0$. The question of when we have $S(n, \chi) \neq 0$ if $(n, m) > 1$ is answered by the following theorem.

Theorem 8.15. *Let χ be a Dirichlet character modulo m and suppose that $S(n, \chi) \neq 0$ for some n such that $(n, m) > 1$. Then there exists a divisor d of m such that $d < m$ and $\chi(a) = 1$ whenever $(a, m) = 1$ and $a \equiv 1 \pmod{d}$.*

Proof. For the given n value, let $g = (m, n)$ and $d = m/g$. We have $d \mid m$ and $d < m$ because $g > 1$. Choose any a such that $(a, m) = 1$ and $a \equiv 1 \pmod{d}$. Since $(a, m) = 1$, in the sum defining $S(n, \chi)$ we can replace k in the exponent and in $\chi(k)$ by ak and get

$$S(n, \chi) = \sum_{k=1}^{m} \chi(k)e^{2\pi i kn/m} = \sum_{k=1}^{m} \chi(ak)e^{2\pi i akn/m}$$

$$= \chi(a) \sum_{k=1}^{\infty} \chi(k)e^{2\pi i akn/m} .$$

Since $a \equiv 1 \pmod{d}$, $d = m/g$, and $g \mid n$, we have

$$akn \equiv kn \pmod{m} .$$

Thus,

$$S(n, \chi) = \chi(a)S(n, \chi) .$$

Since $S(n, \chi) \neq 0$, we have $\chi(a) = 1$. $\square$

This result leads us to consider those characters modulo m for which there is a divisor $d < m$ satisfying the properties in Theorem 8.15.

Definition 8.16. *Let χ be a Dirichlet character modulo m and let d be any positive divisor of m. The number d is called an* induced modulus *for χ if we have*

$$\chi(a) = 1 \text{ whenever } (a, m) = 1 \text{ and } a \equiv 1 \pmod{d} .$$

The smallest induced modulus f_χ for χ is called the conductor *of χ.*

That is to say, d is an induced modulus if the character χ acts like a character modulo d on the elements, which are relatively prime to m, of the residue class of integers congruent to 1 modulo d. Notice that m is always an induced modulus for χ if χ is a Dirichlet character modulo m. If there are no other induced moduli, we say that the character is primitive.

Definition 8.17. *A Dirichlet character modulo m is said to be* primitive *modulo m if it has no induced modulus $d < m$.*

We now derive some results concerning primitive characters; these will be of use to us in the next chapter. It is easy to show that 1 is an induced modulus for χ if and only if $\chi = \chi_0$. Thus, if $m > 1$, the principal character χ_0 is not primitive since it has 1 as an induced modulus. Also, in view of Definitions 8.16 and 8.17 and Theorems 8.14 and 8.15, we have

Theorem 8.18. *If χ is a primitive (Dirichlet) character, then*

$$S(n, \chi) = \overline{\chi}(n)S(1, \chi)$$

for all n.

We also note that the product of primitive characters is always a primitive character.

Theorem 8.19. *Let χ_i $(i = 1, 2)$ be primitive characters modulo m_i (> 1) $(i = 1, 2)$, where $(m_1, m_2) = 1$. Then $\chi = \chi_1\chi_2$ is a primitive character modulo $m = m_1m_2$.*

Proof. By Theorems 8.10 and 8.11, χ is certainly a character and its representation as the product of $\chi_1\chi_2$ is unique. If χ is not primitive, it must have an induced modulus $d < m$ such that if $(a, m) = 1$ and $a \equiv 1 \pmod{d}$, then $\chi(a) = 1$. Let $d_i = (d, m_i)$ $(i = 1, 2)$. Now, suppose that we have any a such that $(a, m_i) = 1$ and $a \equiv 1 \pmod{m_i}$ for i either 1 or 2. As in Theorem 8.11, we define a_i by $a_i \equiv a \pmod{m_i}$ and $a_i \equiv 1 \pmod{m_j}$ $(j \neq i)$. Then $\chi_i(a) = \chi(a_i)$. Since $a_i \equiv a \equiv 1 \pmod{d_i}$ and $a_i \equiv 1 \pmod{d_j}$, we must have $a_i \equiv 1 \pmod{d}$ and $\chi_i(a) = 1$. It follows that d_i must be an induced modulus for χ_i. Since χ_i is primitive, d_i can only be m_i and $d = m$, a contradiction. Thus, χ must be a primitive character modulo m.   □

Another result that must hold for primitive characters is the following.

Theorem 8.20. *If χ is a primitive character modulo m, then*

$$|S(n, \chi)|^2 = m .$$

Proof. We have

$$|S(n, \chi)|^2 = |S(1, \chi)|^2 = S(1, \chi)\overline{S(1, \chi)}$$

$$= S(1, \chi) \sum_{k=1}^{m} \overline{\chi}(k)e^{-2\pi ik/m}$$

$$= \sum_{k=1}^{m} S(k, \chi)e^{-2\pi ik/m} \qquad \text{(by Theorem 8.18)}$$

$$= \sum_{k=1}^{m} \sum_{r=1}^{m} \chi(r)e^{2\pi ikr/m}e^{-2\pi ik/m} ,$$

$$|S(n,\chi)|^2 = |S(1,\chi)|^2 = \sum_{r=1}^{m} \chi(r) \sum_{k=1}^{m} e^{2\pi i k(r-1)/m}$$
$$= m\chi(1) = m .$$

□

Corollary 8.20.1. *If χ is a primitive character modulo m, then*

$$S(1,\overline{\chi})S(1,\chi) = \chi(-1)m .$$

Proof. The proof follows easily on observing that

$$\overline{S(1,\chi)} = \chi(-1)S(1,\overline{\chi}) .$$

□

If we now consider $\Delta_{\mathbb{K}}$, the fundamental discriminant of the maximal order of $\mathbb{K}$, we know by Theorem 8.9 that $(\Delta_{\mathbb{K}}/n)$ is a Dirichlet character modulo $|\Delta_{\mathbb{K}}|$. We will next show that $(\Delta_{\mathbb{K}}/n)$ is primitive, but in order to do this, we need to develop a result concerning the representation of $\Delta_{\mathbb{K}}$ as a product of certain funamental discriminants.

If p is an odd prime, define by p^* the value $(-1)^{(p-1)/2}p$. Notice that $p^* \equiv 1 \pmod 4$ and therefore satisfies the properties of a fundamental discriminant. Notice also that so do the numbers 4, 8, and -8. We will call numbers $4, \pm 8$, and p^* for all primes p the *prime discriminants*.

Theorem 8.21. *A product of distinct relatively prime prime discriminants is a fundamental discriminant. A fundamental discriminant $\Delta_{\mathbb{K}}$ is a unique product of distinct relatively prime prime discriminants.*

Proof. Clearly, the product of distinct relatively prime discriminants is a fundamental discriminant. We now prove the converse by dividing it into three cases. We will also assume that $\Delta_{\mathbb{K}} > 0$, as the proof for $\Delta_{\mathbb{K}} < 0$ is derived by using similar reasoning.

CASE I: $\Delta_{\mathbb{K}} \equiv 1 \pmod 4$:
    In this case, $\Delta_{\mathbb{K}} = p_1 p_2 \cdots p_k$, where the p_i are distinct odd primes. Also, the number of these primes equivalent to 3 (mod 4) must be even because $\Delta_{\mathbb{K}} \equiv 1 \pmod 4$. Thus,

$$\Delta_{\mathbb{K}} = p_1^* p_2^* \cdots p_k^* .$$

CASE II: $\Delta_{\mathbb{K}} \equiv 12 \pmod{16}$:
    In this case,

$$\Delta_{\mathbb{K}} = 4p_1 p_2 \cdots p_k$$

and the number of primes equivalent to 3 (mod 4) must be odd. Thus,

$$\Delta_{\mathbb{K}} = -4p_1^* p_2^* \cdots p_k^* .$$

CASE III: $\Delta_{\mathbb{K}} \equiv 8 \pmod{16}$:

    Here, $\Delta_{\mathbb{K}} = 8p_1p_2 \cdots p_k = \pm 8p_1^* p_2^* \cdots p_k^*$.

□

With this result we can now establish that the Kronecker symbol $(\Delta_{\mathbb{K}}/n)$ is a primitive character.

Theorem 8.22. *If* $\chi(n) = (\Delta_{\mathbb{K}}/n)$, *then* $\chi(n)$ *is a primitive character modulo* $\Delta_{\mathbb{K}}$.

Proof. We first note that if $\chi(n) = (p^*/n)$, where p is a prime, then χ is a primitive character. This is because χ is a non-principal character (Theorem 8.9) and therefore cannot have 1 as an induced modulus. Also, if $\chi(n) = (-4/n)$ is not primitive, then it can only have 2 as an induced modulus. In this case, $\chi(3) = \chi(1) = 1$, but $\chi(3) = -1$; hence, χ is primitive. If $\chi(n) = (\pm 8/n)$, then $\chi(5) = -1$, so neither 2 nor 4 is an induced modulus. If 2 is the induced modulus, then $(\pm 8/n) = 1$ for all odd n, which is not so. If 4 is the induced modulus, then $\chi(n) = 1$ whenever $n \equiv 1 \pmod{4}$; however, $\chi(5) = -1$. Thus, $(\pm 8/n)$ is also a primitive character for either sign. By Theorem 8.21 and the above remarks, we have

$$\chi(n) = \left(\frac{\Delta_{\mathbb{K}}}{n} \right) = \prod_{i=1}^{k} \chi_i(n) \, ,$$

where each of $\chi_i(n)$ is a primitive character. By Theorem 8.19, $\chi(n)$ must be a primitive character modulo $\Delta_{\mathbb{K}}$. □

8.3 The *L*-Function

Definition 8.23. *Let* χ *be a non-principal character modulo* m *and let* $s = \sigma + it$ *($\sigma, t \in \mathbb{R}$) be a complex variable. We define the* Dirichlet *L*-function *by*

$$L(s, \chi) = \sum_{n=1}^{\infty} \frac{\chi(n)}{n^s} \, .$$

We note that if $\sigma > 1$, then $L(s, \chi_\Delta)$ is absolutely convergent. We now investigate the case where $\sigma > 0$. In order to do this, we will need some results which are variations on the theme of partial summation.

Theorem 8.24. *Let* $a, b \in \mathbb{Z}$, $a \leq b$, *and* $C(n)$ *and* $f(n)$ *be real or complex numbers, where* $n \in \mathbb{Z}$ *and* $a \leq n \leq b$. *If*

$$S(n) = \sum_{a \leq t \leq n} C(t) \, ,$$

then

$$\left| \sum_{n=a}^{b} C(n)f(n) \right| \le \max_{a \le n \le b} |S(n)| \left(|f(b)| + \sum_{t=a}^{b-1} |f(t) - f(t+1)| \right) .$$

Proof. Let $S(a-1) = 0$. We have

$$\sum_{n=a}^{b} C(n)f(n) = \sum_{n=a}^{b} (S(n) - S(n-1)) f(n)$$

$$= \sum_{n=a}^{b} S(n)f(n) - \sum_{n=a}^{b-1} S(n)f(n+1)$$

$$= \sum_{n=a}^{b-1} (S(n)(f(n) - f(n+1))) + S(b)f(b) .$$

Thus,

$$\left| \sum_{n=a}^{b} C(n)f(n) \right| \le \sum_{n=a}^{b-1} |S(n)||f(n) - f(n+1)| + |S(b)f(b)|$$

$$\le \max_{a \le n \le b} |S(n)| \left(|f(b)| + \sum_{t=a}^{b-1} |f(t) - f(t+1)| \right) .$$

$\square$

Corollary 8.24.1. *If $f(n)$ is a positive decreasing function of n, then*

$$\left| \sum_{n=a}^{b} C(n)f(n) \right| \le \max_{a \le n \le b} |S(n)||f(a)| .$$

We now suppose that $x \in \mathbb{R}$ and f is a function $f : \mathbb{R} \to \mathbb{R}$.

Theorem 8.25. *Let $x \ge 1$ and let $f(x)$ be a function with a continuous derivative for $x \ge 1$. Let $S(x) = \sum_{1 \le n \le x} C(n)$, where $C(n)$ is a real or complex number for $n \in \mathbb{Z}$. Then*

$$\sum_{1 \le n \le x} C(n)f(n) = S(x)f(x) - \int_{1}^{x} S(t)f'(t)\, dt .$$

Proof. Let $k = \lfloor x \rfloor$; then

$$S(x) = \sum_{n=1}^{k} C(n) = S(k) .$$

Define $S(0) = 0$ and get

$$\sum_{1 \le n \le x} C(n)f(n) = \sum_{n=1}^{k}(S(n) - S(n-1))f(n)$$

$$= \sum_{n=1}^{k} S(n)f(n) - \sum_{n=1}^{k-1} S(n)f(n+1)$$

$$= \sum_{n=1}^{k-1}(f(n) - f(n+1))S(n) + S(k)(f(k) - f(x)) + S(k)f(x)$$

$$= -\sum_{n=1}^{k-1} S(n) \int_{n}^{n+1} f'(t)\,dt - S(k)\int_{k}^{x} f'(t)\,dt + S(x)f(x)$$

$$= S(x)f(x) - \int_{1}^{x} S(t)f'(t)\,dt .$$

$\square$

By Corollaries 8.24.1 and 8.6.1 we get

$$\left| \sum_{n=N}^{M} \chi(n)\frac{1}{n^s} \right| \le \frac{m}{|N|^s}; \qquad (8.2)$$

hence, $L(s,\chi)$ must converge for all $\sigma > 0$. We also can put an upper bound on $L(1,\chi)$, a quantity that will be of much interest later.

Theorem 8.26.
$$|L(1,\chi)| < \log m + 1 .$$

Proof. By Theorem 8.25 we have

$$\sum_{n=1}^{N} \chi(n)\frac{1}{n} = \frac{S(N)}{N} + \int_{1}^{N} \frac{S(x)}{x^2}\,dx ,$$

where

$$S(x) = \sum_{n=1}^{\lfloor x \rfloor} \chi(n) .$$

Thus, since $|S(N)| < m$, we get

$$\left| \sum_{n=1}^{\infty} \chi(n)\frac{1}{n} \right| = \left| \int_{1}^{\infty} \frac{S(x)}{x^2}\,dx \right| \le \int_{1}^{m} \frac{|S(x)|}{x^2}\,dx + \int_{m}^{\infty} \frac{|S(x)|}{x^2}\,dx .$$

Now, $|S(x)| \le x$ for $0 \le x \le m$, and $|S(x)| \le m$ always. Hence

$$\left| \sum_{n=1}^{\infty} \chi(n)\frac{1}{n} \right| \le \int_{1}^{m} \frac{1}{x}\,dx + m\int_{m}^{\infty} \frac{1}{x^2}\,dx = \log m + 1 .$$

$\square$

Also, as we shall see below, $L(1, \chi_\Delta) > 0$ when $\chi_\Delta(n) = (\Delta/n)$.

We should also note the following simple result, which connects the value of $L(1, \chi_\Delta)$ to that of $L(1, \chi_{\Delta_K})$. Here, $\Delta = f^2 \Delta_K$, where f is the conductor of $\mathcal{O}_\Delta$.

Theorem 8.27.

$$L(1, \chi_\Delta) = \prod_{p|f} \left(1 - \frac{(\Delta_K/p)}{p} \right) L(1, \chi_{\Delta_K}),$$

where the product is taken over all the distinct prime divisors p of f.

Proof. It is easy to see that

$$\prod_{p|f} \left(1 - \frac{(\Delta_K/p)}{p} \right) = \sum_{g|f} \mu(g) \frac{(\Delta_K/g)}{g},$$

where $\mu(n)$ is the Möbius[4] μ-function. Thus,

$$\prod_{p|f} \left(1 - \frac{(\Delta_K/p)}{p} \right) L(1, \chi_{\Delta_K}) = \sum_{g|f} \mu(g) \frac{(\Delta_K/g)}{g} \sum_{n=1}^{\infty} \left(\frac{\Delta_K}{n} \right) \frac{1}{n}$$

$$= \sum_{n=1}^{\infty} \sum_{g|(f,n)} \mu(g) \left(\frac{\Delta_K}{g} \right) \frac{1}{g} \left(\frac{\Delta_K}{n/g} \right) \frac{1}{n/g}$$

$$= \sum_{n=1}^{\infty} \sum_{g|(f,n)} \mu(g) \left(\frac{\Delta_K}{n} \right) \frac{1}{n}.$$

Since $\sum_{r|k} \mu(r) = 0$ unless $k = 1$, we get

$$\prod_{p|f} \left(1 - \frac{(\Delta_K/p)}{p} \right) L(1, \chi_{\Delta_K}) = \sum_{\substack{n=1 \\ (n,f)=1}}^{\infty} \left(\frac{\Delta_K}{n} \right) \frac{1}{n}$$

$$= \sum_{n=1}^{\infty} \left(\frac{f^2 \Delta_K}{n} \right) \frac{1}{n} = L(1, \chi_\Delta).$$

$\square$

8.4 Ideal Density

Since we know that any invertible ideal in $\mathcal{O}_\Delta$ is equivalent to an integral ideal $\mathfrak{a}$ of $\mathcal{O}_\Delta$ such that $(N(\mathfrak{a}), f) = 1$, we can confine our attention to such ideals in order to determine h_Δ. Our next objective is to estimate the number of invertible ideals $\mathfrak{a}$ of $\mathcal{O}_\Delta$ that are such that $(N(\mathfrak{a}), f) = 1$ and $N(\mathfrak{a}) < t$ for some $t > 0$. To this end we first define $H(\mathcal{C}, t)$.

Definition 8.28. *Let C be any class of invertible ideals of $\mathcal{O}_\Delta$. We define $H(C,t)$ to be the number of distinct invertible ideals $\mathfrak{a}$ in the ideal class C^{-1} such that $N(\mathfrak{a}) < t$ and $(N(\mathfrak{a}), f) = 1$. We denote by $G(\mathfrak{a}, t)$ the number of distinct principal ideals (α) formed by taking $\alpha \in \mathfrak{a}$ for which $0 < N((\alpha)) < t$ and $(N((\alpha)), f) = 1$.*

We now prove the following simple lemma.

Lemma 8.29. *If $a \in C$, then*

$$H(C, t) = G(\mathfrak{a}, tN(\mathfrak{a})) .$$

Proof. Let $\mathfrak{a} \in C$. If $\alpha \in \mathfrak{a}$ and $(f, N((\alpha))) = 1$, then $(\alpha) = \mathfrak{a}\mathfrak{b}$, $\mathfrak{b} \in C^{-1}$ and $N((\alpha)) = N(\mathfrak{a})N(\mathfrak{b})$ by Theorems 4.35 and 4.36. Also, every $\mathfrak{b} \in C^{-1}$ with $(N(\mathfrak{b}), f) = 1$ defines a principal ideal $(\alpha) = \mathfrak{a}\mathfrak{b}$, where $(N((\alpha)), f) = 1$. Thus, every $\mathfrak{b} \in C^{-1}$ with $(N(\mathfrak{b}), f) = 1$ and $0 < N(\mathfrak{b}) < t$ corresponds uniquely to a principal ideal $(\alpha) \subseteq \mathfrak{a}$ with $N((\alpha)) \leq tN(\mathfrak{a})$ and $(N((\alpha)), f) = 1$. $\square$

From this lemma, we see that the task of estimating $H(C,t)$ is reduced to that of estimating $G(\mathfrak{a}, tN(\mathfrak{a}))$ for some $\mathfrak{a} \in C$. In order to accomplish this we require some further preliminary results.

Lemma 8.30. *Let $F(x, y) = Ax^2 + Bxy + Cy^2$, where $A, B, C \in \mathbb{Z}$, and put $D = B^2 - 4AC$. If $(k, A) = 1$ and $k \mid D$, then there are precisely $k\phi(k)$ distinct pairs (x, y) modulo k such that $(F(x, y), k) = 1$.*

Proof. By the Chinese Remainder Theorem it suffices to show that if $p^m \parallel k$, where p is any prime, then there are $p^m \phi(p^m)$ pairs (x, y) modulo p^m such that $p \nmid F(x, y)$. We consider two cases.

CASE 1: $p > 2$:
  We have
$$4AF(x, y) = (2Ax + By)^2 - Dy^2 .$$

Since $p \nmid 2A$, we can have $p \nmid F(x, y)$ if and only if $p \nmid 2Ax + By$. For each of the possible p^m values of y modulo p^m there are $p - 1$ values of x modulo p such that $p \nmid 2Ax + By$. Thus, there are $p^{m-1}(p - 1) = \phi(p^m)$ values of x modulo p^m.
CASE 2: $p = 2$:
  In this case, $2 \mid B$ and

$$F(x, y) \equiv Ax + By \pmod{2} .$$

By using the same reasoning as above, we see that there are $2^{m-1} = \phi(2^m)$ values of x modulo 2^m for each of the 2^m values of y modulo 2^m such that $2 \nmid F(x, y)$.

$\square$

Our argument for estimating $G(\mathfrak{a}, tN(\mathfrak{a}))$ will depend upon counting lattice points in certain regions of the plane. We will therefore require a simple lemma which allows us to do this.[5]

Lemma 8.31. *Let Γ be a continuous arc such that the radius of curvature at any point of Γ is greater than or equal to $r > 0$, and suppose that about each point of Γ, there is drawn a circle of radius r. Let the resulting domain be denoted by $\Gamma(r)$ and let $|\Gamma|$ be the length of the arc Γ. Then the area $|\Gamma(r)|$ of the domain $\Gamma(r)$ satisfies the inequality*

$$|\Gamma(r)| \leq 2r|\Gamma| + \pi r^2 .$$

Proof. Let (x, y) and (x', y') be two points on Γ and at each of these points construct a normal to the curve Γ. If the limit as (x', y') approaches (x, y), these normals will intersect at the centre of curvature for the point (x, y) of Γ. Let the normals intersect in an angle $d\phi$. If R is the radius of curvature, then

$$ds = R\, d\phi ,$$

and, on the other hand, since $R > r$, the element of area of $\Gamma(r)$ is

$$dA = \frac{2r}{2} \left((R + r)\, d\phi + (R - r)\, d\phi \right) = 2rR\, d\phi = 2r\, ds .$$

If we integrate along Γ and supplement this integral by the areas of the semicircles at the end points of Γ, we get

$$|\Gamma(r)| \leq 2r|\Gamma| + \pi r^2 .$$

$\square$

If Λ is the lattice given by $\Lambda = \{x(0, 1) + y(1, 0) : x, y, \in \mathbb{Z}\}$, we can prove the following corollary to Lemma 8.31.

Corollary 8.31.1. *Let A be a region bounded by a curve Γ consisting of a finite number n of arcs $\Gamma_1, \Gamma_2, \ldots, \Gamma_n$ of curves satisfying the conditions of the lemma with $r_i \geq \sqrt{2}$ $(i = 1, 2, \ldots, n)$. Then if $M(A)$ is the number of points of Λ in A or on Γ_i, then*

$$M(A) = |A| + E ,$$

where $|A|$ is the area of the region A and

$$E = O(|\Gamma|) .$$

Proof. We obtain the number of points of Λ within or on the boundary of A as follows: Around each point of the curve Γ_i we draw a circle of radius $\sqrt{2}$. Since $\sqrt{2}$ is less than or equal to the radius of curvature of any point of Γ_i, we get our result on using the lemma. $\square$

Definition 8.32. *Define the* Dirichlet structure constant κ_Δ *by*

$$
\kappa_\Delta =
\begin{cases}
\dfrac{2\pi}{w\sqrt{|\Delta|}} & \text{if } \Delta < 0 \quad (w \text{ is definied in §4.3}) \\[2em]
\dfrac{2R_\Delta}{\sqrt{\Delta}} & \text{if } \Delta > 0 .
\end{cases}
$$

We are finally ready to prove the main theorem of this section.

Theorem 8.33. *For $H(\mathcal{C}, t)$ defined as in Definition 8.28 we have*

$$
H(\mathcal{C}, t) = \frac{\phi(f)\kappa_\Delta t}{f} + O(\sqrt{t}) .
$$

Proof. For a given invertible $\mathfrak{a}$ we will investigate the problem of evaluating $G(\mathfrak{a}, tN(\mathfrak{a}))$. By Theorem 4.24 and Definition 4.33 we may assume that

$$
\mathfrak{a} = [\alpha_1, \alpha_2] ,
$$

where $\alpha_1 = a$, $\alpha_2 = b + c\omega$, $a, b, c \in \mathbb{Z}$, $N(\mathfrak{a}) = ac$, and $(ac, f) = 1$. Let $\alpha \in \mathfrak{a}$ such that $(N((\alpha)), f) = 1$. We have

$$
\alpha = \alpha_1 x + \alpha_2 y, \quad \overline{\alpha} = \overline{\alpha}_1 x + \overline{\alpha}_2 y \quad (x, y \in \mathbb{Z}) ,
$$

and putting $A = \alpha_1 \overline{\alpha_1}$, $B = \alpha_1 \overline{\alpha_2} + \overline{\alpha}_1 \alpha_2$, $C = \alpha_2 \overline{\alpha}_2$, we get

$$
N((\alpha)) = |\alpha\overline{\alpha}| = \left| Ax^2 + Bxy + Cy^2 \right| .
$$

Thus, to determine $G(\mathfrak{a}, tN(\mathfrak{a}))$ we need to count the number of pairs $(x, y) \in \mathbb{Z}^2$ such that

$$
\left| Ax^2 + Bxy + Cy^2 \right| \leq tN(\mathfrak{a}) \tag{8.3}
$$

and $(f, Ax^2 + Bxy + Cy^2) = 1$. However, we notice that if β is an associate of α, then $(\alpha) = (\beta)$; thus, we must count these pairs (x, y) in such a way that no two corresponding values of α are associates.

In the case of $\Delta > 0$, by Theorem 4.26, we need select only those values of α such that $\alpha > 0$ and $1 < |\alpha/\overline{\alpha}| < \epsilon_\Delta^2$. Thus, we must append to inequality (8.3) the further side conditions:

$$
1 \leq \left| \frac{\alpha_1 x + \alpha_2 y}{\overline{\alpha}_1 x + \overline{\alpha}_2 y} \right| < \epsilon_\Delta^2, \quad \alpha_1 x + \alpha_2 y > 0. \tag{8.4}
$$

Let $\mathcal{P}$ denote the set of all pairs (x, y) modulo f such that $(Ax^2 + Bxy + Cy^2, f) = 1$. By Lemma 8.30 we know that

$$
|\mathcal{P}| = f\phi(f) .
$$

Let $(x_0, y_0) \in \mathcal{P}$ and let $N(\mathfrak{a}, t, x_0, y_0)$ denote the number of points (x, y) of Λ for which both (8.3) and (8.4) hold and $x \equiv x_0 \pmod{f}$ and $y \equiv y_0 \pmod{f}$. Then

$$G(\mathfrak{a}, tN(\mathfrak{a})) = \sum_{(x,y) \in \mathcal{P}} N(\mathfrak{a}, t, x, y) .$$

Suppose $(x_0, y_0) \in \mathcal{P}$ and $x = x_0 + uf$, $y = y_0 + vf$. Let A be the region in the $u - v$ plane subject to the constraints of (8.3) and (8.4) for $u = (x - x_0)/f$, $v = (y - y_0)/f$. Then the area $|A|$ of A is given by

$$|A| = \int \int du \, dv .$$

Putting

$$\begin{cases} \xi = \alpha_1 x_0 + \alpha_2 y_0 + f(\alpha_1 u + \alpha_2 v) , \\ \overline{\xi} = \overline{\alpha}_1 x_0 + \overline{\alpha}_2 y_0 + f(\overline{\alpha}_1 u + \overline{\alpha}_2 v) , \end{cases} \qquad (8.5)$$

we must have

$$|\xi\overline{\xi}| < s \quad (s = tN(\mathfrak{a}))$$
$$1 \leq |\xi/\overline{\xi}| < \epsilon_\Delta^2 ,$$
$$\xi < 0 .$$

These conditions in the $\xi - \overline{\xi}$ plane define two sectors (of equal area) of a hyperbola. We will deal with the sector defined by

$$\xi\overline{\xi} < s , \quad \xi/\overline{\xi} < \epsilon_\Delta^2 , \quad \xi > \overline{\xi} > 0 .$$

The Jacobian of the transformation (8.5) is $f^2(\alpha_1\overline{\alpha}_2 - \overline{\alpha}_1\alpha_2) = f^2 N(\mathfrak{a})\sqrt{\Delta}$; thus,

$$\frac{f^2 \sqrt{\Delta} N(\mathfrak{a})|A|}{2} = \int \int d\xi \, d\overline{\xi} ,$$

where the integration is over $\xi\overline{\xi} \leq s$, $\overline{\xi} > 0$, and $\overline{\xi} < \xi < \epsilon_\Delta^2 \overline{\xi}$. Thus,

$$\frac{f^2 \sqrt{\Delta} N(\mathfrak{a})|A|}{2} = \int_0^{\sqrt{s}} d\xi \int_{\xi/\epsilon_\Delta^2}^{\xi} d\overline{\xi} + \int_{\sqrt{s}}^{\sqrt{s\epsilon_\Delta}} d\xi \int_{\xi/\epsilon_\Delta^2}^{s/\xi} d\overline{\xi} = s \log \epsilon_\Delta .$$

It follows that

$$|A| = \frac{2R_\Delta t}{f^2 \sqrt{\Delta}} .$$

Since the radius of curvature of the hyperbolic arcs bounding A increases with increasing t and the arc length is $O(\sqrt{t})$, we see from Corollary 8.31.1 that

$$N(\mathfrak{a}, t, x_0, y_0) = \frac{2R_\Delta t}{f^2 \sqrt{\Delta}} + O(\sqrt{t}) .$$

Hence,

$$G(\mathfrak{a}, tN(\mathfrak{a})) = \frac{2\phi(f)R_\Delta t}{f\sqrt{\Delta}} + O(\sqrt{t}) \, .$$

In the case of $\Delta < 0$, the region described by (8.3) is an ellipse and there are only w associates to every α. Thus, since the area of the ellipse given by (8.3) is well known[6] to be $2\pi tN(\mathfrak{a})/\sqrt{|\Delta|}$, we get

$$G(\mathfrak{a}, tN(\mathfrak{a})) = \frac{2\pi\phi(f)t}{wf\sqrt{\Delta}} + O(\sqrt{t}) \, .$$

Thus, if $\mathcal{C}$ is any invertible ideal class, we have

$$H(\mathcal{C}, t) = \frac{\kappa_\Delta t\phi(f)}{f} + O(\sqrt{t}) \, .$$

$\square$

Corollary 8.33.1. *If $H(t)$ denotes the number of distinct invertible ideals $\mathfrak{a}$ of $\mathcal{O}_\Delta$ such that $(N(\mathfrak{a}), f) = 1$ and $N(\mathfrak{a}) \leq t$, then*

$$H(t) = \frac{h_\Delta \kappa_\Delta t\phi(f)}{f} + O(\sqrt{t}) \, .$$

Proof. Clearly,

$$H(t) = \sum_{\mathcal{C}} H(\mathcal{C}, t)$$

and our result easily follows from the finiteness of h_Δ.

$\square$

8.5 The Class Number Formula

Our next step is to establish the analytic class number formula:

$$\kappa_\Delta h_\Delta = L(1, \chi_\Delta) \, .$$

In order to derive this from our previous results, we require a simple lemma.

Lemma 8.34. *Let $m > 0$ and $(\Delta, m) = 1$. The number of distinct solutions to the congruence*

$$x^2 \equiv \Delta \pmod{4m} \tag{8.6}$$

is equal to

$$2 \sum_{d\mid m}{}' \left(\frac{\Delta}{d}\right) \, ,$$

where the sum is taken over the squarefree divisors of m.

Proof. If Δ is odd, then $\Delta \equiv 1 \pmod 4$ and $(\Delta, 4m) = 1$. If p is any prime, the number of solutions of

$$x^2 \equiv \Delta \pmod{p^n}$$

is 2 if $p = m = 2$, $2(1 + (\Delta/p))$ if $p = 2$, $m > 2$, and $1 + (\Delta/p)$ if $p > 2$. Thus, by the Chinese Remainder Theorem the number of solutions of (8.6) is

$$2 \prod_{p|m} \left(1 + \left(\frac{\Delta}{p}\right)\right) = 2 \sum_{d|m}{}' \left(\frac{\Delta}{d}\right).$$

If Δ is even, then $4 \mid \Delta$ and m is odd. Hence, the congruence

$$x^2 \equiv \Delta \equiv 0 \pmod 4$$

has two solutions; the remaining part of the proof is similar to that of the first case. $\square$

Note that the number of solutions of (8.6) which are distinct modulo $2m$ is

$$\sum_{d|m}{}' \left(\frac{\Delta}{d}\right).$$

Now, let $\mathfrak{a}$ be any ideal of $\mathcal{O}_\Delta$ such that $N(\mathfrak{a}) = k$ and $(k, f) = 1$. We have seen by Theorem 4.24 that $\mathfrak{a}$ will be such an ideal if and only if

$$\mathfrak{a} = (m) \left[k_0, \frac{l + \sqrt{\Delta}}{2}\right],$$

where $k = m^2 k_0$, $(k, f) = 1$, and $4k_0 \mid \Delta - l^2$. Furthermore, if

$$\mathfrak{b} = (m) \left[k_0, \frac{r + \sqrt{\Delta}}{2}\right],$$

then $\mathfrak{b} = \mathfrak{a}$ if and only if $l \equiv r \pmod{2k_0}$. If by $\nu(k)$ we denote the number of distinct ideals of norm k, we see by the previous result that

$$\nu(k) = \sum_{m^2|k} \sum_{j|k/m^2}{}' \left(\frac{\Delta}{j}\right).$$

If $k = s^2 q$, where q is squarefree and n is written as $n_1 n_2^2$, where n_1 is squarefree, then we can write

$$\sum_{n|k} \left(\frac{\Delta}{n}\right) = \sum_{n_2^2|k} \sum_{n_1|k/n_2^2}{}' \left(\frac{\Delta}{n_1}\right);$$

thus, we have the simple expression for $\nu(k)$:

$$\nu(k) = \sum_{n|k} \left(\frac{\Delta}{n}\right). \tag{8.7}$$

The total number of distinct invertible ideals $\mathfrak{a}$ such that $N(\mathfrak{a}) \le t$ is given by

$$H(t) = \sum_{\substack{1 \le k \le t \\ (k,f)=1}} \nu(k). \tag{8.8}$$

From this observation and (8.7) we can derive another result concerning $H(t)$.

Theorem 8.35.

$$\lim_{t \to \infty} \frac{H(t)}{t} = \frac{\phi(f)}{f} L(1, \chi_\Delta) \,.$$

Proof. Let $M(t; f, n)$ be the number of positive integers not exceeding t/n and relatively prime to f. By (8.7) and (8.8) we have

$$\frac{H(t)}{t} = \frac{1}{t} \sum_{\substack{1 \le k \le t \\ (k,f)=1}} \sum_{n|k} \left(\frac{\Delta}{n}\right)$$

$$= \frac{1}{t} \sum_{1 \le k \le t} \left(\frac{\Delta}{n}\right) \sum_{\substack{1 \le k \le t \\ (k,f)=1 \\ n|k}} 1$$

$$= \frac{1}{t} \sum \left(\frac{\Delta}{n}\right) \sum_{\substack{l \le k \le t/n \\ (k,f)=1}} 1$$

$$= \sum_{n=1}^{\infty} \left(\frac{\Delta}{n}\right) \frac{M(t; f, n)}{t} \,.$$

Now, $M(t; f, n)$ cannot increase as n increases and

$$\frac{M(t; f, n)}{t} \le \frac{1}{n} \,;$$

hence, by Corollary 8.24.1 we see that

$$\sum_{n=1}^{\infty} \left(\frac{\Delta}{n}\right) \frac{M(t; f, n)}{t}$$

will converge uniformly in t. For a fixed value of n,

$$\lim_{n \to \infty} \frac{M(t; f, n)}{t} = \frac{\phi(f)}{nf} \,.$$

Thus,

$$\lim_{t\to\infty} \frac{H(t)}{t} = \sum_{n=1}^{\infty} \left(\frac{\Delta}{n}\right) \lim_{t\to\infty} \frac{M(t; f, n)}{t} = \frac{\phi(f)}{f} L(1, \chi_\Delta).$$

□

Corollary 8.35.1.

$$\kappa_\Delta h_\Delta = L(1, \chi_\Delta).$$

Proof. The proof follows easily from the theorem and Corollary 8.33.1.   □

Let $\Delta = f^2 \Delta_{\mathbb{K}}$, where $\Delta_{\mathbb{K}}$ is the fundamental discriminant of $\mathbb{K}$, and put

$$\alpha_{\mathbb{K}} = \prod_{p|f} \left(1 - \frac{(\Delta_{\mathbb{K}}/p)}{p}\right).$$

By Theorem 8.27 and Corollary 8.35.1 we have

$$h_\Delta = \frac{\alpha_{\mathbb{K}} \kappa_{\Delta_{\mathbb{K}}} h_{\mathbb{K}}}{\kappa_\Delta},$$

where $h_{\mathbb{K}}$ is the class number of the maximal order $[1, \omega_0]$ of $\mathbb{K}$. Thus, we can find h_Δ in terms of $h_{\mathbb{K}}$ by using

$$h_\Delta = \frac{\alpha_{\mathbb{K}} f h_{\mathbb{K}}}{n}, \tag{8.9}$$

where n is the unit index of ϵ_Δ over $\epsilon_{\mathbb{K}}$.

We can also use Corollary 8.35.1 to put an upper bound on h_Δ. In order to do this, we will make use of an idea of Slavutskii.[7] We begin with the case of $\Delta > 0$ and let $\epsilon_\Delta = (x + y\sqrt{\Delta})/2$, where $x, y \in \mathbb{Z}^{>0}$. Since

$$x^2 - y^2 = \pm 4,$$

we get

$$x \geq y\sqrt{\Delta - 4}$$

and

$$\epsilon_\Delta > \frac{y(\sqrt{\Delta - 4} + \sqrt{\Delta})}{2} > \sqrt{\Delta - 4}.$$

Thus,

$$R_\Delta > \log\sqrt{\Delta - 4}. \tag{8.10}$$

If we put $k = \phi(\Delta)/2$ and $S(n) = \sum_{r=1}^{n}(\Delta/r)$, we get

$$|L(1, \chi_\Delta)| = \left|\sum_{n=1}^{\infty} \frac{S(n) - S(n-1)}{n}\right|$$

$$= \left|\sum_{n=1}^{\infty} \frac{S(n)}{n(n+1)}\right|$$

$$< \sum_{n=1}^{k-1} \frac{n}{n(n+1)} + \sum_{n=k}^{\infty} \frac{k}{n(n+1)}$$

by part 2 of Theorem 8.7. Now,

$$\sum_{n=1}^{k-1} \frac{n}{n(n+1)} + k \sum_{n=k}^{\infty} \frac{1}{n(n+1)} = \sum_{n=2}^{k} \frac{1}{n} + k \sum_{n=k}^{\infty} \left(\frac{1}{n} - \frac{1}{n+1} \right) = \sum_{n=1}^{k} \frac{1}{n} .$$

Since it is easily established[8] that

$$\sum_{n=1}^{k} \frac{1}{n} < \log k + \gamma + \frac{1}{k} - \frac{1}{2(k+1)} ,$$

where $\gamma = 0.5772\ldots$ is Euler's constant, we get

$$|L(1, \chi_\Delta)| < \log(\Delta - 1) - \log 2 + \gamma + \frac{\phi(\Delta) + 4}{\phi(\Delta)(\phi(\Delta) + 2)} .$$

If $\Delta > 60$, then $\phi(\Delta) \geq 16$ and $(\phi(\Delta) + 4)/(\phi(\Delta)(\phi(\Delta) + 2)) < 0.07$. Hence, in this case

$$|L(1, \chi_\Delta)| < \log(\Delta - 1) - 0.04 .$$

Since

$$h_\Delta = \frac{L(1, \chi_\Delta)}{\kappa_\Delta} = \frac{\sqrt{\Delta} L(1, \chi_\Delta)}{2R_\Delta} < \frac{\sqrt{\Delta} \log(\Delta - 1) - 0.04}{\log(\Delta - 4)}$$

by (8.10), we see that for $\Delta > 80$, we have

$$h_\Delta < \sqrt{\Delta} .$$

If $\Delta < 80$, it is easy to verify that this inequality also holds by checking each case.

If $\Delta < 0$, we find, by using the same technique, that $|L(1, \chi_\Delta)| < \log |\Delta|$. In this case, we get

$$h_\Delta < \frac{\sqrt{|\Delta|} \log |\Delta|}{3} .$$

In summary, we have[9]

$$h_\Delta < \begin{cases} \sqrt{\Delta} & \text{when } \Delta > 0 \\ \dfrac{\sqrt{|\Delta|} \log |\Delta|}{3} & \text{when } \Delta < 0 . \end{cases} \tag{8.11}$$

Notes and References

[1] This result was proved by Dirichlet in 1839. See [Dir99].

[2] The material in this chapter is classical. Useful references are [Hua82], [Coh62], [Apo98], [Ayo63], and [IK04].

[3] See, for example, the treatment in [Hua82], §12.3.

[4] See, for example, [Apo98], §2.2.

[5] See [Ayo63], p. 283.

[6] See [Coh62], p. 160ff.

[7] [Sla69].

[8] This follows easily from the more precise inequality

$$\sum_{n=1}^{k} \frac{1}{n} < \log k + \gamma + \frac{1}{2k} - \frac{1}{12k^2} + \frac{1}{120k^4} \, ,$$

which can be found in [Knu97], §1.2.11.2.

[9] In the case where Δ is positive and a fundamental discriminant, this result was improved by Le [Le94] to

$$h_{\mathbb{K}} < \sqrt{\Delta}/2 \, .$$

See also [Ram01], p. 248.

Some Additional Analytic Results

9.1 More on Gauss Sums

By (8.9), it is clear that we can evaluate h_Δ for any order $\mathcal{O}$ of discriminant Δ of $\mathbb{K}$, once we know the unit index n and $h_\mathbb{K}$. However, if we refer to Corollary 8.35.1 for determining $h_\mathbb{K}$, we notice that the expression for $L(1, \chi_\Delta)$ is an infinite series. The purpose of this section is to derive a closed-form formula for $h_\mathbb{K}$. To this end we must first consider some properties of a particular Gauss sum.

Let $G(n, k)$ be the Gauss sum defined by

$$G(n, k) = \sum_{j=0}^{k-1} e^{2\pi i j^2 n/k} = \sum_{j=1}^{k} e^{2\pi i j^2 n/k} \, ,$$

where $(k, n) = 1$. We first prove the following simple theorem.

Theorem 9.1. *If p is an odd prime, then*

$$G(n, p) = \sum_{j=1}^{p} \left(\frac{j}{p} \right) e^{2\pi i j n/p} \, .$$

Proof. The number of solutions to

$$x^2 \equiv k \pmod{p}$$

is $1 + (k/p)$; hence,

$$\sum_{j=1}^{p} e^{2\pi i j^2 n/p} = \sum_{k=1}^{p} \left(1 + \left(\frac{k}{p} \right) \right) e^{2\pi i k n/p}$$

$$= \sum_{k=1}^{p} \left(\frac{k}{p} \right) e^{2\pi i k n/p} \, .$$

$\square$

Corollary 9.1.1. $G(n,p) = (n/p)G(1,p)$.

Proof. $\chi(n) = (n/p)$ is a real primitive character modulo p. Hence, the result follows by Theorem 8.18.   □

We now turn our attention to the value of $G(1,p)$.

Corollary 9.1.2.

$$G(1,p)^2 = \left(\frac{-1}{p}\right) p \, .$$

Proof. This follows from Corollary 8.20.1.   □

By this result we see that $G(1,p) = \pm\sqrt{p}$ when $p \equiv 1 \pmod 4$ and $G(1,p) = \pm i\sqrt{p}$ when $p \equiv -1 \pmod 4$. The problem of determining the correct sign is very difficult and was solved by Gauss in 1805 after several attempts. Since that time, many different proofs of the value of $G(1,p)$ have appeared.[1] In what follows, we give the proof presented by Hua.[2] It is convenient for its simplicity and its relative brevity.

Theorem 9.2.

$$G(1,p) = \begin{cases} \sqrt{p} \text{ when } p \equiv 1 \pmod 4 \\ i\sqrt{p} \text{ when } p \equiv -1 \pmod 4 \, . \end{cases}$$

Proof. For a given p, we know that

$$G(1,p) = \eta\sqrt{p}$$

when $p \equiv 1 \pmod 4$ or

$$G(1,p) = \eta i\sqrt{p}$$

when $p \equiv -1 \pmod 4$. Here, $\eta = \pm 1$. We can combine both of these formulas as

$$\frac{1}{2}(1+i^p)(1-i)G(1,p) = \eta\sqrt{p} \, .$$

Thus, if we can show that

$$\Re\left(\frac{1}{2}(1+i^p)(1-i)G(1,p)\right) > -\sqrt{p} \, ,$$

we have $\eta = 1$.

We first note that if f is any function and p is odd, then

$$\sum_{j=1}^{(p-1)/2} f(j) + \sum_{j=1}^{(p-1)/2} f(p/2 - j) = \sum_{j=1}^{p-1} f(j/2) \, .$$

This is easily seen by evaluating the sum on the right over the even values of j and then over the odd values of j. Putting $f(j) = e^{2\pi i j^2/p}$, we see that $f(p/2 - j) = i^p e^{2\pi i j^2/p}$; hence,

$$(1 + i^p) \sum_{j=1}^{(p-1)/2} e^{2\pi i j^2/p} = \sum_{j=1}^{p-1} e^{\pi i j^2/2p} .$$

Now,

$$G(1,p) - 1 = \sum_{j=1}^{p-1} e^{2\pi i j^2/p}$$

$$= \sum_{j=1}^{(p-1)/2} e^{2\pi i j^2/p} + \sum_{j=1}^{(p-1)/2} e^{2\pi i (p-j)^2/p}$$

$$= 2 \sum_{j=1}^{(p-1)/2} e^{2\pi i j^2/p} .$$

Putting

$$W = \sum_{j<\sqrt{p}} e^{\pi i j^2/2p}, \quad Z = \sum_{j>\sqrt{p}}^{p-1} e^{\pi i j^2/2p} ,$$

we get

$$\frac{1}{2}(1 + i^p)(1 - i)G(1,p) = \frac{1}{2}(1 + i^p)(1 - i) + (1 - i)(W + Z) .$$

Since

$$\Re\left(\frac{1}{2}(1 + i^p)(1 - i)\right) = 0, 1$$

and

$$|\Re((1 - i)Z)| \le \sqrt{2}|Z| ,$$

we get

$$\Re\left(\frac{1}{2}(1 + i^p)(1 - i)G(1,p)\right) \ge \Re((1 - i)W) - \sqrt{2}|Z| .$$

Also,

$$\Re((1 - i)W) = \sum_{j<\sqrt{p}} \left(\cos\frac{\pi j^2}{2p} + \sin\frac{\pi j^2}{2p}\right)$$

and $\cos x + \sin x \ge 1$ for $0 < x < \pi/2$; thus, we get

$$\Re((1 - i)W) \ge \lfloor\sqrt{p}\rfloor > \sqrt{p}/2 .$$

It remains to bound $|Z|$. Put

$$v_j = e^{\pi i j(j+1)/2p}, \quad w_j = \frac{1}{\sin(\pi j/2p)}, \quad q = \lfloor\sqrt{p}\rfloor .$$

We have

$$(v_j - v_{j-1})w_j = 2ie^{\pi ij^2/2p} .$$

Consequently,

$$2iZ = \sum_{j=q+1}^{p-1} (v_j - v_{i-1})w_j$$

and

$$2|Z| = \left| \sum_{j=q+1}^{p-1} v_j(w_j - w_{j+1}) + v_{p-1}w_p - v_q w_{q+1} \right| .$$

Now, w_j decreases for increasing values of j in the range of summation. Also, since $\sin x > 2x/\pi$ for $0 \le x \le \pi/2$, we get

$$2|Z| \le \left(\sum_{j=q+1}^{p-1} w_j - w_{j+1} \right) + w_p + w_{q+1} = 2w_{q+1} \le \frac{2p}{q+1} < 2\sqrt{p} .$$

Thus,

$$\Re \left(\frac{1}{2}(1 + i^p)(1 - i)G(1,p) \right) > \left(\frac{1}{2} - \sqrt{2} \right) \sqrt{p} > -\sqrt{p} .$$

$\square$

If $\chi(j)$ is the Kronecker symbol (m/j), define

$$S(n,m) = S(n,\chi) = \sum_{j=1}^{|m|} \left(\frac{m}{j} \right) e^{2\pi ijn/|m|} .$$

Note that if $p^* = (-1)^{(p-1)/2}p$, where p is an odd prime, then by Corollary 9.1.1, Theorem 9.2, and the Law of Quadratic Reciprocity, we get

$$S(n,p^*) = \left(\frac{p^*}{n} \right) \sqrt{p^*} . \tag{9.1}$$

Also, by direct calculation, we have

$$S(n,-4) = \left(\frac{-4}{n} \right) \sqrt{-4} , \tag{9.2}$$

$$S(n,\pm 8) = \left(\frac{\pm 8}{n} \right) \sqrt{\pm 8} . \tag{9.3}$$

9.2 A Closed Formula for $h_{\mathbb{K}}$

In order to derive the formula for $h_{\mathbb{K}}$, we will need to generalize (9.1). Indeed, we will show that

$$S(n,\Delta_{\mathbb{K}}) = \left(\frac{\Delta_{\mathbb{K}}}{n} \right) \sqrt{\Delta_{\mathbb{K}}} .$$

We require two lemmas.

Lemma 9.3. *If m_1 and m_2 are coprime integers and $\chi_i(j) = (m_i/j)$ ($i = 1, 2$), then if one of m_1 or m_2 is congruent to 1 (mod 4), we have*

$$\chi_1(|m_2|)\chi_2(|m_1|) = \begin{cases} -1 & \text{if } m_1, m_2 < 0 \\ 1 & \text{otherwise}. \end{cases}$$

Proof. Suppose that $m_1 \equiv 1$ (mod 4); then

$$\chi_1(|m_2|) = \left(\frac{m_1}{|m_2|}\right) = \left(\frac{|m_2|}{|m_1|}\right).$$

Thus,

$$\chi_1(|m_2|)\chi_2(|m_1|) = \left(\frac{m_2|m_2|}{|m_1|}\right) = 1$$

when $m_2 > 0$. If $m_2 < 0$, then

$$\chi_1(|m_2|)\chi_2(|m_1|) = \left(\frac{-1}{|m_1|}\right) = \begin{cases} 1 & \text{if } m_1 > 0 \\ -1 & \text{if } m_1 < 0. \end{cases}$$

$\square$

Lemma 9.4. *Let m_1 and m_2 be coprime integers such that at least one is congruent to 1 (mod 4) and let $\chi_i(j) = (m_i/j)$ ($i = 1, 2$). If χ is a character mod $|m_1 m_2|$ such that $\chi(j) = \chi_1(j)\chi_2(j)$ and*

$$S(n, m_1) = \chi_1(n)\sqrt{m_1}, \quad S(n, m_2) = \chi_2(n)\sqrt{m_2},$$

then

$$S(n, m_1 m_2) = \chi(n)\sqrt{m_1 m_2}.$$

Proof. By Theorem 8.13, we have

$$\begin{aligned} S(n, m_1 m_2) &= \chi_1(|m_2|)\chi_2(|m_1|)S(n, m_1)S(n, m_2) \\ &= \chi_1(|m_2|)\chi_2(|m_1|)\chi(n)\sqrt{m_1}\sqrt{m_2}. \end{aligned}$$

By Lemma 9.3, we get

$$S(n, m_1 m_2) = \chi(n)\sqrt{m_1 m_2}$$

because[3] $\sqrt{m_1}\sqrt{m_2} = -\sqrt{m_1 m_2}$ when $m_1, m_2 < 0$.

$\square$

We are now able to prove the result mentioned earlier.

Theorem 9.5. *If Δ is a fundamental discriminant, then*

$$S(n, \Delta) = \left(\frac{\Delta}{n}\right)\sqrt{\Delta}.$$

Proof. The proof follows easily from (9.1), (9.2), (9.3), Theorem 8.21, and the preceeding lemma.   □

We will now make use of the result of Theorem 9.5 in order to derive our formula for h_{K}. We will do this by finding a finite sum for $L(1, \chi_{\Delta_{\mathrm{K}}})$. We also require a simple, preliminary lemma.

Lemma 9.6. *If* $0 < \phi < 2\pi$, *then*

$$\sum_{n=1}^{\infty} \frac{\sin n\phi}{n} = \frac{1}{2}(\pi - \phi)$$

and

$$\sum_{n=1}^{\infty} \frac{\cos n\phi}{n} = -\log\left(2\sin\frac{\phi}{2}\right).$$

Proof. It is well known that if $|z| \leq 1$ and $z \neq 1$, then

$$-\log(1 - z) = \sum_{n=1}^{\infty} \frac{z^n}{n},$$

where the logarithm takes its principal value. If we put $z = e^{i\phi}$, where $0 < \phi < 2\pi$, then $\arg(1 - z) = (\phi - \pi)/2$ and $|1 - z| = 2\sin(\phi/2)$. Hence,

$$\sum_{n=1}^{\infty} \frac{e^{in\phi}}{n} = -\log\left(2\sin\frac{\phi}{2}\right) - \frac{(\phi - \pi)i}{2}.$$

The lemma follows on equating real and imaginary parts.   □

Theorem 9.7. *If* Δ_{K} *is a fundamental discriminant, then*

$$L(1, \chi_{\Delta_{\mathrm{K}}}) = \begin{cases} -\dfrac{\pi}{|\Delta_{\mathrm{K}}|^{3/2}} \displaystyle\sum_{j=1}^{|\Delta_{\mathrm{K}}|} \left(\dfrac{\Delta_{\mathrm{K}}}{j}\right) j & \text{when } \Delta_{\mathrm{K}} < 0 \\[4mm] -\dfrac{1}{\sqrt{\Delta_{\mathrm{K}}}} \displaystyle\sum_{j=1}^{\Delta_{\mathrm{K}}} \left(\dfrac{\Delta_{\mathrm{K}}}{j}\right) \log \sin \dfrac{\pi j}{\Delta_{\mathrm{K}}} & \text{when } \Delta_{\mathrm{K}} > 0. \end{cases}$$

Proof. We have

$$\sqrt{\Delta_{\mathrm{K}}} L(1, \chi_{\Delta_{\mathrm{K}}}) = \sum_{n=1}^{\infty} \left(\frac{\Delta_{\mathrm{K}}}{n}\right) \frac{\sqrt{\Delta_{\mathrm{K}}}}{n} = \sum_{n=1}^{\infty} \frac{1}{n} \sum_{j=1}^{|\Delta_{\mathrm{K}}|} \left(\frac{\Delta_{\mathrm{K}}}{j}\right) e^{2\pi i j n/|\Delta_{\mathrm{K}}|}$$

$$= \sum_{j=1}^{|\Delta_{\mathrm{K}}|} \left(\frac{\Delta_{\mathrm{K}}}{j}\right) \sum_{n=1}^{\infty} \frac{e^{2\pi i j n/|\Delta_{\mathrm{K}}|}}{n}$$

by Theorem 9.5. If $\Delta_{\mathbb{K}} < 0$, then on taking imaginary parts of the above formula, we get

$$\sqrt{|\Delta_{\mathbb{K}}|}L(1,\chi_{\Delta_{\mathbb{K}}}) = \sum_{j=1}^{|\Delta_{\mathbb{K}}|}\left(\frac{\Delta_{\mathbb{K}}}{j}\right)\sum_{n=1}^{\infty}\frac{1}{n}\sin\frac{2\pi nj}{|\Delta_{\mathbb{K}}|}$$

$$= \sum_{j=1}^{|\Delta_{\mathbb{K}}|}\left(\frac{\Delta_{\mathbb{K}}}{j}\right)\left(\frac{\pi}{2}-\frac{\pi j}{\Delta_{\mathbb{K}}}\right)$$

$$= -\frac{\pi}{|\Delta_{\mathbb{K}}|}\sum_{j=1}^{|\Delta_{\mathbb{K}}|}\left(\frac{\Delta_{\mathbb{K}}}{j}\right)j .$$

If $\Delta > 0$, by taking real parts we get

$$\sqrt{\Delta_{\mathbb{K}}}L(1,\chi_{\Delta_{\mathbb{K}}}) = \sum_{j=1}^{|\Delta_{\mathbb{K}}|}\left(\frac{\Delta_{\mathbb{K}}}{j}\right)\sum_{n=1}^{\infty}\frac{1}{n}\cos\frac{2\pi nj}{\Delta_{\mathbb{K}}}$$

$$= -\sum_{j=1}^{|\Delta_{\mathbb{K}}|}\left(\frac{\Delta_{\mathbb{K}}}{j}\right)\log\left(2\sin\frac{\pi j}{\Delta_{\mathbb{K}}}\right)$$

$$= -\sum_{j=1}^{\Delta_{\mathbb{K}}}\left(\frac{\Delta_{\mathbb{K}}}{j}\right)\log\sin\frac{\pi j}{\Delta_{\mathbb{K}}} .$$

$\square$

Putting the results of Theorem 9.7 together with that of Corollary 8.35.1 we get

$$h_{\mathbb{K}} = \begin{cases} -\dfrac{w}{2\Delta_{\mathbb{K}}}\displaystyle\sum_{j=1}^{|\Delta_{\mathbb{K}}|}\left(\frac{\Delta_{\mathbb{K}}}{j}\right)j & \text{when } \Delta_{\mathbb{K}} < 0 \qquad (9.4) \\[2em] -\dfrac{1}{2R_{\Delta_{\mathbb{K}}}}\displaystyle\sum_{j=1}^{\Delta_{\mathbb{K}}}\left(\frac{\Delta_{\mathbb{K}}}{j}\right)\log\sin\frac{\pi j}{\Delta_{\mathbb{K}}} & \text{when } \Delta_{\mathbb{K}} > 0 . \quad (9.5) \end{cases}$$

We can modify (9.5) somewhat by writing it as

$$\epsilon_{\mathbb{K}}^{-2h_{\mathbb{K}}} = \frac{\prod_r \sin\frac{\pi r}{\Delta_{\mathbb{K}}}}{\prod_n \sin\frac{\pi n}{\Delta_{\mathbb{K}}}},$$

where by $\prod_r$ $(\prod_n)$ we mean the product taken over all the r (n) such that $1 \le r \le \Delta_{\mathbb{K}}$ $(1 \le n \le \Delta_{\mathbb{K}})$ and $(\Delta_{\mathbb{K}}/r) = 1$ $((\Delta_{\mathbb{K}}/n) = -1)$. If we put $\zeta = e^{2\pi i/\Delta_{\mathbb{K}}}$ and use part 2 of Theorem 8.7, we get

$$\epsilon_{\mathbb{K}}^{-2h_{\mathbb{K}}} = \frac{\prod_r(1-\zeta^r)}{\prod_n(1-\zeta^n)}, \qquad (9.6)$$

another result proved by Dirichlet.

If $\Delta_{\mathbb{K}} = p$, where p is a prime and $p \equiv 1 \pmod 4$, it is a simple matter to deduce from (9.6) that

$$\sqrt{p}\, \epsilon_p^{h_p} = \prod_n (1 - \zeta^n)\,, \tag{9.7}$$

where h_p is the class number of $\mathbb{Q}(\sqrt{p})$. In 1948, Kisilev[4] used (9.7) to show that if $\epsilon_p = (x + y\sqrt{p})/2$ is the fundamental unit of $\mathbb{Q}(\sqrt{p})$, then

$$h_p y \equiv x B_{\frac{p-1}{2}} \pmod p \tag{9.8}$$

and

$$x \equiv (-1)^{(h_p - 1)/2} \left(\frac{p-1}{2}\right)! \pmod p \ .$$

Here, B_n is the nth Bernoulli number defined by

$$\frac{x}{e^x - 1} = \sum_{n=0}^{\infty} \frac{B_n}{n!} x^n \ .$$

In the course of proving some results concerning $\Delta_{\mathbb{K}} = pm$, where p is an odd prime, Ankeny, Artin, and Chowla[5] rediscovered (9.8) for $p \equiv 5 \pmod 8$. They asked[6] if it were always the case that $p \nmid y$. Later, Mordell[7] put this in the form of a conjecture stating that $p \nmid y$, and this has since become known as the Ankeny-Artin-Chowla conjecture.[8] In Chapter 10 we will describe how this conjecture has been verified for all $p < 10^{11}$. Notice[9] that since $h_p < p$, we can have $p \mid y$ if and only if $p \mid B_{(p-1)/2}$.

When $\Delta_{\mathbb{K}} = 4p$, where $p \equiv -1 \pmod 4$, Mordell[10] showed that $p \mid y$ if and only if $p \mid E_{(p-3)/2}$, where E_n is the Euler number defined by

$$\sec x = \sum_{n=0}^{\infty} \frac{(-1)^n E_n x^n}{n!}$$

or

$$\frac{2e^x}{e^{2x} + 1} = \sum_{n=0}^{\infty} \frac{E_n}{n!} x^n \ .$$

All of these results have since been extended to other values of $\Delta_{\mathbb{K}}$ and to other moduli by Kisilev and Slavutskii.[11]

We conclude this section by pointing out that the above results relate to the Pell equation through the following proposition.

Proposition 9.8. *Let $\epsilon_\Delta = (x + y\sqrt{\Delta})/2$ and $\Delta = (2/r)^2 D$. If (t, u) is the fundamental solution of the Pell equation (1.7) and $p\ (> 3)$ is a prime such that $p \mid \Delta$, then $p \mid u$ if and only if $p \mid y$.*

Proof. Put $\alpha = \epsilon_\Delta$, $\beta = \bar{\epsilon}_\Delta$. Then by our results in §1.4 and Table 4.1, we have

$$ru = yu_v(x, n) \; ,$$

where $n = N(\epsilon_\Delta) = \pm 1$. Clearly, if $p \mid y$, then $p \mid u$. Suppose $p \mid u$ and $p \nmid y$. In this case we must have $p \mid u_v(x, n)$, but since $d = \Delta y^2$, we have $p \mid d$. It is easy to see that we must have $p \mid u_p(x, n)$; hence, $\omega(p) = p$ and $p \mid v$. Since $v \in \{1, 2, 3, 6\}$, we can only have $p = 3$, a contradiction.   □

9.3 The Riemann Zeta-Function

As we have seen in Chapter 8, the Dirichlet L-function $L(s, \chi)$ for $\chi(n) = (\Delta/n)$ is of great importance for connecting the value of κ_Δ to that of the class number h_Δ of the order $\mathcal{O}_\Delta$. In this section we shall derive several other properties of $L(s, \chi)$. These will be of considerable importance when we turn our attention to deriving computational techniques for evaluating h_Δ.

We first define the *Riemann zeta-function* $\zeta(s)$ for $s = \sigma + it$, $\sigma > 1$ by

$$\zeta(s) = \sum_{n=1}^{\infty} \frac{1}{n^s} \; . \tag{9.9}$$

An important special value for the zeta-function is $\zeta(2) = \pi^2/6$. If $\sigma \geq \sigma_0 > 1$, then the series on the right of (9.9) converges uniformly and absolutely as

$$\sum_{n=1}^{\infty} \left| \frac{1}{n^{\sigma+it}} \right| = \sum_{n=1}^{\infty} \frac{1}{n^{\sigma}} \leq \sum_{n=1}^{\infty} \frac{1}{n^{\sigma_0}} < 1 + \int_{1}^{\infty} \frac{du}{u^{\sigma_0}} = 1 + \frac{1}{\sigma_0 - 1} \; .$$

Also, each of the terms in (9.9) is an analytic function of s, and, as a consequence, $\zeta(s)$ is analytic for $\Re(s) = \sigma > 1$.

Theorem 9.9. *If $s = \sigma + it$ and $\sigma > 1$, then*

$$\zeta(s) = \prod_{p} \left(1 - \frac{1}{p^s} \right)^{-1} \; ,$$

where the product is taken over all rational prime numbers p.

Proof. Let $x > 2$ and define the function $\zeta_x(s)$ by

$$\zeta_x(s) = \prod_{p \leq x} \left(1 - \frac{1}{p^s} \right)^{-1} \; .$$

Now, for any p,

$$\left(1 - \frac{1}{p^s} \right)^{-1} = \sum_{m=0}^{\infty} \frac{1}{p^{ms}} \; ,$$

and each of these progressions is absolutely convergent and can therefore be multiplied term by term. Thus,

$$\zeta_x(s) = \prod_{p \leq x} \sum_{m=0}^{\infty} \frac{1}{p^{ms}}$$

$$= \sum_{m_1=0}^{\infty} \sum_{m_2=0}^{\infty} \cdots \sum_{m_j=0}^{\infty} \left(\frac{1}{p_1^{m_1} p_2^{m_2} \cdots p_j^{m_j}} \right)^s ,$$

where $2 = p_1 < p_2 < \cdots < p_j$ and $p_1, p_2, \ldots, p_j$ are all the primes less than or equal to x. Also, by the Fundamental Theorem of Arithmetic, all of the terms in this sum are distinct. Hence, we can write this sum as

$$\sum_{n \leq x} \frac{1}{n^s} + \sideset{}{'}\sum_{n > x} \frac{1}{n^s} ,$$

where the $\sum'$ represents summation over the integers $n > x$ whose prime divisors are all less than or equal to x. Now,

$$\left| \sideset{}{'}\sum_{n > x} \frac{1}{n^s} \right| \leq \sideset{}{'}\sum_{n > x} \frac{1}{n^\sigma} < \sum_{n > x} \frac{1}{n^\sigma} \leq \frac{1}{x^\sigma} + \int_x^{\infty} \frac{du}{u^\sigma}$$

$$= \frac{1}{x^\sigma} + \frac{1}{\sigma - 1} x^{1-\sigma} \leq \frac{\sigma}{\sigma - 1} x^{1-\sigma} .$$

Hence,

$$\zeta_x(s) = \sum_{n \leq x} \frac{1}{n^s} + O\left(\frac{\sigma}{\sigma - 1} x^{1-\sigma} \right) .$$

If we let $x \to \infty$, then since $\sigma > 1$, we get $x^{1-\sigma} \to 0$; thus,

$$\zeta(s) = \prod_p \left(1 - \frac{1}{p^s} \right)^{-1} .$$

□

We can extend the definition of $\zeta(s)$, given earlier for $\sigma > 1$ only, to a function in the region for which $\sigma > 0$. We have

$$\zeta(s) = \sum_{n=1}^{\infty} n^{-s} = \sum_{n=1}^{\infty} n \left(n^{-s} - (n+1)^{-s} \right)$$

$$= s \sum_{n=1}^{\infty} n \int_n^{n+1} x^{-s-1} \, dx$$

$$= s \int_1^{\infty} \lfloor x \rfloor x^{-s-1} \, dx .$$

Putting $\lfloor x \rfloor = x - \{x\}$, where $\{x\}$ represents the fractional part of x, we find that

$$\zeta(s) = \frac{s}{s-1} - s \int_1^\infty \{x\} x^{-s-1} \, dx .$$

The integral here is absolutely convergent for $\sigma > 0$ and converges uniformly for $\sigma \geq \sigma_0 > 0$; therefore, it represents an analytic function for $\sigma > 0$. It follows that $\zeta(s)$ is meromorphic for $\sigma > 0$ with only one pole with residue 1 at $s = 1$.

As suggested by results in Chapter 8, there is a similar result for $L(s, \chi)$. If we define

$$S(x) = \sum_{n \leq x} \chi(n) ,$$

then

$$L(s, \chi) = \sum_{n=1}^\infty \frac{\chi(n)}{n^s} = \sum_{n=1}^\infty S(n) \left[n^{-s} - (n+1)^{-s} \right]$$

$$= s \int_1^\infty S(x) x^{-s-1} \, dx$$

for $\sigma > 1$. If χ is not the principal character χ_0, we know by Theorem 8.6 that $S(n)$ is a bounded function of x. Thus, the integral here provides the analytic continuation of $L(s, \chi)$ to that region in the plane for which $\sigma > 0$.

We can generalize the Riemann zeta-function, which can be thought of as being defined over the rational field $\mathbb{Q}$, by defining a similar kind of function over the field $\mathbb{K}$. We define the *Dedekind zeta-function*

$$\zeta_{\mathbb{K}}(s) = \sum_{\mathfrak{a}} \frac{1}{N(\mathfrak{a})^s} \quad (\sigma > 1) ,$$

where the summation extends over all the (integral) ideals $\mathfrak{a}$ of $\mathbb{K}$. We can also write this as

$$\zeta_{\mathbb{K}}(s) = \sum_{n=1}^\infty \frac{\nu(n)}{n^s} ,$$

where $\nu(n)$ is the number of distinct ideals of $\mathbb{K}$ of norm n. It is of some interest to examine $\zeta_{\mathbb{K}}(s)$ from the point of view of possible analytic continuation to $\sigma > 1/2$. As in Chapter 8, we let

$$H(n) = \sum_{t \leq n} \nu(t) .$$

Since $\zeta_{\mathbb{K}}$ is defined over $\mathbb{K}$, we may assume that $\mathbb{K}$ has fundamental discriminant $\Delta_{\mathbb{K}}$, and in this case, the conductor is 1. Thus, by Corollary 8.33.1 we have

$$H(n) = h\kappa n + O(\sqrt{n}) = O(n) ,$$

where $h = h_\mathbb{K}$, $\kappa = \kappa_{\Delta_\mathbb{K}}$. By using the method of partial summation of Theorem 8.24, we see that

$$\sum_{n=1}^{N} \frac{\nu(n)}{n^s} = \sum_{n=1}^{N-1} H(n) \left(n^{-s} - (n+1)^{-s}\right) + H(N)N^{-s} .$$

Since $H(N) = O(N)$, we get

$$\zeta_k(s) = \sum_{n=1}^{\infty} \frac{\nu(n)}{n^s} = \sum_{n=1}^{\infty} H(n) \left(n^{-s} - (n+1)^{-s}\right) \quad (\sigma > 1)$$

$$= s \int_1^\infty H(x)x^{-s-1}\,dx$$

$$= s \int_1^\infty (h\kappa x + d(x))x^{-s-1}\,dx$$

$$= \frac{sh\kappa}{s-1} + s \int_1^\infty d(x)x^{-s-1}\,dx .$$

Since $d(x) = O(\sqrt{x})$, the integral here converges for $\sigma > 1/2$, which means that $\zeta_\mathbb{K}(s)$ can be extended analytically to $\sigma > 1/2$. Thus, $\zeta_\mathbb{K}(s)$ is analytic for $\sigma > 1/2$ except for a single pole at $s = 1$ with residue $h\kappa$; that is, $\lim_{s\to 1}(s-1)\zeta_\mathbb{K}(s) = h\kappa$.

Just as in the case of the Riemann zeta-function, we can express $\zeta_\mathbb{K}(s)$ as an Euler product.

Theorem 9.10. *If $\sigma > 1$, then*

$$\zeta_\mathbb{K}(s) = \prod_{\mathfrak{p}} \left(1 - \frac{1}{N(\mathfrak{p})^s}\right)^{-1} ,$$

where the product is taken over all the prime ideals $\mathfrak{p}$ of $\mathbb{K}$.

Proof. The argument is almost the same as that used in the proof of Theorem 9.9. We use

$$\zeta_{\mathbb{K},x}(s) = \prod_{N(\mathfrak{p})\le x} \left(1 - \frac{1}{N(\mathfrak{p})^s}\right)^{-1}$$

and the fact, established in §4.4, that the decomposition of an ideal into prime ideals is unique.   □

We can now use this result to relate $\zeta_\mathbb{K}(s)$, $\zeta(s)$ and $L(s,\chi)$, where $\chi(n) = (\Delta_\mathbb{K}/n)$.

Theorem 9.11. *If $\sigma > 1$ and $\chi(n) = (\Delta_\mathbb{K}/n)$, then*

$$\zeta_\mathbb{K}(s) = \zeta(s)L(s,\chi) .$$

Proof. We note that

$$\zeta_{\mathbb{K}}(s) = \prod_{p} \prod_{\mathfrak{p}|p} \left(1 - \frac{1}{N(\mathfrak{p})^s}\right)^{-1}.$$

We now investigate the inner product. There are three cases.

CASE 1: $(\Delta_{\mathbb{K}}/p) = 1$:
Since $(\Delta_{\mathbb{K}}/p) = 1$, we have $(p) = \mathfrak{p}\bar{\mathfrak{p}}$ and $N(\mathfrak{p}) = N(\bar{\mathfrak{p}}) = p$. Thus,

$$\prod_{\mathfrak{p}|p} \left(1 - \frac{1}{N(\mathfrak{p})^s}\right)^{-1} = \left(1 - \frac{1}{p^s}\right)^{-2} = \left(1 - \frac{1}{p^s}\right)^{-1} \left(1 - \frac{\chi(p)}{p^s}\right)^{-1}.$$

CASE 2: $(\Delta_{\mathbb{K}}/p) = -1$:
In this case, $(p) = \mathfrak{p}$ and $N(\mathfrak{p}) = p^2$; thus,

$$\prod_{\mathfrak{p}|p} \left(1 - \frac{1}{N(\mathfrak{p})^2}\right)^{-1} = \left(1 - \frac{1}{p^{2s}}\right)^{-1} = \left(1 - \frac{1}{p^s}\right)^{-1} \left(1 - \frac{\chi(p)}{p^s}\right)^{-1}.$$

CASE 3: $(\Delta_{\mathbb{K}}/p) = 0$:
Here we have $p = \mathfrak{p}^2$ and $N(\mathfrak{p}) = p$. Hence,

$$\prod_{\mathfrak{p}|p} \left(1 - \frac{1}{N(\mathfrak{p})^s}\right)^{-1} = \left(1 - \frac{1}{p^s}\right)^{-1} = \left(1 - \frac{1}{p^s}\right)^{-1} \left(1 - \frac{\chi(p)}{p^s}\right)^{-1}.$$

In each case, we have

$$\prod_{\mathfrak{p}|p} \left(1 - \frac{1}{N(\mathfrak{p})^s}\right)^{-1} = \left(1 - \frac{1}{p^s}\right)^{-1} \left(1 - \frac{\chi(p)}{p^s}\right)^{-1};$$

it follows by Theorem 9.9 that

$$\zeta_{\mathbb{K}}(s) = \zeta(s) \prod_{p} \left(1 - \frac{\chi(p)}{p^s}\right)^{-1}.$$

By using the same argument as that employed in the proof of Theorem 9.9, it can be shown that for any Dirichlet character χ modulo m,

$$L(s, \chi) = \prod_{p} \left(1 - \frac{\chi(p)}{p^s}\right)^{-1} \quad (\sigma > 1). \tag{9.10}$$

Thus,

$$\zeta_{\mathbb{K}}(s) = \zeta(s) L(s, \chi).$$

$\square$

9.4 The Euler Product for $L(1, \chi)$

We have seen in Chapter 8 that the value of $L(1, \chi)$ exists. Furthermore, we know by (9.10) that we can express $L(s, \chi)$ as an Euler product as long as $\Re(s) > 1$. In view of the importance of $L(1, \chi)$ when $\chi = (\Delta/n)$, it is of some interest to prove that (9.10) holds when $s = 1$. To this end, we first define the *(von Mangoldt) lambda function*

$$\Lambda(n) = \begin{cases} \log p & \text{if } n \text{ is a power of a rational prime } p \\ 0 & \text{otherwise .} \end{cases}$$

We first note that if

$$n = \prod_{i=1}^{k} p_1^{\alpha_i}$$

is the decomposition of n into the powers of distinct primes, then

$$\sum_{r|n} \Lambda(r) = \sum_{i=1}^{k} \alpha_i \log p_i = \log n . \tag{9.11}$$

If by $\pi(x)$ we denote the number of rational primes less than or equal to x, then by the definition of $\Lambda(n)$ we have

$$\Psi(x) := \sum_{r \leq x} \Lambda(r) \leq \pi(x) \log_2 x . \tag{9.12}$$

To get some idea of the growth rate of $\Psi(x)$, we will require a simple lemma concerning $\pi(x)$.

Lemma 9.12. *There exists a constant C such that for all $x > 1$,*

$$\pi(x) < \frac{Cx}{\log x} .$$

Proof. Put $\theta(x) = \sum_{p \leq x} \log p$. Now,

$$\theta(x) \geq \sum_{\sqrt{x} < p \leq x} \log p \geq \left(\pi(x) - \pi(\sqrt{x}) \right) \log \sqrt{x} .$$

Thus, since $\pi(\sqrt{x}) < \sqrt{x}$, we have

$$\pi(x) < \frac{2\theta(x)}{\log x} + \sqrt{x} . \tag{9.13}$$

Consider the binomial coefficient $N = \binom{2n}{n}$. We note that

$$N < (1+1)^{2n} = 2^{2n} ;$$

but since

$$N = \frac{2n(2n-1)\cdots(n+1)}{n!} ,$$

we see that N is divisible by all the primes p such that $n < p \le 2n$, as each of these primes divides the numerator but does not divide the denominator. Thus,

$$N \ge \prod_{n<p\le 2n} p ,$$

and we get

$$2n \log 2 > \log N \ge \sum_{n<p\le 2n} \log p = \theta(2n) - \theta(n) .$$

Putting $n = 2^{r-1}$ and summing from $r = 1$ to t, we find that

$$\theta\left(2^t\right) < \sum_{k=1}^{t} 2^r \log 2 < 2^{t+1} \log 2 .$$

Hence, if $n > 1$ and t is defined by $2^{t-1} \le n < 2^t$, then

$$\theta(n) \le \theta(2^t) < 2^{t+1} \log 2 \le 4n \log 2 .$$

It follows by (9.13) that

$$\pi(n) < \frac{8n \log 2}{\log n} + \sqrt{n} ,$$

and it follows that there must exist a constant C such that

$$\pi(x) < \frac{Cx}{\log x}$$

for all $x > 1$.   □

In fact, Rosser and Schoenfeld[12] have shown by a much deeper analysis that C can be as small as $C = 1.25506$. By using Lemma 9.12 and (9.12) we also have

$$\Psi(x) = O(x) . \tag{9.14}$$

We are now able to prove the following result.

Theorem 9.13. *If $\chi(n)$ is a non-principal Dirichlet character modulo m, then*

$$\left| \sum_p \frac{\chi(p)}{p} \right| ,$$

where the sum is taken over all the rational primes, converges.

Proof. We consider the sum

$$\sum_{n\leq x} \frac{\chi(n)\log n}{n} = \sum_{k_1,k_2\leq x} \frac{\chi(k_1)\chi(k_2)\Lambda(k_1)}{k_1 k_2}$$

by (9.11). Then

$$\sum_{n\leq x} \frac{\chi(n)\log n}{n} = \sum_{k_1\leq x} \frac{\chi(k_1)\Lambda(k_1)}{k_1} \sum_{k_2\leq x/k_1} \frac{\chi(k_2)}{k_2} .$$

Now,

$$L(1,\chi) = \sum_{n\leq x} \frac{\chi(n)}{n} + \sum_{n>x} \frac{\chi(n)}{n} ,$$

and by (8.2), we have

$$\left| \sum_{n>x} \frac{\chi(n)}{n} \right| = O\left(\frac{1}{x}\right) .$$

Hence,

$$\sum_{n\leq x} \frac{\chi(n)\log n}{n} = \sum_{k_1\leq x} \frac{\chi(k_1)\Lambda(k_1)}{k_1} \left(L(1,\chi) + O\left(\frac{k_1}{x}\right)\right)$$

and

$$\left| \sum_{n\leq x} \frac{\chi(n)\log n}{n} \right| = \left| L(1,\chi) \sum_{k_1\leq x} \frac{\chi(k_1)\Lambda(k_1)}{k_1} \right| + O\left(x^{-1} \sum_{k\leq x} \Lambda(k)\right)$$

$$= \left| L(1,\chi) \sum_{k\leq x} \frac{\chi(k)\Lambda(k)}{k} \right| + O(1)$$

by (9.14). By using Corollary 8.24.1, we see that as $x \to \infty$, the sum

$$\left| \sum_{n\leq x} \frac{\chi(n)\log n}{n} \right|$$

converges; it follows that

$$\left| \sum_{k\leq x} \frac{\chi(k)\Lambda(k)}{k} \right| = O(1) .$$

The contribution of the prime values of k in this sum is

$$\sum_{p\leq x} \frac{\chi(p)\log p}{p}$$

and the contribution of the other terms is easily shown to be $O(1)$; hence, if

$$S(x) := \sum_{p \le x} \frac{\chi(p) \log p}{p} \,,$$

then $|S(x)| = O(1)$. Since

$$\sum_{p \le x} \frac{\chi(p)}{p} = \sum_{2 \le n \le x} \frac{S(n) - S(n-1)}{\log n} \,,$$

we have

$$\left| \sum_{p \le x} \frac{\chi(p)}{p} \right| = O(1)$$

by Corollary 8.24.1.   □

If we consider the product $P(x)$, where

$$P(x) = \prod_{p \le x} \left(1 - \frac{\chi(p)}{p} \right)^{-1} \,,$$

then by using the infinite series expansion for $\log(1-x)$ when $x^2 < 1$, we can deduce that

$$\left| \log P(x) - \sum_{p \le x} \frac{\chi(p)}{p} \right| = O(1) \,.$$

Thus, by Theorem 9.13 we have $P(x) = O(1)$; hence,

$$\prod_{p} \left(1 - \frac{\chi(p)}{p} \right)^{-1}$$

converges. Since $L(s, \chi) = \prod_p (1 - \chi(p)/p^s)^{-1}$ holds for all s such that $\Re(s) > 1$ and $L(1, \chi)$ exists, we must have

$$L(1, \chi) = \prod_{p} \left(1 - \frac{\chi(p)}{p} \right)^{-1} \,, \tag{9.15}$$

where χ is any non-principal Dirichlet character. An important question that we will consider in the sequel is how to use the Euler product (9.15) to obtain an approximate value for $L(1, \chi)$.

9.5 Bounds on $L(1, \chi)$

Let χ be a Dirichlet character modulo m of order r. We have already seen in Theorem 8.26 that

$$|L(1, \chi)| < \log m + 1 \; ;$$

however, in view of the importance of $L(1, \chi)$ as a component of the analytic class number formula, it is of some interest to discuss other bounds on this quantity. As a complete discussion of the material in this section would require a book in itself, we will only provide a brief description, without proofs, of the results that will be of interest to us. Some of the results mentioned below are now considered to be classical and can be found in standard works[13] devoted to analytic number theory. As usual, we will provide citations in the notes at the end of this chapter for those results that may be more difficult to find.

As an example of a more precise upper bound for $L(1, \chi)$ when χ is a primitive character, we mention the following result of Granville and Soundararajan.[14]

Theorem 9.14. *Define* $c_2 = 2 - 2/\sqrt{e} = 0.786938\ldots$, $c_3 = 4/3 - 1/e^{2/3} = 0.819916\ldots$, $c_4 = 0.8296539741\ldots$, *and* $c_k = c_\infty = 34/35$ *for* $k \geq 5$. *For any primitive Dirichlet character* χ (mod m) *of order* k, *we have*

$$|L(1, \chi)| \leq \begin{cases} \dfrac{1}{4}(c_k + o(1)) \log m & \text{if } m \text{ is cube-free, or } \chi \text{ has order } m^{o(1)} \\[2ex] \dfrac{1}{3}(c_k + o(1)) \log m & \text{otherwise}. \end{cases}$$

However, this powerful result is unfortunately not effective. In the case of $\chi = \chi_\Delta = (\Delta/n)$, Hua[15] showed that

$$L(1, \chi_\Delta) < \frac{1}{2} \log |\Delta| + 1 \; . \tag{9.16}$$

In the case when Δ (> 0) is a fundamental discriminant, this has been improved and generalized by Louboutin[16] to

$$|L(1, \chi)| \leq \frac{1}{2}(\log f_\chi + \kappa_0) \; , \tag{9.17}$$

where χ is any primitive even character of conductor f_χ (> 1) and

$$\kappa_0 = 2 + \gamma - \log(4\pi) = 0.046191\ldots \; .$$

In certain cases, this inequality can be improved; for example,

$$\left| \left(1 - \frac{\chi(2)}{2} \right) L(1, \chi) \right| < \frac{1}{4}(\log f_\chi + 5)$$

when Δ is a positive fundamental discriminant. Ramaré[17] has improved (9.17) to

$$|L(1, \chi)| < \frac{1}{2} \log f_\chi \qquad (9.18)$$

when χ is a primitive, even character. Thus, since $h_\Delta \geq 1$, we see by Corollary 8.35.1 and (9.18) that

$$R_{\mathbb{K}} < \frac{\sqrt{\Delta_{\mathbb{K}}}}{4} \log \Delta_{\mathbb{K}} \ .$$

To obtain a lower bound on $|L(1, \chi)|$ is much more problematical. In the case of $\chi = \chi_\Delta$, where Δ is a fundamental discriminant, it is known[18] that if $L(s, \chi)$ has no real zero in the interval

$$\left(1 - \frac{c_1}{\log |\Delta|}, 1 \right) \qquad (9.19)$$

for some constant c_1, then

$$L(1, \chi) > \frac{c_2}{\log |\Delta|} \ ,$$

where the constant c_2 depends on c_1 only. It is also known that $L(s, \chi)$ can have at most one zero in the interval (9.19). Furthermore,[19] for some suitable constant c, there is at most one real primitive χ to a modulus $m \leq v$ for which $L(s, \chi)$ has a real zero β satisfying

$$\beta > 1 - \frac{c}{\log v} \ .$$

Such a zero β is called a *Landau-Siegel zero*. It is not known whether any such zero exists, but it is strongly believed that none does. However, even if such a zero does exist, Siegel[20] was able to show that for any $\epsilon > 0$, we must have

$$L(1, \chi_\Delta) > \frac{c_1(\epsilon)}{|\Delta|^\epsilon} \ , \qquad (9.20)$$

where $c_1(\epsilon)$ is an ineffective (unfortunately) constant depending on ϵ. From this, we see that if Δ is a fundamental discriminant and $\epsilon > 0$, there exists some (ineffective) constant $c_2(\epsilon)$ such that

$$h_\Delta > c_2(\epsilon)|\Delta|^{1/2-\epsilon} \quad \text{for } \Delta < 0 \qquad (9.21)$$

and

$$h_\Delta R_\Delta > c_2(\epsilon)\Delta^{1/2-\epsilon} \quad \text{for } \Delta > 0 \ . \qquad (9.22)$$

To render (9.20) effective, we have to deal with the possibility of the existence of a Landau-Siegel zero; this was done by Tatuzawa,[21] who showed that if $0 < \epsilon < 1/2$ and $|\Delta| \geq \max\{e^{1/\epsilon}, e^{11.2}\}$, then

$$L(1, \chi) > \frac{0.655\epsilon}{|\Delta|^{\epsilon}}$$

with one possible exception. This result has been improved by Hoffstein.[22]

It seems, then, that to obtain an effective lower bound on $L(1, \chi)$, it is important to know something about the location of the zeros of $L(s, \chi)$. In the case of $\zeta(s)$, it was proved by Riemann that if

$$\xi(s) = \frac{1}{2} s(s-1) \pi^{-s/2} \Gamma\left(\frac{s}{2}\right) \zeta(s) ,$$

then $\xi(s)$ is an entire function and we get the functional equation

$$\xi(s) = \xi(1-s) .$$

From this result we can derive some important characteristics of the zeros of $\zeta(s)$.

Theorem 9.15. *The zeros of $\xi(s)$, if any, are all located in the strip $0 \le \sigma \le 1$ and lie symmetrically about the lines $t = 0$ and $\sigma = 1/2$. The zeros of $\zeta(s)$ are identical in position and multiplicity with those of $\xi(s)$, except that $\zeta(s)$ has a simple zero at the points $s = -2, -4, -6, \ldots$.*

The strip $0 \le \sigma \le 1$ is called the *critical strip* and the line $\sigma = 1/2$ is called the *critical line*. The points $s = -2, -4, -6, \ldots$ are called the *trivial zeros* of $\zeta(s)$. In 1860, Riemann conjectured that the only non-trivial zeros of $\zeta(s)$ lie on the critical line. Today this conjecture is called the Riemann hypothesis (RH). The RH is of great importance in number theory, and as a consequence of this importance, there have been many attempts to prove it. In spite of the expense of a great deal of effort and ingenuity, it still remains unproved[23]; however, several attempts have been made to verify the RH numerically for certain ranges of t in the critical strip. In fact, the first 10^{13} zeros have been found by Gourdon,[24] and without exception they are on the critical line. Odlyzko[25] examined the neighbourhood of the 10^{22}th zero of $\zeta(s)$. He found 10 billion zeros in this region, all of which are on the critical line. Although such extensive computations do not prove the RH, they are strongly suggestive of its truth.

For the remainder of this section we will assume that $\chi = \chi_\Delta$ and Δ is a fundamental discriminant. We also have a functional equation for $L(s, \chi)$.

Theorem 9.16. *Let χ be a primitive non-principal character modulo m. Let*

$$a = \begin{cases} 0 & \text{if } \chi(-1) = 1 \\ 1 & \text{if } \chi(-1) = -1 . \end{cases}$$

Then

$$\xi(s, \chi) = \xi(1-s, \chi) ,$$

where

$$\xi(s, \chi) = \left(\frac{\pi}{|\Delta|}\right)^{-(s+a)/2} \Gamma\left(\frac{s+a}{2}\right) L(s, \chi) .$$

Since $L(s, \chi) \neq 0$ for $\sigma > 1$ (see (9.10)), we see that $\xi(s, \chi) \neq 0$ and $\xi(s, \overline{\chi}) \neq 0$ for $\sigma > 1$. By the functional equation, $\xi(s, \chi) \neq 0$ for $\sigma < 0$, so that all the zeros of $\xi(s, \chi)$ must lie in the critical strip. The only zeros $L(s, \chi)$ that lie outside the critical strip are at $s = 0, -2, -4, -6, \ldots$ when $\chi(-1) = 1$ and at $s = -1, -3, -5, \ldots$ when $\chi(-1) = -1$. These are simple zeros and are called the *trivial zeros* of $L(s, \chi)$. Since the L-function behaves, with respect to its zeros, in a manner rather similar to that of the zeta-function, it has been conjectured that all of the non-trivial zeros of $L(s, \chi)$ lie on the critical line. This is called the extended Riemann hypothesis (ERH). Its status, like that of the RH, is unproved; however, numerical trials of the ERH have always confirmed its validity. Extensive numerical testing of the ERH was done by Rumely.[26]

Assuming the ERH, Littlewood[27] showed that as $|\Delta|$ goes to infinity:

$$(1 + o(1))(c_1 \log \log |\Delta|)^{-1} < L(1, \chi) < (1 + o(1))c_2 \log \log |\Delta| , \qquad (9.23)$$

where $c_2 = 2e^\gamma$ and $c_1 = c_2/\zeta(2)$. He also showed (again under the ERH) that there are infinitely many values of Δ such that

$$L(1, \chi) \geq \frac{(1 + o(1))c_2}{2} \log \log |\Delta| \qquad (9.24)$$

and

$$L(1, \chi) \leq \frac{1 + o(1)}{(c_1/2) \log \log |\Delta|} ; \qquad (9.25)$$

later, Chowla[28] established (9.25) unconditionally. Shanks[29] improved (9.23) slightly by noticing that the constants c_1 and c_2 can be changed to $c_2 = re^\gamma$ and $c_1 = 4(r + 1)e^\gamma/\pi^2$. He also carried out some numerical calculations to determine whether (9.23) is ever violated. Such computations have been extended by later investigations,[30] and in every case, no violation of (9.23) has ever been detected.

Nevertheless, while such numerical experiments may constitute some support for (9.23), we still seem to be a very long way from a proof. Other researchers[31] have attempted to use probabilistic methods to examine the distribution of the values of Δ which produce what are called the extreme values of $L(1, \chi)$ (i.e., those values of $L(1, \chi)$ for which either (9.24) or (9.25) holds). As a result of their research into this matter, Montgomery and Vaughn[32] made the following conjecture.

Conjecture 9.17. The proportion P of fundamental discriminants $|\Delta| \leq x$ with $L(1, \chi) \geq (c_2/2) \log \log |\Delta|$ is

$$\exp(-C \log x / \log \log x) < P < \exp(-c \log x / \log \log x)$$

for appropriate finite constants $0 < c < C$.

Similar estimates apply to the proporition of fundamental discriminants $|\Delta| \leq x$ with $L(1, \chi) \leq 2/c_1 \log \log |\Delta|$. A stronger version of this conjecture was recently proved by Granville and Soundararajan.[33] Thus, it would appear that the values of Δ which produce extreme values for $L(1, \chi)$ tend to be very sparsely distributed as $|\Delta|$ becomes large. Indeed, Granville and Soundararjan speculated that possibly

$$\max_{|\Delta| < x} L(1, \chi) = \frac{c_2}{2} \left(\log \log x + \log \log \log x + C_1 + o(1)\right) ,$$

where C_1 is some constant.

If we bear in mind the above results and take into additional consideration the Cohen-Lenstra heuristics mentioned in Chapter 7, it certainly appears that for any ϵ, there must exist families of positive fundamental discriminants $\{D_k\}$ for which

$$R(D_k) \gg D_k^{1/2-\epsilon} .$$

However, as mentioned in §6.5, the best result that we can prove unconditionally is

$$R(D_k) \gg (\log D_k)^3 .$$

This enormous difference between what must be the case and what can be unconditionally established is most remarkable and is indicative of a significant gap in our knowledge concerning the growth of the regulator of a quadratic field. Indeed, because the value of $h_{\mathbb{K}}$ tends to be small, we would expect that

$$R_{\mathbb{K}} \gg \Delta_{\mathbb{K}}^{1/2-\epsilon} \tag{9.26}$$

for a large proportion of all the real quadratic fields. We certainly observe this in calculations, but as mentioned earlier, we are far from being able to prove it.

The situation in the case of $\Delta < 0$ is somewhat better. Of course, in this case we are not concerned about the growth of the regulator but rather that of the class number. As early as 1801, Gauss[34] conjectured that the number of negative discriminants $\Delta < 0$ which have a given class number $h_{\mathbb{K}}$ is finite. By Siegel's result (9.21) we get

$$h_{\mathbb{K}} \gg |\Delta|^{1/2-\epsilon} , \tag{9.27}$$

but the implied constant is not computable. Thus, we cannot use (9.27) to find all imaginary quadratic fields with a fixed class number. For $h_{\mathbb{K}} = 1$ and $h_{\mathbb{K}} = 2$, this problem was solved by Heegner, Stark, and Baker.[35] Later, Gross and Zagier,[36] building on previous work of Goldfeld[37] which made use of L-functions of elliptic curves, showed that

$$h_{\mathbb{K}} \gg \prod_{p|\Delta} \left(1 - \frac{2}{\sqrt{p}}\right) \log |\Delta| ,$$

where the implied constant is effectively computable. Indeed, Oesterlé[38] showed that

$$h_{\mathbb{K}} > \frac{1}{7000} \log |\Delta| \prod_{\substack{p | \Delta \\ p \neq |\Delta|}} \left(1 - \frac{\lfloor 2\sqrt{p} \rfloor}{p + 1} \right) .$$

This allowed for the determination[39] of all the imaginary quadratic fields for which $h_{\mathbb{K}} \leq 100$.

Considering the Cohen-Lenstra heuristics, it is easy to understand that the equivalent problem of obtaining lower bounds on class numbers of $\mathbb{K}$ for the case of $\Delta > 0$ is much more complicated. For example, although we strongly believe it, we still do not know whether there exists an infinitude of fundamental discriminants Δ (> 0) for which $h_{\mathbb{K}} = 1$. In certain cases, however, we can say something. For example, if $\{D_k\}$ is a sleeper, then because the regulator is bounded by $O(\log D_k)$, we see by (9.21) that we must have $\lim_{k \to \infty} h_{\mathbb{K}} = \infty$. If we make use of Tatazawa's result, we can even get effective lower bounds on $h_{\mathbb{K}}$ with one possible exception.[40] Getting rid of this exception is very difficult. In 1976, Chowla[41] conjectured that if $\Delta = M^2 + 1$ is a fundamental discriminant and $h_{\mathbb{K}} = 1$, then $M \leq 26$. It was only quite recently that Biró[42] was able to prove this unconditionally. These techniques have also been successfully applied by him[43] to the case of $\Delta = M^2 + 4$ (Yokoi's conjecture) and by Biró and Granville[44] to $\Delta = M^2 + 4M$.

As their regulators are also bounded, we can also consider similar problems for creepers. In this case, the situation is, somewhat surprisingly, less difficult. In the case of the Shanks sequence

$$S_n = (2^n + 3)^2 - 8 ,$$

it is very easy to show[45] that $h_{\mathbb{K}} = 1$ if and only if $n \in \{1, 2, 3, 4, 5\}$. This is because of the remarkable property that $(S_n/127) = 1$ for any $n \in \mathbb{Z}^{\geq 0}$. By results in §4.4 this means that the ideal (127) splits in $\mathcal{O}_{\mathbb{K}}$ as $(127) = \mathfrak{p}\bar{\mathfrak{p}}$ and $N(\mathfrak{p}) = 127$. If $h_{\mathbb{K}} = 1$, there can only be one cycle of reduced ideals in $\mathcal{O}_{\mathbb{K}}$; hence, if $127 < \sqrt{S_n}/2$, we must find 127 as a value of $Q_i/2$ in the simple continued fraction (SCF) expansion of $(\sqrt{S_n} + 1)/2$. However, we have already seen in §6.4 that these values of $Q_i/2$ are all powers of 2. Thus, $h_{\mathbb{K}} = 1$ if $\sqrt{S_n}/2 < 127$ or $S_n < 64516$ or $n \leq 7$. By evaluating $h_{\mathbb{K}}$ for $n = 1, 2, \ldots, 7$, the result is easily established. This technique has been extended to other creepers.[46]

There is even a more general L-function, called the Hecke L-function. This makes use of a more general character χ than a Dirichlet character, called a *Hecke character*. This character can be defined on the class group of $\mathbb{K}$ and we can define the *Hecke L-function* as

$$L(s, \chi, \mathbb{K}) = \sum_{\mathfrak{a}} \frac{\chi(\mathfrak{a})}{N(\mathfrak{a})^s} ,$$

where, as before, s is a complex variable and the sum is taken over all the integral ideals of $\mathcal{O}_{\mathbb{K}}$. When $\chi = \chi^0$, a special Hecke character for which $\chi^0(\mathfrak{a}) = 1$, we have

$$L(s, \chi^0, \mathbb{K}) = \zeta_{\mathbb{K}}(s) .$$

The Hecke L-function satisfies a functional equation similar to that in Theorem 9.16, and the RH for Hecke L-functions asserts that the real part of its non-trivial zeros must be $1/2$. Under this hypothesis, Bach[47] was able to show that the class group of $\mathcal{O}_{\mathbb{K}}$ is generated by classes which contain the prime ideals $\mathfrak{p}$ of $\mathcal{O}_{\mathbb{K}}$ such that

$$N(\mathfrak{p}) < 6 \log^2 |\Delta_{\mathbb{K}}| . \tag{9.28}$$

Furthermore, the class group of any maximal or non-maximal order $\mathcal{O}$ can be generated by classes which contain the prime ideals $\mathfrak{p}$ of $\mathcal{O}$ such that

$$N(\mathfrak{p}) < 12 \log^2 |\Delta| .$$

There remains the problem of estimating the value of $L(1, \chi)$. In 1903, Lerch[48] converted the $L(1, \chi)$ series when $\Delta > 0$ into the form

$$L(1, \chi) = \Delta^{-1/2} \sum_{n=1}^{\infty} \left(\frac{\Delta}{n} \right) E(An^2) + \sum_{n=1}^{\infty} \left(\frac{\Delta}{n} \right) n^{-1} \operatorname{erfc}(n\sqrt{A}) , \tag{9.29}$$

where

$$A = \frac{\pi}{\Delta}, \quad E(x) = \int_x^{\infty} e^{-t} t^{-1} \, dt, \quad \operatorname{erfc}(x) = \frac{2}{\sqrt{\pi}} \int_x^{\infty} e^{-t^2} \, dt .$$

At first glance, this expression appears to be far more formidable than the simpler formula in §8.3, but we can discuss the convergence of this series much more easily and we do not need any RHs. We point out that

$$0 < \operatorname{erfc}(x) < \frac{e^{-x^2}}{x\sqrt{\pi}}, \quad 0 < E(x) < \frac{e^{-x}}{x} ;$$

hence, if

$$T(m) = \Delta^{-1/2} \sum_{n=m+1}^{\infty} \left(\frac{\Delta}{n} \right) E(An^2) + \sum_{n=m+1}^{\infty} \left(\frac{\Delta}{n} \right) n^{-1} \operatorname{erfc}(n\sqrt{A}) ,$$

then we can show that

$$|T(m)| < \frac{\Delta^{3/2} e^{-Am^2}}{\pi^2 m^3} .$$

Since the value of $T(m)$ becomes small very quickly, we do not have to go very far in summing (9.29) until we get a good estimate for $L(1, \chi)$. Note that each of the transcendental functions $E(x)$ and $\text{erfc}(x)$ can be evaluated to good accuracy by using power series expansions in no more than $O(\log \Delta)$ terms (in practice, Chebychev approximations are more efficient). Indeed, if $m = O(\Delta^{1/2})$, this is sufficient accuracy to find the value of $h_{\mathbb{K}}$ (given $R_{\mathbb{K}}$) unambiguously. This result was used[49] to find all the values of $h_{\mathbb{K}}$ for $\mathbb{K} = \mathbb{Q}(\sqrt{D_0})$ and $D_0 < 1.5 \times 10^5$. There is also a similar analysis that can be performed for $\Delta < 0$.

Recently, Louboutin[50] devised a version of this technique which dispenses with the need to compute approximations to the E and erfc functions. As with the previous technique, this requires $O(\Delta^{1/2+\epsilon})$ operations to compute $h_{\mathbb{K}}$, but it is much more efficient. This procedure and that of simply counting classes are the fastest non-conditional methods known for evaluating h_Δ; the only ways of doing any better involve invoking the ERH.

If we refer back to (9.15), it seems reasonable to attempt to approximate $L(1, \chi)$ by using

$$B(x, \chi) = \prod_{p < x} \left(1 - \frac{\chi(p)}{p}\right)^{-1}$$

and finding some x such that $B(x, \chi)$ is as close as we want to $L(1, \chi)$. If we put

$$\overline{B}(x, \chi) = \prod_{p > x} \left(1 - \frac{\chi(p)}{p}\right)^{-1},$$

then under the ERH it is possible to show that

$$|\log \overline{B}(x, \chi)| = O\left(\frac{\log |\Delta| x}{\sqrt{x}}\right). \tag{9.30}$$

As Bach[51] has pointed out, however, it is possible to improve on this result. To estimate $\overline{B}(x, \chi)$, we introduce weights a_i defined by

$$a_i = \frac{(x + i) \log(x + i)}{\sum_{i=0}^{x-1} (x + i) \log(x + i)} \quad (i = 0, 1, 2, \ldots, x - 1).$$

We note that

$$\sum_{i=0}^{x-1} a_i = 1.$$

Bach showed that under the ERH,

$$\left| \log L(1, \chi) - \sum_{i=0}^{x-1} a_i \log B(x + i, \chi) \right| < \frac{A \log |\Delta| + B}{\sqrt{x} \log x} \tag{9.31}$$

for constants A and B, and he provided a table of values for such constants. Notice that this result is better than (9.30). It has also been found to be of great use in determining values of $h_{\mathbb{K}}$, but it must be borne in mind that these values of $h_{\mathbb{K}}$ are contingent on the truth of the ERH.

Notes and References

[1] See, for example, [Lan50], pp. 158–171.

[2] [Hua82], pp. 165–166.

[3] For a historical perspective on this observation, see [Mar07].

[4] [Kis48].

[5] [AAC52].

[6] [AAC52], p. 480.

[7] [Mor60].

[8] In fact, it was Mordell [Mor60] who put Ankeny, Artin, and Chowla's result in the form (9.8). He also noted that this had been proved by them only for $p \equiv 5 \pmod 8$. Later, Ankeny and Chowla [AC62] established it for all $p \equiv 1 \pmod 4$, but, as we pointed out earlier, this had been done previously by Kisilev.

[9] This was noticed by Ankeny and Chowla [AC60] and earlier by Carlitz [Car53] and Kisilev [Kis48].

[10] [Mor61].

[11] See [Sla69] and, in particular, [Sla65] for references. In fact, in [Sla65] it is mentioned that Kisilev anticipated Mordell's result for $p \equiv -1 \pmod 4$.

[12] [RS62].

[13] See, for example, [Dav00] and [IK04].

[14] [GS02].

[15] [Hua82], §12.13. In fact, Hua almost proved this result in 1942, but made a small error. See the discussion by Louboutin in [Lou02b], p. 12.

[16] [Lou02b].

[17] [Ram01].

[18] See [Lan18] and [Dav00], Ch. 21.

[19] [Dav00], Ch. 14.

[20] [Sie35] and [Dav00], Ch. 21.

[21] [Tat51].

[22] [Hof80].

[23] For a historical account of investigations into this famous problem, see [Edw74]. Also, see the interesting account of more recent investigations in [Con03].

[24] [Gou04]. He also computed two billion zeros at height 10^{24}.

[25] [Odl01]. In more recent unpublished work, Odlyzko computed 20 billion zeros at height 10^{25}.

[26] [Rum93].

[27] [Lit28].

[28] [Cho49].

[29] [Sha73].

[30] See [JLW95], [Jac98], and [JRW06].

[31] See [Ell69], [Ell73], [Ell70], [Ell80], Ch. 22, and [MV99].

[32] [MV99].

[33] [GS03].

[34] [Gau86], §303.

[35] The case of $h = 1$ was solved by arithmetical means in [Hee52] and [Sta67] and by transcendental techniques in [Bak66]. For some interesting historical commentary on this problem, see [Sta69a] and [Sta69b]. The case of $h = 2$ was solved in [MW74] and [Sta75].

[36] [GZ86].

[37] [Gol76].

[38] [Oes85]. For a very readable account of all of this work, see [Gol85].

[39] [Wat04].

[40] See, for example, [MW91b]. More information on this can be found in [Mol95].

[41] [CF76], p. 48.

[42] [Bir03a].

[43] [Bir03b].

[44] [BG07].

[45] [MW91a].

[46] [MW94].

[47] [Bac90].

[48] See [Dic19], Vol. III, p. 164.

[49] [WB76].

[50] [Lou02a].

[51] [Bac95].

10

Some Computational Techniques

10.1 Introduction

In the previous chapters, various methods for computing the regulator were presented. Using Shanks' infrastructure, we have seen in Chapter 7 how to improve the continued fraction algorithm for computing the regulator using the baby-step giant-step method, resulting in an algorithm that computes R_Δ unconditionally in time $O(\Delta^{1/4+\epsilon})$ and, using improvements due to Buchmann, Williams, and Vollmer, in time $O(R_\Delta^{1/2} \Delta^\epsilon)$. In Chapter 8 we have seen how the class number and regulator are intimately connected to $L(1, \chi)$ via the analytic class number formula (Corollary 8.35.1), and in Chapter 9 methods for efficiently computing estimates of h_Δ or $h_\Delta R_\Delta$ using the analytic class number formula were discussed.

The next step in improving these baby-step giant-step algorithms for computing the regulator is to combine the infrastructure techniques with analytic methods for approximating $h_\Delta R_\Delta$. The resulting algorithm, due to Lenstra, computes an unconditionally correct approximation of R_Δ which, under the assumption of the extended Riemann hypothesis (ERH), has complexity $O(\Delta^{1/5+\epsilon})$. We will present a description of Lenstra's algorithm in §10.2.

Once the regulator is computed, analytic methods for approximating $L(1, \chi)$ can also be used to compute the class number efficiently. Some purely analytical algorithms were described in Chapter 9. In §10.3 we describe an algorithm due to Shanks and Lenstra for computing the class number that combines the algebraic baby-step giant-step approach with analytic methods, resulting in a running time $O(\Delta^{1/5+\epsilon})$. Given the class number, a method of Buell can then be used to determine the structure of the class group. This method is surveyed in §10.4, along with algorithms of Buchmann, Jacobson, and Teske and Buchmann and Schmidt that compute the structure of the class group without prior knowledge of the class number in time roughly $O(h_\Delta^{1/2})$.

Finally, in §10.5 we will survey some large-scale computations for numerical verification of a variety of conjectures that make use of the algorithms described in this chapter. These will include the most recent numerical verifications of the Cohen-Lenstra heuristics, Hooley's conjecture, and the Ankeney-Artin-Chowla conjecture. The methods used to compute these tables of class numbers, including a recent method of Jacobson, Ramachandran, and Williams for verifying unconditionally tables of class numbers of imaginary quadratic fields, will also be discussed.

10.2 Computing the Regulator

After Shanks' discovery of the infrastructure and the application of the baby-step giant-step method to computing the regulator of a real quadratic field $\mathbb{Q}(\sqrt{\Delta})$, the next significant advance was due to Lenstra,[1] who invented an algorithm for computing the regulator R_Δ or a real quadratic order $\mathcal{O}_\Delta$ in time $O(\Delta^{1/5+\epsilon})$. Lenstra's algorithm has been used extensively in practice; some examples will be discussed at the end of this chapter. The improvement over Shanks' algorithm described in §7.4 is a result of the combination of Shanks' baby-step giant-step method with analytic results from Chapters 8 and 9. The main idea is to compute an approximation of $h_\Delta R_\Delta$ using the analytic class number formula and an appropriately accurate approximation of $L(1, \chi)$. Bounds on the error of this approximation are used to obtain an interval in which $h_\Delta R_\Delta$ is known to lie. Using the infrastructure, the baby-step giant-step method is then employed to find an integer multiple $h^* R_\Delta$ of the regulator by searching this interval. Finally, once again using infrastructure, the integer multiplier h^* is computed, yielding the regulator R_Δ.

We now describe a recent version[2] of Lenstra's method in detail. The first step is to compute an approximation of $\log L(1, \chi)$ using Bach's method (equation (9.31) from Chapter 9). Any sufficiently accurate and efficient method for approximating $L(1, \chi)$ can be used; we use Bach's method because it is the most accurate and efficient in practice. Recall that, given a bound Q, Bach's method computes

$$S(Q, \Delta) = \sum_{i=0}^{Q-1} a_i \log B(Q + i, \chi) , \tag{10.1}$$

an approximation of $\log L(1, \chi)$, which from (9.31), satisfies

$$|\log L(1, \chi) - S(Q, \Delta)| < A(Q, \Delta) := \frac{A \log |\Delta| + B}{\sqrt{Q} \log Q} \tag{10.2}$$

for explicit constants A and B determined by Bach.[3] If we put

$$E = \frac{\sqrt{\Delta} \exp(S(Q, \Delta))}{2} , \tag{10.3}$$

then $h_\Delta R_\Delta \approx E$, and by using the analytic class number formula and (10.2), we have, assuming the ERH, that

$$|E - h_\Delta R_\Delta| = \frac{\sqrt{\Delta}}{2} \, |\exp(S(Q, \Delta)) - L(1, \chi)|$$
$$= E \left| \frac{L(1, \chi)}{\exp(S(Q, \Delta))} - 1 \right|$$
$$\leq E \, |\exp(A(Q, \Delta)) - 1| \, .$$

Thus,

$$|E - h_\Delta R_\Delta| < L^2 \, , \qquad (10.4)$$

where we set

$$L^2 = E \, |\exp(A(Q, \Delta)) - 1| \, .$$

Taking $Q \approx \Delta^{1/5}$ is optimal for Lenstra's algorithm, as we will prove below.

Given the approximation E of $h_\Delta R_\Delta$ from (10.3), we proceed to compute an integer multiple of R_Δ using the baby-step giant-step method in the infrastructure. We first compute a list $\mathcal{L}$ of baby steps consisting of all reduced principal ideals with distance less than $L + \log 2$ by developing the continued fraction expansion corresponding to $\mathfrak{a}_1 = \mathcal{O}_\Delta$. As in the baby-step giant-step algorithm described in Chapter 7, the list $\mathcal{L}$ is usually stored as a hash table in order to enable fast searching and insertion of new elements. If during the computation of the list we find an ideal $\mathfrak{a}_{n+1} = [Q_n/r, (P_n + \sqrt{D})/r]$ with $P_n = P_{n+1}$, then, as in Shanks' algorithm described in §7.4, we can immediately set

$$R_\Delta = 2\delta(\mathfrak{a}_{n+1}) + \log \frac{Q_0}{Q_n} \, ,$$

and if $Q_n = Q_{n+1}$, we set

$$R_\Delta = 2\delta(\mathfrak{a}_{n+1}) + \log \frac{Q_0 \psi_{n+1}}{Q_n} \, .$$

In either case, we terminate the algorithm.

We next compute a reduced principal ideal $\mathfrak{a}_m = [Q_{m-1}/r, (P_{m-1} + \sqrt{D})/r]$ with $\delta(\mathfrak{a}_m) \approx E$. Recall that, as described in §7.4, if $2^k < E < 2^{k+1}$, then we can use k "doubling" steps to find $\mathfrak{a}_m$ by selecting some $\mathfrak{a}_s (= \mathfrak{b}_1)$ with $\delta_s \approx E/2^k$ and computing $\mathfrak{b}_{j+1} = \mathfrak{b}_j * \mathfrak{b}_j$ until we get $\mathfrak{b}_k = \mathfrak{a}_m$, where by $\mathfrak{a} * \mathfrak{b}$ we again denote the result of either reducing $\mathfrak{a}\mathfrak{b}$ or computing a reduced ideal equivalent to $\mathfrak{a}\mathfrak{b}$ using NUCOMP as described in §5.4. At each step, after we compute the product $\mathfrak{a} = \mathfrak{b}_j * \mathfrak{b}_j$ we move through the cycle of principal ideals from $\mathfrak{a}$ until we find some $\mathfrak{a}_i$ such that $\delta_i \leq E/2^{k-j} < \delta_{i+1}$ and set $\mathfrak{b}_{j+1} = \mathfrak{a}_i$. We do this so that each ideal $\mathfrak{b}_{j+1}$ has distance as close to $E/2^{k-j}$ as possible.

We then use this ideal $\mathfrak{a}_m$ to compute a multiple $h^* R_\Delta$ of R_Δ, where h^* is some positive integer, as follows. Select the reduced principal ideal $\mathfrak{a}_t$ from the list $\mathcal{L}$ such that $\delta_t < L < \delta_{t+1}$. Set $\mathfrak{c}_1 = \mathfrak{a}_m$, $\mathfrak{d}_1 = \bar{\mathfrak{a}}_m$ and compute

$c_{i+1} = c_i * a_t$ and $\mathfrak{d}_{i+1} = \mathfrak{d}_i * a_t$ until we find some c_j or $\mathfrak{d}_j$ in $\mathcal{L}$. If $c_j = a_k \in \mathcal{L}$ has distance $\delta(c_j)$ and a_k has distance $\delta(a_k)$, then

$$h^* R_\Delta = \delta(c_j) - \delta(a_k) \ .$$

If $\mathfrak{d}_j = a_k$, where $\mathfrak{d}_j$ has distance $\delta(\mathfrak{d}_j, \mathfrak{d}_1)$ from $\mathfrak{d}_1$, then

$$h^* R_\Delta = \delta(a_m) - (\delta(\mathfrak{d}_j, \mathfrak{d}_1) - \delta(a_k)) - \log \frac{Q_{m-1}}{r} \ .$$

Notice that h^* is not necessarily the class number of $\mathcal{O}_\Delta$ but is at least a rational integer. By (10.4) and the fact that $Q_{m-1} < 2\sqrt{D}$, we know that this process must terminate after the performance of $O(L)$ giant steps.

Using the value of $h^* R_\Delta$, we must now compute R_Δ. We first check whether $R_\Delta < E/\sqrt{L}$ by using a technique similar to that for finding $h^* R_\Delta$. As above, select the reduced principal ideal a_t from $\mathcal{L}$ such that $\delta(a_t) < L < \delta(a_{t+1})$. Set $c_1 = a_t$ and compute $c_{i+1} = c_i * a_t$. If for some c_j we get $\delta(c_j) \geq E/\sqrt{L}$, then we know that $R_\Delta \geq E/\sqrt{L}$. However, if $c_j = a_k$, where $a_k \in \mathcal{L}$, then, as in the baby-step giant-step algorithm described in Chapter 7,

$$R_\Delta = \delta(c_j) - \delta(a_k) \ .$$

If $\bar{c}_j = a_k$, then

$$R_\Delta = \delta(c_j) + \delta(a_k) - \log \frac{Q_{j-1}}{r} \ .$$

In either case we terminate the algorithm.

In the case where $R_\Delta \geq E/\sqrt{L}$, we must find h^*. The main idea is to check for all primes q that could possibly divide h^*; that is, we check whether the ideal a at distance $h^* R_\Delta/q$ from (1) is such that $a = (1)$. If so, then we know that $h^* R_\Delta/q$ is also a multiple of the regulator, so $q \mid h^*$. We check the ideals at distance $h^* R_\Delta/q^2$, $h^* R_\Delta/q^3, \ldots$, until we find one equal to (1) at distance $h^* R_\Delta/q^\alpha$ but not at $h^* R_\Delta/q^{\alpha+1}$. Then we have q^α as the highest power of q that divides h^*. If there are n primes $q_1, \ldots, q_n$ that were found to divide h^*, then

$$h^* = \prod_{i=1}^{n} q_i^{\alpha_i}$$

and R_Δ can be computed easily. Because we know that $R_\Delta > E/\sqrt{L}$, we need only consider primes q such that $h^* R_\Delta/q > E/\sqrt{L}$, so $q < h^* R_\Delta/(E/\sqrt{L})$. By (10.4), $h^* R_\Delta < L^2 + E$, so we have that

$$q < \sqrt{L} + L^2 \sqrt{L}/E$$

suffices.

The following algorithm summarizes Lenstra's method.

Algorithm 10.1: Regulator of a Real Quadratic Order (Lenstra)

Input: discriminant $\Delta > 0$ of a real quadratic order
Output: R_Δ

    /* *Compute approximation of* $h_\Delta R_\Delta$*/

1: Set $Q = \Delta^{1/5}$, compute $E = \sqrt{\Delta}\exp(S(Q,\Delta))/2$ using (10.1).
2: Compute $L = \sqrt{E}\,|\exp(A(Q,\Delta)) - 1|$ using (10.2).
    /* *Compute integer multiple* R' *of* R_Δ*/

3: Set $\mathfrak{a}_1 = (1)$ and compute $\mathcal{L} = \{\mathfrak{a}_1, \mathfrak{a}_2, \ldots, \mathfrak{a}_t, \mathfrak{a}_{t+1}, \ldots, \mathfrak{a}_s\}$ where $\mathfrak{a}_i = \rho(\mathfrak{a}_{i-1})$, $\delta(\mathfrak{a}_{t+1}) > L > \delta(\mathfrak{a}_t)$, and $\delta(\mathfrak{a}_{s+1}) > L + \log 2 > \delta(\mathfrak{a}_s)$.
4: **if** $\mathfrak{a}_{n+1} = [Q_n/r, (P_n + \sqrt{D})/r] \in \mathcal{L}$ such that $P_n = P_{n+1}$ **then**
5:     Set $R_\Delta = 2\delta(\mathfrak{a}_{n+1}) + \log(Q_0/Q_n)$.
6: **else if** $\mathfrak{a}_{n+1} = [Q_n/r, (P_n + \sqrt{D})/r] \in \mathcal{L}$ such that $Q_n = Q_{n+1}$ **then**
7:     Set $R_\Delta = 2\delta(\mathfrak{a}_{n+1}) + \log(Q_0\psi_{n+1}/Q_n)$.
8: **else**
9:     Compute $\mathfrak{a}_m$ with $\delta(\mathfrak{a}_m) \approx E$.
10:     Set $\mathfrak{c}_1 = \mathfrak{a}_m$, $\mathfrak{d}_1 = \bar{\mathfrak{a}}_m$, and $i = 1$.
11:     **while** $\mathfrak{c}_i \notin \mathcal{L}$ and $\mathfrak{d}_i \notin \mathcal{L}$ **do**
12:         Set $i = i + 1$ and compute $\mathfrak{c}_i = \mathfrak{c}_{i-1} * \mathfrak{a}_t$, $\mathfrak{d}_i = \mathfrak{d}_{i-1} * \mathfrak{a}_t$, and $\delta(\mathfrak{d}_i, \mathfrak{d}_1)$.
13:     **end while**
14:     **if** $\mathfrak{c}_i = \mathfrak{a}_k \in \mathcal{L}$ **then**
15:         Set $R' = \delta(\mathfrak{c}_i) - \delta(\mathfrak{a}_k)$.
16:     **else if** $\mathfrak{d}_i = \mathfrak{a}_k \in \mathcal{L}$ **then**
17:         Set $R' = \delta(\mathfrak{a}_m) - (\delta(\mathfrak{d}_i, \mathfrak{d}_1) - \delta(\mathfrak{a}_k)) - \log(Q_{m-1}/r)$.
18:     **end if**
        /* *Check whether* $R_\Delta < E/\sqrt{L}$*/
19:     Set $\mathfrak{c}_1 = \mathfrak{a}_t$, $\mathfrak{c}_2 = \mathfrak{c}_1 * \mathfrak{a}_t$, and $i = 2$.
20:     **while** $\mathfrak{c}_i \notin \mathcal{L}$ and $\delta(\mathfrak{c}_i) < E/\sqrt{L}$ **do**
21:         Set $i = i + 1$ and compute $\mathfrak{c}_i = \mathfrak{c}_{i-1} * \mathfrak{a}_t$.
22:     **end while**
23:     **if** $\mathfrak{c}_i = \mathfrak{a}_k \in \mathcal{L}$ **then**
24:         Set $R_\Delta = \delta(\mathfrak{c}_i) - \delta(\mathfrak{a}_k)$.
25:     **else**
        /* *Compute* h^**/
26:         Set $h^* = 1$ and $p = 2$.
27:         **while** $p < \sqrt{L} + L^2\sqrt{L}/E$ **do**
28:             Set $\alpha = 1$ and compute $\mathfrak{a}$ such that $\delta(\mathfrak{a}) \approx R'/q$.
29:             **while** $\mathfrak{a} = (1)$ **do**
30:                 Set $\alpha = \alpha + 1$ and compute $\mathfrak{a}$ such that $\delta(\mathfrak{a}) \approx R'/(q^\alpha)$.
31:             **end while**
32:             Set $h^* = h^*p^{\alpha-1}$ and p to the next prime larger than p.
33:         **end while**
34:         Set $R_\Delta = R'/h^*$.
35:     **end if**
36: **end if**

We now derive the complexity of Lenstra's method.[4]

Theorem 10.1. *Lenstra's algorithm as described above computes the regulator R_Δ of $\mathcal{O}_\Delta$ in time $O(\Delta^{1/5+\epsilon})$ assuming the ERH.*

Proof. Recall that there are four main steps in this algorithm, namely computing an approximation E of $h_\Delta R_\Delta$ such that $|E - h_\Delta R_\Delta| < L^2$, computing the multiple $h^* R_\Delta$ of R_Δ using the baby-step giant-step method, checking whether $R_\Delta < E/\sqrt{L}$, and computing h^* in the case that $R_\Delta > E/\sqrt{L}$. We will estimate the cost of each of these steps in terms of Q, E, and L, and at the end, we will determine optimal values of these quantities that minimize the total runtime.

The dominant part of the first step is computing the approximation $S(Q, \Delta)$ of $L(1, \chi)$ using Bach's method, costing $O(Q)$ evaluations of Legendre symbols.[5] As the evaluation of a Legendre symbol is polynomial in $\log \Delta$, the cost of this step is $O(Q\Delta^\epsilon)$.

The second step, computing $h^* R_\Delta$, requires three main parts. Computing the baby-step list containing reduced principal ideals with distance less than $L + \log 2$ requires $O(L)$ continued fraction steps, each of which has runtime $O(\Delta^\epsilon)$, resulting in an overall cost of $O(L\Delta^\epsilon)$. The second part, computing a reduced principal ideal $\mathfrak{a}_m$ with $\delta(\mathfrak{a}_m) \approx E$, requires $O(\log E)$ ideal multiplications and reductions, all of which have polynomial complexity in $\log \Delta$, for an overall cost of $O((\log E)\Delta^\epsilon)$. Finally, $O(L)$ giant steps are required to find a match in the baby step list (and hence $h^* R_\Delta$), and as each of these has complexity $O(\Delta^\epsilon)$, the cost of this step is $O(L\Delta^\epsilon)$. Thus, the total cost for computing $h^* R_\Delta$ given E is

$$O(L\Delta^\epsilon) + O((\log E)\Delta^\epsilon) + O(L\Delta^\epsilon) = O(L\Delta^\epsilon).$$

Determining whether $R_\Delta < E/\sqrt{L}$ is done with the baby-step giant-step method using the same list of baby steps as before and the same giant step of distance roughly L. Thus, the cost of this step is $E/(L\sqrt{L})$ giant steps, for a total complexity of $O(E/(L\sqrt{L})\Delta^\epsilon)$.

The last step requires computing ideals at distances $h^* R_\Delta/q$ for all primes $q < \sqrt{L} + L^2\sqrt{L}/E$. Each of these ideals can be computed in time $O(\log(h^* R_\Delta)\Delta^\epsilon) = O(\Delta^\epsilon)$ using the same method used to compute $\mathfrak{a}_m$ described above, as all these distances are bounded by $h^* R_\Delta$. Thus, the overall complexity of the last step is $O(\Delta^\epsilon(\sqrt{L} + L^2\sqrt{L}/E))$.

Putting the above results together indicates that the running time of Lenstra's algorithm is

$$O(Q\Delta^\epsilon) + O(L\Delta^\epsilon) + O(E/(L\sqrt{L})\Delta^\epsilon) + O(\Delta^\epsilon(\sqrt{L} + L^2\sqrt{L}/E)). \quad (10.5)$$

It remains to find optimal values for Q, L, and E and to express (10.5) as a function of Δ.

Assume that $Q = \Delta^\alpha$ for $\alpha > 0$; we will show below that $\alpha = 1/5$ optimizes the complexity of the algorithm. From the definition of $S(Q, \Delta)$ (see §9.5 and (10.1)), it is easy to show that

$$\exp(S(Q, \Delta)) \leq \prod_{q < 2Q} \frac{q}{q-1}$$

and by Mertens' Theorem we know that

$$\prod_{q < 2Q} \frac{q}{q-1} = O(\log Q) \, .$$

Thus, we have

$$E = O(\Delta^{1/2+\epsilon}) \, .$$

Recall that, using (10.4), we have $|E - h_\Delta R_\Delta| \leq L^2$, with

$$L^2 = E \, |\exp(A(Q, \Delta)) - 1| \, .$$

Assuming that $Q = \Delta^\alpha$ and the ERH holds, we have, from (9.31),

$$A(Q, \Delta) = O(\Delta^{-\alpha/2})$$

and one can show that

$$|\exp(A(Q, \Delta)) - 1| = O(\Delta^{-\alpha/2+\epsilon}) \, , \tag{10.6}$$

yielding

$$L^2 = O(\Delta^{1/2-\alpha/2+\epsilon}) \, .$$

Finally, we have that

$$L = O(\Delta^{1/4-\alpha/4+\epsilon}) \, ,$$
$$E/(L\sqrt{L}) = O(\Delta^{1/8+3\alpha/8+\epsilon}) \, ,$$

and

$$\sqrt{L} + L^2\sqrt{L}/E = O(\Delta^{1/8-\alpha/8+\epsilon}) \, .$$

The complexity of Lenstra's algorithm is therefore

$$O(\Delta^{\alpha+\epsilon}) + O(\Delta^{1/4-\alpha/4+\epsilon}) + O(\Delta^{1/8+3\alpha/8+\epsilon}) \, ,$$

and this is minimized when $\alpha = 1/4 - \alpha/4 = 1/8 + 3\alpha/8$ (i.e., $\alpha = 1/5$). Hence, under the ERH, Lenstra's algorithm has complexity $O(\Delta^{1/5+\epsilon})$. $\quad\square$

A method to compute h^* that is more efficient in practice was proposed by Jacobson, Lukes, and Williams.[6] We first compute a list $\mathcal{I}$ of reduced principal ideals $\mathfrak{a}_{t_0}, \mathfrak{a}_{t_1}, \ldots, \mathfrak{a}_{t_n}$, where $\mathfrak{a}_{t_0} = \mathfrak{a}_t$, $\mathfrak{a}_{t_j} = \mathfrak{a}_{t_{j-1}} * \mathfrak{a}_{t_{j-1}}$, and $\delta_{t_{n-1}} < h^* R_\Delta / 2 < \delta_{t_n}$. We then produce a list of all primes $q < \sqrt{L} + L^2 \sqrt{L}/E$ in decreasing order. For each prime q_s, we must find a reduced principal ideal $\mathfrak{a}_e$ such that

$$\frac{h^* R_\Delta}{q_s} < \delta(\mathfrak{a}_e) < \frac{h^* R_\Delta}{q_s} + \delta(\mathfrak{a}_t) .$$

From the preceding prime q_{s+1} ($> q_s$) we have an ideal $\mathfrak{a}_m$, with

$$\frac{h^* R_\Delta}{q_{s+1}} < \delta(\mathfrak{a}_m) < \frac{h^* R_\Delta}{q_{s+1}} + \delta(\mathfrak{a}_t) .$$

Notice that if we find an ideal $\mathfrak{a}_s$ such that

$$\delta(\mathfrak{a}_s) \approx \frac{h^* R_\Delta}{q_s} - \delta(\mathfrak{a}_m)$$

and

$$\frac{h^* R_\Delta}{q_s} < \delta(\mathfrak{a}_s) + \delta(\mathfrak{a}_m) \leq \frac{h^* R_\Delta}{q_s} + \delta(\mathfrak{a}_t) ,$$

then we can set $\mathfrak{a}_e = \mathfrak{a}_s * \mathfrak{a}_m$, with $\delta(\mathfrak{a}_e) \approx \delta(\mathfrak{a}_s) + \delta(\mathfrak{a}_m)$. To find $\mathfrak{a}_s$, we first put $r\delta(\mathfrak{a}_t) = h^* R_\Delta / q_s - \delta(\mathfrak{a}_m)$ for some real number r. We then have $\delta(\mathfrak{a}_s) \approx q\delta(\mathfrak{a}_t)$, where $q = [r] + 1$. If we represent q in binary as

$$q = b_k 2^k + b_{k-1} 2^{k-1} + \cdots + b_0 ,$$

where $b_k = 1$ and $b_j \in \{0, 1\}$ for $j < k$, then we have

$$q\delta_t = b_k 2^k \delta(\mathfrak{a}_t) + b_{k-1} 2^{k-1} \delta(\mathfrak{a}_t) + \cdots + b_0 \delta(\mathfrak{a}_t) .$$

In the list $\mathcal{I}$, we have $\delta(\mathfrak{a}_{t_k}) \approx 2^k \delta(\mathfrak{a}_t)$, so we can find $\mathfrak{a}_s$ with distance $\delta(\mathfrak{a}_s) \approx q\delta(\mathfrak{a}_t)$ by simply computing a reduced ideal equivalent to

$$\prod_{j=0}^{k} \mathfrak{a}_{t_j}^{b_j} .$$

Once $\mathfrak{a}_e$ has been determined, we check whether there is an ideal $\mathfrak{a}_j \in \mathcal{L}$ such that $\mathfrak{a}_e = \mathfrak{a}_j$ and $\delta(\mathfrak{a}_e) = h^* R_\Delta / q_s + \delta(\mathfrak{a}_j)$. If so, then the ideal at distance $h^* R_\Delta / q_s$ must be (1), implying that $q_s \mid h^*$, and we repeat the above process to determine the precise power α_s of q_s that divides h^*. After the above process has been performed for all $q_s < \sqrt{L} + L^2 \sqrt{L}/E$, we have

$$h^* = \prod_{i=1}^{s} q_i^{\alpha_i}$$

and we can compute R_Δ.

Lenstra's algorithm has been implemented and used for a variety of applications, one of the first being Schoof's real quadratic field-based integer factorization algorithm.[7] It is feasible, using currently available technology, to use this algorithm to compute R_Δ for discriminants as large as 10^{40}. The main obstacle in pushing this further is the fact that the amount of storage required for the baby-step list becomes prohibitively large.

It must be emphasized that although the complexity of Lenstra's algorithm is conditional on the ERH, the regulator computed is unconditionally correct. The ERH is only required for estimating the error in the approximation of $h_\Delta R_\Delta$. Given any such approximation, the algorithm will, using the baby-step giant-step method, compute an integer multiple of R_Δ and from that correctly determine R_Δ itself. The estimate of the error in approximating $h_\Delta R_\Delta$ only plays a role in determining the complexity of the baby-step giant-step stage of the algorithm.

Indeed, Srinivasan[8] has presented a variation of this algorithm in which the approximation of $h_\Delta R_\Delta$ is determined using a novel technique called random summation, as opposed to Bach's method. The idea of random summation is to approximate $L(1, \chi)$ by taking random terms in its Euler product expansion instead of a weighted average of truncated Euler products. The error in the approximation can be estimated probabilistically without appealing to the ERH, but there is a small chance that the approximation is incorrect and may need to be computed again. The result is an algorithm that, like Lenstra's algorithm described above, computes an unconditionally correct approximation of R_Δ in expected time $O(\Delta^{1/5+\epsilon})$, where, although the algorithm is not deterministic, the complexity result does not depend on the ERH.[9]

10.3 Computing the Class Number

The simplest method for computing the class number of an imaginary quadratic order is to enumerate every reduced ideal. Because the coefficients of reduced ideals are bounded by $\sqrt{|\Delta|}$, this strategy amounts to simply looping over all possible values of the coefficients and counting the number of (Q, P) pairs that represent valid reduced ideals of $\mathcal{O}_\Delta$. Although the complexity of this method is $O(|\Delta|^{1+\epsilon})$ for a single discriminant, it has the advantages of being unconditionally correct and not involving any operations more complicated than basic integer arithmetic; in particular, no ideal arithmetic is required. In addition, as only coefficients of reduced ideals are used, the operands are small, bounded by $\sqrt{|\Delta|}$. This method can be quite effective when computing tables of class numbers and has been used extensively by Buell.[10] When looping over the ideal coefficients, a separate reduced ideal counter is maintained corresponding to each discriminant in the tabulation interval. The result is a tabulation algorithm that requires only $O(|\Delta|^{1/2+\epsilon})$ operations per discriminant.

This method can also be adapted to work in real quadratic orders. The difference is that we have multiple reduced ideals per equivalence class, so instead of simply counting reduced ideals, one has to count the number of cycles of reduced ideals. It is possible to construct an algorithm based on this approach that runs in time $O(\Delta^{1/2+\epsilon})$ that, as in the imaginary case, computes h_Δ unconditionally.[11]

Assuming that the regulator has been computed, even faster algorithms for computing the class number exist that make use of the analytic class number formula. A few of these have been described in Chapter 8 and Chapter 9 which work by computing a sufficiently accurate approximation of $L(1, \chi)$ or by using a closed-form formula. These algorithms are all unconditionally correct but have complexity $O(|\Delta|^{1+\epsilon})$ or, at best, $O(|\Delta|^{1/2+\epsilon})$ and are thus unsuitable for large discriminants. In the following, we will describe an algorithm due to Shanks[12] that computes h_Δ in time $O(|\Delta|^{1/5+\epsilon})$. This algorithm makes it possible, with currently available technology, to compute class numbers for discriminants as large as 10^{40} in absolute value, but, unfortunately, the correctness of the class numbers is conditional on the ERH. In general, algorithms whose complexity breaks the $O(\Delta^{1/2+\epsilon})$ barrier sacrifice the unconditionality of the computed class numbers, the sole exception being a recent algorithm due to Booker[13] that we will describe in Chapter 15.

We will describe Shanks' algorithm in terms of real quadratic orders; modifications to the imaginary case are relatively straightforward.[14] The main idea behind Shanks' algorithm is to compute an estimate of h_Δ using the analytic class number formula. Assuming that R_Δ is known, a sufficiently accurate approximation of $L(1, \chi)$ is used to verify that the estimate is in fact equal to h_Δ.

The algorithm proceeds as follows. First, assume that h_Δ has a known factor h_1 and put $h_2 = h_\Delta / h_1$. We can take $h_1 = 1$, but the larger h_1 is the faster the algorithm will execute, and a sufficiently large divisor h_1 is required in order to prove the $O(\Delta^{1/5+\epsilon})$ complexity. We will discuss how to compute a suitable value of h_1 below, but for now we assume that this is provided. Given an approximation $S(Q, \Delta)$ of $\log L(1, \chi)$, we have, by the analytic class number formula, that

$$\tilde{h}_2(Q) = \left\lfloor \frac{\sqrt{\Delta}\exp(S(Q, \Delta))}{2R_\Delta h_1} \right\rceil$$

is an approximation of h_2. If we put

$$\kappa(Q) = \sqrt{\Delta}\exp(S(Q, \Delta))/(2R_\Delta h_1) - \tilde{h}_2(Q), \tag{10.7}$$

then, clearly, $|\kappa(Q)| < 1/2$.

Given R_Δ and h_1, we need to find a sufficiently large value of Q such that we can prove $\tilde{h}_2(Q) = h_2$ (i.e., $h_\Delta = h_1\tilde{h}_2(Q)$). We need the following two elementary lemmas noted by Mollin and Williams.[15]

Lemma 10.2. *If*

$$|\log x| < \log\left(\frac{k+1}{k+|y|}\right),$$

where $k \geq 0$, $x > 0$, *and* $|y| < 1/2$, *then*

$$\frac{k}{k+1+y} < x < \frac{k+1}{k+y}.$$

Proof. The proof is easy to show using elementary algebraic manipulations.
□

Lemma 10.3. *If* $k \in \mathbb{Z}$ *and*

$$\frac{k}{k+1+\kappa(Q)} < T(Q,\Delta) < \frac{k+1}{k+\kappa(Q)},$$

for $T(Q,\Delta) = L(1,\chi)/\exp(S(Q,\Delta))$, *then* $h_2 = k$ *if and only if* $\tilde{h}_2(Q) = k$.

Proof. The proof follows easily from the fact that, by (10.7), we have

$$\kappa(Q) = h_2 \exp(S(Q,\Delta))/L(1,\chi) - \tilde{h}_2(Q),$$

so we can write

$$T(Q,\Delta) = \frac{h_2}{\tilde{h}_2(Q) + \kappa(Q)}. \tag{10.8}$$

First, assume that $h_2 = k$. Then we have $T(Q,\Delta) = k/(\tilde{h}_2(Q) + \kappa(Q))$
and

$$\frac{k}{k+1+\kappa(Q)} < \frac{k}{\tilde{h}_2(Q) + \kappa(Q)} < \frac{k+1}{k+\kappa(Q)}.$$

The first inequality yields

$$\tilde{h}_2(Q) + \kappa(Q) < k+1+\kappa(Q),$$

so

$$\tilde{h}_2(Q) < k+1. \tag{10.9}$$

The second inequality yields

$$\tilde{h}_2(Q) > \frac{k(k+\kappa(Q))}{k+1} - \kappa(Q) = \frac{k^2 - \kappa(Q)}{k+1} > \frac{k^2 - 1}{k+1} = k-1 \tag{10.10}$$

because $|\kappa(Q)| < 1/2$. Thus, by (10.9) and (10.10) we have $\tilde{h}_2(Q) = k$.
Next, assume that $\tilde{h}_2(Q) = k$. Then $T(Q,\Delta) = h_2/(k+\kappa(Q))$ and

$$\frac{k}{k+1+\kappa(Q)} < \frac{h_2}{k+\kappa(Q)} < \frac{k+1}{k+\kappa(Q)}.$$

The first inequality yields

$$h_2 > \frac{k(k+\kappa(Q))}{k+1+\kappa(Q)} = \frac{k(k+1+\kappa Q) - k}{k+1+\kappa(Q)} = k - \frac{k}{k+1+\kappa(Q)} > k-1 \,,$$

$$(10.11)$$

again because $|\kappa(Q)| < 1/2$. The second inequality yields

$$h_2 > k+1 \,, \qquad (10.12)$$

and it follows from (10.11) and (10.12) that $h_2 = k$. □

From these two lemmas and (10.2) it follows that $h_2 = \tilde{h}_2(Q)$ whenever Q is sufficiently large that

$$A(Q, \Delta) < \log \left(\frac{\tilde{h}_2(Q) + 1}{\tilde{h}_2(Q) + |\kappa(Q)|} \right) . \qquad (10.13)$$

Thus, given R_Δ and h_1, we can use the following algorithm to compute an approximation of $L(1, \chi)$ to sufficient accuracy to determine $\tilde{h}_2(Q)$ such that $h_\Delta = h_1 \tilde{h}_2(Q)$.

1. Set $Q = \Delta^\beta$ (we will prove below that $\beta = 1/5$ is optimal)
2. $F = \sqrt{\Delta} \exp(S(Q, \Delta))/(2R_\Delta h_1)$
3. $\tilde{h}_2 = \lfloor F \rceil$
4. $\kappa = F - \tilde{h}_2$
5. If $A(Q, \Delta) \ge \log \left(\frac{\tilde{h}_2+1}{\tilde{h}_2+|\kappa|} \right)$, increase Q (eg. $Q = Q+5000$) and go to step 2
6. $h_\Delta = \tilde{h}_2 h_1$

Given R_Δ and h_1, this algorithm will run in time $O(Q\Delta^\epsilon)$, the cost of computing the approximations $\tilde{h}_2$ of h_Δ. We have already seen that R_Δ can be computed in time $O(\Delta^{1/5+\epsilon})$. In order to prove that we can compute h_Δ in the same time, the next step is to determine how large h_1 needs to be in terms of β.

Theorem 10.4. *Assuming the ERH, the algorithm described above computes the class number h_Δ of $\mathcal{O}_\Delta$ in time $O(\Delta^{\beta+\epsilon})$, given the regulator R_Δ and a divisor h_1 of h_Δ such that $h_1 > \Delta^\alpha/R_\Delta$ with $0 < \alpha < 1/2$ and $\beta = 1 - 2\alpha$. The correctness of the computed class number is also conditional on the ERH.*

Proof. First, observe that for $|y| < 1/2$,

$$\log \left(\frac{1+x}{|y|+x} \right) > \log \left(1 + \frac{1}{2x+1} \right) > \frac{1}{2x+2} \,,$$

so we see that (10.13) will hold if

$$A(Q, \Delta) < \frac{1}{2\tilde{h}_2(Q) + 2} . \qquad (10.14)$$

If $h_1 > \Delta^\alpha/R_\Delta$, where $0 < \alpha < 1/2$, then from (10.7), and $|\kappa(Q)| < 1/2$, we have

$$2\tilde{h}_2(Q) + 1 < \Delta^{1/2-\alpha} \exp(S(Q, \Delta)) .$$

Now, we know that

$$\exp(S(Q, \Delta)) \leq \prod_{q < 2Q} \frac{q}{q-1}$$

and by Mertens' Theorem we also know that

$$\prod_{q < 2Q} \frac{q}{q-1} \sim e^{\gamma} \log 2Q .$$

Thus, there exists some positive constant c_1 such that

$$2\tilde{h}_2(Q) + 2 < c_1 \Delta^{1/2-\alpha} \log 2Q$$

and

$$\frac{1}{2\tilde{h}_2(Q) + 2} > \frac{\Delta^{\alpha-1/2}}{c_1 \log 2Q} .$$

From (10.2) we see that, assuming the ERH, there exists a positive constant c_2 such that

$$A(Q, \Delta) < \frac{c_2 \log \Delta}{\sqrt{Q} \log Q} .$$

It follows that (10.14) will certainly hold if

$$\frac{\Delta^{\alpha-1/2}}{c_1 \log 2Q} > \frac{c_2 \log \Delta}{\sqrt{Q} \log Q}$$

$$\Delta^{\alpha-1/2} > \frac{c_1 c_2 \log \Delta \log 2Q}{\sqrt{Q} \log Q}$$

$$\sqrt{Q} > \frac{c_1 c_2 \log \Delta \log 2Q}{\log Q} \Delta^{1/2-\alpha}$$

$$\sqrt{Q} > c_3 \Delta^{1/2-\alpha+\epsilon} ,$$

where $c_3 > 0$. Finally, if $\beta/2 = 1/2 - \alpha$, i.e., $\beta = 1 - 2\alpha$, then (10.14) holds with $Q = \Delta^{\beta+\epsilon}$ and the result follows. Note that, as claimed, the ERH is required for both the value of Q (hence the complexity of the algorithm) and the correctness of the error approximation corresponding to h_2 (hence the correctness of the algorithm). □

Let $h_\Delta < \Delta^l$. From the analytic class number formula and Littlewood's bound in (9.24), there exists a positive constant c_4 such that

$$R_\Delta h_\Delta > c_4 \sqrt{\Delta} / \log \log \Delta .$$

Thus, even if $h_1 = 1$, we have

$$R_\Delta h_1 > c_4 \Delta^{1/2-l}/\log\log\Delta \ .$$

Putting $\alpha = 1/2 - l$, we see that the above algorithm evaluates h_Δ in $O(\Delta^{2l+\epsilon})$ operations. Hence, if $h_\Delta < \Delta^{1/10}$, which one can expect to occur quite frequently for real quadratic fields due to the Cohen-Lenstra heuristics, this method will work well, even if the best divisor of h_Δ we can obtain is $h_1 = 1$. For $h_\Delta > \Delta^{1/10}$, this method will determine h in $O(\Delta^{1/5+\epsilon})$ operations provided we can obtain a divisor h_1 of h_Δ such that $h_1 > \Delta^{2/5}/R_\Delta$ (i.e., taking $\alpha = 2/5$ in Theorem 10.4).

The following method[16] can be used to find a suitable value of h_1 when $h_\Delta > \Delta^{1/10}$. The idea is to use, once again, Shanks' baby-step giant-step method combined with an approximation of h_Δ found via the analytic class number formula to compute the order of a random ideal class. If this integer, which is clearly a divisor of h_Δ, is greater than $\Delta^{2/5}/R_\Delta$, then we are finished; otherwise, we select a second ideal class and compute the order of the subgroup generated by both classes, repeating this process until the order of this subgroup, also a divisor of h_Δ, is greater than $\Delta^{2/5}/R_\Delta$.

Given an approximation $S(Q, \Delta)$ of $\log L(1, \chi)$ as before, we have, by the analytic class number formula, that

$$\tilde{h}(Q) = \left\lfloor \frac{\sqrt{\Delta}\exp(S(Q,\Delta))}{2R_\Delta} \right\rceil$$

is an approximation of h_Δ and, by (10.2),

$$\left| h_\Delta - \tilde{h}(Q) \right| < B_2(Q) \ ,$$

where

$$B_2(Q) = 1/2 + \sqrt{\Delta}\exp(S(Q,\Delta))(\exp(A(Q,\Delta)) - 1)/(2R_\Delta) \ .$$

If we set $Q = \Delta^{1/5}$, $B_2 = B_2(Q)$, and $\tilde{h} = \tilde{h}(Q)$, we have from (10.6) (assuming the ERH),

$$R_\Delta B_2 = O(\Delta^{2/5+\epsilon}) \ , \tag{10.15}$$

and it follows that there exists some integer m such that $|m| < B_2$ and $h_\Delta = \tilde{h} - m$. Thus, for any ideal $\mathfrak{a}$, we have

$$\mathfrak{a}^{\tilde{h}} \sim \mathfrak{a}^m \ , \tag{10.16}$$

and we can use the baby-step giant-step method to find m such that (10.16) holds. The result is that $|\tilde{h} - m|$ is a multiple of the order of $[\mathfrak{a}] \in Cl_\Delta$.

Given $\tilde{h}$ and B_2, we find integers i and j such that $|m| = ki + j$ with $0 \le j \le k$ for some fixed integer $k \ge 1$ to be determined below. We first select a random prime ideal $\mathfrak{a}$ and compute ideals $\mathfrak{d} \sim \mathfrak{a}^{\tilde{h}}$ and $\mathfrak{g} \sim \mathfrak{a}^k$. If (10.16) holds for $m \ge 0$, then there exist i and j such that $\mathfrak{d} \sim \mathfrak{g}^i \mathfrak{a}^j$ and

$$(\overline{\mathfrak{g}})^i \mathfrak{d} \sim \mathfrak{a}^j \qquad (10.17)$$

holds. On the other hand, if (10.16) holds for $m < 0$, then $\mathfrak{a}^{-\tilde{h}} \sim \overline{\mathfrak{d}} \sim \mathfrak{g}^i \mathfrak{a}^j$ and

$$(\overline{\mathfrak{g}})^i \overline{\mathfrak{d}} \sim \mathfrak{a}^j \qquad (10.18)$$

holds.

To find i and j, we proceed in one of two ways. If the regulator is small, then testing equivalence of ideals is easy, so we can find i and j using a simple adaptation of the baby-step giant-step method. On the other hand, if the regulator is large, then equivalence testing is difficult, but we know that the value of m will be small. In that case, we set $k = 1$ and $j = 0$ in (10.17) and (10.18) and simply find the smallest integer i such that $\mathfrak{a}^i \sim \mathfrak{a}^{\tilde{h}}$ or $\mathfrak{a}^i \sim \mathfrak{a}^{-\tilde{h}}$. The trick is to determine a cut off point to decide between these two approaches in such a way that the entire computation of h_1 still has complexity $O(\Delta^{1/5+\epsilon})$. The value B_2 is precisely the value we need.

If $B_2 \geq R_\Delta$, or if $\mathcal{O}_\Delta$ is imaginary, then the number of reduced ideals in any equivalence class is sufficiently small that we can use the baby-step giant-step method to compute i and j and handle equivalence testing by exhaustively enumerating all equivalent reduced ideals corresponding to each of the baby steps. Specifically, we set $k = [\sqrt{B_2/R_\Delta}] + 1$ and compute the cycles of reduced ideals in the classes of the baby steps $\mathcal{O}_\Delta, \mathfrak{a}, \mathfrak{a}^2, \ldots, \mathfrak{a}^{k-1}$. As usual, these ideals are all stored in a list $\mathcal{J}$ that admits fast searching such as a hash table. To find i, we compute the giant steps $\mathfrak{b}_n$ and $\mathfrak{c}_n$ using $\mathfrak{b}_{t+1} = \mathfrak{b}_t * \overline{\mathfrak{g}}$ and $\mathfrak{c}_{t+1} = \mathfrak{c}_t * \overline{\mathfrak{g}}$, where $\mathfrak{b}_0 = \mathfrak{d}$ and $\mathfrak{c}_0 = \overline{\mathfrak{d}}$, until we find either $\mathfrak{b}_i$ or $\mathfrak{c}_i$ in $\mathcal{J}$. If $\mathfrak{b}_i \in \mathcal{J}$, then $\mathfrak{b}_i \sim (\overline{\mathfrak{g}})^i * \mathfrak{d} \sim \mathfrak{a}^j$ and we have $m = ik + j$. If $\mathfrak{c}_i \in \mathcal{J}$, then $\mathfrak{c}_i \sim (\overline{\mathfrak{g}})^i * \overline{\mathfrak{d}} \sim \mathfrak{a}^j$ and we have $m = -ik - j$.

If $B_2 < R_\Delta$, then we set $k = 1$ and $j = 0$ and compute the ideals $\mathfrak{b}_n$ and $\mathfrak{c}_n$ as above and find the smallest n for which one of $\mathfrak{b}_n$ or $\mathfrak{c}_n$ is principal. If $\mathfrak{b}_n$ is principal, then we have $\mathfrak{b}_n \sim (\overline{\mathfrak{g}})^n * \mathfrak{d} \sim \mathcal{O}_\Delta$, so $\mathfrak{d} \sim \mathfrak{a}^{\tilde{h}} \sim \mathfrak{a}^n$ and we take $m = n$. Similarly, if $\mathfrak{c}_n$ is principal, then we have $\mathfrak{c}_n \sim (\overline{\mathfrak{g}})^n * \overline{\mathfrak{d}} \sim \mathcal{O}_\Delta$, so $\overline{\mathfrak{d}} \sim \mathfrak{a}^{-\tilde{h}} \sim \mathfrak{a}^n$ and we take $m = -n$. Thus, in this case, the main difficulty is testing $\mathfrak{b}_n$ and $\mathfrak{c}_n$ for principality. Here, we use the baby-step giant-step method as mentioned in §7.4 with a step size of $S = \sqrt{R_\Delta B_2}$.

We now prove that in either case we compute m in time $O(\Delta^{1/5+\epsilon})$.

Theorem 10.5. *Given an ideal $\mathfrak{a}$ and the regulator R_Δ of $\mathcal{O}_\Delta$, the algorithm described above computes an integer m such that $\mathfrak{a}^{\tilde{h}} \sim \mathfrak{a}^m$ in time $O(\Delta^{1/5+\epsilon})$ assuming the ERH.*

Proof. First, suppose that $B_2 \geq R_\Delta$. The total number of ideals stored in the baby-step list is $O(kR_\Delta)$, so by (10.15) the complexity of computing the baby steps is $O(R_\Delta(\sqrt{B_2/R_\Delta} + 1)\Delta^\epsilon) = O(\sqrt{B_2 R_\Delta} + R_\Delta)\Delta^\epsilon = O(\Delta^{1/5+\epsilon})$. Because $|m| < B_2$ and $j \geq 0$, the value of i found by this method must satisfy $i < B_2/k = O(\Delta^{1/5+\epsilon})$, and the overall complexity is $O(\Delta^{1/5+\epsilon})$.

Otherwise, if $B_2 < R_\Delta$, the algorithm consists of n baby-step giant-step principality tests using a step size of $S = \sqrt{R_\Delta B_2}$. To compute the baby steps, the first $O(S)$ reduced ideals in the principal cycle requires $O(SD^\epsilon)$ operations. This step need only be done once, as all the principality tests we require can make use of this same baby-step list. Each principality test will require R_Δ/S giant steps, at a cost of $O(R_\Delta \Delta^\epsilon/S)$ each. Finally, as we know that $n < B_2$, the total cost in this case is

$$O(S\Delta^\epsilon) + O(B_2 R_\Delta \Delta^\epsilon/S) = O(\Delta^{1/5+\epsilon})$$

by (10.15).

In both cases, the ERH is required to bound $R_\Delta B_2$ in (10.15), so the complexity is dependent on the ERH.   □

The value of m found here often satisfies $h_\Delta = |\tilde{h}-m|$, but this need not be the case. However, $|\tilde{h}-m|$ is unconditionally (i.e., without assuming the ERH) a multiple of e_1, the order of $[\mathfrak{a}]$ in the class group. After factoring $|\tilde{h}-m|$, we use one of the two methods for testing principality outlined above to find, for each prime divisor p_i, the largest positive integer α_i such that $\mathfrak{a}^{|\tilde{h}-m|/p_i^{\alpha_i}}$ is principal, yielding $e_1 = |\tilde{h}-m|/\left(\prod p_i^{\alpha_i}\right)$. If $e_1 > \Delta^{2/5}/R_\Delta$, then we take $h_1 = e_1$ and we are finished. Otherwise, we select another ideal $\mathfrak{b}$ and find the order e_2 of the subgroup generated by $\mathfrak{a}$ and $\mathfrak{b}$. This process is repeated until the order e_i of the subgroup generated by i ideals satisfies $e_i > \Delta^{2/5}/R_\Delta$. Details of how the baby-step giant-step method can be adapted to compute the e_i have been presented by a number of authors.[17] However, as predicted by the Cohen-Lenstra heuristics, most class groups tend to be cyclic or close to cyclic, so e_1 is likely to suffice. In general, it can be shown[18] that, assuming the generalized Riemann hypothesis (GRH) for Hecke L-functions and the ERH, a suitable value $e_n > \Delta^{2/5}/R_\Delta$ can be found in $O(\Delta^{1/5+\epsilon})$ operations. The assumption of the GRH guarantees that, by a result of Bach (9.28), the number of prime ideals required to generate the class group is polynomial in $\log \Delta$, so $n = O(\Delta^\epsilon)$. Thus, we obtain the following theorem.

Theorem 10.6. *Assuming the ERH, Shanks' algorithm as described above computes the class number h_Δ of $\mathcal{O}_\Delta$ in time $O(\Delta^{1/5+\epsilon})$. The correctness of the class number computed is also dependent on the ERH.*

Proof. The proof follows from Theorems 10.1, 10.4, 10.5, and the preceding remarks.   □

It must be emphasized that although an unconditionally correct approximation of the regulator can be computed in time $O(\Delta^{1/5+\epsilon})$ using Lenstra's algorithm (Theorem 10.1), the class number cannot be computed unconditionally in the same time using Shanks' algorithm described above. The reason for this is that, in Shanks' algorithm, the estimate $A(Q, \Delta)$ of the error in our approximation of $L(1, \chi)$ is conditional on the ERH, and the condition for determining that our approximation of h_Δ is correct depends on this estimate.

However, note that the $O(\Delta^{1/5+\epsilon})$ complexity depends on the ERH in both cases.

As with Lenstra's algorithm for computing the regulator, Srinivasan's random summation technique[19] can be used in place of Bach's ERH-dependent method to approximate $L(1,\chi)$, resulting in an algorithm for computing h_Δ in expected time $O(\Delta^{1/5+\epsilon})$ without assuming any Riemann hypotheses. Using random summation to approximate $L(1,\chi)$ removes the need to assume the ERH, and a novel method of selecting random ideals and an accompanying analysis of the probability that each newly selected ideal enlarges the subgroup generated by the previous ideals remove the need to assume the GRH. The output of Srinivasan's algorithm is unconditionally correct in the case of real quadratic fields, as an incorrect approximation of $L(1,\chi)$ will be detected when computing the regulator R_Δ. In the imaginary case, it is not possible to make this check, so the output is only correct with high probability.

10.4 Computing the Class Group

In some cases, it is desirable to compute the structure of the class group in addition to the class number and regulator. Recall that by the structure of the class group, we mean the elementary divisors $m_1, \ldots, m_s$ with $m_1 \geq 1$, $m_{j+1} \mid m_j$ for $1 \leq j < s$, such that

$$Cl_\Delta \cong C(m_1) \times C(m_2) \times \cdots \times C(m_s) ,$$

is the canonical decomposition of Cl_Δ as a direct product of cyclic subgroups. For example, as described in Chapter 7, a number of the Cohen-Lenstra heuristics predict properties of the structure when Δ is fundamental, such as the probability that the odd part is cyclic and the probability that the p-rank is equal to a given value. Tabulating class group structures can thus provide evidence in support of these conjectures.

One approach to computing the class group structure is to first compute the class number and then deduce the structure. This approach is described in detail by Buell[20] and used extensively in his tabulations of class groups of imaginary quadratic fields.[21] The idea is to first factor h_Δ and determine which p-Sylow subgroups could possibly be non-cyclic (i.e., those primes p for which $p^2 \mid h_\Delta$). For each of these primes, the p-Sylow subgroup is explicitly determined. An advantage of this approach is that, as predicted by the Cohen-Lenstra heuristics, the odd part of the class group is cyclic with very high probability, so in most cases, even if some power of $p \mid h_\Delta$, the structure can be resolved with very little additional work beyond computing h_Δ and factoring it.

Another approach is to compute the class number and class group simultaneously. The idea is to use a generic method to compute the subgroup generated by a single ideal and extend this by adding additional generators until the entire class group is obtained. This method has been used by Jacobson[22] for tabulating class groups of real quadratic fields and by Jacobson, Ramachandran, and Williams[23] for tabulating class groups of imaginary quadratic fields.

Suppose that we have computed the subgroup G_{l-1} of Cl_Δ generated by $\mathfrak{g}_1, \ldots, \mathfrak{g}_{l-1}$ (i.e., $G_{l-1} = \langle \mathfrak{g}_1, \ldots, \mathfrak{g}_{l-1} \rangle$). In order to compute $G_l = \langle \mathfrak{g}_1, \ldots, \mathfrak{g}_l \rangle$ given another generator $\mathfrak{g}_l$, compute the smallest integer $v_{l,l} > 0$ such that $[\mathfrak{g}_l^{v_{l,l}}] \in G_{l-1}$. Then there exist integers $v_{1,l}, v_{2,l}, \ldots, v_{l-1,l}$ such that $\mathfrak{g}_1^{v_{1,l}} \cdots \mathfrak{g}_{l-1}^{v_{l-1,l}} \mathfrak{g}_l^{v_{l,l}} \sim \mathcal{O}_\Delta$. Note that we can take $0 \leq v_{i,l} < v_{i,i}$ and $v_{i,j} = 0$ if $i > j$. Thus, the matrix $A = (v_{i,j})_{l \times l}$ is in Hermite normal form, and the Smith normal form[24] of A yields the structure of G_l; the diagonal elements are precisely the elementary divisors of G_l.

The reason this method works is as follows.[25] The columns of A form a Hermite normal form basis of the lattice $\ker \phi$, where ϕ is the surjective homomorphism

$$\phi : \mathbb{Z}^l \to G_l, \quad \mathbf{e} = (e_1, \ldots, e_l) \mapsto g_1^{e_1} \cdots g_l^{e_l} .$$

The vectors $\mathbf{v} \in \ker \phi$ are referred to as *relations*. The fundamental theorem of algebra states that $\mathbb{Z}^l / \ker \phi \cong G_l$; thus, the columns of A provide a representation of G_l corresponding to the generators $\mathfrak{g}_1, \ldots, \mathfrak{g}_l$. In particular, we have $\det A = h_\Delta$. The Smith normal form of A yields another representation of G_l corresponding to another set of generators obtained by applying the row operations applied to A to the original generators. This method, essentially computing relations corresponding to a set of generators of a group, is a standard technique for representing a finite abelian group.

There are two main issues to be addressed in applying this method in practice. The first is how to compute the required relations—in particular, determining the integer $v_{l,l}$ such that $\mathfrak{g}_l^{v_{l,l}} \in G_{l-1}$ given some representation of G_{l-1}. The second issue is how to determine when we have $G_l = Cl_\Delta$.

The first problem is typically solved by a generalization of a method used to compute the order of an element in a group such as the baby-step giant-step or Pollard's rho method. For example, suppose that $v_{l,l} \leq H$ for some known upper bound H. Then Shanks' original baby-step giant-step method can be employed as follows. One can show[26] that there exist integers $v_{1,l}, \ldots, v_{l,l}$ such that $\mathfrak{g}_1^{v_{1,l}} \cdots \mathfrak{g}_l^{v_{l,l}} \sim \mathcal{O}_\Delta$ with $v_{i,l} = f_i B_i + e_i$ for $1 \leq i \leq l$ and $0 \leq e_i, f_i < B_i = \lceil \sqrt{v_{i,i}} \rceil$, $B_l = \lceil \sqrt{H} \rceil$. Assuming that we have computed the sets

$$R = \{\mathfrak{g}_1^{-e_1} \cdots \mathfrak{g}_l^{-e_l} \mid 0 \leq e_i \leq B_i\}$$

$$Q = \{\mathfrak{g}_1^{f_1 B_1} \cdots \mathfrak{g}_l^{f_l B_l} \mid 0 \leq f_i \leq B_i, 1 \leq f_l \leq B_l\} ,$$

we search for a match $\mathfrak{g}_Q = \mathfrak{g}_R$ with $\mathfrak{g}_R \in R$ and $\mathfrak{g}_Q \in Q$. In this case, we have

$$\mathfrak{g}_Q \mathfrak{g}_R^{-1} = \mathfrak{g}_1^{f_1 B_1 + e_1} \cdots \mathfrak{g}_l^{f_l B_l + e_l} \sim \mathcal{O}_\Delta$$

as required, and if f_l is minimal, then $v_{l,l} = f_l B_l + e_l$ is minimal. Furthermore, we have $|G_l| = v_{1,1} v_{2,2} \cdots v_{l,l}$, $B_i \approx \sqrt{v_{i,i}}$ and $B_l \approx \sqrt{H}$, so we would expect that $|R| \approx \sqrt{|G_l|}$ and $|Q| \approx \sqrt{|G_l|}$, thereby yielding an algorithm that should cost roughly $O(\sqrt{|G_l|}\Delta^\epsilon)$ operations.

There are three noteworthy improvements and refined analyses of this method. The first, due to Buchmann, Jacobson, and Teske,[27] is an adaptation of a baby-step giant-step algorithm for computing the order of an element in a finite abelian group whose complexity depends on the order of the element as opposed to an upper bound. The main idea is the same as that used by Buchmann and Williams in their refinement of Shanks' baby-step giant-step algorithm for computing R_Δ discussed in Chapter 7. For each l, the required value of $v_{l,l}$ is found by using the baby-step giant-step method to search the interval 0 to v^2 for some small initial step width v and repeatedly doubling v until the minimal value of $v_{l,l}$ is found. The resulting algorithm can be shown to have complexity $O(2^l \sqrt{|G_l|}\Delta^\epsilon)$, where the exponential dependence on l only occurs for highly non-cyclic groups, the worst of all being groups isomorphic to $C(2) \times \cdots \times C(2)$. Thus, when applied to class group computation where highly non-cyclic groups are expected to be very rare, this method performs well.

The second improvement, due to Buchmann and Schmidt,[28] is also based on the baby-step giant-step method and eliminates the exponential dependence on l, resulting in a complexity of $O(\sqrt{|G_l|}\Delta^\epsilon)$. The main idea here is, instead of the doubling strategy used by Buchmann, Jacobson, and Teske, to use Terr's method—in particular, Lemma 7.17 as stated in our discussion of Buchmann and Vollmer's method for computing R_Δ in Chapter 7. Using this method allows the sizes of the sets R and Q to be balanced better, thereby eliminating the exponential complexity worst case. Although of great interest in terms of complexity, empirical results of Ramachandran[29] indicate that, for imaginary quadratic fields, the algorithm of Buchmann, Jacobson, and Teske is in the majority of cases faster due to the extra overhead required to balance the sizes of the sets R and Q in the Buchmann-Schmidt algorithm. However, it is possible that the Buchmann-Schmidt algorithm may have some advantages when applied to real quadratic fields, where groups of the form $C(2) \times \cdots \times C(2)$ occur more frequently.

The main drawback of both of these methods is that they require storage proportional to $\sqrt{|G_l|}$ due to the fact that they employ variations of the baby-step giant-step method. Clearly, this storage requirement becomes a limiting factor for large discriminants. Teske[30] described an algorithm that uses a variation of Pollard's rho algorithm to find the required relations. The result is a probabilistic algorithm for which the expected number of operations is $O(\sqrt{|G_l|}\Delta^\epsilon)$ that requires constant storage at the cost of additional overhead for minimizing the coefficients of the relations returned.

In the case of imaginary quadratic fields, any of these strategies can be applied directly, as equivalence testing is easy. When applied to real quadratic fields, it is necessary to incorporate a baby-step giant-step strategy for equivalence testing, essentially using a similar method to that described above for computing h_1. To the best of the authors' knowledge, no such method, in particular based on the Buchmann-Jacobson-Teske or Buchmann-Schmidt algorithms, has been described and analyzed for real quadratic fields. However, a simple strategy involving such an extension of the Buchmann-Jacobson-Teske algorithm was employed by Jacobson,[31] and such an algorithm should be fairly straightforward to develop and analyze. Generalizing Teske's probabilistic method to incorporate equivalence testing in the real quadratic case seems to be much more difficult, due to the complications related to equivalence testing.

The second issue, determining when we have $G_l = Cl_\Delta$, is resolved using, once again, the analytic class number formula. We compute an approximation h^* of h_Δ such that $h^* < h_\Delta < 2h^*$; that is, no other divisor or multiple of h_Δ lies between h^* and $2h^*$. This is done by computing an approximation of $L(1, \chi)$ with error less than $\sqrt{2}$ and applying the analytic class number formula. More specifically, we compute $\overline{L(1, \chi)} = \exp(S(Q, \Delta))$ using (10.1) such that $|L(1, \chi) - \overline{L(1, \chi)}| < \sqrt{2}$. In order to ensure that our approximation has the required accuracy, we need to select Q such that $A(Q, \Delta) < \log \sqrt{2}$, using the table of constants given by Bach.[32] Then Düllmann[33] showed that

$$h^* = \frac{\sqrt{2|\Delta|}}{\pi} \overline{L(1, \chi)} \tag{10.19}$$

suffices for the imaginary case and Abel[34] showed that

$$h^* = \sqrt{\frac{\Delta}{2}} \overline{L(1, \chi)} \tag{10.20}$$

suffices for the real case. As soon as $|G_l| > h^*$, we know that $G_l = Cl_\Delta$, as G_l is always guaranteed to be a subgroup of Cl_Δ.

Note that this method will only compute the class group unconditionally if the accuracy of the approximation of $L(1, \chi)$ does not depend on the ERH. Thus, if we use Bach's method for the approximation, we obtain an algorithm which computes the class group deterministically in time $O(h_\Delta^{1/2} \Delta^\epsilon) = O(\Delta^{1/4+\epsilon})$ for which the correctness of the output depends on the ERH. Using unconditionally correct approximations, the fastest of which is due to Louboutin,[35] results an algorithm with complexity $O(\Delta^{1/2+\epsilon})$ where the running time is dominated by the cost of computing the approximation.[36] In all these cases, the complexity results require the assumption of the GRH to ensure, via Bach's theorem, that the number of prime ideals required to generate the class group is $O(\log^2 |\Delta|)$.

10.5 Numerical Results

The methods described above have been used for a number of large-scale tabulations of class numbers and class groups of quadratic fields. We will highlight a few of the most recent results in this area below.

10.5.1 Imaginary Quadratic Fields

One of the pioneers in tabulating class groups of imaginary quadratic fields is Buell, who, in a series of papers from 1976 to 1999, tabulated the class group structures of all imaginary quadratic fields with discriminants less than 2.2×10^9 in absolute value.[37] The algorithm of enumerating all reduced ideals was used to compute the class numbers, and the group structures were determined by computing the structures of the p-Sylow subgroups[38] for each prime p such that $p^2 \mid h_\Delta$. Thus, the class groups computed are all correct without assuming any Riemann hypotheses such as the ERH.

The most recent efforts in this area are due to Jacobson, Ramachandran, and Williams[39] and Ramachandran[40] in her masters thesis. Jacobson, Ramachandran, and Williams presented the results of a tabulation for $|\Delta_{\mathbb{K}}| < 10^{11}$, and this bound is extended to 2×10^{11} by Ramachandran. In both cases, the class groups are unconditionally correct and, in particular, do not depend on the ERH. The main idea used to extend Buell's tables was to use a faster conditional algorithm, namely the generic algorithm of Buchmann, Jacobson, and Teske mentioned above combined with a novel batch verification algorithm applied as a postprocessing step. The result is a tabulation that costs only $O(\Delta^{1/4+\epsilon})$ operations per field as opposed to $O(\Delta^{1/2+\epsilon})$.

The verification algorithm is based on an idea of Booker[41] to use the Eichler-Selberg trace formula,[42] a formula relating the trace of the Hecke operator T_n acting on the space $S_k(\Gamma_0(N), \chi)$ of cusp forms of weight k, level N, and character χ to a sum of class numbers of imaginary quadratic fields. As the space $S_2(\Gamma_0(1), 1)$ has dimension zero, the traces of the corresponding Hecke operators are also zero, yielding an identity relating a sum of class numbers to an easily computable expression.[43]

The idea for verifying a table of class numbers for all fundamental discriminants with $|\Delta_{\mathbb{K}}| < B$ is to compute a sum involving the Hurwitz class numbers[44] $H(\Delta) = \sum_f h_\omega(\Delta/f^2)$ for values of $\Delta = t^2 - 4n$ for a fixed integer $n \geq 1$. The Eichler-Selberg trace formula tells us that this is equal to another sum involving all divisors of n that can be computed rapidly. Thus, we simply compute both parts of the identity and check whether they are equal. This is done for sufficiently many n such that each class number in our table has appeared in at least one identity, after which we know that all the computed class numbers are correct. Note that the group structures will also be correct, as the algorithm of Buchmann, Jacobson, and Teske correctly produces the structure of some subgroup of Cl_Δ without assuming the ERH, so if the class numbers are verified, these are also correct.

The version of the trace formula that we require is derived as follows.[45] For $\Delta = \Delta_{\mathbb{K}} f^2$, where $\Delta_{\mathbb{K}}$ is fundamental, let

$$H(\Delta) = h_w(\Delta_{\mathbb{K}})K(\Delta) \tag{10.21}$$

denote the Hurwitz class number of the quadratic order $\mathcal{O}_\Delta$, where

$$h_w(\varDelta) = \begin{cases} h_\varDelta & \text{if } |\varDelta| > 4 \\ 1/2 & \text{if } \varDelta = -4 \\ 1/3 & \text{if } \varDelta = -3 \end{cases}$$

and

$$K(\varDelta) = \sum_{t|f} t \prod_{q|t} \left(1 - \frac{\left(\frac{\varDelta}{q} \right)}{q} \right).$$

Using results of Schoof and van der Vlugt[46], we have equality

$$Tr(T_n) = A_1 + A_2 + A_3 + A_4 = 0 , \tag{10.22}$$

where

$$A_1 = \frac{1}{12} \sigma(n) ,$$

$$A_2 = -\frac{1}{2} H(-4n) - \sum_{t=1}^{\lceil \sqrt{4n} \rceil - 1} H(t^2 - 4n) ,$$

$$A_3 = - \left(\sum_{\substack{d|n \\ d < \sqrt{n}}} d \right) - \frac{1}{2} \sigma(n)\sqrt{n} ,$$

$$A_4 = \sum_{d|n} d ,$$

and $\sigma(n) = 1$ if n is a square and zero otherwise. Rearranging (10.22) gives us

$$H(-4n) + 2 \sum_{t=1}^{\lceil \sqrt{4n} \rceil - 1} H(t^2 - 4n)$$

$$= 2 \left(\sum_{d|n} d \right) - 2 \left(\sum_{\substack{d|n \\ d < \sqrt{n}}} d \right) - \sigma(n)\sqrt{n} + \frac{1}{6} \sigma(n) . \tag{10.23}$$

To verify all the class numbers for discriminants $\varDelta$ with $|\varDelta| < B$, we need to verify that (10.23) is satisfied for a certain set of n values that ensures every fundamental discriminant appears in (10.23) for at least one value of n. One possibility is to use a preprocessing step to select an appropriate set of n values, but this is somewhat costly.

A better approach is to evaluate (10.23) for $1 \leq n \leq X = \lceil B/4 \rceil$ by summing both sides of (10.23), yielding the identity

$$\sum_{n=1}^{X} \left(H(-4n) + 2 \sum_{t=1}^{\lceil \sqrt{4n} \rceil - 1} H(t^2 - 4n) \right)$$

$$= \sum_{n=1}^{X} \left(2 \left(\sum_{d|n} d \right) - 2 \left(\sum_{\substack{d|n \\ d < \sqrt{n}}} d \right) - \sigma(n)\sqrt{n} + \frac{1}{6}\sigma(n) \right). \quad (10.24)$$

Clearly, every fundamental discriminant Δ with $|\Delta| \leq B$ will occur at least once in (10.24). The trick in making this efficient is to realize that we can count precisely how many times a particular discriminant Δ, not necessarily fundamental, appears in (10.24). This allows us to rewrite (10.24) as a sum over discriminants Δ where each discriminant appears exactly once, as opposed to evaluating (10.23) separately for each value of n. If we define $r(\Delta, X)$ as the number of different representations of Δ as $t^2 - 4n$ for integers t and n satisfying $1 \leq n \leq X$ and $1 \leq t \leq \lceil\sqrt{4X}\rceil$ [i.e., the number of times Δ appears in (10.24)], then (10.24) can be rewritten as

$$\left(\sum_{\substack{\Delta \equiv 0 \ (\mathrm{mod}\ 4) \\ |\Delta| \leq 4X}} H(\Delta) \right) + 2 \left(\sum_{\substack{\Delta \equiv 0,1 \ (\mathrm{mod}\ 4) \\ |\Delta| \leq 4X}} r(\Delta, X) H(\Delta) \right)$$

$$= \sum_{n=1}^{X} \left(2 \left(\sum_{d|n} d \right) - 2 \left(\sum_{\substack{d|n \\ d < \sqrt{n}}} d \right) - \sigma(n)\sqrt{n} + \frac{1}{6}\sigma(n) \right). \quad (10.25)$$

The first sum in the left-hand side accounts for representations of Δ of the form $\Delta = t^2 - 4n$ with $t = 0$. If is easy to verify[47] that

$$r(\Delta, X) = \begin{cases} \left\lfloor \dfrac{\lfloor \sqrt{4X + \Delta} \rfloor}{2} \right\rfloor & \text{if } \Delta \equiv 0 \ (\mathrm{mod}\ 4) \\[4mm] \left\lfloor \dfrac{\lfloor \sqrt{4X + \Delta} \rfloor + 1}{2} \right\rfloor & \text{if } \Delta \equiv 1 \ (\mathrm{mod}\ 4) \end{cases} \quad (10.26)$$

by counting the number of square values of $4n + \Delta$ for $1 \leq n \leq X$.

The verification algorithm consists of evaluating the left- and right-hand sides of (10.25) separately and checking that they are equal. Equality will be violated if one or more of the tabulated class numbers is incorrect. Due to the nature of our algorithm for computing Cl_Δ, if the number we computed is not equal to the class number, it is always a divisor of h_Δ and hence less than or equal equal to h_Δ. Thus, any number of incorrect class numbers would cause the left-hand side of (10.25) to be strictly less than the right-hand side, so if (10.25) holds, we are able to unconditionally verify our results.

The right-hand side of (10.25) is evaluated by processing each value of n separately, allowing for a trivial parallel implementation. The main computational task is computing the set of divisors of each n, Although this could be done by factoring each value of n individually, it is more efficient to factor all required values of n simultaneously using the Sieve of Eratosthenes.[48]

The left-hand side of (10.25) is evaluated by looping over all fundamental discriminants $\Delta_{\mathbb{K}}$ with $|\Delta_{\mathbb{K}}| \leq B$, (i.e., every discriminant for which a class number was tabulated). As (10.25) contains Hurwitz class numbers of fundamental and non-fundamental discriminants, we need to find, for each fundamental discriminant $\Delta_{\mathbb{K}}$, all non-fundamental discriminants $\Delta_f = \Delta_{\mathbb{K}} f^2$ that appear in (10.24). Thus, for every discriminant Δ_f such that $|\Delta_f| \leq 4X$ for $1 \leq f \leq \lfloor \sqrt{4X/\Delta_{\mathbb{K}}} \rfloor$, we compute $H(\Delta_f)$ (using (10.21)) and $r(\Delta_f, X)$ (using (10.26)) for each Δ_f and add the term $2r(\Delta_f, X)H(\Delta_f)$ to a running total. We also add $H(\Delta_f)$ or each $\Delta_f \equiv 0 \pmod 4$ to account for the representation of $\Delta_f = -4n$. Note that it is also trivial to parallelize this step, by having each processor handle a distinct interval of fundamental discriminants.

This method was used to compute all Cl_Δ for all fundamental Δ with $|\Delta| < 2 \times 10^{11}$, a total of 60792710179 fields.[49] The total running time, using a cluster of 256 2.4-GHz Xeon processors with 1 GB of RAM each, was roughly 2 weeks for computing the class groups and 3 days for the verification. The data were used to test a number of conjectures, including the relevant Cohen-Lenstra heuristics from Chapter 7 and Littlewood's bounds on $L(1, \chi)$ presented in Chapter 9; not surprisingly, all the data support the truth of these conjectures. In addition, following Buell, first occurrences of various "exotic" class groups were recorded. Table 10.1 lists a few new examples discovered during this tabulation of the smallest $|\Delta|$ for which Cl_Δ has p-rank equal to 3.

Table 10.1. Minimal Examples of Imaginary Quadratic Fields $\mathbb{Q}(\sqrt{\Delta})$ with p-Rank Equal to 3

p	Δ	Cl_Δ
11	-23235125867	$C(264) \times C(11) \times C(11)$
13	-38630907167	$C(1131) \times C(13) \times C(13)$
19	-136073793499	$C(190) \times C(19) \times C(19)$

10.5.2 Real Quadratic Fields

There have been many efforts to tabulate class numbers and class groups of real quadratic fields.[50] The most recent effort for tabulating class groups is due to Jacobson,[51] for which Cl_Δ was computed for all $\Delta < 10^9$. Lenstra's algorithm as described above was used to compute the regulators, and a generalization of the generic algorithm of Buchmann, Jacobson, and Teske mentioned above was used to compute the class groups. Unfortunately, as it is currently an open problem to devise a batch verification algorithm as used for the imaginary case, the correctness of the class groups is conditional on the ERH.

As in the class group tables of imaginary quadratic fields, the data were used to test a variety of conjectures, including the Cohen-Lenstra heuristics presented in Chapter 7 and Littlewood's bounds on $L(1, \chi)$; all the data completely supported these conjectures. First occurrences of exotic class groups were also recorded, including the smallest discriminant $(\Delta = 999790597)$ for which the class group has 5-rank equal to 3 $(Cl_\Delta \cong C(40) \times C(5) \times C(5))$.

The most extensive tabulation of class numbers at the time of writing this book is due to te Riele and Williams,[52] for which all class numbers were computed for quadratic fields $\mathbb{Q}(\sqrt{p})$ of prime discriminant $p \equiv 1 \pmod 4$ and $p < 2 \times 10^{11}$. Versions of Lenstra's and Shanks' algorithms described above were used to compute the regulators and class numbers, so the data are conditional on the ERH. These algorithms were modified to make use of an improvement to Bach's $L(1, \chi)$ approximation algorithm and to take advantage of the fact that for discriminants in this range, the approximations obtained are usually quite close to the actual class number, thereby improving the computation of a divisor of h_Δ. The data were used to test the Cohen-Lenstra heuristics involving only the class numbers (and not the class group structures) as well as Hooley's conjecture, and, as usual, the data supported all the conjectures.

The tabulation of class numbers described by Jacobson, Lukes, and Williams,[53] although not as extensive as that of te Riele and Williams, is noteworthy in that the data were applied to a few conjectures not considered in the latter case. The authors computed all class numbers for real quadratic fields with discriminant $\Delta < 10^8$ and for prime discriminants less than 10^9. Similar algorithms were used to compute the regulators and class groups, but in addition to testing the Cohen-Lenstra heuristics and Hooley's conjecture, the data were applied to Littlewood's bounds on $L(1, \chi)$ in order to provide evidence in support of what is believed about the magnitude of R_Δ, namely that there exists an infinitude of real quadratic fields for which $R_\Delta > \sqrt{\Delta} \log \log \Delta$. However, as in the case of te Riele and Williams, the correctness of the data is conditional on the ERH.

Finally, we mention the numerical verification of the Ankeney-Artin-Chowla (AAC) conjecture by van der Poorten, te Riele, and Williams.[54] As mentioned in §9.2, the AAC conjecture asserts that if $(t + u\sqrt{p})/2, t, u \in \mathbb{Z}$, is the fundamental unit of $\mathbb{Q}(\sqrt{p})$ for any prime $p \equiv 1 \pmod 4$, then $p \nmid u$. This conjecture was verified for all such $p < 10^{11}$. For each prime p, a modification of the first part of Lenstra's algorithm was used to compute a multiple $k\mathcal{R}_\Delta$ of $\mathcal{R}_\Delta = \log_2 \epsilon_p$. It was shown that in order to verify that $p \nmid u$, it is sufficient to verify that $p \nmid Y$ for $\epsilon_\Delta^k = (T + Y\sqrt{p})/2$ for any integer k not divisible by p. Thus, for the values of p considered, if $kR_2 < 8p$, then $p \nmid k$ and the distance $kR_2 = \log_2 \epsilon_\Delta^k$ can be used, using infrastructure techniques, to compute $u \bmod p$ and verify that it is not equal to zero.

The method used to approximate distances is a crucial component of any algorithm using infrastructure computations in practice. One has to ensure that sufficient accuracy is maintained to guarantee an accurate output, but, at the same time, arithmetic with distances needs to be as fast as possible, so excessive amounts of precision are undesirable. One possibility is to use basic floating point arithmetic with as large a precision as necessary to ensure the numerical stability of the results, but this method, although simple to implement, is difficult to analyze in terms of the precision required. A variation of this approach was introduced by te Reile and Williams and van der Poorten, te Riele, and Williams for their tabulations mentioned above, in which base-2 logarithms are used for distances as opposed to base-e, allowing part of the floating point arithmetic to be performed with integers. Another possibility is to use explicit representations of the relative generators as opposed to their logarithms, as in the algorithm of Buchmann and Vollmer[55] for computing R_Δ. This method clearly does not suffer from problems with numerical accuracy, but the amount of storage required is somewhat large and, unless great care is taken, arithmetic with them is more time-consuming.

In the following chapters, we will explore two additional methods for representing and manipulating distances. The first of these, (f, p) representations, is similar to that of van der Poorten, te Riele, and Williams but with extensions that enable fairly easily derived error estimates, allowing the required precision guaranteeing accurate output to be determined explicitly. This method will be used for cryptographic applications described in Chapter 14 and an algorithm for unconditionally computing R_Δ in Chapter 15. The second, compact representations, is a method for explicitly representing and computing with elements in a quadratic field $\mathbb{Q}(\sqrt{\Delta})$ in such a way that the bit length of the representation is polynomial in $\log \Delta$. In addition to infrastructure computations, this provides the capability of explicitly representing units and solutions of Pell's equation using polynomially many bits.

Notes and References

[1][Len82].

[2][JLW95].

[3]The constants A and B are found in [Bac95], Table 3.

[4]The proof presented here follows the exposition in [MW92a].

[5]See ß 2 of [JLW95] for a description of a method to compute $S(Q, \Delta)$.

[6]See [JLW95]. This method is based on ideas of Fung [Fun90] for pure cubic fields but is easily adapted to real quadratic fields.

[7][Sch83].

[8][Sri98].

[9]The authors are unaware of any implementations of Srinivasan's method [Sri98]. It would be interesting to compare its performance in practice to Lenstra's algorithm using Bach's method for approximating $L(1, \chi)$.

[10]See, for example, [Bue76], [Bue87], and [Bue99].

[11]According to [MW92a], footnote 11, p. 273, the existence of this algorithm was communicated to the authors of that paper by H.W. Lenstra, Jr. To the best of our knowledge, a formal description has not appeared in the literature.

[12][Sha71].

[13][Boo06].

[14]For a complete description of Shanks' algorithm for computing the class number of an imaginary quadratic field, see [Sch83] or [Coh93].

[15]The first of these Lemmas is Lemma 8.1 of [MW92a]. The second is Lemma 8.2, adapted to the situation where Bach's method is used to approximate $L(1, \chi)$.

[16]This method, based on that proposed by Lenstra [Len82] and Schoof [Sch83], is described by Mollin and Williams in [MW92a].

[17]See [Len82], [Sch83], and [Sri98] for some examples.

[18][MW92a].

[19][Sri98].

[20][Bue89].

[21][Bue76], [Bue87], and [Bue99].

[22][Jac98].

[23][JRW06] and [Ram06].

[24]See [Coh93] for definitions of the Hermite and Smith normal forms of an integer matrix as well as algorithms to compute them.

[25]This is described in detail in [Hun74], Ch. 7 Appendix.

[26]See, for example, [Ram06].

[27][BJT97].

[28][BS05].

[29][Ram06].

[30][Tes98].

[31][Jac98].

[32][Bac95].

[33][Dül91].

[34][Abe94].

[35][Lou02a].

[36]We will see in Chapter 15 an algorithm due to Booker for approximating $L(1, \chi)$ unconditionally in time $O(\Delta^{1/4+\epsilon})$, which, when combined with the methods in this

chapter, results in an algorithm for computing Cl_Δ unconditionally in the same time.

[37] See [Bue76], [Bue87], and [Bue99].

[38] As described in [Bue89].

[39] [JRW06].

[40] [Ram06].

[41] A. Booker, private communication.

[42] [SvdV91], Theorem 2.2.

[43] This formula is also suggested for computing individual class numbers in [Coh93].

[44] See [Coh93], Definition 5.3.6.

[45] [JRW06].

[46] The result follows from Theorems 2.2 and 2.5 of [SvdV91] with $k = 2$.

[47] See [Ram06].

[48] See, for example, [CP05], Section 3.2.

[49] See [Ram06] for details.

[50] See [MW92a] for a partial survey up to 1992.

[51] [Jac98].

[52] [tRW03].

[53] [JLW95].

[54] [vdPtRW01].

[55] [BV06] and [BV07], Section 10.2.

11

(f, p) Representations of $\mathcal{O}$-ideals

11.1 Basic Concepts and Definitions

We assume in this chapter and the next that $\mathcal{O}$ is an order of $\mathbb{K}$ with positive discriminant Δ. As we have seen in the previous chapter, we can greatly improve the speed of determining R_Δ when we make use of the infrastructure technique of Shanks. Unfortunately, however, this requires that we compute distances, and as such quantities are logarithms of quadratic irrationals, they must be transcendental numbers.[1] This means, of course, that we cannot compute them to full accuracy but must instead be content with approximations to a fixed number of figures. When Δ is small, this is not likely to cause many difficulties, but when Δ becomes large, we have no real handle on how much round-off or truncation error might accumulate. Numerical analysts pay a great deal of attention to this problem, but, frequently, computational number theorists ignore it, hoping or believing that their techniques are sufficiently robust that serious deviations of their results from the truth will not occur. It must be admitted that this is usually what happens, but if a computational algorithm is to produce a numerical answer that is to be formally accepted as correct, it must contain within it the same aspects of rigour that one would expect within any mathematical proof. This means that we must provide provable bounds on the possible errors in our results.

In the procedures that we describe below,[2] we deal with this problem of error accumulation by making use of what we call (f, p) representations of ideals.[3] After defining these representations and deriving a number of their properties, we will produce a number of core algorithms essential for our subsequent work. Because of the fundamental importance of these procedures, we will provide them in considerable detail in the Appendix. It must be emphasized that there are many other ways in which we could approach these same processes; however, we have found through our experience in implementing such techniques that the algorithms given below, however tedious, are the most efficient means of producing the required output.[4]

Suppose we are given some $\theta \in \mathbb{R}^+$ and some $p \in \mathbb{Z}^{>0}$. There exists some $k, q, r \in \mathbb{Z}$ such that

$$2^k < \theta \le 2^{k+1}$$

and

$$\lceil 2^{2p-k}\theta \rceil = 2^p q + r\ ,$$

where $-2^p + 1 \le r \le 0$. If we put $\eta = \lceil 2^{2p-k}\theta \rceil - 2^{2p-k}\theta$, then $0 \le \eta < 1$ and

$$|2^{2p-k}\theta - 2^p q| = |r - \eta| \le |r| + |\eta| < 2^p - 1 + 1 = 2^p\ .$$

Hence,

$$2^p q < 2^{2p-k}\theta + 2^p \le 2^{2p+1} + 2^p$$

and

$$q < 2^{p+1} + 1\ .$$

Since

$$2^p q = 2^{2p-k}\theta - r + \eta > 2^{2p} + \eta \ge 2^{2p}\ ,$$

we also have $q > 2^p$. Furthermore,

$$|2^{2p-k}\theta - 2^p q| < 2^p < q\ .$$

It follows that for any $\theta \in \mathbb{R}^+$ and $p \in \mathbb{Z}^{>0}$, there always exists some $q, k \in \mathbb{Z}$ such that $2^p < q \le 2^{p+1}$ and

$$\left| \frac{2^p \theta}{2^k q} - 1 \right| < \frac{1}{2^p}\ .$$

In the case of $\theta = (a + b\sqrt{D})/c$, where $a, b, c \in \mathbb{Z}$ and $\theta \notin \mathbb{Q}$, we can compute $\lceil m\theta \rceil$ ($m \in \mathbb{Z}^{>0}$) by assuming with no loss of generality that $c > 0$ and using

$$\left\lceil \frac{m(a + b\sqrt{D})}{c} \right\rceil = \left\lceil \frac{\lceil m(a + b\sqrt{D}) \rceil}{c} \right\rceil\ .$$

Since $\lceil m(a + b\sqrt{D}) \rceil = ma + \lceil mb\sqrt{D} \rceil$, we are left with the problem of evaluating $\lceil d\sqrt{D} \rceil$ for $d = mb$. If $d < 0$, then $\lceil d\sqrt{D} \rceil = -\lceil -d\sqrt{D} \rceil + 1$; thus, we may now assume that $d > 0$. We develop the simple continued fraction (SCF) expansion of $\sqrt{D}$ and compute A_i, B_i until $B_i > d$ and $2 \nmid i$. In this case, we have by (3.27) and (3.29),

$$0 < \frac{A_i}{B_i} - \sqrt{D} < \frac{1}{B_i^2}\ .$$

Hence,

$$0 < d\frac{A_i}{B_i} - d\sqrt{D} < \frac{d}{B_i^2} < \frac{1}{B_i}\ .$$

Since $I = \lceil dA_i/B_i \rceil$ satisfies

$$0 < I - d\sqrt{D} < 1 - \frac{1}{B_i} + \frac{1}{B_i} = 1 \,,$$

we have

$$\lceil d\sqrt{D} \rceil = I \,.$$

Thus, since $B_i \geq F_{i+1} > \tau^{i-1}$, we can compute $\lceil m\theta \rceil$ in $O(\log(m|b|))$ steps in the SCF expansion of $\sqrt{D}$. If $\lceil \theta \rceil > 1$, it is easy to find k from

$$2^k < \lceil \theta \rceil \leq 2^{k+1} \,.$$

If $\lceil \theta \rceil \leq 1$, we find $\lceil \theta^{-1} \rceil$ and compute t such that

$$2^{t-1} \leq \lceil \theta^{-1} \rceil < 2^t \,;$$

in this case, $k = -t$.

Definition 11.1. *Let $p \in \mathbb{Z}^{>0}$, $f \in \mathbb{R}$ with $f \geq 1$ and let $\mathfrak{a}$ be an $\mathcal{O}$-ideal. An (f, p) representation of $\mathfrak{a}$ is a triple $(\mathfrak{b}, d, k)$ where the following hold:*

1. $\mathfrak{b}$ *is an $\mathcal{O}$-ideal equivalent to $\mathfrak{a}$, $d \in \mathbb{N}$ with $2^p < d \leq 2^{p+1}$, $k \in \mathbb{Z}$.*
2. *There exists a $\theta \in \mathbb{K}$ with $\mathfrak{b} = \theta\mathfrak{a}$ and*

$$\left| \frac{2^{p-k}\theta}{d} - 1 \right| < \frac{f}{2^p} \,. \tag{11.1}$$

An (f, p) representation of $\mathfrak{a}$ is said to be *reduced* if $\mathfrak{b}$ is a reduced $\mathcal{O}$-ideal. Note that $(\mathfrak{a}, 2^{p+1}, -1)$ or $(\mathfrak{a}, 2^p + 1, 0)$ is an (f, p) representation of $\mathfrak{a}$ ($\theta = 1$) for any $f \geq 1$. The symbol f here should not be confused with that used for the conductor of $\mathcal{O}$.

Although it is obvious by (11.1), it is nevertheless important to observe that if $(\mathfrak{b}, d, k)$ is an (f, p) representation of an $\mathcal{O}$-ideal $\mathfrak{a}$, then it is also an (f', p) representation of $\mathfrak{a}$ for any $f' \geq f$. By our earlier remarks, we see that, given any $f, \theta \in \mathbb{R}$ and $p \in \mathbb{Z}^{>0}$ such that $f \geq 1$, there always exist $k, d \in \mathbb{Z}$ such that $2^p < d \leq 2^{p+1}$ and

$$\left| \frac{2^p\theta}{2^kd} - 1 \right| < \frac{f}{2^p} \,.$$

If f, p, k, and θ satisfy (11.1) and $f < 2^p$, then

$$0 < \frac{d}{2^p}\left(1 - \frac{f}{2^p}\right) < \frac{\theta}{2^k} < \frac{d}{2^p}\left(1 + \frac{f}{2^p}\right) < 4 \,. \tag{11.2}$$

Since

$$1 > 1 - \left(\frac{f}{2^p}\right)^2 = \left(1 - \frac{f}{2^p}\right)\left(1 + \frac{f}{2^p}\right) \,,$$

we see that

$$\left|\log\left(\frac{2^p\theta}{2^kd}\right)\right| < \left|\log\left(1-\frac{f}{2^p}\right)\right|.$$

Hence,

$$|p - \log_2 d + \log_2 \theta - k| < \left|\log_2\left(1-\frac{f}{2^p}\right)\right|. \tag{11.3}$$

Since $-1 \le p - \log_2 d < 0$ and $|\log(1-f/2^p)|$ is small when f is not too close to 2^p, we see that $k \approx \log_2 \theta$. When $f/2^p < \sqrt{2}-1$, we can be more precise than this. If we recall that $1 < d/2^p \le 2$ and take logarithms across (11.2), we see that

$$\log_2\left(1-\frac{f}{2^p}\right) < \log_2 \theta - k < 1 + \log_2\left(1+\frac{f}{2^p}\right).$$

Since $f/2^p < \sqrt{2}-1$, we have

$$\log_2\left(1+\frac{f}{2^p}\right) < \frac{1}{2}.$$

Also, $1-(f/2^p)^2 > 1/2$, and this means that

$$-\log_2\left(1-\frac{f}{2^p}\right) < 1 + \log_2\left(1+\frac{f}{2^p}\right) < \frac{3}{2}.$$

Hence,

$$|\log_2 \theta - k| < \frac{3}{2}. \tag{11.4}$$

Thus, if we are given an (f,p) representation $(\mathfrak{b},d,k)$ of an unknown $\mathcal{O}$-ideal $\mathfrak{a}$, then $2^{k-p}d$ is an approximation to the value of the (unknown) relative generator θ of $\mathfrak{b}$ with respect to $\mathfrak{a}$ to accuracy $f/2^p$, and p can be thought of as the precision of the approximation. In working with reduced ideals we will use k as our measure of distance.

Suppose we are given $(\mathfrak{b}',d',k')$ and $(\mathfrak{b}'',d'',k'')$, which are respectively an (f',p) representation of $\mathfrak{a}'$ and an (f'',p) representation of $\mathfrak{a}''$, where $\mathfrak{a}'$ and $\mathfrak{a}''$ are both $\mathcal{O}$-ideals. We now consider the problem of finding f and $(\mathfrak{b},d,k)$, a reduced (f,p) representation of $\mathfrak{a}'\mathfrak{a}''$. We begin with the following result for dealing with products of (f,p) representations.[5]

Theorem 11.2. *Let $(\mathfrak{b}',d',k')$ be an (f',p) representation of an $\mathcal{O}$-ideal $\mathfrak{a}'$ and let $(\mathfrak{b}'',d'',k'')$ be an (f'',p) representation of an $\mathcal{O}$-ideal $\mathfrak{a}''$. If $d'd'' \le 2^{2p+1}$, put $d = \lceil d'd''/2^p \rceil$ and $k = k'+k''$. If $d'd'' > 2^{2p+1}$, put $d = \lceil d'd''/2^{p+1} \rceil$ and $k = k'+k''+1$. Then $(\mathfrak{b}'\mathfrak{b}'',d,k)$ is an (f,p) representation of the product ideal $\mathfrak{a}'\mathfrak{a}''$, where $f = 1 + f' + f'' + 2^{-p}f'f''$.*

Proof. By the bounds on d' and d'' and the definition of d in the theorem, it is easy to see that $2^p < d \le 2^{p+1}$. By assumption, we may let $\mathfrak{b}' = \theta'\mathfrak{a}'$ and $\mathfrak{b}'' = \theta''\mathfrak{a}''$ with $\theta',\theta'' \in \mathbb{K}^{\ge 0}$ and

$$\left|\frac{2^{p-k'}\theta'}{d'}-1\right|<\frac{f'}{2^p}, \quad \left|\frac{2^{p-k''}\theta''}{d''}-1\right|<\frac{f''}{2^p}.$$

By (11.2) we get

$$\left(1-\frac{f'}{2^p}\right)\left(1-\frac{f''}{2^p}\right)<\frac{2^{2p-k'-k''}\theta'\theta''}{d'd''}<\left(1+\frac{f'}{2^p}\right)\left(1+\frac{f''}{2^p}\right).$$

If we put $f^* = f' + f'' + f'f''/2^p$, then $(1+f'/2^p)(1+f''/2^p) = 1 + f^*/2^p$ and $(1-f'/2^p)(1-f''/2^p) > 1 - f^*/2^p$. Hence,

$$1-\frac{f^*}{2^p}<\frac{2^{2p-k'-k''}\theta'\theta''}{d'd''}<1+\frac{f^*}{2^p}.$$

Suppose $d'd'' \leq 2^{2p+1}$. Since $d = d'd''/2^p + \eta$ $(0 \leq \eta < 1)$, we get

$$1-\frac{f^*}{2^p}<\frac{2^{p-k}\theta'\theta''}{d-\eta}<1+\frac{f}{2^p}$$

and

$$\left(1-\frac{\eta}{d}\right)\left(1-\frac{f^*}{2^p}\right)<\frac{2^{p-k}\theta'\theta''}{d}<\left(1-\frac{\eta}{d}\right)\left(1+\frac{f^*}{2^p}\right)<1+\frac{f^*}{2^p}.$$

Since $d > \eta 2^p$, we have

$$\left(1-\frac{\eta}{d}\right)\left(1-\frac{f^*}{2^p}\right)>\left(1-\frac{f}{2^p}\right),$$

and it follows that

$$1-\frac{f}{2^p}<\frac{2^{p-k}\theta'\theta''}{d}<1+\frac{f}{2^p};$$

since $\mathfrak{b}'\mathfrak{b}'' = \theta'\theta''\mathfrak{a}'\mathfrak{a}''$, we get our result. The theorem follows in similar manner when $d'd'' > 2^{2p+1}$. □

We can now use this result to produce the following algorithm.

Algorithm 11.1: NUMULT

Input: $(\mathfrak{b}', d', k')$, $(\mathfrak{b}'', d'', k'')$, p, where $(\mathfrak{b}', d', k')$ is a reduced (f', p) representation of an invertible $\mathcal{O}$-ideal $\mathfrak{a}'$ and $(\mathfrak{b}'', d'', k'')$ is reduced (f'', p) representation of an invertible $\mathcal{O}$-ideal $\mathfrak{a}''$. Here,

$$\mathfrak{b}' = \left[\frac{Q'}{r}, \frac{P'+\sqrt{D}}{r}\right], \quad \mathfrak{b}'' = \left[\frac{Q''}{r}, \frac{P''+\sqrt{D}}{r}\right], \quad Q' \geq Q'' > 0.$$

Output: A reduced (f, p) representation $(\mathfrak{b}, d, k)$ of $\mathfrak{a}'\mathfrak{a}''$, where

$$\mathfrak{b} = \left[\frac{Q}{r}, \frac{P+\sqrt{D}}{r}\right],$$

$(P+\sqrt{D})/Q > 1$, $-1 < (P-\sqrt{D})/Q < 0$, $k \leq k'+k''+1$, $f = f^*+17/8$ with $f^* = f' + f'' + 2^{-p}f'f''$. (Optional output: $a, b \in \mathbb{Z}$, where $\nu = (a+b\sqrt{D})/r \in \mathcal{O}$, and $\mathfrak{b} = \nu\mathfrak{b}'\mathfrak{b}''/(N(\mathfrak{b}')N(\mathfrak{b}''))$.)

11.2 w-Near Representations

A reduced (f,p) representation $(\mathfrak{b}, d, k)$ of an $\mathcal{O}$-ideal $\mathfrak{a}$ is said to be w-near for some $w \in \mathbb{Z}^{\geq 0}$ if the following conditions hold:

1. $k < w$.
2. If $\rho(\mathfrak{b}) = \psi\mathfrak{b}$, $\mathfrak{b}_1 = \bar{\mathfrak{b}}$, and $\mathfrak{b}_2 = \rho(\mathfrak{b}_1) \,(= \psi\mathfrak{b})$, then there exist integers d' and k' with $k' \geq w$, $2^p < d' \leq 2^{p+1}$ such that

$$\left| \frac{2^{p-k'}\theta\psi}{d'} - 1 \right| < \frac{f}{2^p} \,.$$

If $(\mathfrak{b}, d, k)$ is a w-near (f,p) representation of some $\mathcal{O}$-ideal $\mathfrak{a}$ and f is not too large, then the parameters θ and k will not be far from 2^w and w, respectively. We can be more precise about this in the following lemma.

Lemma 11.3. *Let $(\mathfrak{b}, d, k)$ be a w-near (f,p) representation of some $\mathcal{O}$-ideal $\mathfrak{a}$ with $p > 4$ and $f < 2^{p-4}$. If θ and ψ have the meaning assigned to them above, then*

$$\frac{15N(\mathfrak{b})}{16\sqrt{\Delta}} < \frac{15}{16\psi} < \frac{\theta}{2^w} < \frac{17}{16} \quad \text{and} \quad 0 > k - w > -\log_2\left(\frac{34\psi}{15}\right) \,.$$

Proof. By (11.1) and Condition 2 we have

$$1 - \frac{f}{2^p} < \frac{2^{p-k}\theta}{d} < 1 + \frac{f}{2^p}, \quad 1 - \frac{f}{2^p} < \frac{2^{p-k'}\theta\psi}{d} < 1 + \frac{f}{2^p} \,.$$

Since $k < w$ and $d \leq 2^{p+1}$, we get $2^{k-p}d \leq 2^{w-p-1} \cdot 2^{p+1} = 2^w$; hence

$$\frac{\theta}{2^w} < \frac{2^{k-p}d}{2^w}\left(1 + \frac{f}{2^p}\right) < 1 + \frac{2^{p-4}}{2^p} = \frac{17}{16} \,. \tag{11.5}$$

Also, since $k' \geq w$ and $d' > 2^p$, we get $2^{k'-p}d' > 2^w$ and

$$\frac{\theta\psi}{2^w} > 1 - \frac{2^{p-4}}{2^p} = \frac{15}{16} \,. \tag{11.6}$$

From (3.33) we have

$$N(\mathfrak{b})\psi < \sqrt{\Delta} \,;$$

hence,

$$\frac{15}{16\psi} > \frac{15}{16}\frac{N(\mathfrak{b})}{\sqrt{\Delta}} \,.$$

By (11.5) and (11.6) we have

$$2^{-k+w} < \frac{2^{-p}d2^w}{\theta}\left(1 + \frac{f}{2^p}\right) < \frac{34}{15}\psi \,.$$

Hence,

$$0 > k - w > -\log_2\left(\frac{34\psi}{15}\right).$$

$\square$

Corollary 11.3.1. *Under the conditions of Lemma 11.3, we have*

$$\frac{15}{16\sqrt{\Delta}} < \frac{\theta}{2^w} < \frac{17}{16} \quad and \quad 0 > k - w > -\log_2\left(\frac{34}{15}\sqrt{\Delta}\right).$$

Furthermore, if $\mathfrak{a}$ is a reduced ambiguous $\mathcal{O}$-ideal and $b \in \mathbb{R}$, $b \geq 1$, then

$$\theta > \frac{15}{16b} \quad and \quad w - k < \log_2\left(\frac{34}{15}b\right)$$

with probability approximately $1 - \log_2(1 + b^{-1})$.

Proof. The first set of inequalities follows from $1 < \psi < \sqrt{\Delta}$. In the second case, we have $\psi = \psi_j$ in the continued fraction expansion of $(P + \sqrt{D})/Q$, where $\mathfrak{a} = [Q/r, (P + \sqrt{D})/r]$. By Theorem 5.21, we have $\psi_j = \phi_{n+1-j}$, where n is the period length of the continued fraction. By the Gauss-Kuz'min theorem (Theorem 3.18), the probability that $\psi_i > b$ for any $b \geq 1$ is about $\log_2(1 + b^{-1})$. Hence, $\psi_j \leq b$ with approximate probability $1 - \log_2(1 + b^{-1})$, in which case $\theta/2^w > 15/16\psi \geq 15/16b$ and $w - k < \log_2(34\psi/15) \leq \log_2(34b/15)$. $\square$

For example, setting $b = 60/17 = 3.529\ldots$, we would expect that $\theta/2^w > 17/64$ and $w - k \leq 2$ about 64% of the time.

If $(\mathfrak{b}, d, k)$ and $(\mathfrak{c}, e, h)$ are two w-near (f, p) representations of some $\mathcal{O}$-ideal $\mathfrak{a}$, it is not necessarily the case that $\mathfrak{b} = \mathfrak{c}$. However, we can provide the following theorem.

Theorem 11.4. *Let $(\mathfrak{b}, d, k)$ and $(\mathfrak{c}, e, h)$ be two w-near (f, p) representations of some $\mathcal{O}$-ideal $\mathfrak{a}$ with $p > 4$ and $f < 2^{p-4}$. Then*

$$\mathfrak{b} \in \left\{\rho^{-2}(\mathfrak{c}), \rho^{-1}(\mathfrak{c}), \mathfrak{c}, \rho(\mathfrak{c}), \rho^2(\mathfrak{c})\right\}.$$

Proof. Since $\mathfrak{c}$ and $\mathfrak{b}$ are equivalent reduced $\mathcal{O}$-ideals, by the results in §5.3 we must have $\mathfrak{c} = \rho^i(\mathfrak{b}) = \mathfrak{b}_{i+1} = \theta_{i+1}\mathfrak{b}_1$ for $\mathfrak{b}_1 = \mathfrak{b}$. Also, $\rho(\mathfrak{b}) = \psi_1\mathfrak{b}_1$ and $\rho(\mathfrak{c}) = \mathfrak{b}_{i+2} = \psi_{i+1}\mathfrak{b}_{i+1}$. By the definition of an (f, p) representation, we have $\theta, \gamma \in \mathbb{K}$, where $\theta, \gamma > 0$, $\mathfrak{b} = \theta\mathfrak{a}$, and $\mathfrak{c} = \gamma\mathfrak{a}$. Furthermore, by Lemma 11.3

$$\frac{15}{16\psi_1} < \frac{\theta}{2^w} < \frac{17}{16} \tag{11.7}$$

and

$$\frac{15}{16\psi_{i+1}} < \frac{\gamma}{2^w} < \frac{17}{16}. \tag{11.8}$$

Since $c = (\theta_{i+1}\theta/\gamma)c$, we must also have

$$\epsilon_\Delta^j = \frac{\theta_{i+1}\theta}{\gamma} \, ,$$

where $\epsilon_\Delta = \theta_{n+1}$, $\rho^n(\mathfrak{b}) = \mathfrak{b}$ and n is positive and minimal.

We may assume that $i < n$; suppose $3 \le i \le n - 3$. We get

$$\epsilon_\Delta^j = \frac{\theta\psi_1}{\gamma}(\psi_2 \cdots \psi_i) > \frac{2\theta\psi_1}{\gamma}$$

by Proposition 3.16. By (11.7) and (11.8), we get $\theta\psi_1/\gamma > 15/17$; hence, $\epsilon_\Delta^j > 1$. Also, by (11.7) and (11.8),

$$\epsilon_\Delta^j = \frac{\theta_{i+2}\theta}{\psi_{i+1}\gamma} < \frac{17}{15}\theta_{n-1} = \frac{17}{15}\frac{\theta_{n+1}}{\psi_{n-1}\psi_n} < \frac{17}{30}\epsilon_\Delta < \epsilon_\Delta \, .$$

However, $1 < \epsilon_\Delta^j < \epsilon_\Delta$ is impossible; hence, $i \in \{2, 1, 0, n-2, n-1\}$. It follows that

$$\mathfrak{b} \in \{\rho^{-2}(c), \rho^{-1}(c), c, \rho(c), \rho^2(c)\} \, .$$

$\square$

Very frequently, however, it turns out that $\mathfrak{b} = c$. Some explanation for this is provided in the next theorem.

Theorem 11.5. *Let $(\mathfrak{b}, d, k)$ and (c, e, h) be two w-near reduced (f, p) representations of an $\mathcal{O}$-ideal $\mathfrak{a}$. Suppose $(\rho(\mathfrak{b}), d', k')$ and $(\rho(c), e', h')$ $(h', k' \ge 0)$ are also (f, p) representations of $\mathfrak{a}$. If*

$$\min\left\{\frac{e'}{d}, \frac{d'}{e}\right\} \ge \frac{1}{2} + \frac{f}{2^p - f} \, , \tag{11.9}$$

then $\mathfrak{b} = c$.

Proof. There must exist $\theta_1, \theta_2 \in \mathbb{K}$ such that $\mathfrak{b} = \theta_1\mathfrak{a}$ and $c = \theta_2\mathfrak{a}$; hence, $\mathfrak{b} = (\theta_1/\theta_2)c$. Furthermore,

$$\frac{d2^k}{2^p}\left(1 - \frac{f}{2^p}\right) < \theta_1 < \frac{d2^k}{2^p}\left(1 + \frac{f}{2^p}\right) \, ,$$
$$\frac{e2^h}{2^p}\left(1 - \frac{f}{2^p}\right) < \theta_2 < \frac{e2^h}{2^p}\left(1 + \frac{f}{2^p}\right) \, .$$

Also, $k, h < w$, $2^p < e$, and $d \le 2^{p+1}$. If $\mathfrak{b} \ne c$, then $\theta_1 \ne \theta_2$. Suppose $\theta_1 > \theta_2$. Now, $\rho(c) = (\psi)c$ and therefore $\theta_1 \ge \psi\theta_2$; but

$$\frac{e'2^{h'}}{2^p}\left(1 - \frac{f}{2^p}\right) < \psi\theta_2 < \frac{e'2^{h'}}{2^p}\left(1 + \frac{f}{2^p}\right) \, .$$

Hence,

$$\frac{d2^k}{2^p}\left(1+\frac{f}{2^p}\right) > \frac{e'2^{h'}}{2^p}\left(1-\frac{f}{2^p}\right),$$

$$2^{k-h'} > \frac{e'}{d}\left(\frac{1-f/2^p}{1+f/2^p}\right),$$

and

$$\frac{e'}{d} < 2^{k-h'}\left(\frac{1+f/2^p}{1-f/2^p}\right) \le \frac{1}{2}\left(\frac{1+f/2^p}{1-f/2^p}\right) = \frac{1}{2}+\frac{f}{2^p-f}.$$

Similarly, if $\theta_1 < \theta_2$, then

$$\frac{d'}{e} < \frac{1}{2}+\frac{f}{2^p-f}.$$

$\square$

Since $1/2 < e'/d$ and $d'/e < 2$, we would certainly expect (11.9) to hold very frequently, particularly when f is significantly smaller than 2^p. We can justify this statement by using the following argument.

Theorem 11.6. *Let* $\mu \in \mathbb{R}$ *such that* $1/2 \le \mu < 1$, $a \in \mathbb{Z}^{>0}$ *and* $S = \{a+1, a+2,\ldots,2a\}$. *If* i *and* j *are selected at random from* S, *the probability that*

$$i/j > \mu \tag{11.10}$$

is given by

$$P := \Pr\left(\frac{i}{j} > \mu\right) = 3 - 2\mu - (2\mu)^{-1} + \gamma, \tag{11.11}$$

where $|\gamma| < 3/a$.

Proof. We first observe that there are a^2 possible pairs such that $i, j \in S$. If $T \subseteq S \times S$ such that for $(i,j) \in T$, we have (11.10); then $P = |T|/a^2$. Select any $j \in S$. If $\lfloor \mu j \rfloor < a$, then any $i \in S$ satisfies (11.10). If $\lfloor \mu j \rfloor \ge a$, the values of $i \in S$ for which (11.10) holds are

$$\lfloor \mu j \rfloor + 1, \lfloor \mu j \rfloor + 2,\ldots,2a.$$

It follows that

$$|T| = a\sum_{\substack{\lfloor \mu j\rfloor < a \\ j=a+1}}^{2a} 1 + \sum_{\substack{\lfloor \mu j\rfloor \ge a \\ j=a+1}}^{2a}(2a - \lfloor \mu j\rfloor). \tag{11.12}$$

Let N denote the maximum value of $j \in \mathbb{Z}$ such that $\lfloor \mu j \rfloor < a$. Clearly, we may write $N = a/\mu - \eta$, and it is easy to show that $0 < \eta \le 1$. We can now write (11.12) as

$$|T| = a(N - a) + 2a(2a - N) - \sum_{j=N+1}^{2a} \lfloor \mu j \rfloor . \qquad (11.13)$$

Since $\lfloor \mu j \rfloor = \mu j - \eta_j$ and $0 \leq \eta_j < 1$, we get

$$\sum_{i=N+1}^{2a} \lfloor \mu j \rfloor = \mu \sum_{j=N+1}^{2a} j - H ,$$

where $H = \sum_{j=N+1}^{2a} \eta_j$. Thus, $0 \leq H < 2a - N$ and

$$\sum_{j=N+1}^{2a} j = \frac{2a(2a + 1)}{2} - \frac{N(N + 1)}{2} .$$

If we substitute these results back into (11.13), we get

$$|T| = 3a^2 - aN - 2\mu a^2 - \mu a + \mu N^2/2 + \mu N/2 + H .$$

Replacing N by $a/\mu - \eta$, we get

$$|T| = 3a^2 - \frac{a^2}{2\mu} - 2\mu a^2 + \gamma a ,$$

where $|\gamma| < 3$. □

Put $a = 2^p$, $g = f/(2^p - f)$, $f < 2^{p-4}$, and $\mu = 1/2 + g$. Then $0 < g < 1/15$. If we assume that for a large number of (f, p) representations of ideals in $\mathcal{O}$, the values for e, d, e', and d' are randomly distributed in S, then[6]

$$\mathrm{Pr}\left(\frac{e'}{d} > \frac{1}{2} + g\right) = \mathrm{Pr}\left(\frac{d'}{e} > \frac{1}{2} + g\right) .$$

Also, if 2^p is much larger than 3, we get by Theorem 11.6 that

$$\mathrm{Pr}\left(\frac{e'}{d} > \frac{1}{2} + g\right) \approx 3 - (1 + 2g) - (1 + 2g)^{-1}$$

$$= 3 - (1 + 2g) - (1 - 2g + 4g^2 - 8g^3 + \cdots)$$

$$\approx 1 - 4g^2 .$$

Hence,

$$\mathrm{Pr}\left(\min\left\{\frac{e'}{d}, \frac{d'}{e}\right\} > \frac{1}{2} + g\right) \approx (1 - 4g^2)^2 \approx 1 - 8g^2 .$$

When g is small, this probability is very close to 1.

We also have an algorithm for producing a w-near representation for an $\mathcal{O}$-ideal $\mathfrak{a}$, given a reduced (f, p) representation $(\mathfrak{b}, d, k)$ of $\mathfrak{a}$. We will assume that $\mathfrak{b} = [Q/r, (P + \sqrt{D})/r]$, where $(P + \sqrt{D})/Q > 1$ and $-1 < (P - \sqrt{D})/Q < 0$.

Algorithm 11.2: WNEAR

Input: $(\mathfrak{b}, d, k), w, p$, where $(\mathfrak{b}, d, k)$ is a reduced (f, p) representation of some
$\mathcal{O}$-ideal $\mathfrak{a}$. Here $\mathfrak{b} = [Q/r, (P+\sqrt{D})/r]$, where $P+\lfloor\sqrt{D}\rfloor \geq Q, 0 \leq \lfloor\sqrt{D}\rfloor - P \leq Q$.

Output: $(\mathfrak{c}, g, h)$ a w-near $(f + 9/8, p)$ representation of $\mathfrak{a}$. (Optional output: a reduced $(f + 9/8, p)$ representation $(\rho(\mathfrak{c}), g', h')$ of $\mathfrak{a}$.)

11.3 Exponentiation of Ideals and Computation of $\mathfrak{a}[x]$

In this section we will first develop an algorithm, EXP, to determine, given an invertible $\mathcal{O}$-ideal $\mathfrak{a}$, a w-near representation $(\mathfrak{b}, d, k)$ of $\mathfrak{a}^n$ for a positive integer n. In the case of computing $b \equiv a^n \pmod{m}$, where $a, b, m, n \in \mathbb{Z}^{>0}$, we can put

$$n = b_0 2^k + b_1 2^{k-1} + \cdots + b_k$$

$(b_0, b_1, \ldots, b_k \in \{0, 1\})$. If $s_0 = b_0 = 1$ and $s_{i+1} = 2s_i + b_{i+1}$, then $s_k = n$. We let $r_i \equiv a^{s_i} \pmod{m}$; then

$$r_{i+1} \equiv a^{s_{i+1}} = a^{2s_i + b_{i+1}} = a^{b_{i+1}} r_i^2 \pmod{m} .$$

Hence,

$$r_{i+1} = \begin{cases} r_i^2 \pmod{m} & \text{when } b_{i+1} = 0 \\ a r_i^2 \pmod{m} & \text{when } b_{i+1} = 1 . \end{cases}$$

Thus, we can compute $b \equiv r_k \pmod{m}$ in $O(k) = O(\log n)$ elementary arithmetic operations.[7]

We can do the same thing with ideals, but we use ideal reduction as opposed to reduction modulo m. This, of course, returns us to the problem of ideal multiplication, which we have discussed in §5.4. We can now incorporate the previous two algorithms into one operation, which on input of two w-near representations outputs a w-near representation of the product of the two ideals represented by the inputs.

Algorithm 11.3: WMULT

Input: $(\mathfrak{b}', d', k'), (\mathfrak{b}'', d'', k''), w, p$ where $(\mathfrak{b}', d', k')$ is a w-near (f', p) representation of an invertible $\mathcal{O}$-ideal $\mathfrak{a}'$ and $(\mathfrak{b}'', d'', k'')$ is a w-near (f'', p) representation of an invertible $\mathcal{O}$-ideal $\mathfrak{a}''$. Here $\mathfrak{b}' = [Q'/r, (P' + \sqrt{D})/r]$, $\mathfrak{b}'' = [Q''/r, (P'' + \sqrt{D})/r]$.

Output: A w-near $(f^* + 13/4, p)$ representation $(\mathfrak{c}, g, h)$ of $\mathfrak{a}'\mathfrak{a}''$ with $f^* = f' + f'' + 2^{-p}f'f''$. (Optional output: a w-near $(f^* + 13/4, p)$ representation $(\rho(\mathfrak{c}), g', h')$ of $\mathfrak{a}$.)

1: **if** $Q' \geq Q''$ **then**
2:     $(\mathfrak{b}, d, k) = \text{NUMULT}((\mathfrak{b}', d', k'), (\mathfrak{b}'', d'', k''), p)$
3: **else**

4:         $(\mathfrak{b}, d, k) = \text{NUMULT}((\mathfrak{b}'', d'', k''), (\mathfrak{b}', d', k'), p)$
5: **end if**
6: $(\mathfrak{c}, g, h) = \text{WNEAR}((\mathfrak{b}, d, k), w, p).$
    $(((\mathfrak{c}, g, h), (\rho(\mathfrak{c}), g', h')) = \text{WNEAR}((\mathfrak{b}, d, k), w, p).)$

From this point forward we will regard the addition, subtraction, comparison, multiplication, and division of integers of $O(p + \log_2 \Delta)$ bits to be elementary operations. After step 1 of WMULT, we have $k \leq k' + k'' + 1 \leq 2w - 1$. Also, $k \geq k' + k'' - t$, where $t = O(\log \mu) = O(\log \Delta)$ by (5.45). Hence, $k' + k'' - k = O(\log \Delta)$. Since by Lemma 11.3 and (3.33), $w - k' = O(\log \Delta)$ and $w - k'' = O(\log \Delta)$, we find that

$$-w < w - k < -w + O(\log \Delta) \ .$$

It follows that WMULT will execute in $O(\max\{w, \log \Delta\})$ elementary operations.

Clearly, there are any number of possible selections for a value of w in WMULT. If we select w such that

$$2^{w-1} < \Delta^{1/4} < 2^w \ , \tag{11.14}$$

then by (5.47) we expect $|\log_2 \mu - w|$ to be small much of the time NUMULT is executed and, therefore, $|k - w|$ will be small much of the time. This follows on noting that after NUMULT is executed,

$$k \approx k' + k'' - \log_2 \mu$$
$$\approx k' + k'' - w$$
$$\approx w + w - w = w \ .$$

In fact, we have discovered by empirical studies that the amount of time needed to find a w-near reduced representation $(\mathfrak{b}, d, k)$ for this value[8] of w from the result produced by NUMULT takes between 10% and 18% of the time required for NUMULT to execute.

We can now use the standard binary exponentiation technique to do "exponentiation" on w-near (f, p) representations.

Algorithm 11.4: EXP

Input: $(\mathfrak{b}_0, d_0, k_0), n, w, p$, where $n \in \mathbb{N}$ and $(\mathfrak{b}_0, d_0, k_0)$ is a w-near (f_0, p) representation of some invertible $\mathcal{O}$-ideal $\mathfrak{a}$.
Output: A w-near (f, p) representation of $(\mathfrak{b}, d, k)$ of $\mathfrak{a}$ for suitable $f \in [1, 2^p)$.
1: Compute the binary representation of n, say $n = \sum_{i=0}^{l} b_i 2^{l-i}$ ($b_0 = 1, b_i \in \{0, 1\}$ for $1 \leq i \leq l$, $l = \lfloor \log_2 n \rfloor$).
2: Set $(\mathfrak{b}, d, k) = (\mathfrak{b}_0, d_0, k_0)$.
3: **for** $i = 1$ to l **do**
4:      $(\mathfrak{b}, d, k) = \text{WMULT}((\mathfrak{b}, d, k), (\mathfrak{b}, d, k), w, p).$
5:      **if** $b_i = 1$ **then**

6:          $(\mathfrak{b}, d, k) = \text{WMULT}((\mathfrak{b}, d, k), (\mathfrak{b}_0, d_0, k_0), w, p)$.
7:      **end if**
8: **end for**

By our previous remark, we see that EXP executes in $O(\log n \log \Delta)$ elementary operations when $w < \log \Delta$.

There remains the problem of determining an upper bound on f after EXP has executed. We will require a preliminary lemma.

Lemma 11.7. *Let* $a_0 \in \mathbb{Z}^{\geq 0}$, $p, k \in \mathbb{Z}^{>0}$, *and* $c, h \in \mathbb{R}^+$ *with* $p \geq 8$ *and* $h \geq \max\{16, k\}$. *Define the sequence* $\{a_i\}$ $(i \geq 0)$ *by*

$$a_i = 2c + \left(\left(1 + \frac{1}{h}\right)^2 + \frac{c}{2^p} \right) a_0 + \left(2 + \frac{1}{h}\right) a_{i-1} \quad (i = 1, 2, 3, \dots, k) .$$

Then

$$a_k < 2^k e^{1/2} \left(\frac{(528 + c)a_0}{256} + 2c \right) .$$

Proof. Set $g = 2 + h^{-1}$ (> 2). It is easy to verify that a closed form for a_i is given by

$$a_i = \left(g^{i+1} - g + 1 + \frac{g^i - 1}{g - 1} \frac{c}{2^p} \right) a_0 + 2c \frac{g^i - 1}{g - 1}$$

for $i \in \mathbb{Z}^{>0}$. Now, $(g^i - 1)/(g - 1) < g^i$ and, thus, the multiple of a_0 in the above formula is bounded above by $g^{i+1} + cg^i/256$. Since $h \geq 16$, we get

$$a_i < g^i \left(\left(2 + \frac{1}{16} + \frac{c}{256}\right) a_0 + 2c \right)$$

$$= 2^i \left(1 + \frac{1}{2h}\right)^i \left(\frac{528 + c}{256} a_0 + 2c \right)$$

$$< 2^i \exp\left(\frac{i}{2h}\right) \left(\frac{(528 + c)a_0}{256} + 2c \right) .$$

Since $h \geq k$, we have $\exp(i/2h) \leq e^{1/2}$ for $i \leq k$. $\square$

Theorem 11.8. *Suppose* $p \geq 8$ *and* $h \in \mathbb{R}^+$ *with* $h \geq \max\{16, \log_2 n\}$. *Put* $m = 3.43 f_0 + 10.72$. *If* $hmn < 2^p$, *then the value of* f *after EXP has executed satisfies* $f < mn$ *and, therefore,* $f < 2^p/h$.

Proof. After the ith iteration of step 3 of EXP, put $\mathfrak{b}_i = \mathfrak{b}$, $d_i = d$, and $k_i = k$. If we set $s_0 = b_0 = 1$ and $s_i = 2s_{i-1} + b_i$ for $1 \leq i \leq l$, then $(\mathfrak{b}_i, d_i, k_i)$ is a w-near (f_i, p) representation of $\mathfrak{a}^{s_i}$ where $c = 13/4$ and

$$f_i = c + f_0 + \left(c + 2f_{i-1} + \frac{f_{i-1}^2}{2^p} \right) + f_0 \frac{(c + 2f_{i-1} + f_{i-1}^2/2^p)}{2^p} \qquad (11.15)$$

for $i = 1, 2, \ldots, l$. We can rewrite this difference equation as

$$f_i = 2c + \left(\left(1 + \frac{f_{i-1}}{2^p}\right)^2 + \frac{c}{2^p}\right) f_0 + \left(2 + \frac{f_{i-1}}{2^p}\right) f_{i-1} . \tag{11.16}$$

Put $f = f_l$. Since $s_l = n$, algorithm EXP produces a w-near (f, p) representation $(\mathfrak{b}_l, d_l, k_l)$ of $\mathfrak{a}^n$.

Put $a_0 = f_0$. If we define a_i $(i = 1, 2, \ldots, l)$ as in Lemma 11.7, then since $h \geq l$, we have $a_l < m2^l \leq mn$. Since a_i is a strictly increasing function of i and $hmn < 2^p$, we must have $ha_i < 2^p$ for $i = 0, 1, 2, \ldots, l$. Hence, $hf_0 < 2^p$, and it follows inductively from (11.16) that $f_i \leq a_i$ $(i = 0, 1, 2, \ldots, l)$. Hence, $f_i < mn$ and $hf_i < 2^p$ $(i = 0, 1, 2, \ldots, l)$. In particular, $f < mn$ and $hf < 2^p$. $\qquad \square$

Suppose we are given p and f wth $f < 2^{p-4}$. Let $\mathfrak{a}$ $(= \mathfrak{a}_1)$ be any reduced $\mathcal{O}$-ideal. By our results in Chapter 5, we can use the SCF expansion of $(P + \sqrt{D})/Q$, where $\mathfrak{a} = [Q/r, (P + \sqrt{D})/r]$, to produce a sequence of reduced ideals

$$\mathfrak{a}_1, \mathfrak{a}_2, \mathfrak{a}_3, \ldots, \mathfrak{a}_j, \ldots , \tag{11.17}$$

with $\mathfrak{a}_j = \theta_j \mathfrak{a}_1$ $(j = 1, 2, \ldots)$. We may also assume that for each $\mathfrak{a}_j$ we have $d_j, k_j \in \mathbb{Z}$ such that $(\mathfrak{a}_j, d_j, k_j)$ is a reduced (f, p) representation of $\mathfrak{a}$. Since

$$\left| \frac{2^p \theta_j}{2^{k_j} d_j} - 1 \right| < \frac{1}{16}$$

and $2^p < d_j \leq 2^{p+1}$, we get

$$\frac{15}{16} 2^{k_j} < \theta_j < \frac{17}{8} 2^{k_j} . \tag{11.18}$$

By Proposition 3.16, we have $\theta_{j+2} > 2\theta_j$, $\theta_{j+i} > 3\theta_j$ $(i \geq 3)$. Thus, if $i \geq 3$, then

$$2^{k_{j+i}} > \frac{8}{17} \theta_{j+i} > \frac{8 \cdot 3}{17} \theta_j > 2^{k_j} .$$

Hence, $k_{j+i} > k_j$ when $i \geq 3$. If $j = 2$, then

$$2^{k_{j+2}} > \frac{15}{17} 2^{k_j} > 2^{k_j - 1} ;$$

consequently, $k_{j+2} \geq k_j$.

Now suppose $(\mathfrak{a}_j, d_j, k_j)$ and $(\mathfrak{a}_h, d_h, k_h)$ are both w-near (f, p) representations of $\mathfrak{a}$. Since $\mathfrak{a}_{j+1} = \rho(\mathfrak{a}_j)$ and $\mathfrak{a}_{h+1} = \rho(\mathfrak{a}_h)$, we must have $k_j < w$, $k_{j+1} \geq w$, and $k_h < w$, $k_{h+1} \geq w$. We will assume with no loss of generality that $h > j$. Clearly, we cannot have $h = j + 1$. If $h = j + i$, where $i \geq 3$, then

$$k_h = k_{j+1+i-1} \geq k_{j+1} \geq w ,$$

a contradiction. Thus, if we have distinct $\mathcal{O}$-ideals $\mathfrak{a}_j$ and $\mathfrak{a}_h$ such that both $(\mathfrak{a}_j, d_j, k_j)$ and $(\mathfrak{a}_h, d_h, k_h)$ are w-near (f, p) representations of $\mathfrak{a}$, then $|h - j| = 2$. It follows that there can be at most two distinct $\mathcal{O}$-ideals which can occur in any w-near (f, p) representation of $\mathfrak{a}$. We will use the notation $\mathfrak{a}[w]$ to denote any one of these ideals if there are two; certainly, there must be at least one such ideal. That $\mathfrak{a}[w]$ need not be unique will not be a problem in our applications of this concept.[9]

We will now develop an algorithm that can be used to compute an $\mathcal{O}$-ideal $\mathfrak{a}[x]$ in the important special case when $\mathfrak{a} = (1)$ and x is a positive integer. Our first algorithm ADDXY gives us the ability to determine, given $\mathcal{O}$-ideals $\mathfrak{a}[x]$ and $\mathfrak{a}[y]$, an $\mathcal{O}$-ideal $\mathfrak{a}[x + y]$. This will enable us to jump quickly through the cycle of reduced principal ideals in $\mathcal{O}$.

Algorithm 11.5: ADDXY

Input: $(\mathfrak{a}[x], d', k')$, $(\mathfrak{a}[y], d'', k'')$, x, y, p, where $(\mathfrak{a}[x], d', k')$ and $(\mathfrak{a}[y], d'', k'')$ are respectively x- and y-near (f', p) and (f'', p) representations of the $\mathcal{O}$-ideal $\mathfrak{a} = (1)$.
Output: $(\mathfrak{a}[x + y], d, k)$, an $(x + y)$-near (f, p) representation of $\mathfrak{a}$, where $f = 13/4 + f' + f'' + f'f''/2^p$.
1: Put $(\mathfrak{c}, g, h) = \text{NUMULT}((\mathfrak{a}[x], d', k'), (\mathfrak{a}[y], d'', k''), p)$.
2: Put $(\mathfrak{c}', g', h') = \text{WNEAR}((\mathfrak{c}, g, h), x + y, p)$.
3: Put $\mathfrak{a}[x + y] = \mathfrak{c}'$, $d = g'$, $k = h'$.

We remark here that after step 1 has executed, we have $h \le k' + k'' + 1 \le x + y - 1$. Thus, only case 1 of WNEAR need be executed in step 2. Also, by Lemma 11.3 and (3.33), we may use the same reasoning as that employed in the remark concerning the computational complexity of WMULT to deduce that ADDXY will execute in $O(\log \Delta)$ elementary operations. This is because $x + y - h = O(\log \Delta)$ and $h < x + y$. Finally, it is important to observe that since $\mathfrak{a} = (1)$, we have $\mathfrak{a}^2 = \mathfrak{a} = (1)$; hence, $\mathfrak{a}[x + y]$ as determined in the algorithm is principal.

The next algorithm, AX, finds for a given x and the $\mathcal{O}$-ideal $\mathfrak{a} = (1)$, an x-near (f, p) representation of $\mathfrak{a}$ for a certain value of f.

Algorithm 11.6: AX

Input: $x \in \mathbb{Z}^{>0}$ and $p \in \mathbb{Z}^{>0}$.
Output: $(\mathfrak{a}[x], d, k)$ an x-near (f, p) representation of $\mathfrak{a} = (1)$ for a suitable $f \in [1, 2^p)$.
1: Put $l = \lfloor \log_2 x \rfloor$ and compute the binary representation of x, say

$$x = \sum_{i=0}^{l} b_i 2^{l-i}$$

$(b_0 = 1, b_i \in \{0, 1\}$ for $1 \le i \le l)$.

2: Let $Q = r$, $P = r\lfloor(\lfloor\sqrt{D}\rfloor - r + 1)/r\rfloor + r - 1$, $\mathfrak{b} = [1, (P + \sqrt{D})/r]$,
   $d = 2^p + 1$, $k = 0$, $i = 0$, $s_0 = 1$.
3: Put $(\mathfrak{b}_0, d_0, k_0) = \text{WNEAR}((\mathfrak{b}, d, k), 1, p)$
4: **while** $i < l$ **do**
5:     Put $(\mathfrak{b}_{i+1}, d_{i+1}, k_{i+1}) = \text{ADDXY}((\mathfrak{b}_i, d_i, k_i), (\mathfrak{b}_i, d_i, k_i), s_i, s_i, p)$.
6:     Put $s_{i+1} = 2s_i$
7:     **if** $b_{i+1} = 1$ **then**
8:         Put $s_{i+1} = 2s_i + 1$ and

$$(\mathfrak{b}_{i+1}, d_{i+1}, k_{i+1}) \leftarrow \text{WNEAR}((\mathfrak{b}_{i+1}, d_{i+1}, k_{i+1}), s_{i+1}, p).$$

9:     **end if**
10:     $i \leftarrow i + 1$.
11: **end while**
12: Put $\mathfrak{a}[x] = \mathfrak{b}_l$ $d = d_l$, $k = k_l$.

Clearly, Algorithm AX will execute in $O(\log x \log \Delta)$ elementary operations. That the algorithm is correct follows easily by observing that

$$\mathfrak{b}_j = \mathfrak{a}[s_j] \sim \mathfrak{a} \quad (j = 0, 1, 2, \dots, l).$$

As in the case of EXP, we must now find an upper bound on f. We do this in the next theorem.

Theorem 11.9. *Suppose $p \geq 8$ and $h \in \mathbb{R}^+$ with $h \geq \log_2 x$. Put $m = 11.2$. If $hmx < 2^p$, then the value of f after AX has executed satisfies $f < mx$ and therefore $f < 2^p/h$.*

Proof. After step 11.6, we see that $(\mathfrak{b}_{i+1}, d_{i+1}, k_{i+1})$ is an s_{i+1}-near (f_{i+1}, p) representation of $\mathfrak{a}$, where

$$f_{i+1} = \frac{9}{8} + \frac{13}{4} + 2f_i + \frac{f_i^2}{2^p} \quad (1 \leq i + 1 \leq l) \tag{11.19}$$

and $f_0 = 1 + 9/8 = 17/8$. We put $f = f_l$. Since $s_l = x$, algorithm AX produces an x-near (f, p) representation $(\mathfrak{b}_l, d_l, k_l)$ of $\mathfrak{a}$. We now define $a_0 = f_0$, $c = 37/8$ and

$$a_{i+1} = \left(2 + \frac{1}{h}\right) a_i + c.$$

If $g = 2 + 1/h$, a closed-form representation for a_i is given by $a_i = g^i a_0 + c(g^i - 1)/(g - 1)$; hence, an analysis similar to that employed in the proof of Lemma 11.7 yields

$$a_l < g^l(a_0 + c) < 2^l e^{1/2}(a_0 + c) < 2^l m \leq mx,$$

where $m = 11.2$. As in the proof of Theorem 11.8, we have $ha_i < 2^p$ $(i = 0, 1, 2, \dots, l)$ and $hf_0 < 2^p$. Thus, by using induction on (11.19), we can show that $f_i \leq a_i$ $(i = 0, 1, 2, \dots, l)$. It follows that $f < mn$ and $hf < 2^p$. $\qquad\square$

Suppose now that we are given some $x \in \mathbb{R}$ and $a \in \mathbb{R}^{\geq 0}$ such that

$$|x - \log_2 \theta_j| \leq a ,$$

where $\mathfrak{a}_j = \theta_j \mathfrak{a}_1$ in (11.17). If a is not too large, we would expect that if $\mathfrak{a}_i = \mathfrak{a}[x]$ in (11.17), then i and j should be close in value. However, just how close would they be? In order to answer this question we will begin by defining $c(m)$.

Definition 11.10. *For a fixed $m \in \mathbb{R}$, we define $c(m) = \max\{m_1, m_2\}$, where m_1 and m_2 are respectively the largest integers such that the Fibonacci numbers F_{m_1} and F_{m_2+1} satisfy:*

$$F_{m_1} < \frac{16}{15} 2^m \text{ and } F_{m_2+1} < \frac{17}{16} 2^{m+1} .$$

Notice that $m_1 \geq 0$, $m_2 \geq -1$. For example, if $m = -3/2$ then $m_1 = 0$, $m_2 = -1$, and $c(-3/2) = 0$. A short table of values for $c(m)$ is given in Table 11.1.

Table 11.1. Some Values of $c(m)$

m	$c(m)$	m	$c(m)$
≤ -2	0	2	5
-1	1	3	6
0	2	4	8
1	3	5	9

It is easy to show that if $m' \leq m$, then $c(m') \leq c(m)$. We can also find an upper bound on $c(m)$.

Proposition 11.11. *If $m \geq 1$, then*

$$c(m) < 3 + \frac{3m}{2} .$$

Proof. We have seen in §3.4 that if $m \geq 1$, then $F_m > \tau^{m-2} > (8/5)^{m-2}$. Also, $(8/5)^{3/2} > 2$. Suppose that $m_1 \geq 2.14 + (3/2)m$. Then

$$F_{m_1} > \left(\frac{8}{5}\right)^{m_1-2} \geq \left(\frac{8}{5}\right)^{0.14} \left(\frac{8}{5}\right)^{(3/2)m} > \left(\frac{8}{5}\right)^{0.14} 2^m > \frac{16}{15} 2^m ,$$

which is impossible. Next, suppose $m_2 \geq 2.61 + (3/2)m$. Then

$$F_{m_2+1} > \left(\frac{8}{5}\right)^{m_2-1} > \left(\frac{8}{5}\right)^{1.61} 2^{(3/2)m} = \frac{(8/5)^{1.61}}{2} 2^{(3/2)m+1} > \frac{17}{16} 2^{m+1},$$

which is also impossible. Thus, we must have

$$m_1 < 2.14 + \frac{3}{2}m \text{ and } m_2 < 2.61 + \frac{3}{2}m$$

and, hence,

$$c(m) < 3 + \frac{3}{2}m \ .$$

$\square$

We can now use $c(m)$ to bound the value of $|i - j|$.

Theorem 11.12. *Let* $x \in \mathbb{Z}$, *where* $x \geq 1$. *Suppose* $a, b \in \mathbb{R}$ *and*

$$a < \log_2 \theta_j - x < b \ .$$

If $\mathfrak{a}_i = \mathfrak{a}[x]$, *then*

$$j - c(b) \leq i \leq j + c(-a - 1) \ .$$

Proof. We must have

$$2^{x+a} < \theta_j < 2^{x+b} \ . \tag{11.20}$$

By Lemma 11.3 we know that

$$\theta_i < \frac{17}{16}2^x, \quad \theta_{i+1} > \frac{15}{16}2^x \ . \tag{11.21}$$

By Proposition 3.16 when $n \geq 1$ and $i \geq 0$, we have

$$\theta_{i+n} \geq F_n \theta_{i+1} > F_n \left(\frac{15}{16}\right) 2^x \ .$$

By (11.21) and Definition 11.10, we find that for $m = b$,

$$\theta_{i+m_1+1} > F_{m_1+1}\theta_{i+1} > F_{m_1+1}\left(\frac{15}{16}\right)2^x > 2^{m+x} = 2^{b+x} > \theta_j \ .$$

It follows that $j < i + m_1 + 1 \leq i + 1 + c(m)$. Hence, $i \geq j - c(b)$.

Also, if $i > n$, by Proposition 3.16, (11.20), and (11.21), we get for $n \geq 0$ and $m = -a - 1$,

$$F_{n+1}\theta_{i-n} \leq \theta_i < \left(\frac{17}{16}\right)2^x = \left(\frac{17}{16}\right)2^{-a}2^{x+a} < \left(\frac{17}{16}\right)2^{m+1}\theta_j \ .$$

Putting $n = m_2 + 1$ and noting that $F_{m_2+2} > (17/16)2^{m+1}$, we get

$$\theta_{i-m_2-1} < \theta_j \ .$$

Thus, $j > i - m_2 - 1 \geq i - c(m) - 1$ and $i \leq j + c(-a - 1)$. If $i \leq n = m_2 + 1$, then $i \leq c(m) + 1 \leq j + c(-a - 1)$.

$\square$

Notes and References

[1] This is an easy consequence of the Gelfond-Schneider Theorem. See, for example, [Niv56], Ch. 10.

[2] Much of this material on (f, p) representations was developed in [JSW01] and [JSW06b].

[3] This idea represents a refinement of the representations introduced in [HP00].

[4] Of course, we are assuming here that the values of our ideal norms are not so large that Schönhage's reduction algorithm would be of greater efficiency than the simple reduction technique that we will employ. See the discussion at the beginning of §5.2.

[5] This technique provides a somewhat less precise result on f than that given in Theorem 5.1 of [JSW06b], but it is easier to present.

[6] Extensive numerical testing of this by the authors tends to support this assumption.

[7] There is a lengthy literature on this problem containing many improvements to this basic idea for fast exponentiation. For references, see [Gor98]. More recent information can be found in [MS06] and [BS07]. Several of these techniques can also be applied to the problem of exponentiating ideals, but we will only describe the simplest of these here.

[8] The idea of using this value of w is an adaptation of an idea that occurs in a similar context involving hyperelliptic curves in [JSS07a].

[9] The use of $\mathfrak{a}[x]$ was introduced in [dHJW07], but we have adopted the notation $\mathfrak{a}[x]$ here instead of the $\mathfrak{a}(x)$ used there in order to avoid functional notation which would imply a unique $\mathfrak{a}(x)$.

Compact Representations

12.1 Compact Representation of θ_j

As we have seen in Chapter 9 (cf. (9.26)), we would expect that

$$R_{\mathbb{K}} \gg \Delta_{\mathbb{K}}^{1/2-\epsilon}$$

holds for a large proportion of all the real quadratic fields. It follows that for such fields,

$$\epsilon_\Delta \gg \exp(\Delta_{\mathbb{K}}^{1/2-\epsilon}) \, .$$

If $\epsilon_\Delta = (x + y\sqrt{\Delta})/2$, where $x, y \in \mathbb{Z}$, then $x = \epsilon_\Delta + \bar{\epsilon}_\Delta$, $y = (\epsilon_\Delta - \bar{\epsilon}_\Delta)/\sqrt{\Delta}$. Also, since $\epsilon_\Delta |\bar{\epsilon}_\Delta| = 1$ and $\epsilon_\Delta > 1$, we see that

$$x, y \gg \exp(\Delta^{1/2-\epsilon}) \, .$$

When Δ is large, this implies that x and y could be so enormous that it would not be possible to write them out in conventional decimal representation. Indeed, it is clear that the problem of doing so is of exponential complexity in $\Delta^{1/2}$. For example, at the end of §3.3 we mentioned the 30-digit

$$D = 990676090995853870156271607886 \, .$$

By using the ideas mentioned in Chapter 10, it was shown[1] that

$$R_\Delta = 4770372955851343.43 \, .$$

From this information, we see that if $\epsilon_\Delta = (x + y\sqrt{\Delta})/2$ with $x, y \in \mathbb{Z}$, then $x, y > 10^{2 \cdot 10^{15}}$. This means that it would require over 6,000,000 books, each of 1000 pages, of the same small format used by Nelson[2] to record the solution of the Cattle Problem, to write out x and y. However, it requires less than one page to write ϵ_Δ as a compact representation.

In this chapter we define what is meant by a compact representation[3] of a quadratic integer α (> 1) and show how it can be computed when we are given an approximation to $\log_2 |\alpha|$ and a representation like (4.9) for the ideal (α) in $\mathcal{O}$. We will begin with the problem of representing some θ, where $\mathfrak{a}[x] = (\theta)$ for some given $x \in \mathbb{Z}^{>0}$.

We next require two simple algorithms. The first of these, which will be needed below, is a modification of WNEAR. WNEAR is executed on an $\mathcal{O}$-ideal $\mathfrak{b}$ to produce an equivalent ideal $\mathfrak{c}$. In the case where $k < w$, EWNEAR computes $(\mathfrak{c}, g, h)$, a w-near representation of $\mathfrak{a}$ and κ, where $\mathfrak{c} = \kappa \mathfrak{b}$.

Algorithm 12.1: EWNEAR

Input: $(\mathfrak{b}, d, k)$, w, p, where $(\mathfrak{b}, d, k)$ is a reduced (f, p) representation of some $\mathcal{O}$-ideal $\mathfrak{a}$ and $k < w$.

Output: $(\mathfrak{c}, g, h)$ a w-near $(f + 9/8, p)$ representation of $\mathfrak{a}$ and a, b, where $N(\mathfrak{b})\kappa = (a + b\sqrt{D})/r$ and $\mathfrak{c} = \kappa \mathfrak{b}$ $(\kappa > 1)$.

Next, suppose we have $\mathcal{O}$-ideals $\mathfrak{a}$, $\mathfrak{b}$, and $\mathfrak{c}$ and $N = N(\mathfrak{b})$, $M \in \mathfrak{a} \cap \mathbb{Z}$, where

$$M\mathfrak{b} = \gamma_1 \mathfrak{a}, \quad N\mathfrak{c} = \gamma_2 \mathfrak{b},$$

and $\gamma_1, \gamma_2 \in \mathcal{O}$. If

$$\gamma_1 = \frac{a_1 + b_1\sqrt{D}}{r}, \quad \gamma_2 = \frac{a_2 + b_2\sqrt{D}}{r},$$

where $a_1, b_1, a_2, b_2 \in \mathbb{Z}$, then

$$M\mathfrak{c} = \gamma \mathfrak{a},$$

and $\gamma = (a + b\sqrt{D})/r \in \mathcal{O}$. Indeed, $\gamma = \gamma_1 \gamma_2 / N$. The process of determining γ, given γ_1, γ_2, and N, is given in the following algorithm. At this point we will also start putting most of our algorithms in the body of this book rather than in the Appendix.

Algorithm 12.2: IMULT

Input: $\gamma_1 = (a_1 + b_1\sqrt{D})/r$, $\gamma_2 = (a_2 + b_2\sqrt{D})/r$, N, where $a_1, b_1, a_2, b_2 \in \mathbb{Z}$.
Output: γ, $a, b \in \mathbb{Z}$, where $\gamma = \gamma_1 \gamma_2 / N = (a + b\sqrt{D})/r$.
1: Put

$$a = (a_1 a_2 + D b_1 b_2)/rN,$$
$$b = (a_1 b_2 + a_2 b_1)/rN.$$

We can now employ these algorithms to produce a simple modification of ADDXY.

Algorithm 12.3: EADDXY

Input: $(\mathfrak{a}[x], d', k')$, $(\mathfrak{a}[y], d'', k'')$, x, y, p, where $(\mathfrak{a}[x], d', k')$, $(\mathfrak{a}[y], d'', k'')$ are respectively x (y)-near $(f', p), (f'', p)$ representations of the $\mathcal{O}$-ideal $\mathfrak{a} = (1)$.

Output: $(\mathfrak{a}[x+y], d, k)$, a, b, where $(\mathfrak{a}[x+y], d, k)$ is a $x+y$-near (f,p) representation of $\mathfrak{a}$, $f = 13/4 + f' + f'' + f'f''/2^p$,

$$\mathfrak{a}[x+y] = \left(\frac{\lambda\theta'\theta''}{N(\mathfrak{a}[x])N(\mathfrak{a}[y])} \right) \mathfrak{a} .$$

Here

$$\lambda = \frac{a + b\sqrt{D}}{r} \text{ and } \mathfrak{a}[x] = \theta'\mathfrak{a}, \mathfrak{a}[y] = \theta''\mathfrak{a} .$$

1: Put $((\mathfrak{c}, g, h), a', b') = \text{NUMULT}((\mathfrak{a}[x], d', k'), (\mathfrak{a}[y], d'', k''), p)$.
2: Put $((\mathfrak{c}', g', h'), a'', b'') = \text{EWNEAR}((\mathfrak{c}, g, h), x+y, p)$.
3: Put $\mathfrak{a}[x+y] = \mathfrak{c}'$, $d = g'$, $k = h'$,

$$(a, b) = \text{IMULT}(a', b', a'', b'', N(\mathfrak{c})) .$$

In step 12.3, we use the optional output feature of NUMULT to find a' and b' where

$$\mathfrak{c} = \left(\frac{\nu}{N(\mathfrak{a}[x])N(\mathfrak{a}[y])} \right) \mathfrak{a}[x]\mathfrak{a}[y]$$

and $\nu = (a' + b'\sqrt{D})/r \in \mathcal{O}$. In step 12.3 we find

$$\mathfrak{c}' = \kappa\mathfrak{c} ,$$

where $\kappa = (a'' + b''\sqrt{D})/r \in \mathcal{O}$. Hence,

$$\lambda = \frac{\nu\kappa}{N(\mathfrak{c})} = \frac{a + b\sqrt{D}}{r} \in \mathcal{O}$$

and

$$\mathfrak{a}[x+y] = \mathfrak{c}' = \left(\frac{\lambda}{N(\mathfrak{a}[x])N(\mathfrak{a}[y])} \right) \mathfrak{a}[x]\mathfrak{a}[y] = \theta\mathfrak{a} ,$$

where $\theta = \lambda\theta'\theta''/(N(\mathfrak{a}[x])N(\mathfrak{a}[y]))$. Also,

$$\left| \frac{\theta 2^p}{2^k d} - 1 \right| < \frac{f}{2^p} .$$

Suppose we are now given some integer $x \geq 1$. With these routines we can now produce an algorithm that will find a certain representation of some $\theta \in \mathcal{O}$ where $\mathfrak{a}[x] = (\theta)$ ($\mathfrak{a}_1 = \mathfrak{a} = (1)$). This is again a modification of a previous algorithm (AX).

Algorithm 12.4: CRAX

Input: x, p, where $x \in \mathbb{Z}^{>0}$ and $2^p > 11.2x \max\{16, \log_2 x\}$.
Output: $(\mathfrak{a}[x], d, k)$, an x-near (f,p) representation of $\mathfrak{a} = (1)$ where $f < 2^{p-4}$ and a set of integer pairs (m_i, n_i) and $L_i \in \mathbb{Z}^{>0}$, $i = 0, 1, 2, \ldots, l$, where $l = \lfloor \log_2 x \rfloor$.

1: Compute the binary representation of x with

$$x = \sum_{i=0}^{l} b_i 2^{l-i}, \quad b_0 = 1, \quad b_i \in \{0, 1\} \quad (1 \le i \le l) .$$

2: Put

$$Q = r, \quad P = r \left\lfloor \frac{\lfloor \sqrt{D} \rfloor - r + 1}{r} \right\rfloor + r - 1, \quad \mathfrak{b} = \left[1, \frac{P + \sqrt{D}}{r} \right] ,$$

$d = 2^p + 1, \ k = 0, \ i = 0, \ s_0 = 1, \ L_0 = 1.$
3: Put $((\mathfrak{b}_0, d_0, k_0), m_0, n_0) = \text{EWNEAR}((\mathfrak{b}, d, k), 1, p).$
4: **while** $i < l$ **do**
5:     Put $L_{i+1} = N(\mathfrak{b}_i)$ and

$$((\mathfrak{b}_{i+1}, d_{i+1}, k_{i+1}), m_{i+1}, n_{i+1})$$
$$= \text{EADDXY}((\mathfrak{b}_i, d_i, k_i), (\mathfrak{b}_i, d_i, k_i), s_i, s_i, p) .$$

6:     Put $s_{i+1} = 2s_i.$
7:     **if** $b_{i+1} = 1$ **then**
8:         Put $s_{i+1} \leftarrow s_{i+1} + 1, \ N = N(\mathfrak{b}_{i+1}),$ and

$$((\mathfrak{b}_{i+1}, d_{i+1}, k_{i+1}), m', n')$$
$$\leftarrow \text{EWNEAR}((\mathfrak{b}_{i+1}, d_{i+1}, k_{i+1}), s_{i+1}, p) .$$

9:         Put $(m_{i+1}, n_{i+1}) = \text{IMULT}(m_{i+1}, n_{i+1}, m', n', N).$
10:     **end if**
11:     $i \leftarrow i + 1.$
12: **end while**
13: Put $\mathfrak{a}[x] = \mathfrak{b}_l, \ d = d_l, \ k = k_l.$

Notice that CRAX will execute in $O(\log x \log \Delta)$ elementary operations.

As this algorithm executes, it finds reduced principal $\mathcal{O}$-ideals $\mathfrak{b}_i = \mathfrak{a}[s_i] = \mu_i \mathfrak{a}$, where

$$\left| \frac{2^p \mu_i}{2^{k_i} d_i} - 1 \right| < \frac{f}{2^p} \quad (i = 0, 1, 2, \ldots, l) .$$

Since $2^p > 11.2x \max\{16, \log_2 x\}$, we know by Theorem 11.9 that $f < 2^{p-4}$. If we put $\lambda_i = (m_i + n_i \sqrt{D})/r$, then

$$\mu_{i+1} = \left(\frac{\lambda_{i+1}}{L_{i+1}^2} \right) \mu_i^2 \quad (i = 0, 1, 2, \ldots, l-1) \tag{12.1}$$

where $\mu_0 = \lambda_0$. Also, $\mu_i \in \mathcal{O}$ and

$$|N(\mu_i)| = N(\mathfrak{a}[s_i]) = N(\mathfrak{b}_i) = L_{i+1} . \tag{12.2}$$

Now, for any fixed i $(1 \leq i \leq l)$, we must have some θ_j in the simple continued fraction (SCF) expansion of $(P + \sqrt{D})/r$ such that $\mu_i = \theta_j$. Also, by Lemma 11.3,

$$\frac{15N(b_i)}{16\sqrt{\Delta}} 2^{s_i} < \theta_j < \frac{17}{16} 2^{s_i} . \tag{12.3}$$

Hence, by (12.3), we have

$$\frac{15L_{i+1}}{16\sqrt{\Delta}} 2^{s_i} < \mu_i < \frac{17}{16} 2^{s_i} \quad (i = 1, 2, 3, \ldots, l) . \tag{12.4}$$

Since, by (12.1),

$$\lambda_i = \frac{L_i^2 \mu_i}{\mu_{i-1}^2} \quad (i = 1, 2, 3, \ldots, l) ,$$

we get

$$0 < \lambda_i < \left(\frac{16\sqrt{\Delta}}{15} 2^{-s_{i-1}} \right)^2 \frac{17}{16} 2^{s_i}$$

$$= \frac{16 \cdot 17}{15^2} \Delta 2^{s_i - 2s_{i-1}}$$

$$< \frac{5}{2}\Delta .$$

Also, since

$$\overline{\lambda}_i = \frac{L_i^2 \overline{\mu}_i}{\overline{\mu}_{i-1}^2}$$

and $|\mu_i \overline{\mu}_i| = L_{i+1}$, we get

$$|\overline{\lambda}_i| = \frac{L_{i+1}\mu_{i-1}^2}{\mu_i} < \frac{16\sqrt{\Delta}}{15} 2^{-s_i} \left(\frac{17}{16} \right)^2 2^{2s_i - 1}$$

$$\leq \frac{17^2}{15 \cdot 16}\sqrt{\Delta} < \frac{7}{5}\sqrt{\Delta} .$$

We now define $H(\alpha)$ for $\alpha \in \mathcal{O}$ to be $\max\{|\alpha|, |\overline{\alpha}|\}$. Notice that if $\alpha, \beta \in \mathcal{O}$, then $H(\alpha\beta) \leq H(\alpha)H(\beta)$. Also, since $|N(\alpha)| = |\alpha\overline{\alpha}| \geq 1$, $H(\alpha)$ cannot be arbitrarily small; in fact, $H(\alpha) \geq 1$. By our previous results we have

$$H(\lambda_i) < \frac{5}{2}\Delta . \tag{12.5}$$

Futhermore, since $\lambda_i = (m_i + n_i\sqrt{D})/r$, we see that $\lambda_i = (x_i + y_i\sqrt{\Delta})/2$, where $x_i = 2m_i/r$ and $y_i = n_i$. It follows that

$$|x_i|, |y_i\sqrt{\Delta}| < \lambda_i + |\overline{\lambda}_i| < \frac{5}{2}\Delta + \frac{7}{2}\sqrt{\Delta} \tag{12.6}$$

for $i = 1, 2, \ldots, l$. We also have

$$\frac{15L_1}{16\sqrt{\Delta}} < \lambda_0 < \frac{17}{8} \quad (s_0 = 1)$$

and $\lambda_0|\overline{\lambda}_0| = N(\mathfrak{b}_0) = L_1$. Thus, (12.5) and (12.6) also hold for $i = 0$.

If we define $L_0 = 1$, from (12.1), we get

$$\mu_j = \prod_{i=0}^{j} \left(\frac{\lambda_i}{L_i^2}\right)^{2^{j-i}} \quad (j = 0, 1, 2, \ldots, l) .$$

If we put $d_i = L_{i+1}$ $(i = -1, 0, 1, \ldots, l)$, then

$$\mu_j = \lambda_j \prod_{i=0}^{j-1} \left(\frac{\lambda_i}{d_i}\right)^{2^{j-i}} \quad (j = 0, 1, \ldots, l) . \tag{12.7}$$

Also, since $d_i = L_{i+1}$ and L_{i+1} is the norm of a reduced $\mathcal{O}$-ideal, we have $0 < d_i < \sqrt{\Delta}$. When $j = l$, we get $s_l = x$, $\mathfrak{a}[x] = \mathfrak{b}_l = (\mu_l)$; hence, $\mathfrak{a}[x] = (\theta)$, where

$$\theta = \lambda \prod_{i=0}^{l} \left(\frac{\lambda_i}{d_i}\right)^{2^{l-i}} . \tag{12.8}$$

(In this case, $\lambda = d_l$.) Since $l = \lfloor \log_2 x \rfloor$ and $2^x < (16\sqrt{\Delta}/15)\theta$, we find that

$$l = O(\log_2 \log_2 \theta)$$

when $\theta > 16\sqrt{\Delta}/15$.

Hence, for any $\mathcal{O}$ and any θ such that $(\theta) = \mathfrak{a}[x] \in \mathcal{O}$, we can get a representation (12.8) of θ with the following properties:

1. $l = O(\log \log \theta)$ for large θ.
2. $\lambda, \lambda_i \in \mathcal{O}$, $d_i \in \mathbb{Z}$ $(0 \le i \le l)$.
3. $0 < d_i \le \Delta^{1/2}$, $H(\lambda) = O(\Delta^{1/2})$, and $H(\lambda_i) = O(\Delta)$ $(0 \le i \le l)$.
4. Also, μ_j given by (12.7) is in $\mathcal{O}$, $|N(\mu_j)| = d_j$, μ_j generates a reduced $\mathcal{O}$-ideal $\mathfrak{b}_j$, where $\mathfrak{b}_0 = \mathfrak{a}[1]$ and $d_{i+1}^2 \mathfrak{b}_{i+1}^2 = \lambda_{i+1} \mathfrak{b}_i^2$ $(i = 0, 1, \ldots, l-1)$.

A representation (12.8) of θ satisfying the properties (1)–(4) above is an example of a *compact representation* of θ. The total number of bits needed to represent θ by this means is $O(l \log_2 \Delta) = O(\log \log \theta \log \Delta)$, which, for large θ, is considerably smaller than the $O(\log \theta)$ bits needed to represent θ in conventional decimal representation. In the next section, we will produce the definition of a compact representation for any $\gamma \in \mathcal{O}$. In order to do this, we will first show how to compute what we will call a compact representation of γ.

12.2 Compact Representation of Quadratic Integers

Suppose we are given some $c \in \mathbb{Z}^{\ge 0}$, a reduced (f, p) representation $(\mathfrak{b}, d, k)$ of an $\mathcal{O}$-ideal $\mathfrak{a}$, and a reduced $\mathcal{O}$-ideal $\mathfrak{c}$ such that if $\mathfrak{b} = \mathfrak{a}_i$, then $\mathfrak{c} = \mathfrak{a}_j$, where

$$i \le j \le i + c .$$

We can easily devise an algorithm, based on the SCF expansion of a quadratic irrational, which will find $(\mathfrak{c}, g, h)$, a reduced $(f + 1/4, p)$ representation of $\mathfrak{a}$ and the quadratic integer $\lambda \in \mathcal{O}$ such that

$$N(\mathfrak{b})\mathfrak{c} = \lambda\mathfrak{b} .$$

Algorithm 12.5: FIND

Input: $c \in \mathbb{Z}^{\ge 0}$; $(\mathfrak{b}, d, k)$ a reduced (f, p) representation of an $\mathcal{O}$-ideal $\mathfrak{a}$
with $\mathfrak{b} = \mathfrak{a}_i = [Q_{i-1}/r, (P_{i-1} + \sqrt{D})/r]$, where $(P_{i-1} + \sqrt{D})/Q_{i-1} > 1$,
$-1 < (P_{i-1} - \sqrt{D})/Q_{i-1} < 0$; $\mathfrak{c} = [Q_{j-1}/r, (P_{j-1} + \sqrt{D})/r]$ with

$$i \le j \le i + c .$$

Output: $(\mathfrak{c}, g, h)$ a reduced $(f + 1/4, p)$ representation of $\mathfrak{a}$ (optional output
$m, n \in \mathbb{Z}$, where $\lambda = (m + n\sqrt{D})/r \in \mathcal{O}$ and $N(\mathfrak{b})\mathfrak{c} = \lambda\mathfrak{b}$.)

As pointed out in the Appendix, we have

$$1 \le \frac{\lambda}{N(\mathfrak{b})} < \epsilon_\Delta \tag{12.9}$$

for the λ produced by FIND, where ϵ_Δ is the fundamental unit of $\mathcal{O}$.

Remark 12.1. Suppose we are given a reduced principal $\mathcal{O}$-ideal $\mathfrak{b} = (\theta)$ such
that $H(\theta) < B$ for some bound B and we wish to find $|\theta|$. Since at least one
of $|\theta|$ or $|\overline{\theta}|$ must exceed 1, we may assume that

$$1 < |\theta| < B \text{ or } 1 < |\overline{\theta}| < B .$$

Thus, we can find $|\theta|$ (or $|\overline{\theta}|$) by using the SCF algorithm on $\mathfrak{a}_1 = [1, \omega]$ to
find θ_i and $\mathfrak{a}_i$ such that $(\theta_i) = \mathfrak{a}_i = \mathfrak{b}$ (or $\overline{\mathfrak{b}}$). Then $|\theta| = \theta_i$ (or $|\overline{\theta}| = \theta_i$). In
the latter case, we can easily produce $|\theta|$ by conjugation of θ_i. This process
executes in $O(\log B)$ elementary operations on numbers of $O(\log B)$ bits.

At this point, we will assume that we have been given some reduced prin-
cipal $\mathcal{O}$-ideal $\mathfrak{b}$ and some $y, q \in \mathbb{Q}$ such that $y \ge q + 2$, $\mathfrak{b} = (\theta)$, and

$$|\log_2 \theta - y| < q . \tag{12.10}$$

Certainly, we have $\log_2 \theta > y - q \ge 2$. We will also assume that q is not very
large, say $q < 10$, and $\epsilon_\Delta > 2^{2q+3}\sqrt{\Delta}$. When q is small, this is not a very
great restriction because we know by (6.28) that ϵ_Δ must exceed $\sqrt{\Delta}/2$. Also,
if $\epsilon_\Delta < 2^{2q+3}\sqrt{\Delta}$, then finding θ is a very simple problem because

$$\theta = \epsilon_\Delta^n \gamma \quad (n \in \mathbb{Z}^{\ge 0}) \tag{12.11}$$

and $1 \le \gamma < \epsilon_\Delta$. Thus, γ can be easily determined by a direct application of the continued fraction algorithm on $\mathfrak{a}$ to find $\mathfrak{b}$. From (12.10) and (12.11) we see that with good rational approximations of R_Δ and $\log_2 \gamma$, we can easily compute the integer n in (12.11).

If we put $x = \lfloor y - q \rfloor - 1$, then

$$1 < \log_2 \theta - x < 2q + 2 . \tag{12.12}$$

If $\mathfrak{a}_i = \mathfrak{a}[x]$, by Theorem 11.12 we must have

$$\mathfrak{b} \in \{\mathfrak{a}_i, \mathfrak{a}_{i+1}, \dots, \mathfrak{a}_{i+c}\} ,$$

as $c(-2) = 0$ and $c = c(2q + 2) < 6 + \lceil 3q \rceil$ by Proposition 11.11. With this information we can now produce an algorithm for determining a compact representation of θ.

Algorithm 12.6: CR

Input: A reduced principal $\mathcal{O}$-ideal $\mathfrak{b} = [Q/r, (P + \sqrt{D})/r]$, where $(P + \sqrt{D})/Q > 1$ and $-1 < (P - \sqrt{D})/Q < 0$. Here $\mathfrak{b} = (\theta)$ for $\theta \in \mathcal{O}$; $y, q \in \mathbb{Q}^+$; $y \ge q + 2$ and

$$|\log_2 \theta - y| < q .$$

Output: A compact representation of θ, where

$$\theta = \lambda \prod_{i=0}^{l} \left(\frac{\lambda_i}{d_i} \right)^{2^{l-i}} .$$

(Optional output: $(\mathfrak{b}, g, h)$, a reduced $(f + 1/4, p)$ representation of $\mathfrak{a}$.)
1: Put $c = 6 + \lceil 3q \rceil$, $x = \lfloor y - q \rfloor - 1$ and find p such that $2^p > 11.2x \max\{16, \log_2 x\}$.
2: Execute $\text{CRAX}(x, p)$ to find an (f, p) representation $(\mathfrak{a}[x], d, k)$ of $\mathfrak{a}$, where $\mathfrak{a}[x] = [Q'/r, (P' + \sqrt{D})/r]$ and a compact representation

$$\mu = d_l \prod_{i=0}^{l} \left(\frac{\lambda_i}{d_i} \right)^{2^{l-i}}$$

of μ, where $(\mu) = \mathfrak{a}[x]$.
3: Execute $\text{FIND}(c, (\mathfrak{a}[x], d, k), \mathfrak{b})$ to produce $(\mathfrak{b}, g, h)$, a reduced $(f + 1/4, p)$ representation of $\mathfrak{a}$, and m, n, where $\lambda = (m + n\sqrt{D})/r \in \mathcal{O}$ and

$$N\mathfrak{b} = \lambda \mathfrak{a}[x], \quad N = N(\mathfrak{a}[x]) = d_l .$$

Proof (of correctness of CR). Clearly, $(\theta) = (\lambda\mu/N)$ and

$$\frac{\lambda\mu}{N} = \lambda \prod_{i=0}^{l} \left(\frac{\lambda_i}{d_l}\right)^{2^{l-i}}.$$

We now need to show that $\theta = \lambda\mu/N$. By (12.12) we have

$$2 < \frac{\theta}{2^x} < 2^{2q+2} \tag{12.13}$$

and by (12.4) we have

$$\frac{15N}{16\sqrt{\Delta}} < \frac{\mu}{2^x} < \frac{17}{16}.$$

Hence,

$$1 < \frac{\theta}{\mu} < \frac{2^{2q+2}16\sqrt{\Delta}}{15N} < \frac{2^{2q+3}\sqrt{\Delta}}{N} < \frac{\epsilon_\Delta}{N}. \tag{12.14}$$

Since $\theta = \epsilon_\Delta^n \lambda\mu/N$, we get

$$1 < \frac{\theta}{\mu} = \frac{\epsilon_\Delta^n \lambda}{N} < \frac{\epsilon_\Delta}{N};$$

however, by (12.9) we have $\epsilon_\Delta^{-1} < N/\lambda \le 1$. It follows, then, that $n = 0$ and $\theta = \lambda\mu/N$. It remains to bound $H(\lambda)$. Since $\lambda/N = \theta/\mu$, we deduce from (12.14) that $\lambda > N$ and $\lambda < 2^{2q+3}\sqrt{\Delta}$. Since $|\lambda\bar{\lambda}| = NN(\mathfrak{b})$, we also have $|\bar{\lambda}| < N(\mathfrak{b}) < \sqrt{\Delta}$. Hence, $H(\lambda) < 2^{2q+3}\sqrt{\Delta}$. This algorithm will execute in $O(\log y \log \Delta + q)$ elementary operations. $\qquad\square$

Naturally, if we have a value of y' such that $|\log_2 \theta - y'| < q'$, where q' is smaller than q, then CR will execute faster and $H(\lambda)$ will be smaller. We can find such a value for y' by simply executing AX on x and p, where $2^p > 11.2x \max\{16, \log_2 x\}$. We then get an x-near (f, p) representation of $\mathfrak{a}$, where $f < 2^{p-4}$. If we next execute FIND on c, $(\mathfrak{a}[x], d, k)$, and $\mathfrak{b}$, we get $(\mathfrak{b}, g, h)$, a reduced $(f + 1/4, p)$ representation of $\mathfrak{a}$. By (11.3)

$$|p - \log_2 g + \log_2 \theta - h| < \log_2\left(1 - \frac{f + 1/4}{2^p}\right).$$

Now, if $0 < t < 1$, then

$$|\log_2(1 - t)| < 1.45|\log_e(1 - t)|$$
$$= 1.45\left(t + \frac{t^2}{2} + \frac{t^3}{3} + \cdots\right)$$
$$= 1.45t\left(1 + \frac{t}{2} + \frac{t^2}{3} + \cdots\right)$$
$$< 1.45t\left(1 + \frac{t}{2}(1 - t)^{-1}\right);$$

hence,

$$\left| \log_2 \left(1 - \frac{f'}{2^p} \right) \right| < \frac{3f'}{2^{p+1}} < 0.1 \ ,$$

when $f' < 2^{p-4} + 1/4$ and $p > 10$. If, for example, we compute $r = \lfloor 3 \log_2 g \rfloor$ (i.e., find $r \in \mathbb{Z}$ such that $2^r < g^3 \le 2^{r+1}$), then

$$-\frac{1}{3} \le \frac{r}{3} - \log_2 g < 0 \ .$$

Putting $y' = -p + h + r/3 \in (1/3)\mathbb{Z} \subseteq \mathbb{Q}$, we get

$$|y' - \log_2 \theta| < \frac{1}{3} + 0.1 < \frac{1}{2} \ .$$

(Of course, by Theorem 11.9, it is possible to select a sufficiently large value of p such that $3f'/2^{p+1}$ is as small as we like. After doing this and determining a sufficiently accurate[4] approximation to $\log_2 g$, we can produce a value of y' which is as close as we wish to $\log_2 \theta$.)

We now turn to the problem of finding a compact representation of γ, where $\mathfrak{c} = (\gamma)$ is not necessarily a reduced $\mathcal{O}$-ideal. We assume that we are given $\sigma = \operatorname{sign} \gamma$, $\mathfrak{c} = S[Q/r, (P + \sqrt{D})/r]$, and some $z, q' \in \mathbb{Q}$ such that

$$|\log_2 |\gamma| - z| < q' \ .$$

We first use the reduction algorithm of §5.2 on the primitive $\mathcal{O}$-ideal $\mathfrak{c}_1 = [Q/r, (P + \sqrt{D})/r]$ to find

$$\theta_{i+2} = \frac{G_i + B_i \sqrt{D}}{Q}$$

such that

$$S\mathfrak{c}_{i+2} = \theta_{i+2}\mathfrak{c} \ ,$$

and $\mathfrak{c}_{i+2}$ is a reduced $\mathcal{O}$-ideal. Put $\beta = S/|\theta_{i+2}|$, $\mathfrak{b} = \mathfrak{c}_{i+2}$. We can compute, by the technique of §11.1, values for h and e such that

$$\left| \frac{2^p |\theta_{i+2}|/S}{e2^h} - 1 \right| < \frac{1}{2^p} \ ;$$

hence, by our earlier observations, we can compute $y', q'' \in \mathbb{Q}$, $q'' < 1/2$, such that

$$\left| \frac{\log_2 |\theta_{i+2}|}{S} - y' \right| < q'' \ .$$

It follows that if we put $y = z + y'$ and $q = q' + q''$, we get

$$|\log_2 \theta - y| < q$$

for $\theta = |\theta_{i+2}\gamma|/S$ or $|\gamma| = \beta\theta$ when $\mathfrak{b} = (\theta)$. If $y \geq q + 2$, we can use CR to find a compact representation of θ. If $y \leq t - q - 5/2$, where $t \in \mathbb{Q}$ and $|t - \log_2 N(\mathfrak{b})| < 1/2$, then since $\log_2 |\bar{\theta}| + \log_2 \theta = \log_2 N(\mathfrak{b})$, we get

$$\left|\log_2 |\bar{\theta}| - (t - y)\right| < q + \frac{1}{2}$$

and $t - y \geq q + 1/2 + 2$. We can now use CR to find a compact representation of $|\bar{\theta}|$, where $\mathfrak{b} = (\bar{\theta})$. From this we can easily produce a compact representation of $|\theta|$ by conjugating the compact representation of $|\bar{\theta}|$.

If

$$\theta = \lambda \prod_{i=0}^{l} \left(\frac{\lambda_i}{d_i}\right)^{2^{l-i}} , \tag{12.15}$$

then

$$|\gamma| = \beta\lambda \prod_{i=0}^{l} \left(\frac{\lambda_i}{d_i}\right)^{2^{l-i}} .$$

Now, in the process of producing (12.15) we get

$$d_l \mathfrak{b} = \lambda \mathfrak{a}[x]$$

for some $x \in \mathbb{Z}^{>0}$ and $N(\mathfrak{a}[x]) = d_l$. Hence,

$$d_l \mathfrak{c} = \beta\lambda \mathfrak{a}[x] ,$$

and we see that $\beta\lambda \in \mathfrak{c} \subseteq \mathcal{O}$. If we put $\nu = N(\mathfrak{b})\beta \in \mathcal{O}$, then

$$\nu = \frac{S|G_i - B_i\sqrt{D}|}{r} .$$

Since $|G_i| < \sqrt{Q}\sqrt[4]{D}$ and $|B_i| < \sqrt{Q}/\sqrt[4]{D}$ (see §5.2), we get

$$|\nu|, |\bar{\nu}| < \frac{2S\sqrt{Q}\sqrt[4]{D}}{r}$$

and

$$H(\nu) = \frac{2S\sqrt{Q}\sqrt[4]{D}}{r} = \sqrt{2}\Delta^{1/4}\sqrt{N(\mathfrak{c})} .$$

As $H(\lambda) < 2^{2q+4}\sqrt{\Delta}$, we must have

$$H(\nu\lambda) < 2^{2q+4}\sqrt{2}\Delta^{3/4}\sqrt{N(\mathfrak{c})} ,$$

and, therefore,

$$H(\beta\lambda) < \frac{2^{2q+4}\sqrt{2}\Delta^{3/4}\sqrt{N(\mathfrak{c})}}{N(\mathfrak{b})} .$$

Thus, if $\lambda' = \sigma\beta\lambda$, we get a compact representation of γ:

$$\gamma = \lambda' \prod_{i=0}^{l} \left(\frac{\lambda_i}{d_i}\right)^{2^{l-i}} .$$

This algorithm executes in $O(\log \log H(\gamma) \log \Delta + q + \log N(\mathfrak{c}))$ elementary operations.

There remains the problem of finding a compact representation of γ when

$$t - q - \frac{5}{2} < y < q + 2 .$$

Since $|G_i| < \sqrt{Q}\sqrt[4]{D}$ and $|B_i| < \sqrt{Q}/\sqrt[4]{D}$, we get

$$\left|\frac{\theta_{i+2}}{S}\right|, \left|\frac{\overline{\theta}_{i+2}}{S}\right| \le \frac{S(|G_i| + |B_i|\sqrt{D})}{rN(\mathfrak{c})} < \frac{\sqrt{2}\Delta^{1/4}}{\sqrt{N(\mathfrak{c})}} . \qquad (12.16)$$

Also, from (12.16) and

$$S^2 N(\mathfrak{b}) = |\theta_{i+2}\overline{\theta}_{i+2}|N(\mathfrak{c}) ,$$

we have

$$\left|\frac{\theta_{i+2}}{S}\right| > \frac{N(\mathfrak{b})}{\Delta^{1/4}\sqrt{2N(\mathfrak{c})}} . \qquad (12.17)$$

Now,

$$\log_2 |\gamma| < z + q' = y - y' + q'$$

$$< y - \log_2 \left|\frac{\theta_{i+2}}{S}\right| + q$$

$$< 2q + 2 + \frac{\log_2 \Delta}{4} + \log_2 \frac{\sqrt{2N(\mathfrak{c})}}{N(\mathfrak{b})}$$

by (12.17). Also,

$$\log_2 |\gamma| > z - q' = y - y' - q' > y - \log_2 \left|\frac{\theta_{i+2}}{S}\right| - q .$$

Hence

$$\log_2 |\overline{\gamma}| = \log_2 N(\mathfrak{c}) - \log_2 |\gamma| < \log_2 N(\mathfrak{c}) + q - y + \log_2 \left|\frac{\theta_{i+2}}{S}\right|$$

$$< -t + 2q + \frac{5}{2} + \log_2 \left|\frac{N(\mathfrak{c})\theta_{i+2}}{S}\right|$$

$$< 2q + 3 + \frac{\log_2 \Delta}{4} + \log_2 \frac{\sqrt{2N(\mathfrak{c})}}{N(\mathfrak{b})}$$

by (12.16). It follows that

$$H(\gamma) < \frac{\sqrt{2N(\mathfrak{c})}}{N(\mathfrak{b})} \Delta^{1/4} 2^{2q+3} .$$

If q is small, we can now find a compact representation of γ by using the technique described in Remark 12.1.

Thus, for any $\mathcal{O}$ and any $\gamma \in \mathcal{O}$, we can now produce the following *definition* for a compact representation[5] of γ. This is a representation of γ as

$$\gamma = \lambda \prod_{i=0}^{l} \left(\frac{\lambda_i}{d_i} \right)^{2^{l-i}} ,$$

where

1. $l = O(\log \log H(\gamma))$.
2. $\lambda, \lambda_i \in \mathcal{O}$, $d_i \in \mathbb{Z}$ $(0 \le i \le l)$.
3. $0 < d_i \le \Delta^{1/2}$, $H(\lambda) = O(\Delta |N(\gamma)|)$, $H(\lambda_i) = O(\Delta)$ $(0 \le i \le l)$.
4. If

$$\mu_j = \lambda_j \prod_{i=0}^{j-1} \left(\frac{\lambda_i}{d_i} \right)^{2^{j-i}} ,$$

then $\mu_j \in \mathcal{O}$ and $|N(\mu_j)| = d_j$. Also, μ_j is a generator of a principal $\mathcal{O}$-ideal $\mathfrak{b}_j$ and

$$d_i^2 \mathfrak{b}_{i+1} = \lambda_{i+1} \mathfrak{b}_i^2 \quad (i = 0, 1, \dots, l-1) .$$

Furthermore, we have shown how such a compact representation of γ can be computed when we are given $\mathfrak{c} = (\gamma)$, the sign of γ, and some small q such that $|\log_2 |\gamma| - z| < q$.

12.3 The Arithmetic of Compact Representations

We have seen that we can represent a very large real quadratic integer $\gamma \in \mathcal{O}$ in a notation which requires only $O(\log \log H(\gamma) \log \Delta + \log |N(\gamma)|)$ bits. All that is needed in order to do this is the $\mathcal{O}$-ideal (γ) represented as $[Q/r, (P + \sqrt{D})/r]$, the sign of γ, and a reasonably close estimate of the value of $\log_2 |\gamma|$. Certainly, this can be done for ϵ_Δ when we have an estimate of R_Δ.

In this section we will describe techniques[6] that can be used to solve certain problems concerning quadratic integers which are represented by compact representations. All of the techniques described will execute in polynomial time; that is, the algorithms are polynomial in $\log \Delta$, l, and $\log |N(\gamma)|$. Here, l is the length of the compact representation, which is $O(\log |\log |\gamma||) = O(\log \log H(\gamma))$.

We remind the reader that if we are given any $\lambda \in \mathcal{O}$, we could use the technique of Theorem 4.35 to determine values of $a, b, c \in \mathbb{Z}$ such that $(\lambda) = [a, b + c\omega]$. However, the following proposition makes this more explicit.

Proposition 12.2. *Let* $\lambda = m + n\omega$, *where* $m, n \in \mathbb{Z}$. *If* $a, b, c \in \mathbb{Z}$ *such that* $c = (m, n)$, $a = |N(\lambda)|/c$, *and* $b = cb'$, *where* $b' \equiv (m/c)(n/c)^{-1} \pmod{a/c}$, *then* $(\lambda) = [a, b + c\omega]$.

Proof. Since (λ) is an ideal of $\mathcal{O}$, we may write $(\lambda) = [a, b + c\omega]$ for some $a, b, c \in \mathbb{Z}$. Now, $\lambda \in [a, b + c\omega]$, and $c \mid b$, and $c \mid a$; thus, we must have $c \mid m$ and $c \mid n$. If $d = (m, n)$, then since $b + c\omega \in (\lambda)$, we have $b + c\omega = x\lambda + y\lambda\omega$ for some $x, y \in \mathbb{Z}$. It follows that $d \mid c$ and therefore $d = c$. Since $N([a, b + c\omega]) = ac$, we also have $|N(\lambda)| = ac$. Finally, since $b = cb'$ and $[a, b + c\omega] = c[a', b' + \omega]$, where $a' = a/c$, and $(\lambda') = (\lambda/c) = [a', b' + \omega]$, we must have $b' \equiv (m/c)(n/c)^{-1} \pmod{a/c}$. □

Thus, if we are only given the compact representation

$$\gamma = \lambda \prod_{i=0}^{l} \left(\frac{\lambda_i}{d_i} \right)^{2^{l-i}} , \qquad (12.18)$$

then we can put $\mathfrak{b}_0 = (\lambda_0)$, and we can recover the ideals $\mathfrak{b}_1, \mathfrak{b}_2, \ldots, \mathfrak{b}_l$ by using Proposition 12.2 and

$$d_i^2 \mathfrak{b}_{i+1} = \lambda_{i+1} \mathfrak{b}_i^2 . \qquad (12.19)$$

Also,

$$(\gamma) = \mathfrak{c} = \left(\frac{\lambda}{d_l} \right) \mathfrak{b}_l .$$

The sign of γ can be easily deduced from the sign of λ.

By our remarks at the beginning of §11.1, we can compute for each λ_i/d_{i-1}^2, values of e_i and k_i such that

$$\left| \frac{\lambda_i 2^p}{d_{i-1}^2 e_i 2^{k_i}} - 1 \right| < \frac{1}{2^p} .$$

We also compute e and k such that

$$\left| \frac{\lambda 2^p}{d_l e 2^k} - 1 \right| < \frac{1}{2^p} .$$

From this information and (12.19), we can use the ideas in the proof of Theorem 11.2 to produce d_i and k_i such that $(\mathfrak{b}_i, d_i, k_i)$ is an (f_i, p) representation of $\mathfrak{a} = (1)$. Here, f_i satisfies (11.15) with $c = 1 = f_0$. It follows from the argument used in the proof of Theorem 11.8 that if $h \geq \max\{16, l\}$, then $f_l < 2^{p-4}$ when $2^p > 7h2^l$. (Note that $7 > e^{1/2}(529/256 + 2)$.) By Theorem 11.2 we can then compute an (f, p) representation $(\mathfrak{c}, d, k)$ of $\mathfrak{a}$ such that $f < 2 + f_l + 2^{-p} f_l$. Furthermore, as described earlier, we can use this representation to find some $g \in \mathbb{Q}$ such that $|\log_2 |\gamma| - g| < 1/2$. These processes execute in $O(\log \log H(\gamma) \log \Delta + \log N(\mathfrak{c}))$ elementary operations.

So far we have only discussed the problem of recovering information that could be used to produce (12.18). We now examine how we may solve the following problems, given (12.18):

1. Determine $\text{sign}(\gamma)$.
2. Determine $N(\gamma)$.

Since $\text{sign}\,\gamma = \text{sign}\,\lambda \cdot \text{sign}\,\lambda_l$ and $N(\gamma) = N(\lambda)/d_l$, these problems are easily solved in polynomial time.

Suppose we have compact representations for α and $\beta \in \mathcal{O}$. In polynomial time we can find $a, b \in (1/3)\mathbb{Z}$ and $\mathcal{O}$-ideals $\mathfrak{a}$ and $\mathfrak{b}$ such that

$$|\log_2 |\alpha| - a| < \frac{1}{2}, \quad |\log_2 |\beta| - b| < \frac{1}{2}$$

and $\mathfrak{a} = (\alpha)$ and $\mathfrak{b} = (\beta)$. Here are some problems concerning α and β that we can also solve.

3. Find a compact representation of $\alpha\beta$.
4. Is $\alpha = \beta$?
5. Is $|\alpha| > |\beta|$?
6. Does $\beta \mid \alpha$? If so, find a compact representation of α/β.

Problem 3 is easily solved by finding $\mathfrak{c} = \mathfrak{ab}$. Then $\mathfrak{c} = (\gamma)$, where $\gamma = \alpha\beta$. Also,

$$|\log_2 |\gamma| - (a + b)| < 1.$$

Hence, we can find a compact representation of γ in $O(\log\log H(\gamma)\log\Delta + N(\mathfrak{c}))$ elementary operations.

We next observe that since

$$a - \frac{1}{2} < \log_2 |\alpha| < a + \frac{1}{2}$$

and

$$b - \frac{1}{2} < \log_2 |\beta| < b + \frac{1}{2},$$

then if $b \geq a + 1$, we have $|\beta| > |\alpha|$ and if $a \geq b + 1$, we have $|\alpha| > |\beta|$. Thus, the only difficulty in solving Problems 4 and 5 arises when $|a - b| < 1$.

We see that if $\alpha = \beta$, then $\mathfrak{a} = \mathfrak{b}$ and $|a - b| < 1$. Also, if $\mathfrak{a} = \mathfrak{b}$, then $|\alpha| = \epsilon_\Delta^n |\beta|$ for $n \in \mathbb{Z}$. Hence,

$$|n\log_2 \epsilon_\Delta - (a - b)| < 1$$

and

$$|n| < \frac{2}{\log_2 \epsilon_\Delta}.$$

Since $\log \epsilon_\Delta > 2$ for $\Delta > 64$, we must have $|n| < 1$ or $n = 0$. Hence, $|\alpha| = |\beta|$ if and only if $\mathfrak{a} = \mathfrak{b}$ and $|a - b| < 1$. The determination of whether $\alpha = \beta$ can now be easily settled by examining the signs of α and β.

Suppose $|\alpha| \neq |\beta|$ and $|a - b| < 1$. Consider $\mathfrak{c} = \mathfrak{a\bar{b}}$ and let $\gamma = \alpha\bar{\beta}$, where $\mathfrak{c} = (\gamma)$. Since

$$|\alpha\bar{\beta}| = \left| \frac{\alpha N(\mathfrak{b})}{\beta} \right|,$$

we see that $|\alpha/\beta| < 1$ if and only if $|\alpha\overline{\beta}| < N(\mathfrak{b})$. Now,

$$\log_2 |\gamma| = \log_2 |\alpha| + \log_2 |\overline{\beta}|$$
$$= \log_2 |\alpha| + \log_2 N(\mathfrak{b}) - \log_2 |\beta|$$
$$\log_2 |\overline{\gamma}| = \log_2 |\beta| + \log_2 N(\mathfrak{a}) - \log_2 |\alpha| \,.$$

Since $|\log_2 |\alpha| - \log_2 |\beta|| < |a - b| + 1$, we get

$$\log |\gamma| < 2 + \log_2 N(\mathfrak{b}), \quad \log_2 |\overline{\gamma}| < 2 + \log_2 N(\mathfrak{a}) \,.$$

Hence, $|\gamma| < 4N(\mathfrak{b})$, $|\overline{\gamma}| < 4N(\mathfrak{a})$, and $H(\gamma) < 4\max\{N(\mathfrak{a}), N(\mathfrak{b})\}$. We can then find $|\gamma|$ by using the technique of Remark 12.1 and then compare $|\gamma|$ to $N(\mathfrak{b})$. If $|\gamma| < N(\mathfrak{b})$, then $|\alpha| < |\beta|$, and if $|\gamma| > N(\mathfrak{b})$, we have $|\alpha| > |\beta|$.

To solve Problem 6, we note that $\beta \mid \alpha$ if and only if $\mathfrak{a}\overline{\mathfrak{b}} = N(\mathfrak{b})\mathfrak{c}$, where $\mathfrak{c}$ is an $\mathcal{O}$-ideal. To see this, we first suppose that $\alpha = \beta\gamma$ where $\gamma \in \mathcal{O}$. Then $\mathfrak{a} = \mathfrak{b}\mathfrak{c}$, where $\mathfrak{c} = (\gamma)$ and $\mathfrak{a}\overline{\mathfrak{b}} = N(\mathfrak{b})\mathfrak{c}$. If $\mathfrak{a}\overline{\mathfrak{b}} = N(\mathfrak{b})\mathfrak{c}$, then $\mathfrak{a} = \mathfrak{b}\mathfrak{c}$ and $(\alpha) = (\beta)\mathfrak{c}$. Hence, $\mathfrak{c}$ is a principal $\mathcal{O}$-ideal and $\mathfrak{c} = (\gamma)$, where $\gamma \in \mathcal{O}$. Thus, we can determine whether or not $\beta \mid \alpha$ by checking whether or not the $\mathcal{O}$-ideal $\mathfrak{a}\overline{\mathfrak{b}} = N(\mathfrak{b})\mathfrak{c}$, where $\mathfrak{c}$ is an $\mathcal{O}$-ideal. If $\beta \mid \alpha$, then if $\gamma = \alpha/\beta$, we get $\mathfrak{c} = (\gamma)$ and

$$\log_2 |\gamma| = \log_2 |\alpha| - \log_2 |\beta| \,,$$

with

$$|\log_2 |\gamma| - (a - b)| < 1 \,.$$

Thus, we can find a compact representation of $|\gamma|$ in polynomial time, and from this and the signs of α, β, it is easy to find a compact representation of γ.

We now turn our attention to how, given some $m \in \mathbb{Z}^{>0}$, we can use (12.18) to determine the values of $x, y \pmod{m}$, where $\gamma = x + y\omega$. We point out that the problem of computing[7] $\gamma \pmod{m}$ is very easy if $(m, d_i) = 1$ for all i such that $1 \leq i \leq l$. In this case, we put $\Gamma_0 = \lambda_0$ and $D_0 = d_0$ and compute

$$D_{i+1} \equiv d_{i+1}D_i^2 \pmod{m} \quad (i = 0, 1, 2, \ldots, l - 1)$$

and

$$\Gamma_{i+1} \equiv \lambda_{i+1}\Gamma_i^2 \pmod{m} \quad (i = 0, 1, 2, \ldots, l - 1) \,,$$

where $\Gamma_{i+1} \in \mathcal{O}$. Then

$$\gamma \equiv \lambda D_l^{-1}\Gamma_l \pmod{m} \,.$$

However, if some of the values of the d_i $(i = 0, 1, 2, \ldots, l)$ are not relatively prime to m, we must make a modification to this simple process. We begin by introducing a proposition which is similar to Proposition 5.15.

Proposition 12.3. *If $\mathfrak{a}$ is an invertible ideal of $\mathcal{O}$ and $\alpha \in \mathfrak{a}$, then there exists an invertible ideal $\mathfrak{b}$ of $\mathcal{O}$ such that $(N(\mathfrak{b}))\mathfrak{a} = (\alpha)\mathfrak{b}$.*

Proof. Since $\alpha \in \mathfrak{a}$, we have $\mathfrak{a} \supseteq (\alpha)$. Thus,

$$(N(\mathfrak{a})) = \mathfrak{a}\bar{\mathfrak{a}} \supseteq (\alpha)\bar{\mathfrak{a}} ,$$

and, therefore, every element of $(\alpha)\bar{\mathfrak{a}}$ is divisible by $N(\mathfrak{a})$. It follows that

$$(\alpha)\bar{\mathfrak{a}} = (N(\mathfrak{a}))\mathfrak{c} ,$$

where $\mathfrak{c}$ is an ideal of $\mathcal{O}$. From this we deduce that

$$(\alpha) = \mathfrak{c}\mathfrak{a} .$$

Since $(N(\alpha)) = \mathfrak{c}\mathfrak{a}(\bar{\alpha})$, we see that $\mathfrak{c}$ is invertible; hence,

$$(\alpha)\bar{\mathfrak{c}} = (N(\mathfrak{c}))\mathfrak{a} .$$

Putting $\mathfrak{b} = \bar{\mathfrak{c}}$, we see that $\mathfrak{b}$ is invertible and $(\alpha)\mathfrak{b} = (N(\mathfrak{b}))\mathfrak{a}$. $\qquad\square$

Our next step will be to use this proposition to show that there are many representations of the form (12.18) for γ. We know that $\mathfrak{b}_i = (\mu_i)$, where

$$\mu_j = \lambda_j \prod_{i=0}^{j-1} \left(\frac{\lambda_i}{d_i}\right)^{2^{j-i}} .$$

We select any $\beta_i \in \mathfrak{b}_i$. By Proposition 12.3 we know that there must exist some principal $\mathcal{O}$-ideal $\mathfrak{c}_i$ such that

$$N(\mathfrak{c}_i)\mathfrak{b}_i = (\beta_i)\mathfrak{c}_i \tag{12.20}$$

and $\mathfrak{c}_i = (\gamma_i)$, where $\gamma_i = \bar{\beta}_i\mu_i/d_i \in \mathcal{O}$. If we put $\beta_{-1} = 1$, then $\mathfrak{c}_{-1} = \mathfrak{b}_{-1} = (1)$ and $\gamma_{-1} = 1$. Since

$$\mu_i = \lambda_i \left(\frac{\mu_{i-1}}{d_{i-1}}\right)^2 \quad (i = 0, 1, 2, \ldots, l) ,$$

we get

$$\gamma_i = \nu_i \left(\frac{\gamma_{i-1}}{N(\mathfrak{c}_{i-1})}\right)^2 \quad (i = 0, 1, 2, \ldots, l) , \tag{12.21}$$

where

$$\nu_i = \frac{\bar{\beta}_i\lambda_i N(\mathfrak{c}_{i-1})^2}{d_i\bar{\beta}_{i-1}^2} = \frac{\bar{\beta}\lambda_i\beta_{i-1}^2}{d_{i-1}^2}$$

$$= \frac{\pm N(\mathfrak{c}_i)\lambda_i\beta_{i-1}^2}{d_{i-1}^2\beta_i} .$$

By (12.19) and (12.20), we have

$$\left(N(\mathfrak{c}_{i-1})^2 d_{i-1}^2 \beta_i\right) \mathfrak{c}_i = \left(N(\mathfrak{c}_i)\lambda_i \beta_{i-1}^2\right) \mathfrak{c}_{i-1}^2 \; ;$$

hence, $\nu_i \in \mathfrak{c}_i \subseteq \mathcal{O}$. If we select $\beta_l = N(\mathfrak{b}_l) = d_l$, then $\gamma_l = \mu_l$ and $\mathfrak{b}_l = \mathfrak{c}_l$, and by (12.21), we get

$$\gamma = \frac{\lambda \mu_l}{d_l} = \frac{\lambda \gamma_l}{d_l} = \lambda \prod_{i=1}^{l} \left(\frac{\nu_i}{N(\mathfrak{c}_i)}\right)^{2^{l-i}} , \qquad (12.22)$$

another representation of the form (12.18).

In order to compute $\gamma \pmod{m}$ easily, we need to be able to find β_i in each $\mathfrak{b}_i$ such that $(N(\mathfrak{c}_i), m) = 1$. By (12.20) and the multiplication property of the norm, we must find β_i such that

$$\left(\frac{N(\beta_i)}{N(\mathfrak{b}_i)}, m\right) = 1 .$$

If we let $\mathfrak{b} = [a, b + \omega]$ be any primitive principal ideal in $\mathcal{O}$ and let $\beta \in \mathfrak{b}$, then $\beta = xa + y(b + \omega)$ for $x, y \in \mathbb{Z}$ and

$$\frac{N(\beta)}{N(\mathfrak{b})} = ax^2 + T(b + \omega)xy + \left(\frac{N(b + \omega)}{a}\right) y^2 .$$

Since $\mathfrak{b}$ is invertible, we know that if

$$a_1 = a, \quad a_2 = T(b + \omega), \quad a_3 = \frac{N(b + \omega)}{a} ,$$

then $(a_1, a_2, a_3) = 1$ by our results in §4.5. Put $h_1 = a$, $h_2 = a_1 + a_2 + a_3$, and $h_3 = a_3$. Since $(a_1, a_2, a_3) = 1$, we must have $(h_1, h_2, h_3) = 1$. We now present the following algorithm.

Algorithm 12.7: SPLIT

Input: $h \in \mathbb{Z}$, $m \in \mathbb{Z}^{>0}$.
Output: $r, s \in \mathbb{Z}^{>0}$ such that $m = rs$, $(r, h) = 1$ and any prime divisor of s must divide h.
 1: Put $g_1 = (m, h)$, $r_1 = m/g_1$, $s_1 = g_1$, $i = 1$.
 2: **while** $g_i > 1$ **do**
 3:     Put

$$g_{i+1} = (r_i, g_i)$$
$$r_{i+1} = r_i/g_{i+1}$$
$$s_{i+1} = g_{i+1}s_i$$
$$i \leftarrow i + 1 .$$

 4: **end while**
 5: Put $r = r_i$, $s = s_i$.

Proof (of correctness of SPLIT). We note that

$$r_j = \frac{m}{s_j}, \quad s_j = \prod_{i=1}^{j} g_i .$$

It follows that since $r_j \in \mathbb{Z}$, we must find some j such that $g_j = 1$ and $j = O(\log m)$. Also, $r_j s_j = m$, $r_j \mid r_i$ $(i \le j)$, $g_j \mid g_i$ $(i \le j)$. We now show that $(r_j, g_1) = 1$. Certainly, $(r_j, g_j) = 1$; suppose $(r_j, g_k) = 1$ for some k such that $2 \le k \le j$. Since $g_k = (r_{k-1}, g_{k-1})$, we get $(r_k, g_{k-1}/g_k) = 1$. Also, $r_j \mid r_k$, and therefore $(r_j, g_{k-1}/g_k) = 1$. Since $(r_j, g_k) = 1$, we must have $(r_j, g_{k-1}) = 1$; thus, we may conclude by induction that $(r_j, g_1) = 1$. Since $(r_1, h/g_1) = 1$ and $r_j \mid r_1$, we get $(r_j, h/g_1) = 1$. Since $(r_j, g_1) = 1$, we must have $(r_j, h) = 1$. If p is any prime such that $p \mid s_j$, then $p \mid g_i$ for some $i \le j$; it follows that $p \mid g_1$ and $p \mid h$. Thus, $r = r_j$ and $s = s_j$ satisfy the requirements of the algorithm and it executes in $O(\log m)$ arithmetic operations on numbers of $O(\log \max\{m, h\})$ bits.

We now use SPLIT to put

$$m = r_1 s_1, \quad s_1 = r_2 s_2, \quad s_2 = r_3 s_3 ,$$

where $(r_i, h_i) = 1$ and any prime which divides s_i must divide h_i $(i = 1, 2, 3)$. We get $m = r_1 r_2 r_3 s_3$ and $(r_1, r_2) = (r_2, r_3) = (r_3, r_1) = 1$. If p is a prime and $p \mid s_3$, then $p \mid s_2$ and $p \mid s_1$, but this means that $p \mid (h_1, h_2, h_3)$, which is impossible. Thus,

$$m = r_1 r_2 r_3 .$$

We next use the Chinese Remainder Theorem to find x, y such that $0 < x, y < m$, $x \equiv 1 \pmod{r_1 r_2}$, $x \equiv 0 \pmod{r_3}$, $y \equiv 1 \pmod{r_2 r_3}$, and $y \equiv 0 \pmod{r_1}$. If $\beta = xa + y(b + \omega)$ and p is a prime divisor of $(m, N(\beta)/a)$, then since $N(\beta)/a \equiv h_i \pmod{r_i}$ $(i = 1, 2, 3)$, we must have $p \mid h_i$ for some $i \in \{1, 2, 3\}$, which is contrary to the construction of the r_i values. It follows then that

$$\left(\frac{N(\beta)}{N(\mathfrak{b})}, m \right) = 1 . \tag{12.23}$$

Thus, given m and an $\mathcal{O}$-ideal $\mathfrak{b} = [a, b + \omega]$, we can compute a value of $\beta \in \mathfrak{b}$ such that (12.23) holds and this process requires only $O(\log m)$ arithmetic operations on numbers of $O(\log \max\{m, a, |T(b + \omega)|, |N(b + \omega)|/a\})$ bits.

To compute $\gamma \pmod{m}$, we need to find a sequence $\beta_1, \beta_2, \ldots, \beta_l$ such that $\beta_i \in \mathfrak{b}_i$ and $(N(\beta_i)/d_i, m) = 1$. We can do this by putting $\beta_i = N(\mathfrak{b}_i) = d_i$ whenever $(m, d_i) = 1$ and by using the process described above whenever $(m, d_i) > 1$. We can use

$$\nu_i = \frac{\overline{\beta}_i \lambda_i \beta_{i-1}^2}{d_{i-1}^2}$$

to compute ν_i and we must compute ν_i exactly, not just modulo m. We can then use (12.22) to determine $\gamma \pmod{m}$. The overall complexity of this process is $O(l \log m)$ arithmetic operations on numbers of $O(\log \max\{m, \Delta, |N(\gamma)|\})$ bits. We emphasize that, in practice, it is usually not necessary to invoke SPLIT to compute a β_i such that $(N(\mathfrak{c}_i), m) = 1$. Finding $\mathfrak{c}_i$ by applying ρ repeatedly to $\mathfrak{b}_i$ will rapidly result in a β_i for which $(N(\beta_i)/d_i, m) = 1$.

Notes and References

[1][Wil02], p. 418.

[2][Nel81].

[3]Although to some degree this idea was anticipated in work of Lagarias ([Lag79], [Lag81]), the term was first mentioned in [Coh93], p. 274, and the basic idea is described there on pp. 280–282. The concept was extended and formalized in [BTW95] and an improved, but much briefer, version appears in [BV07], pp. 251–256. Our approach here is different from that of [BTW95] and [BV07] in that it makes use of (f, p) representation theory. This allows us to avoid trying to approximate logarithms and produces somewhat better results than those in [BV07].

[4]Fast methods of determining accurate approximations of the values produced by elementary functions such as log can be found in [Bre76]. For our purposes, however, we will not require such precision.

[5]This is a different definition from that found in [BTW95] or [BV07], but it is certainly in the spirit of the definitions that appear in these works.

[6]Similar techniques are described in [BTW95].

[7]This problem was addressed in [BTW95], but the technique given there is not very practical. We give here the method described in [JW02].

13

The Subexponential Method

13.1 Introduction

Up to this point, all the algorithms we have presented for computing the regulator R_Δ of the real quadratic order $\mathcal{O}_\Delta$, and, hence, for solving Pell's equation, have exponential complexity in the size of the discriminant Δ. The most exciting recent development has certainly been the discovery of a Las Vegas algorithm[1] by Buchmann for computing R_Δ and h_Δ whose expected running time is subexponential in $\log |\Delta|$. This algorithm has enabled the computation of R_Δ for discriminants Δ as large as 101 decimal digits, a dramatic improvement over what had been attainable previously. Unfortunately, as we will discuss in more detail below, this improvement comes at a price. Like Lenstra's algorithm for computing R_Δ described in Chapter 10, the complexity result is conditional on the generalized Riemann hypothesis (GRH) for Hecke L-functions and the extended Riemann hypothesis (ERH), but in the case of the subexponential algorithm, the correctness of the output is conditional as well. Nevertheless, the fact that R_Δ can be computed in subexponential time, even assuming the GRH and ERH, remains an important breakthrough.

The subexponential algorithm is based on the widely used index-calculus method.[2] This probabilistic strategy has been employed with great success for integer factorization and computing discrete logarithms in finite fields,[3] the most recent algorithm being the number field sieve.[4] The work of Hafner, McCurley, and Buchmann showed that the index-calculus approach can also be applied to computational problems in quadratic fields—in particular, the discrete logarithm problem in the class group, computing the class number and structure of the class group, computing the regulator, and solving the principal ideal problem. In addition, some of the ideas used in this algorithm have been used to reason about the complexity classes to which these problems belong. In particular, the work of McCurley, Buchmann, and Williams shows that, under the assumption of the GRH, these problems are in the complexity class $\mathcal{NP} \cap \text{co-}\mathcal{NP}$.

The main ideas behind applying index-calculus to computations in number fields were first suggested by Lenstra and Lenstra[5] in 1987. Seysen[6] used these ideas in his integer factorization algorithm based on using index-calculus to find ambiguous ideals in an imaginary quadratic order. These ideas were soon elaborated by Hafner and McCurley[7] in 1989 for imaginary quadratic fields and by Buchmann[8] in 1989 for real quadratic fields. Various people have contributed further improvements,[9] and Vollmer's recent work[10] describes variations of these algorithms with the best known complexity. The resulting algorithms all have expected running time subexponential in the bit length of Δ. The most efficient variation for solving these problems in practice, due to Jacobson,[11] uses ideas from the self-initializing quadratic sieve factoring algorithm to improve significantly the relation generation stage of the algorithm.

In this chapter, we will describe the index-calculus algorithm for computing the regulator R_Δ of a real quadratic order. We will begin with a discussion of Vollmer's method for solving the discrete logarithm problem in the ideal class group of an imaginary quadratic order, as this is the one of the most straightforward applications of the index-calculus method to quadratic orders. We will then show how this approach can be modified to compute the class number and group structure of an imaginary quadratic order, essentially giving the algorithm of Hafner and McCurley, followed by Buchmann's extensions for computing the class group and regulator of a real quadratic order. Vollmer's Monte Carlo algorithm for computing R_Δ will also be presented, as well as an index-calculus algorithm for solving the principal ideal problem. A complete analysis of these algorithms is presented in the recent book by Buchmann and Vollmer,[12] so we will present these algorithms as they are typically used in practice and only provide sketches of the complexity analysis, referring the interested reader to this source for more details. This will be followed by a discussion of known complexity results, including the fact that the problem of computing R_Δ, and hence solving the Pell equation, is in $\mathcal{NP} \cap \text{co-}\mathcal{NP}$. We will also describe modifications to these algorithms that work especially well in practice, including the use of sieving for generating relations and the significant computational results obtained with them. Finally, we will conclude with an outlook toward further potential improvements

13.2 Solving the Discrete Logarithm Problem in Cl_Δ

We begin by describing how to use the index calculus approach to solve the discrete logarithm problem (DLP) in the class group of an imaginary quadratic order.

Definition 13.1. *The* imaginary quadratic order discrete logarithm problem *is, given $\mathcal{O}_\Delta$-ideals $\mathfrak{a}$ and $\mathfrak{g}$, to compute the minimal positive integer x such that $\mathfrak{a} \sim \mathfrak{g}^x$ or prove that no such integer exists.*

Note that this is nothing more than the usual finite abelian group discrete logarithm problem set in the ideal class group. Throughout this section, we will assume that we are given ideals $\mathfrak{a}$ and $\mathfrak{g}$ of an imaginary quadratic order $\mathcal{O}_\Delta$ and wish to find a solution to the discrete logarithm problem.

The first index-calculus algorithm for solving this problem is due to Mc-Curley.[13] It uses a strategy similar to that of computing discrete logarithms in finite fields and requires that the class number be computed ahead of time. We will present a modified version of a more recent algorithm due to Vollmer[14] that does not require the class number and solves the discrete logarithm problem in expected subexponential time in $\log|\Delta|$. In particular, it runs in expected time $L_\Delta[1/2, \sqrt{2} + o(1)]$, where

$$L_\Delta[a, b] = \exp\left(b(\log|\Delta|)^a (\log\log|\Delta|)^{1-a}\right) .$$

Notice that if $a = 0$, then $L_\Delta[0, b] = (\log|\Delta|)^b$ is a polynomial in $\log|\Delta|$, and if $a = 1$, then $L_\Delta[1, b] = |\Delta|^b$ is exponential in $\log|\Delta|$. Thus, when $0 < a < 1$, $L_\Delta[a, b]$ is a function that is larger asymptotically than a polynomial function but smaller than an exponential function, and it is said to be subexponential.

In general, index-calculus algorithms consist of two main stages: generating random relations and solving a linear algebra problem. *Relations* correspond to objects that are smooth in some sense, and the key to the fast running times enjoyed by many index-calculus algorithms is the probability that a random object of a particular size is smooth. For example, integer factorization algorithms search for certain smooth integers whose prime divisors are all less than some bound. The linear algebra problem varies by context. For integer factorization, one solves a linear system modulo 2 in an effort to find squares modulo the integer to be factored, and for computing discrete logarithms in finite fields, a linear system must be solved modulo the size of the field.

In our setting, relations correspond to smooth principal ideals (i.e., principal ideals that factor into a product of prime ideals whose norms are all less than some bound). In particular, let $p_1, p_2, \ldots, p_k$ be the smallest k rational primes for which the Kronecker symbol $(\Delta/p_i) \neq -1$, so $(p_i) = \mathfrak{p}_i\overline{\mathfrak{p}}_i$, where $\mathfrak{p}_i$ is an invertible prime ideal. The *factor base* is defined to be the set

$$FB = \{\mathfrak{p}_1, \mathfrak{p}_2, \ldots, \mathfrak{p}_k\} .$$

Define for $\mathbf{v} = (v_1, v_2, \ldots, v_k) \in \mathbb{Z}^k$,

$$FB^{\mathbf{v}} = \prod_{i=1}^{k} \mathfrak{p}_i^{v_i} ,$$

where we use the fractional ideal $\mathfrak{p}^{-1}$ if v_i is negative. We say that $\mathbf{v}$ is a relation if $FB^{\mathbf{v}} = (\gamma)$; that is, the principal ideal (γ) factors completely over the factor base. Relations have the following properties:

1. If $\mathbf{v}$ is a relation, then $a\mathbf{v}$ is also a relation for every $a \in \mathbb{Z}$. To see this, note that $FB^{a\mathbf{v}} = (FB^{\mathbf{v}})^a = (\gamma)^a \sim \mathcal{O}_\Delta$.
2. If $\mathbf{v}$ and $\mathbf{w}$ are relations, then $\mathbf{v} + \mathbf{w}$ is a relation. To see this, note that $FB^{\mathbf{v}+\mathbf{w}} = FB^{\mathbf{v}}FB^{\mathbf{w}} = (\gamma_1)(\gamma_2) \sim \mathcal{O}_\Delta$.

It follows that any integer linear combination of a set of relations is also a relation, so the set of all relations $\Lambda = \{\mathbf{v} \in \mathbb{Z}^k \mid FB^{\mathbf{v}} = (\gamma)\}$ is a sublattice of $\mathbb{Z}^k$ called the *relation lattice*.

The use of relations to solve the discrete logarithm problem is as follows. We first compute a set of random relations corresponding to the extended factor base $FB^* = \{\mathfrak{g}, \mathfrak{a}^{-1}, \mathfrak{p}_1, \ldots, \mathfrak{p}_k\}$, where $\mathfrak{a}^{-1}$ is the fractional ideal inverse of $\mathfrak{a}$. Suppose that $\mathbf{v}_1, \ldots, \mathbf{v}_n$ are relations over FB^* and consider the *relation matrix*

$$B = (\mathbf{v}_1^T \ldots \mathbf{v}_n^T) ,$$

the matrix whose columns consist of the relations. If $\mathfrak{a} \sim \mathfrak{g}^x$, then $\mathfrak{g}^x \mathfrak{a}^{-1} \sim \mathcal{O}_\Delta$ and the vector $(x, 1, 0, \ldots, 0) \in \mathbb{Z}^{k+2}$ is a relation with respect to FB^*. Thus, if the columns of B generate the entire extended relation lattice Λ^*, there exists a vector $\mathbf{x} \in \mathbb{Z}^n$ such that $B\mathbf{x} = (x, 1, 0, \ldots, 0)$. We cannot find $\mathbf{x}$ by solving this linear system directly because x is unknown. Instead, we use the following approach that finds a non-minimal solution to the discrete logarithm problem, provided that a solution exists.

Algorithm 13.1: Imaginary Quadratic Order DLP

Input: discriminant $\Delta < 0$ of an imaginary quadratic order, reduced ideals $\mathfrak{a}$ and $\mathfrak{g}$ such that $[\mathfrak{a}] \in \langle[\mathfrak{g}]\rangle$.
Output: $x \in \mathbb{Z}$ such that $\mathfrak{a} \sim \mathfrak{g}^x$.
1: Compute a factor base $FB = \{\mathfrak{p}_1, \ldots, \mathfrak{p}_k\}$ consisting of all non-inert prime ideals $\mathfrak{p}_i$ with $N(\mathfrak{p}_i) < P$ for some bound P.
2: Compute $n > k$ random relations $\mathbf{v}_1, \ldots, \mathbf{v}_n$ over FB.
3: Compute $\mathbf{v}_a, \mathbf{v}_g$ such that $\mathfrak{a} \sim FB^{\mathbf{v}_a}$ and $\mathfrak{g}^{-1} \sim FB^{\mathbf{v}_g}$. Note that $\mathfrak{a}^{-1} FB^{\mathbf{v}_a}$ and $\mathfrak{g} FB^{\mathbf{v}_g}$ are both principal, so $(1, 0, \mathbf{v}_g)$ and $(0, 1, \mathbf{v}_a)$ are relations on the extended factor base $FB^* = \{\mathfrak{g}, \mathfrak{a}^{-1}, \mathfrak{p}_1, \ldots, \mathfrak{p}_k\}$.
4: Form the relation matrices A and B with

$$A = (\mathbf{v}_1^T \ldots \mathbf{v}_n^T), \quad B = \begin{pmatrix} 1 & 0 & \mathbf{0} \\ 0 & 1 & \mathbf{0} \\ \mathbf{v}_g^T & \mathbf{v}_a^T & A \end{pmatrix} = \begin{pmatrix} \mathbf{b} \\ B' \end{pmatrix} .$$

Thus, the columns of A are relations on FB and the columns of B are relations on FB^*.
5: Solve $B'\mathbf{y} = (1, 0, \ldots, 0)$ over $\mathbb{Z}$. If no solution exists, increase n and go to step 13.1.
6: $x = \mathbf{b} \cdot \mathbf{y}$, the first entry in $\mathbf{y}$.

Notice that the solution $\mathbf{y}$ of $B'\mathbf{y} = (1, 0, \ldots, 0)$ also satisfies $B\mathbf{y} = (x, 1, 0, \ldots, 0)$ with $x = \mathbf{b} \cdot \mathbf{y}$. The following proposition states that the value of x computed is in fact a solution of the discrete logarithm problem.

Proposition 13.2. *Assume that the columns of B generate Λ^*, the relation lattice corresponding to the extended factor base FB^*. Then $\mathfrak{a} \sim \mathfrak{g}^x$ is solvable if and only if $\exists \mathbf{x} \in \mathbb{Z}^{n+2}$ such that $B\mathbf{x} = (x, 1, 0, \ldots, 0)$.*

Proof. First, if $\mathbf{x}$ exists, then $\mathbf{z} = (x, 1, 0, \ldots, 0)$ is a relation with respect to the extended factor base FB^*, because this vector is a linear combination of the columns of B, which are, in turn, relations over FB^*. Thus, $(FB^*)^{\mathbf{z}} = \mathfrak{g}^x \mathfrak{a}^{-1} \sim \mathcal{O}_\Delta$ and we have $\mathfrak{g}^x \sim \mathfrak{a}$, as required.

Conversely, if $\mathfrak{g}^x \sim \mathfrak{a}$ is solvable, then $\mathfrak{g}^x \mathfrak{a}^{-1} \sim \mathcal{O}_\Delta$ and $\mathbf{z} = (x, 1, 0, \ldots, 0)$ is a relation over FB^*. If the columns of B generate Λ^*, then every relation over FB^* can be expressed as a linear combination of the columns of B. Thus, $\mathbf{z}$ has to be equal to some linear combination of the columns of B; that is, $\mathbf{x}$ exists as required. $\square$

A necessary condition for the columns of B to generate Λ^* is that the prime ideals $\mathfrak{p}_1, \ldots, \mathfrak{p}_k$ generate the cyclic group $\langle [\mathfrak{g}] \rangle$. If this is the case, then the two relations corresponding to factorizations of $\mathfrak{a}$ and $\mathfrak{g}^{-1}$ over these prime ideals will exist. Also, linear combinations of these two relations and the other relations only involving the prime ideals $\mathfrak{p}_i$ can be used to generate any extended relation involving $\mathfrak{a}$ and $\mathfrak{g}$ as long as the columns of A generate all of Λ.

It is possible to select the factor base in such a way that it generates all of Cl_Δ and, hence, the subgroup generated by $[\mathfrak{g}]$. Assuming the GRH, the size of the factor base will be polynomial in $\log |\Delta|$ by Bach's theorem (9.28), and in order to analyze the running time, we require an even larger factor base. However, much smaller factor bases typically suffice to generate Cl_Δ in practice,[15] so the bound P can be selected according to other considerations, such as limitations on the size of the relation matrix that can be handled in practice, and the algorithm will still almost certainly terminate successfully.

In order to make use of this algorithm, we need to be able to solve three main computational tasks. First, we need to be able to rapidly compute random relations in order to find a relation matrix that will hopefully yield solutions to the required linear system. Second, we need to be able to factor an arbitrary ideal class over the factor base. Finally, we need to solve linear systems over $\mathbb{Z}$.

The following idea for generating relations is due to Hafner and McCurley.[16] Although this is not the fastest method in practice, it is fairly simple and clearly demonstrates the feasibility of finding relations. In addition, this method does randomly sample relations and is thus easy to incorporate into an analysis of the algorithm. We will discuss more efficient methods for finding relations in §13.7.

For a random $\mathbf{v} \in \mathbb{Z}^k$ with each entry $v_i \in \{0, 1, \ldots, |\Delta| - 1\}$, compute a reduced ideal $\mathfrak{a} \sim FB^{\mathbf{v}}$. Note that in general $FB^{\mathbf{v}}$ is not reduced, so $\mathfrak{a} \neq FB^{\mathbf{v}}$. If $\mathfrak{a}$ can be factored over the factor base as $FB^{\mathbf{w}}$, then we have

$$\mathfrak{a} = FB^{\mathbf{w}} \sim FB^{\mathbf{v}} \implies FB^{\mathbf{v} - \mathbf{w}} \sim \mathcal{O}_\Delta \,,$$

and $\mathbf{v} - \mathbf{w}$ is a relation. The condition that the entries of $\mathbf{v}$ be sampled randomly from the integers $\{0, 1, \ldots, |\Delta| - 1\}$ ensures that the probability of successfully finding a relation using this strategy is sufficiently high.[17] As we will see in §13.7, this condition can typically be relaxed in practice in order to speed the search for relations significantly.

To factor the ideal $\mathfrak{a}$, we recall from Theorem 4.44 that $\mathfrak{a}$ factors uniquely into a product of prime ideal powers and by Theorem 4.36 that the ideal norm is multiplicative. Thus, to factor $\mathfrak{a}$, we factor the integer $N(\mathfrak{a})$ and check whether all integer primes in the factorization are norms of prime ideals in the factor base. This yields the prime ideals that divide $\mathfrak{a}$ and the magnitude of the multiplicities, but it remains to determine whether $\mathfrak{p}$ or $\mathfrak{p}^{-1}$ divides $\mathfrak{a}$ for each $\mathfrak{p}$ (i.e., the sign of the exponent in the prime ideal factorization of $\mathfrak{a}$). If a prime ideal $\mathfrak{p} = [p/r, (t + \sqrt{D})/r]$ divides $\mathfrak{a} = [Q/r, (P + \sqrt{D})/r]$, then we must have $P \equiv t \pmod{4p}$. So, given $\mathfrak{a}$ with $p \mid N(\mathfrak{a})$, we simply check whether $P \equiv t \pmod{4p}$. If so, then $\mathfrak{p} \mid \mathfrak{a}$; otherwise, $\mathfrak{p}^{-1} \mid \mathfrak{a}$.

The second task is finding factorizations of the ideal classes $[\mathfrak{a}]$ and $[\mathfrak{g}^{-1}]$ over the factor base. If the ideals $\mathfrak{a}$ and $\mathfrak{g}^{-1}$ themselves are not smooth, we attempt to find equivalent ideals that are and proceed as follows. Suppose we are trying to factor $[\mathfrak{a}]$. For a random $\mathbf{v} \in \mathbb{Z}^k$, compute a reduced ideal $\mathfrak{c} \sim \mathfrak{a}FB^{\mathbf{v}}$. Suppose $\mathfrak{c}$ can be factored over the factor base as $\mathfrak{c} = FB^{\mathbf{w}}$. Then we have

$$\mathfrak{a}FB^{\mathbf{v}} \sim FB^{\mathbf{w}} \implies \mathfrak{a} \sim FB^{\mathbf{w}-\mathbf{v}},$$

and we have factored $[\mathfrak{a}]$ over the factor base.

The third task is solving a linear system over the integers. This problem[18] can be solved in time $O(n^{3+\epsilon})$ for $n \times m$ matrices assuming that $m = O(n)$ and the coefficients of the matrix are in $O(n^\epsilon)$. Note that this step could be done modulo the class number h_Δ if it were known, and it would result in a minimal solution to the discrete logarithm problem. However, as discussed in Chapter 7, computing h_Δ appears to be a difficult problem, so we cannot assume that it is known in general.

We now present an example illustrating the use of Algorithm 13.1. In this example, and in rest of the chapter, we will use the shorthand notation (Q, P) to denote the ideal $[Q/r, (P + \sqrt{D})/r]$.

Example 13.3. We compute $x \in \mathbb{Z}$ such that $\mathfrak{g}^x \sim \mathfrak{a}$ in $\mathcal{O}_\Delta$ with $\Delta = -32003$, $\mathfrak{g} = (78, -17)$, and $\mathfrak{a} = (58, -19)$. We follow the algorithm described above.

1. We set the factor base to be the smallest three prime ideals in terms of norm, yielding
$$FB = \{(6, 1), (14, 1), (26, 9)\}.$$

2. We compute five random relations, hoping that having two more relations than factor base elements (two more columns than rows in the relation matrix) will be enough to find a solution to the discrete logarithm problem.
   a) The random vector $\mathbf{v} = (1, 9, 4)$ gives us
   $$FB^{\mathbf{v}} \sim (6915236217162, 1237094462899).$$
   Reducing yields $(182, -69) \sim FB^{\mathbf{w}}$ with $\mathbf{w} = (0, 1, 1)$. Thus, $\mathbf{v} - \mathbf{w} = (1, 8, 3)$ is a relation.

b) The random vector $\mathbf{v} = (5, 7, 8)$ gives us

$$FB^{\mathbf{v}} \sim (326489612029948458, 133483783758025177).$$

Reducing yields $(182, 69) \sim FB^{\mathbf{w}}$ with $\mathbf{w} = (0, -1, -1)$. Thus, $\mathbf{v} - \mathbf{w} = (5, 8, 9)$ is a relation.

c) The random vector $\mathbf{v} = (5, 5, 3)$ gives us

$$FB^{\mathbf{v}} \sim (17945539794, 13474508797).$$

Reducing yields $(6, 1) \sim FB^{\mathbf{w}}$ with $\mathbf{w} = (1, 0, 0)$. Thus, $\mathbf{v} - \mathbf{w} = (4, 5, 3)$ is a relation.

d) The random vector $\mathbf{v} = (6, 5, 7)$ gives us

$$FB^{\mathbf{v}} \sim (1537627686169302, 735260185408771).$$

Reducing yields $(26, 9) \sim FB^{\mathbf{w}}$ with $\mathbf{w} = (0, 0, 1)$. Thus, $\mathbf{v} - \mathbf{w} = (6, 5, 6)$ is a relation.

e) The random vector $\mathbf{v} = (4, 8, 7)$ gives us

$$FB^{\mathbf{v}} \sim (58600699595118954, 33025441594964113).$$

Reducing yields $(78, -17) \sim FB^{\mathbf{w}}$ with $\mathbf{w} = (1, 0, 1)$. Thus, $\mathbf{v} - \mathbf{w} = (3, 8, 6)$ is a relation.

3. Next, we factor $\mathfrak{a}$ and $\mathfrak{g}^{-1}$ over the factor base.
   a) Using $\mathbf{v} = (9, 3, 5)$, we have

$$\mathfrak{a}FB^{\mathbf{v}} \sim (145388537407386, 3489781931749).$$

Reducing yields $(26, 9) \sim FB^{\mathbf{w}}$ with $\mathbf{w} = (0, 0, 1)$. Thus, $\mathfrak{a} \sim FB^{\mathbf{u}}$ with $\mathbf{u} = \mathbf{w} - \mathbf{v} = (-9, -3, -4)$.

   b) $\mathfrak{g}^{-1}$ factors immediately over the factor base with $\mathbf{u} = (-1, 0, -1)$.

4. The relation matrix corresponding to the extended factor base that includes $\mathfrak{g}$ and $\mathfrak{a}^{-1}$ is

$$B = \begin{pmatrix} 1 & 0 & 0 & 0 & 0 & 0 & 0 \\ 0 & 1 & 0 & 0 & 0 & 0 & 0 \\ -1 & -9 & 1 & 5 & 4 & 6 & 3 \\ 0 & -3 & 8 & 8 & 5 & 5 & 8 \\ -1 & -4 & 3 & 9 & 3 & 6 & 6 \end{pmatrix} = \begin{pmatrix} \mathbf{b} \\ B' \end{pmatrix}, \quad B' = \begin{pmatrix} 0 & 1 & 0 & 0 & 0 & 0 & 0 \\ -1 & -9 & 1 & 5 & 4 & 6 & 3 \\ 0 & -3 & 8 & 8 & 5 & 5 & 8 \\ -1 & -4 & 3 & 9 & 3 & 6 & 6 \end{pmatrix}.$$

5. We solve the linear system $B'\mathbf{y} = (1, 0, 0, 0)$ over $\mathbb{Z}$. One solution is $\mathbf{y} = (2429, 1, -888, 0, 311, 0, 694)$.

6. Finally, we compute $x = \mathbf{b} \cdot \mathbf{y} = 2429$. It can be verified that $\mathfrak{g}^{2429} \sim \mathfrak{a}$.

Analysis

The goal in analyzing this algorithm is to find the optimal size of the factor base that balances the time required for finding relations with that needed to perform the linear algebra, thereby minimizing the overall running time. The larger the factor base one uses, the easier it is to find relations, as smooth ideals are more likely to be found. On the other hand, using a smaller factor base speeds the linear algebra because the dimensions of the relation matrix will be smaller.

To begin, we need to know the probability that a random reduced ideal is smooth with respect to a given factor base, allowing us to determine the expected number of trials required to find a relation. Given that an ideal is smooth if its norm (an integer) is smooth with respect to the norms of the prime ideals in the factor base (a bounded set of prime numbers), one might expect that this probability would be similar to the probability that a random integer of a certain size is smooth with respect to some bound. This is indeed the case, as proved by Seysen for imaginary quadratic orders and by Abel for real quadratic orders.[19] In particular, we have the following.[20]

Proposition 13.4. *For any $\epsilon > 0$ there is a positive real number $c(\epsilon)$ such that for any $x, y \in \mathbb{R}_{>0}$ and any discriminant Δ with*

$$\max\{(\log x)^{1+\epsilon}, (\log |\Delta|)^{2+\epsilon}\} \leq y \leq \exp((\log x)^{1-\epsilon}) ,$$

the number of primitive $\mathcal{O}_\Delta$-ideals with y-smooth norm $\leq x$ is at least

$$x \exp(-u(\log u + \log \log u + c(\epsilon))) ,$$

where $u = (\log x)/(\log y)$.

In a similar manner to that employed to analyze index-calculus algorithms for integer factorization, one can derive the probability that a single attempt at finding a relation succeeds.[21]

Lemma 13.5. *Suppose that the factor base contains all prime ideals with norm less than $L_\Delta[1/2, z]$ for some constant z. Then the probability that the relation generation strategy described above successfully finds a relation is bounded from below by $L_\Delta[1/2, -1/(4z)] - o(1)$.*

Second, we need to know how many relations we expect to require before the linear system $B'\mathbf{y} = (1, 0, \ldots, 0)$ has a solution over $\mathbb{Z}$. Assuming that the discrete logarithm problem instance $\mathfrak{g}^x \sim \mathfrak{a}$ has a solution, the linear system will have a solution if the factor base FB generates the class group and if the set of relations $\mathbf{v}_1, \ldots, \mathbf{v}_n$ generates the entire relation lattice Λ. One approach is to first compute a set of relations that generate a full-rank sublattice of Λ and then to determine the probability that additional relations lie outside of this sublattice.

To ensure that the relations generated form a full-rank sublattice of Λ, we use an idea of Seysen[22] of ensuring that the relation matrix whose columns consist of the first $k = |FB|$ relations is strictly diagonally dominant.

Definition 13.6. *A matrix* $A = (a_{ij}) \in \mathbb{Z}^{k \times k}$ *is* strictly diagonally dominant *if*

$$|a_{ii}| > \sum_{j \neq i} |a_{ji}|$$

for $1 \leq i \leq k$.

If A is strictly diagonally dominant, then it can be shown that it has full rank.[23] We can ensure that A is strictly diagonally dominant if, when finding the ith relation for each $1 \leq i \leq k$, the random exponent vector v is selected with the ith coefficient $v_i \geq (k-1)|\Delta| + \log|\Delta|$. Assuming that the factor base contains all non-inert prime ideals of norm less than $L_\Delta[1/2, z]$, the prime number theorem implies that $k = |FB| = L_\Delta[1/2, z + o(1)]$. With this assumption on k, we expect that $L_\Delta[1/2, 1/(4z) + o(1)]$ random exponents will be required to find each of the k relations. Each trial requires time $L_\Delta[1/2, z + o(1)]$, because the computation of $\mathfrak{p}_i^{v_i}$ with $v_i = O(|\Delta|)$ requires polynomial time in $\log|\Delta|$ for each of the $L_\Delta[1/2, z + o(1)]$ prime ideals $\mathfrak{p}_i \in FB$, and factoring a reduced ideal over FB, even using trial division, can be done in time $L_\Delta[1/2, z + o(1)]$. Thus, we have the following.[24]

Proposition 13.7. *Assuming that* $k = L_\Delta[1/2, z + o(1)]$, *the running time required to generate a full-rank sublattice of* Λ *is bounded from above by* $L_\Delta[1/2, 2z + 1/(4z) + o(1)]$.

Once a full-rank sublattice of Λ has been found, additional relations are generated until we can say with constant probability that the collection of relations generates the full relation lattice Λ. Buchmann and Vollmer[25] show that every random relation lies outside of a proper sublattice of Λ with probability 2^{-17} and that $L_\Delta[1/2, z + 1/(4z) + o(1)]$ additional relations are required to generate all of Λ with constant probability. An analogue of Proposition 13.7 shows that these can be found in the same expected time as the initial full-rank sublattice, and, thus, we obtain the following.[26]

Proposition 13.8. *The expected running time for finding a set of relations that generates* Λ *is* $L_\Delta[1/2, 2z + 1/(4z) + o(1)]$.

Once we have a set of relations that we expect generates Λ, it remains to solve the linear system $B'\mathbf{y} = (1, 0, \ldots, 0)$. Using, for example, the algorithm of Giesbrecht, Jacobson, and Storjohann,[27] this can be done in time $L_\Delta[1/2, 3z + o(1)]$. Thus, the discrete logarithm problem can be solved in expected time $L_\Delta[1/2, \max(2z + 1/4z, 3z) + o(1)]$. As this is optimized when $z = 1/2$, we obtain the following theorem.

Theorem 13.9. *Given* $\mathcal{O}_\Delta$ *ideals* $\mathfrak{a}$ *and* $\mathfrak{g}$ *such that* $[\mathfrak{a}] \in \langle[\mathfrak{g}]\rangle$, *the imaginary quadratic order discrete logarithm problem can be solved in expected time* $L_\Delta[1/2, 3/2 + o(1)]$ *for* $|\Delta| > 157$ *assuming the GRH.*

The condition that $|\Delta| > 157$ is required to ensure that, under the GRH, the factor base generates the class group; this will be discussed in more detail below.

This algorithm is fine for scenarios in which the discrete logarithm is known to exist; for example, when attempting to cryptanalyze certain public-key cryptosystems based on class groups of imaginary quadratic orders. However, there are two issues that must be addressed if this algorithm is to be applied to an arbitrary instance of the imaginary quadratic order discrete logarithm problem.

1. In general, the value of x computed will not be minimal. For some applications that may be adequate; for example, when cryptanalyzing imaginary quadratic order based public-key cryptosystems.
2. This algorithm gives no way of certifying that an instance of the imaginary quadratic order discrete logarithm problem does not have a solution. If after generating n relations the resulting linear system does not have a solution, it is not possible (without extending the algorithm) to determine whether there is no solution or if the relations generated so far do not generate the full relation lattice Λ.

Both of these issues can be addressed by extending the algorithm to compute the class number. If the class number is known, then the computed value of x can be reduced modulo h_Δ to find the minimal solution. In order to certify that an instance of the DLP has no solution, it is necessary to verify that the relations generate the entire relation lattice Λ. We will describe how computing the class number from the relations can be used to accomplish this under the assumption of the GRH and ERH.

13.3 Computing the Class Number and Class Group

Recall that the relation lattice $\Lambda = \{\mathbf{v} \in \mathbb{Z}^k \mid FB^{\mathbf{v}} = (\gamma)\}$ is a sublattice of $\mathbb{Z}^k$. Let θ be the homomorphism

$$\theta : \mathbb{Z}^k \to Cl_\Delta$$
$$\mathbf{v} \mapsto [FB^{\mathbf{v}}].$$

If we assume that the prime ideals in FB generate the class group, then θ is surjective and Λ is its kernel, so $\mathbb{Z}^k/\Lambda \cong Cl_\Delta$ and $\det(\Lambda) = h_\Delta$. This was first observed by Pohst and Zassenhaus[28] in 1979 and presented in a more fully developed form[29] in 1985. The algorithm described below is based on this observation and was described in 1989 by Hafner and McCurley.[30]

In order to guarantee that FB generates the class group, we take all the prime ideals with norm less than a certain bound P. Unconditionally, we can use $P = \sqrt{|\Delta|}$, due to the fact that every ideal equivalence class contains a reduced ideal with norm less than $\sqrt{|\Delta|}$ (see §5.1), but, unfortunately, this bound is exponential in the bit length of Δ, resulting in an algorithm of exponential complexity at best. If we assume the GRH, then a theorem of Bach,[31] stated as (9.28) for maximal orders, implies that we can take $B = c \log^2 |\Delta|$

with $c = 6$ if Δ is the discriminant of a maximal order and $c = 12$ otherwise. Thus, in order to obtain an algorithm with subexponential complexity, we have to rely on Bach's result, meaning that the factor base generates Cl_Δ only assuming the GRH. Hence, the correctness of the class group obtained is also conditional on the GRH.

Assuming that the factor base FB generates Cl_Δ, we compute a set of random relations $L = \{v_1, \dots, v_n\}$ that we hope generates Λ, as described in the previous section. We know that the relations produced generate Λ with constant probability, but in order to find the class number, we need to verify this. In particular, we know that L generates Λ as soon as the determinant of the lattice generated by L is equal to h_Δ. We compute $\det(L)$ by finding a Hermite normal form (HNF) basis of the lattice generated by L. We form the relation matrix A whose columns consist of the random relations in L and, via elementary column operations, reduce A to its Hermite normal form $\text{HNF}(A) = [0 \mid H]$, where H is upper triangular and all elements in a row of H to the right of the diagonal are strictly less than the diagonal element of that row. Then $\det(L) = \det(H)$ can be computed easily. In addition, the columns of H form a basis for the sublattice generated by L, given in a canonical form. There are a number of algorithms for computing the HNF of an integer matrix; one of the most promising for this particular application is due to Giesbrecht, Jacobson, and Storjohann.[32]

In order to determine whether $\det(L) = h_\Delta$, we compute an approximation h^* of h_Δ such that $h^* < h_\Delta < 2h^*$, so the only integral multiple of h_Δ in the interval $(h^*, 2h^*)$ is h_Δ itself. If the lattice generated by L is a proper sublattice of Λ, then its determinant will be a multiple of h_Δ and lie outside this interval. Thus, as soon as $\det(L) < 2h^*$, we know that L generates Λ and $\det(L) = \det(\Lambda) = h_\Delta$. The approximation h^* can be computed by approximating $L(1, \chi)$ and using the analytic class number formula to approximate h_Δ as described in (10.19) and (10.20). Unfortunately, in order to be able to approximate $L(1, \chi)$ to sufficient accuracy in polynomial time, Bach's method relies on the ERH. Thus, the correctness of h^* is also conditional on ERH, as is the class number computed by our algorithm.

Once we have computed h_Δ, we can also compute the structure of Cl_Δ by computing the Smith normal form (SNF) of the HNF matrix H. As the columns of A form a generating system of Λ, the diagonal entries of $\text{SNF}(A)$ are precisely the elementary divisors of Cl_Δ. In practice, it is more efficient to compute $\text{SNF}(H)$ because it is already in upper-triangular form, and, as the columns of H are a basis of Λ, this gives the same result.

The following algorithm uses the approach described above to compute h_Δ and Cl_Δ.

Algorithm 13.2: Class Group of an Imaginary Quadratic Order

Input: discriminant $\Delta < 0$ of an imaginary quadratic order

Output: h_Δ and elementary divisors $m_1, \dots, m_l$ of Cl_Δ such that $Cl_\Delta = C(m_1) \times \cdots \times C(m_l)$.

1: Compute a factor base $FB = \{\mathfrak{p}_1, \ldots, \mathfrak{p}_k\}$ consisting of all non-inert prime ideals $\mathfrak{p}_i$ with $N(\mathfrak{p}_i) < P$ for some bound P.
2: Compute h^* such that $h^* < h_\Delta < 2h^*$ using (10.19) or (10.20).
3: Compute $n > k$ random relations $\mathbf{v}_1, \ldots, \mathbf{v}_n$.
4: Compute $[0 \mid H] = \text{HNF}(A)$ where $A = [\mathbf{v}_1^T \ldots \mathbf{v}_n^T]$ and compute $h = \det(H)$.
5: If $h = 0$ or $h > 2h^*$, increase n and go to step 3.
6: Compute $S = \text{SNF}(H)$.
7: Set $h_\Delta = h$ and $Cl_\Delta = C(m_1) \times \cdots \times C(m_l)$, where $m_1, \ldots, m_l$ are the diagonal elements of S that are greater than 1.

Example 13.10. We continue the earlier example to compute the structure of Cl_Δ for $\Delta = -32003$. We first need a bound on h_Δ in order to determine whether the set of relations generate Λ. Using Bach's method with $Q = 1427$, we get that $45 < h_\Delta < 90$ by (10.19). The relation matrix is

$$A = \begin{pmatrix} 1\ 5\ 4\ 6\ 3 \\ 8\ 8\ 5\ 5\ 8 \\ 3\ 9\ 3\ 6\ 6 \end{pmatrix}$$

and its HNF is

$$\text{HNF}(A) = \begin{pmatrix} 0\ 0\ 21\ 13\ 2 \\ 0\ 0\ 0\ \ 1\ 0 \\ 0\ 0\ 0\ \ 0\ 3 \end{pmatrix}.$$

The determinant is 63, and as $63 < 2h^* = 90$ we know that $h_\Delta = 63$. Finally, we compute the Smith normal form (SNF) of $\text{HNF}(A)$ and obtain

$$\text{SNF}(A) = \begin{pmatrix} 63\ 0\ 0 \\ 0\ \ 1\ 0 \\ 0\ \ 0\ 1 \end{pmatrix}.$$

Thus, Cl_Δ is cyclic of order 63 under the assumption of the ERH and the assumption that the factor base used generates Cl_Δ. Although Bach's bound does not imply that this factor base generates Cl_Δ, it does in fact suffice for this example; we will discuss how this can be proved in §13.7 when discussing practical improvements. Also, note that we can compute a minimal solution to the DLP instance from the previous example by reducing the non-minimal solution modulo 63, yielding $x = 2429 \bmod 63 = 35$.

Analysis

The analysis of this algorithm is almost the same as that for the imaginary quadratic order DLP described above. In particular, we assume that $k = |FB| = L_\Delta[1/2, z + o(1)]$ for some $z \leq 1$ and that the factor base FB generates Cl_Δ, which will be the case[33] if $|\Delta| > 157$. We use the same strategy to generate relations; by Proposition 13.8, this requires expected time

$L_\Delta[1/2, 2z + 1/(4z) + o(1)]$. It can be seen from (10.1), (10.2), and (10.19) that Bach's method computes the bound on h_Δ in time polynomial in $\log |\Delta|$. Finally, Storjohann[34] has shown that the HNF and SNF of the relation matrix can be computed in time $L_\Delta[1/2, 4z + o(1)]$ as long as the number of relations $n = O(k)$. Thus, the expected running time of the entire algorithm is $L_\Delta[1/2, \max(2z + 1/4z, 4z) + o(1)]$, and as this is optimized when $z = 1/\sqrt{8}$, we obtain the following theorem.[35]

Theorem 13.11. *The class number and class group of an imaginary quadratic order can be computed in expected time $L_\Delta[1/2, \sqrt{2} + o(1)]$ for $|\Delta| > 157$ assuming the GRH and ERH.*

By using this method to compute the class number, it is possible to certify that an arbitrary instance of the imaginary quadratic order DLP does not have a solution. Once the class number has been computed, we know that the relations generate the entire relation lattice Λ. Thus, by Proposition 13.2 we know that the given instance of the imaginary quadratic order DLP has a solution if and only if the linear system $B'\mathbf{y} = (1, 0, \ldots, 0)$ has a solution and we obtain the following theorem.

Theorem 13.12. *The imaginary quadratic order DLP can be solved in expected time $L_\Delta[1/2, \sqrt{2} + o(1)]$ for $|\Delta| > 157$ assuming the GRH and ERH.*

Recall that, according to Theorem 13.9, we can compute discrete logarithms in expected time $L_\Delta[1/2, 3/2 + o(1)]$ if a solution is known to exist. In practice, this algorithm, which does not certify whether a solution exists, would be expected to be faster than that described above, as the relation generation stages would be essentially the same, but solving a linear system is faster than computing a HNF. While it may seem counterintuitive that this algorithm would have a worse asymptotic complexity, this should be seen as a quirk of the asymptotic nature of these estimates. Note that the function $2z + 1/4z$ has a minimum value of $\sqrt{2}$ at $z = 1/\sqrt{8}$. When optimizing the entire algorithm, this minimal value is only achieved when using a HNF algorithm of complexity $L_\Delta[1/2, 4z + o(1)]$. Any other complexity for the linear algebra, such as that used for the version that only solves the DLP, will result in a larger value than $\sqrt{2}$. In order to break the $\sqrt{2}$ barrier, it is necessary to use a faster linear algebra algorithm *and* to improve the $L_\Delta[1/2, 2z + 1/4z + o(1)]$ running time for finding relations.

In fact, it should be possible[36] to improve the expected running time for solving the DLP and computing the class group to $L_\Delta[1/2, 1.013 + o(1)]$. The idea comes from the fact that the set of prime ideals required to generate the class group given by Bach's bound is significantly smaller (polynomial in $\log |\Delta|$) than the subexponential-sized factor base required to obtain the complexities given above. We can make use of this fact to find relations in expected time $L_\Delta[1/2, 1/(4z) + o(1)]$ by selecting the random vectors $\mathbf{v}$ with only $O(\log^2 |\Delta|)$ non-zero terms and factoring the reduced ideal norms with an asymptotically fast algorithm such as the elliptic curve method. If it can be shown that the same number of such "sparse" relations is required to generate the full relation lattice as the "dense" relations used in the previous analysis, then these relations could be found[37] in expected time $L_\Delta[1/2, z + 1/(4z) + o(1)]$. Note that the function $z + 1/4z$ has a minimum value of 1 at $z = 1/2$, so the overall complexity could be improved to $L_\Delta[1/2, 1 + o(1)]$ with the discovery of an algorithm for computing the HNF of the relation matrix with complexity $L_\Delta[1/2, 2 + o(1)]$. The fastest known algorithm in terms of asymptotic complexity, due to Storjohann and Labahn,[38] computes the HNF of the relation matrix in time $L_\Delta[1/2, 2.38z + o(1)]$. Putting these observations together yields an optimal value of $z = 1/\sqrt{5.52} \approx 0.4257$ and an overall complexity of $L_\Delta[1/2, 2.38/\sqrt{5.52} + o(1)] = L_\Delta[1/2, 1.013 + o(1)]$. Unfortunately, to the best of our knowledge, the required result on the number of sparse relations needed to generate the entire relation lattice has not been proved, so this running time remains conjectural for now.

Once the class group structure has been computed, it is sometimes of interest to compute a set of generators, namely reduced ideals $\mathfrak{g}_1, \ldots, \mathfrak{g}_l$ that generate the disjoint subgroups isomorphic to $C(m_1), \ldots, C(m_l)$. More generally, as observed by Buchmann and Düllmann,[39] if we have an explicit version of the isomorphism $\phi : \mathbb{Z}^k \mapsto Cl_\Delta$, allowing us to represent any ideal equivalence class $\mathfrak{a}$ that factors over FB in the group $C(m_1) \times \cdots \times C(m_l)$, then we can use it to solve instances of the DLP by mapping the input ideals to this representation of the class group. One advantage of this approach is that, given representations of the input ideals over $C(m_1) \times \cdots \times C(m_l)$, the DLP can be solved easily using the extended Euclidean algorithm to find solutions modulo each of the m_i followed by a generalized version of the Chinese Remainder Theorem to compute the solution modulo $m_1 \times \cdots \times m_l$. Note that a generalized version of the Chinese Remainder Theorem is required because the m_i are not relatively prime.[40] A second advantage is that once the class group is computed, individual instances of the DLP can be solved quite easily, the most expensive part being factoring the input ideals over the factor base, essentially equivalent to finding two relations.

Suppose that H is a HNF basis of the relation lattice Λ. Then, as argued above, S, the SNF of H, has the elementary divisors of Cl_Δ as its diagonal elements. During the computation of S, we can also compute the corresponding unimodular transformation matrices $U, V \in GL_k(\mathbb{Z})$ such that

$$S = UHV .$$

If $U^{-1} = (u'_{ij})_{k \times k}$, then

$$\mathfrak{g}_i \sim \prod_{j=1}^{k} \mathfrak{p}_j^{u'_{ji}}, \quad 1 \leq i \leq k ,$$

form a system of generators of Cl_Δ if we ignore those $g_i \sim \mathcal{O}_\Delta$ corresponding to the trivial elementary divisors $m_i = 1$. Conversely, if $U = (u_{ij})_{k \times k}$, then for each factor base element $\mathfrak{p}_j$ we have

$$\mathfrak{p}_j \sim \prod_{i=1}^{k} \mathfrak{g}_i^{u_{ij}} \, .$$

Thus, if $\mathfrak{a} \sim FB^{\mathsf{v}}$, it can be represented over the system of generators by

$$\mathfrak{a} \sim \prod_{i=1}^{k} \mathfrak{g}_i^{(\sum_{j=1}^{k} v_j u_{ij})} \, , \tag{13.1}$$

and the equivalence class $[\mathfrak{a}]$ maps to

$$\left(\sum_{j=1}^{k} v_j u_{ij}, \ldots, \sum_{j=1}^{k} v_j u_{lj} \right) \in C(m_1) \times \cdots \times C(m_l) \, .$$

In practice, we avoid computing the entire transformation matrix $U \in \mathbb{Z}^k$. Instead, we work with the essential part of H, the entries remaining after removing all rows and columns corresponding to diagonal elements equal to 1. This matrix will have much smaller dimensions than H, because the diagonal entries not equal to 1 correspond roughly to the number of non-cyclic components of Cl_Δ which, by the Cohen-Lenstra heuristics presented in Chapter 7, is expected to be small. It is possible[41] to convert a factorization of an ideal $\mathfrak{a}$ over the entire factor base to one over only those prime ideals corresponding to entries in the essential part of H, allowing the method described above to work using the essential part of H in place of H.

It is important to emphasize the conditional nature of the output and running time of the index-calculus algorithm for computing Cl_Δ. In particular, versions of the Riemann hypothesis are required in two places. We require the ERH on L-functions in order to guarantee the fact that the value h^* computed using Bach's algorithm for approximating $L(1,\chi)$ satisfies $h^* < h_\Delta < 2h^*$. As this value is used to determine when the full relation lattice has been generated, the correctness of the computed value of h_Δ is conditional on the ERH. If the ERH is false and the algorithm terminates incorrectly, then the best we can say is that the class number computed is a multiple of the actual class number, as in that case it will be the determinant of a proper sublattice of Λ.

The GRH on Hecke L-functions is required to ensure that the prime ideals contained in the factor base generate the class group. This assumption is essential in proving the subexponential complexities, as without it, the best available bound is exponential in $\log |\Delta|$. It is also required for the correctness of the output, because if the factor base does not generate the entire class group, the best we can hope to find is the size and structure of a subgroup.

Thus, both the running time and the correctness of the subexponential algorithm for computing the class group of an imaginary quadratic order depend on the assumptions of the GRH and ERH. It remains an open problem to find an unconditionally correct algorithm with subexponential running time.

Note that although the GRH and ERH are required for the running time of the algorithm for solving the imaginary quadratic order DLP, these are only required to certify correctness when a problem instance does not have a solution, as any solution x produced can be verified unconditionally by checking that $\mathfrak{g}^x \sim \mathfrak{a}$.

13.4 Computing the Regulator

Not long after Hafner and McCurley's algorithm for computing the class group of an imaginary quadratic field was announced, it was generalized to compute class groups and regulators of real quadratic fields, and in fact arbitrary algebraic number fields, by Buchmann.[42] This Las Vegas algorithm runs in expected time $L_\Delta[1/2, 1.7+o(1)]$ for a real quadratic field of discriminant Δ, and as with the algorithm for imaginary quadratic orders described above, both the complexity and correctness of the output are conditional on the GRH and ERH. Abel[43] subsequently showed that the same algorithm could be used for real quadratic orders and improved the complexity to $L_\Delta[1/2, 5\sqrt{3}/6 + o(1)]$.

In this section, we will describe two algorithms for computing R_Δ due to Vollmer[44] that currently have the best known complexity. The first is a Las Vegas algorithm which computes R_Δ and Cl_Δ in expected time $L_\Delta[1/2, \sqrt{2} + o(1)]$, where both the running time and the correctness of the output are conditional on the GRH and ERH. The second is a Monte Carlo algorithm which computes R_Δ in time $L_\Delta[1/2, 3\sqrt{2}/4 + o(1)]$. The complexity result is conditional on the GRH, and the output is correct with probability greater than some fixed value p selected at runtime. Once again, the analysis of these algorithms is presented in detail by Buchmann and Vollmer elsewhere,[45] so we will only give a sketch of these results here.

Although the algorithms described above also work for real quadratic orders in principle, there are two main issues that need to be addressed. The first is the problem of finding random relations. The complication is that in the real case there are multiple reduced ideals in any particular ideal equivalence class. The method described above finds a random ideal equivalence class, but the reduced ideal produced also needs to be randomly selected from that class. We will describe a strategy that accomplishes this below.

The second issue is determining when the relations found generate the entire relation lattice. Recall that by approximating $L(1, \chi)$, we can obtain an estimate of the product of the class number and regulator. The method described above only produces a multiple of the class number, so it must be extended to produce an approximation of the regulator as well.

Buchmann's idea, as specialized to the real quadratic case, was to compute relations of the form $(\mathbf{v}, \log|\gamma|)$, where $FB^{\mathbf{v}} = (\gamma)$; that is, γ generates the principal ideal $FB^{\mathbf{v}}$. It can be verified that integer linear combinations of such extended relations are themselves extended relations, so the set of all extended relations forms a sublattice of $\mathbb{Z}^k \times \mathbb{R}$. We will refer to

$$\Lambda' = \{(\mathbf{v}, \log|\gamma|) \in \mathbb{Z}^k \times \mathbb{R} \mid FB^{\mathbf{v}} = (\gamma)\} \subset \mathbb{Z}^k \times \mathbb{R}$$

as the *extended relation lattice*.

The key to Buchmann's algorithm is the following result.[46]

Proposition 13.13. *The set Λ' is a $(k+1)$-dimensional lattice. If the ideal classes of the elements of FB generate the class group, then $\det(\Lambda') = h_\Delta R_\Delta$.*

To see this, let Λ be the part of Λ' in $\mathbb{Z}^k$. As before, we have that Λ is k dimensional and $Cl_\Delta \cong \mathbb{Z}^k/\Lambda$ as long as FB generates Cl_Δ, so $\det(\Lambda) = h_\Delta$. Let $\{\mathbf{b}_1, \ldots, \mathbf{b}_k\}$ be a basis of Λ. Then Buchmann shows that

$$\{(\mathbf{b}_1, \log|\gamma_1|), \ldots, (\mathbf{b}_k, \log|\gamma_k|), (0, \ldots, 0, R_\Delta)\},$$

where

$$\prod_{i=1}^{k} \mathfrak{p}_i^{b_{i,j}} = (\gamma_j)$$

for $1 \le j \le k$, is a basis of Λ'. It can be verified that this set of extended relations is linearly independent, that any linear combination of them is in Λ', and that an arbitrary element of Λ' can be expressed as a linear combination of these extended relations.

Similar to Hafner and McCurley's algorithm, we produce a generating system

$$L' = \{(\mathbf{v}_1, \log|\gamma_1|), (\mathbf{v}_2, \log|\gamma_2|), \ldots, (\mathbf{v}_n, \log|\gamma_n|)\}$$

of random relations that we hope will generate all of the extended relation lattices Λ'. In order to determine whether L' generates Λ', we use the idea described in Chapter 10, namely (10.20), to compute an approximation h^* such that $h^* < h_\Delta R_\Delta < 2h^*$ and check whether $\det(L') < 2h^*$. As soon as this condition holds, the SNF of the relation matrix $A = (\mathbf{v}_1, \ldots, \mathbf{v}_n)$ yields Cl_Δ, $\det(A) = h_\Delta$, and $R_\Delta = \det(L')/h_\Delta$.

To compute $\det(L')$, we first compute the HNF of A, giving us $\det(L)$. We also need to compute the multiple of R_Δ corresponding to the real part of L'. Let $\mathbf{r} = (\log|\gamma_1|, \log|\gamma_2|, \ldots, \log|\gamma_n|)$. If $\mathbf{x} \in \mathbb{Z}^n$ is such that $A\mathbf{x} = \mathbf{0}$, then $\mathbf{r} \cdot \mathbf{x}$ is some integer multiple of R_Δ. To see this, note that $\mathbf{x}$ corresponds to a linear combination of the columns of A that yields the extended relation $(\mathbf{0}, \log|\gamma|)$ with

$$\gamma = \prod_{i=1}^{n} \gamma_i^{x_i}.$$

Because $FB^{\mathbf{0}} = \mathcal{O}_\Delta$, γ must be a unit and

$$\mathbf{r} \cdot \mathbf{x} = x_1 \log|\gamma_1| + \cdots + x_n \log|\gamma_n| = \log|\gamma|$$

is a multiple of R_Δ. If we compute a basis $\{\mathbf{x}_1, \ldots, \mathbf{x}_l\}$ of the null space of A, then we have $\det(L') = \det(L)R'$, where $R' = \mathrm{rgcd}(\mathbf{r} \cdot \mathbf{x}_1, \ldots, \mathbf{r} \cdot \mathbf{x}_l) = mR_\Delta$.

The function rgcd here denotes a "real" gcd of integer multiples of the same real number (in our case R_Δ). In particular, we define

$$\text{rgcd}(xR, yR) = \gcd(x, y)R ,$$

where $x, y \in \mathbb{Z}$ and $R \in \mathbb{R}$.

One method to compute the rgcd of two multiples of R_Δ, as described by Cohen, Diaz y Diaz, and Olivier,[47] is to use an analogue of the Euclidean algorithm for integers, terminating when the "remainder" is less than $\log((1 + \sqrt{5})/2)$, the minimum possible value of R_Δ. Maurer[48] described another method that computes $\text{rgcd}(l_1, l_2)$ with $l_1 = x_1 R_\Delta$ and $l_2 = x_2 R_\Delta$, $x_1, x_2 \in \mathbb{Z}$, by computing the continued fraction expansion of $l_1/l_2 = x_1 R_\Delta/(x_2 R_\Delta) = x_1/x_2$. The result is integers y_1 and y_2 such that $y_1 x_1 + y_2 x_2 = \gcd(x_1, x_2)$, allowing one to compute $y_1 l_1 + y_2 l_2 = (y_1 x_1 + y_2 x_2)R_\Delta = \gcd(x_1, x_2)R_\Delta$ as required. Maurer's method works well in practice and has the important advantage that the required precision to ensure the numerical accuracy of the output can be determined given explicit representations of $\gamma_1, \ldots, \gamma_n$ and the kernel vectors associated to the regulator multiples.

The following algorithm uses the approach described above to compute R_Δ, h_Δ, and Cl_Δ of a real quadratic order $\mathcal{O}_\Delta$.

Algorithm 13.3: Regulator and Class Group of a Real Quadratic Order

---

Input: discriminant $\Delta > 0$ of a real quadratic order
Output: R_Δ, h_Δ, and elementary divisors $m_1, \ldots, m_l$ of Cl_Δ such that $Cl_\Delta = C(m_1) \times \cdots \times C(m_l)$.

1: Compute a factor base $FB = \{\mathfrak{p}_1, \ldots, \mathfrak{p}_k\}$ consisting of all non-inert prime ideals $\mathfrak{p}_i$ with $N(\mathfrak{p}_i) < P$ for some bound P.
2: Compute h^* such that $h^* < h_\Delta R_\Delta < 2h^*$.
3: Compute $n > k$ random extended relations $(\mathbf{v}_1, \log|\gamma_1|), \ldots, (\mathbf{v}_n, \log|\gamma_n|)$.
4: Compute $[0 \mid H] = \text{HNF}(A)$ where $A = [\mathbf{v}_1^T \ldots \mathbf{v}_n^T]$ and compute $h = \det(H)$.
5: Compute a basis $\{\mathbf{x}_1, \ldots, \mathbf{x}_{n-k}\}$ of $\ker(A)$.
6: Compute $R = \text{rgcd}(\mathbf{r} \cdot \mathbf{x}_1, \ldots, \mathbf{r} \cdot \mathbf{x}_{n-k})$ where $\mathbf{r} = (\log|\gamma_1|, \ldots, \log|\gamma_n|)$.
7: If $h = 0$ or $hR > 2h^*$, increase n and go to step 3.
8: Compute $S = \text{SNF}(H)$.
9: Set $R_\Delta = R$, $h_\Delta = h$ and $Cl_\Delta = C(m_1) \times \cdots \times C(m_l)$, where $m_1, \ldots, m_l$ are the diagonal elements of S that are greater than 1.

If desired, a compact representation of the fundamental unit can be computed from R_Δ using the methods from Chapter 12.

The "random exponents" method of Hafner and McCurley can be generalized[49] to randomly sample extended relations from Λ'. We first select uniformly at random $\mathbf{v} \in \mathbb{Z}^k$ with each entry $v_i \in \{0, 1, \ldots, |\Delta| - 1\}$ and compute a reduced ideal $\mathfrak{b} = (\alpha)FB^{\mathbf{v}}$. In order to ensure that we randomly sample from all of Λ', we randomly select a reduced ideal $\mathfrak{a}$ equivalent to $\mathfrak{b}$ by using infrastructure techniques to find a reduced ideal with distance $\delta(\mathfrak{b}, \mathfrak{a}) \approx \delta$ for a random integer $\delta \in \{0, \ldots, B_d\}$ for some bound $B_d > R_\Delta$. This yields $\mathfrak{a} = (\gamma)FB^{\mathbf{v}}$ for $\gamma = \alpha\beta \in \mathbb{Q}(\sqrt{\Delta})$. If $\mathfrak{a}$ can be factored over the factor base as $(d)FB^{\mathbf{w}}$, where d denotes the denominator of the fractional ideal $FB^{\mathbf{w}}$, then we have

$$\mathfrak{a} = (d)FB^{\mathbf{w}} = (\gamma)FB^{\mathbf{v}} \implies FB^{\mathbf{w}-\mathbf{v}} = (\gamma/d) ,$$

and $(\mathbf{w} - \mathbf{v}, \log|\gamma/d|)$ is an extended relation.

Note that the real part of these relations can be computed using floating point approximations, but in practice it is better to store the generators γ explicitly, using a standard representation or a compact representation as described in Chapter 12, if necessary. Having explicit representations allows Maurer's method mentioned above to be used to compute a multiple of R_Δ with sufficient precision to ensure the numerical accuracy of the result.

Example 13.14. We compute the class group and regulator of $\mathcal{O}_\Delta$ with $\Delta = 12301$.

1. We set the factor base to be the smallest three prime ideals in terms of norm, yielding

$$FB = \{(6, 109), (10, 101), (14, 101)\} .$$

2. We need a bound on $h_\Delta R_\Delta$ in order to determine whether the set of relations generate Λ. Using Bach's method with $Q = 1289$ and (10.20), we get that $61 < h_\Delta R_\Delta < 122$.

3. We compute five random relations, hoping that having two more relations than factor base elements (two more columns than rows in the relation matrix) will be enough to generate Λ'.

   a) The random vector $\mathbf{v} = (5, 0, 1)$ gives us

   $$FB^{\mathbf{v}} = (3402, -1103) .$$

   Reducing yields $\mathfrak{a} = (70, 101) = FB^{\mathbf{w}}$, with $\mathbf{w} = (0, 1, 1)$. The reduction algorithm yields $FB^{\mathbf{v}}(\gamma_1) = FB^{\mathbf{w}}$, with

   $$\gamma_1 = \frac{131 + \sqrt{12301}}{486} ,$$

   so $(\mathbf{w} - \mathbf{v}, \log|\gamma_1|) = (-5, 1, 0, -0.6976433914)$ is an extended relation.

   b) The random vector $\mathbf{v} = (5, 2, 0)$ gives us

   $$FB^{\mathbf{v}} = (12150, 5701) .$$

   Reducing yields $\mathfrak{b} = (10, 109) = (5)FB^{\mathbf{w}}$, with $\mathbf{w} = (0, -1, 0)$. The reduction algorithm yields $FB^{\mathbf{v}}(\gamma_2) = FB^{\mathbf{w}}$, with

   $$\gamma_2 = \frac{2527 + 23\sqrt{12301}}{60750} ,$$

   so $(\mathbf{w} - \mathbf{v}, \log|\gamma_2|) = (-5, -3, 0, -2.481863958)$ is an extended relation.

c) The random vector $\mathbf{v} = (9, 4, 0)$ gives us

$$FB^{\mathbf{v}} = (24603750, 4719901) \,.$$

Reducing yields $\mathfrak{b} = (14, 109) = (7)FB^{\mathbf{w}}$, with $\mathbf{w} = (0, 0, -1)$. The reduction algorithm yields $FB^{\mathbf{v}}(\gamma_3) = FB^{\mathbf{w}}$, with

$$\gamma_3 = \frac{102293 + 907\sqrt{12301}}{172226250} \,,$$

so $(\mathbf{w} - \mathbf{v}, \log|\gamma_3|) = (-9, -4, -1, -6.743908938)$ is an extended relation.

d) The random vector $\mathbf{v} = (2, 4, 0)$ gives us

$$FB^{\mathbf{v}} = (11250, -5099) \,.$$

Reducing yields $(182, 94)$, which unfortunately does not factor over the factor base. However, two applications of the reduction operator yields $\mathfrak{b} = (126, 95) = (3^2 7)FB^{\mathbf{w}}$, with $\mathbf{w} = (-2, 0, -1)$. We have, again from the reduction algorithm, $FB^{\mathbf{v}}(\gamma_4) = FB^{\mathbf{w}}$, with

$$\gamma_4 = \frac{93061 + 839\sqrt{12301}}{708750} \,,$$

so $(\mathbf{w} - \mathbf{v}, \log|\gamma_4|) = (-4, -4, -1, -1.337141404)$ is an extended relation.

e) The random vector $\mathbf{v} = (9, 9, 0)$ gives us

$$FB^{\mathbf{v}} = (76886718750, -5531123849) \,.$$

Reducing yields $\mathfrak{b} = (150, 101) = (3)FB^{\mathbf{w}}$, with $\mathbf{w} = (-1, 2, 0)$. The reduction algorithm yields $FB^{\mathbf{v}}(\gamma_5) = FB^{\mathbf{w}}$, with

$$\gamma_5 = \frac{153187 + 1238\sqrt{12301}}{4613203125} \,,$$

so $(\mathbf{w} - \mathbf{v}, \log|\gamma_5|) = (-10, -7, 0, -9.672852058)$ is an extended relation.

4. The relation matrix corresponding to these extended relations is

$$A = \begin{pmatrix} -5 & -5 & -9 & -4 & -10 \\ 1 & -3 & -4 & -4 & -7 \\ 0 & 0 & -1 & -1 & 0 \end{pmatrix}$$

and we set

$$\begin{aligned} \mathbf{r} &= (\log|\gamma_1|, \log|\gamma_2|, \dots, \log|\gamma_5|) \\ &= (-0.6976433914, -2.481863958, -6.743908938, \\ &\quad -1.337141404, -9.672852058) \,. \end{aligned}$$

5. The HNF of the relation matrix A is

$$\mathrm{HNF}(A) = \begin{pmatrix} 0\,0\,5\,0\,4 \\ 0\,0\,0\,1\,0 \\ 0\,0\,0\,0\,1 \end{pmatrix}.$$

The determinant is 5, so we know that $h = 5$ is a multiple of the class number.

6. A basis for the kernel of A is

$$\ker(A) = (\mathbf{x}_1 \mathbf{x}_2) = \begin{pmatrix} -3 & 7 \\ -1 & 0 \\ 4 & -9 \\ -4 & 9 \\ 0 & 1 \end{pmatrix}.$$

To find multiples of R_Δ, we compute the dot product of $\mathbf{r}$ with each of the two kernel vectors. The first kernel vector yields

$$R'_1 = \mathbf{r} \cdot \mathbf{x}_1 = -17.05227601$$

and the second yields

$$R'_2 = \mathbf{r} \cdot \mathbf{x}_2 = 34.10455201\,.$$

It can be seen by inspection that the "real gcd" of R'_1 and R'_2 is

$$R = 17.05227601\,,$$

and this yields a multiple of the regulator.

7. Finally, we have that $hR = 85.26138005 < 121$, so we conclude that $h_\Delta = 5$ and $R_\Delta = 17.05227601$. Computing the SNF of $\mathrm{HNF}(A)$ yields

$$\mathrm{SNF}(A) = \begin{pmatrix} 5\,0\,0 \\ 0\,1\,0 \\ 0\,0\,1 \end{pmatrix},$$

so Cl_Δ is cyclic of order 5.

Note that these results hold under the assumption of the ERH and the assumption that the factor base used generates Cl_Δ. Although Bach's bound does not imply that this factor base generates Cl_Δ, it does in fact suffice for this example; we will discuss how this can be proved in §13.7 when discussing practical improvements.

Analysis

As in the imaginary case, the complexity of this algorithm depends on the probability that a random ideal is smooth. Using Proposition 13.4, one can show that the strategy described above for computing extended relations succeeds with the same probability as in the imaginary case.[50]

Lemma 13.15. *Suppose that the factor base contains all prime ideals with norm less than $L_\Delta[1/2, z]$ for some constant z. Then the probability that the relation generation strategy described above successfully finds a relation is bounded from below by $L_\Delta[1/2, -1/(4z) - o(1)]$.*

Also, as in the imaginary case, we generate the first k relations in such a way that the relation matrix formed by the integer vector parts of the extended relations is strictly diagonally dominant, and hence non-singular. By using compact representations as described in Chapter 12 to compute and represent the generators of each relation, it can be shown[51] that each trial of the relation generation strategy requires time $L_\Delta[1/2, z+o(1)]$. By applying Lemma 13.15, we can bound the running time for finding a full-rank relation matrix.[52]

Proposition 13.16. *Assuming that $k = O(L_\Delta[1/2, z])$, the running time required to generate a full-rank sublattice of Λ' is bounded from above by $L_\Delta[1/2, 2z + 1/(4z) + o(1)]$.*

Once we have a full-rank sublattice of Λ', it remains to estimate how many additional extended relations are required to conclude that, with constant probability, we have a complete generating system of Λ'. Buchmann and Vollmer[53] show that every random extended relation found is outside of a proper sublattice of Λ' with probability $\gg 1/(1 + \log \Delta)$. Arguing in a similar fashion to the imaginary case, one can show that additional $L_\Delta[1/2, z + 1/(4z) + o(1)]$ relations are required to generate Λ' with constant probability and we obtain the following.[54]

Proposition 13.17. *The expected running time for finding a set of extended relations that generates Λ' is in $L_\Delta[1/2, 2z + 1/(4z) + o(1)]$.*

The next step is to compute a multiple of the regulator by computing the real gcd of a set of regulator multiples, each of which corresponds to a basis vector of the kernel of the relation matrix. Another strategy due to Vollmer,[55] that results in a better asymptotic running time is as follows. Compute two additional extended relations $(\mathbf{v}_{n+1}, \log |\gamma_{n+1}|)$ and $(\mathbf{v}_{n+2}, \log |\gamma_{n+2}|)$. Assuming that the first n extended relations generate Λ', there exist vectors $\mathbf{x}_1, \mathbf{x}_2 \in \mathbb{Z}^n$ such that

$$A\mathbf{x}_i = \mathbf{v}_{n+i}, \quad i = 1, 2, \tag{13.2}$$

where, as before, $A \in \mathbb{Z}^{k \times n}$ is the matrix whose columns consist of the integer vector parts of the extended relations. Equation (13.2) implies that

$$\prod_{j=1}^{n}(\gamma_i^{x_{ij}}) = (\gamma_{n+i})$$

and

$$\epsilon_i = \left(\prod_{j=1}^{n}(\gamma_i^{x_{ij}})\right) / \gamma_{n+i}$$

are units for $i = 1, 2$. Thus,

$$R_i = \log |\gamma_{n+i}| - \mathbf{r} \cdot \mathbf{x_i}, \quad i = 1, 2 \,,$$

are both integer multiples of R_Δ. The observation that the gcd of two random integers is likely to be 1 suggests that $\mathrm{rgcd}(R_1, R_2) = R_\Delta$ with non-negligible probability, and in fact, Vollmer[56] proved that this holds with probability greater than $1/2$. As the linear system (13.2) can be solved deterministically[57] in time $L_\Delta[1/2, 3z + o(1)]$ and Lemma 13.15 implies that the expected time to find a single relation is $L_\Delta[1/2, z + 1/(4z) + o(1)]$, we obtain the following.[58]

Proposition 13.18. *The expected time required to find the regulator by the method described above is $L_\Delta[1/2, \max(z + 1/(4z), 3z) + o(1)]$.*

The proof of this result also requires the use of Maurer's algorithm for computing $\mathrm{rgcd}(R_1, R_2)$, which, according to Buchmann and Vollmer,[59] requires time $L_\Delta[1/2, 2z + o(1)]$ in this context.

The analysis of the complete algorithm is similar to that of computing the class group of an imaginary quadratic order. We assume that $k = |FB| = O(L_\Delta[1/2, z])$ for some $z \leq 1$ and that the factor base FB generates Cl_Δ, which will be the case[60] if $\Delta > 41$. By Proposition 13.17, finding a complete generating set of Λ' requires expected time $L_\Delta[1/2, 2z + 1/(4z) + o(1)]$. It can be seen from (10.1) that Bach's method computes the bound on $h_\Delta R_\Delta$ in time polynomial in $\log \Delta$ as long as $Q = O(\Delta^\epsilon)$, and (10.2) and (10.20) imply that Q can be chosen of this size. The expected time for computing the regulator is $L_\Delta[1/2, \max(z + 1/(4z), 3z) + o(1)]$ by Proposition 13.18, and by using Storjohann's algorithms,[61] the HNF and SNF of the relation matrix can be computed in time $L_\Delta[1/2, 4z + o(1)]$. Thus, the expected running time of the entire algorithm is in $L_\Delta[1/2, \max(2z + 1/4z, 4z) + o(1)]$, and as this is optimized when $z = 1/\sqrt{8}$, we obtain the following theorem.[62]

Theorem 13.19. *The regulator, class number, and class group of a real quadratic order $\mathcal{O}_\Delta$ can be computed in expected time $L_\Delta[1/2, \sqrt{2} + o(1)])$ for $\Delta > 41$ assuming the GRH and ERH.*

As in the imaginary case, both the running time and the correctness of this algorithm depend on the assumptions of the GRH and ERH. The GRH is required to ensure that the factor base generates Cl_Δ even when it is chosen to have subexponential size in $\log \Delta$, and the ERH is required to determine that the set of extended relations produced generates all of Λ'. Without the assumption of the GRH, we can only guarantee that some subgroup of Cl_Δ is produced. Without the assumption of the ERH, the best we can say is that the class number and regulator produced are multiples of their actual values.

It should be possible to improve the expected running time for computing the class group and regulator to $L_\Delta[1/2, 1.013 + o(1)]$ using the same ideas as in the imaginary case. Here, we also use the fact that Storjohann and Labahn's HNF algorithm computes a unimodular transformation matrix U

such that $AU = \mathrm{HNF}(A) = [0 \mid H]$ in the same time as computing $\mathrm{HNF}(A)$ alone. The first $\dim(\ker(A))$ columns of U form a basis of the kernel of A, which can be used to compute a multiple of R_Δ. Similar to the imaginary case, this complexity only holds under the assumption that $O(L_\Delta[1/2, z])$ "sparse" relations suffice to generate the entire extended relation lattice.

If one is only interested in computing the regulator, then it is possible to obtain a faster Monte Carlo algorithm that computes an unconditionally correct approximation of R_Δ with high probability. The idea is to use the approach of producing integer multiples of R_Δ outlined above. One first generates a full-rank sublattice of Λ' as described above, followed by a sequence of additional random extended relations. Multiples of R_Δ corresponding to two of these extra extended relations are computed, also as described above, and their real gcd is computed using Maurer's algorithm. Repeating this process for different pairs of random extended relations yields independent multiples of R_Δ, and after each application we can keep the smallest of the regulator multiples found. As each regulator multiple produced is equal to R_Δ with constant probability, repeating this process a sufficient number of times ensures that the regulator multiple obtained is equal to R_Δ with any probability $p < 1$ we wish.

The key observation in deriving an improved running time for this method is that it is not necessary to ensure that the extended relations generate Λ'; it is enough for the sublattice they generate to have full rank, in order for the linear systems (13.2) to have solutions. Thus, the method for finding relations can be adjusted by selecting the initial random exponent vectors in such a way that, except for the ith entry that is selected to be large in order to ensure strict diagonal dominance, only the first k' entries are non-zero, where $FB' = \{\mathfrak{p}_1, \ldots, \mathfrak{p}_{k'}\}$ contains all the non-inert prime ideals with norm bounded by $12 \log^2 \Delta$. We cannot prove that relations produced in this manner are randomly sampled from the set of all relations, but as argued by Vollmer,[63] this assumption is not necessary to ensure that the regulator multiplies obtained are random.

The advantage of this method is that the expected time required to find a relation is reduced from $L_\Delta[1/2, z + 1/(4z) + o(1)]$ to $L_\Delta[1/2, 1/(4z) + o(1)]$, provided that an asymptotically fast factoring algorithm such as the elliptic curve method is used to factor norms of ideals when testing for smoothness. Therefore, the required $L_\Delta[1/2, z + o(1)]$ extended relations can be found in expected time $L_\Delta[1/2, z + 1/(4z) + o(1)]$, and following the analysis of the regulator algorithm described above, we see that the expected time to compute a multiple of R_Δ is $L_\Delta[1/2, \max(z + 1/(4z), 3z) + o(1)]$. As this is again minimized for $z = 1/\sqrt{8}$, we obtain the following theorem.[64]

Theorem 13.20. *Given p with $0 < p < 1$, the regulator of a real quadratic order $\mathcal{O}_\Delta$ can be computed in expected time $L_\Delta[1/2, 3\sqrt{2}/4 + o(1)]$ under the GRH. The output is unconditionally correct with probability p.*

Thus, we can remove the GRH and ERH assumptions for the correctness of the regulator and obtain a faster algorithm, but at the price of having a Monte Carlo algorithm which only guarantees correctness up to a given probability. It is unknown whether it is possible to certify that the regulator is correct in subexponential time, without assuming the GRH and ERH and using the previous algorithm to compute h_Δ as well.

13.5 Principality Testing

Once we have verified that L' generates Λ', we can also solve the principal ideal problem.

Definition 13.21. *Given an $\mathcal{O}_\Delta$-ideal $\mathfrak{a}$, the* principal ideal problem **P** *is to determine whether $\mathfrak{a}$ is principal and, if so, compute an approximation of $\log |\alpha|$ where $\mathfrak{a} = (\alpha)$.*

The problem of computing an approximation of $\log |\alpha|$ is also known as the *infrastructure discrete logarithm problem*, in analogy to the DLP in a cyclic group.

The algorithm proceeds as follows. Given a set of extended relations L' that generate Λ', we know that every principal ideal of $\mathcal{O}_\Delta$ can be represented by a vector $\mathbf{v} \in \mathbb{Z}^k$, where $\mathbf{v}$ is a linear combination of the vectors in L'. Thus, in order to determine whether an ideal $\mathfrak{a}$ is principal, we search for an equivalent ideal that is smooth over the factor base, yielding $\mathfrak{a} = (\gamma)FB^{\mathbf{v}}$ for some $\gamma \in \mathbb{Q}(\sqrt{\Delta})$. We then test whether there exists a solution $\mathbf{x} \in \mathbb{Z}^n$ of $A\mathbf{x} = \mathbf{v}$, where A is the relation matrix as defined above. If not, then $\mathfrak{a}$ is not principal. Otherwise, we have $\mathfrak{a} = (\alpha)$ with

$$\alpha = \gamma \prod_{i=1}^{n} \gamma_i^{x_i}$$

and $\delta(\mathfrak{a}) = \log |\alpha| = \log |\gamma| + (\mathbf{r} \cdot \mathbf{x})$. In practice, we compute an approximation of $\log |\alpha| \bmod R_\Delta$ using methods of Maurer,[65] which also determine the floating point precision required in order to ensure that the result is numerically accurate. Given $\delta(\mathfrak{a})$, we can compute a compact representation of α using the methods of Chapter 12, if desired.

Example 13.22. We solve the principal ideal problem for the ideal $\mathfrak{a} = (50, 101)$ in $\mathcal{O}_\Delta$ with $\Delta = 12301$. Recall, from the previous example, that we have $h_\Delta = 5$, $R_\Delta = 17.05227601$, and that a generating system of Λ' is given by the relation matrix

$$A = \begin{pmatrix} -5 & -5 & -9 & -4 & -10 \\ 1 & -3 & -4 & -4 & -7 \\ 0 & 0 & -1 & -1 & 0 \end{pmatrix}$$

with corresponding distance vector

$$\mathbf{r} = (-0.6976433914, -2.481863958, -6.743908938,$$
$$- 1.337141404, -9.672852058).$$

To test $\mathfrak{a}$ for principality, we proceed as follows.

1. The first step is to find an equivalent ideal that factors over the factor base. In this case, $\mathfrak{a}$ itself is smooth with $\mathbf{v} = (0, 2, 0)$.
2. We solve the linear system $A\mathbf{x} = \mathbf{v}$. One solution is $\mathbf{x} = (1, 2, -1, 1, -1)$.
3. Given $\mathbf{x}$, we compute

$$\delta(\mathfrak{a}) = \mathbf{r} \cdot \mathbf{x} \bmod R_\Delta = 9.418248286.$$

We can verify that $\mathfrak{a} = (\alpha)$, where

$$\alpha = \prod_{i=1}^{5} \gamma_i^{x_i} = \frac{12311 + 111\sqrt{12301}}{2}$$

and that $\log|\alpha| = 9.418248286$, so this result is unconditionally correct.

Example 13.23. We solve the principal ideal problem for the ideal $\mathfrak{a} = (30, 101)$ in $\mathcal{O}_\Delta$ with $\Delta = 12301$.

1. The first step is to find an equivalent ideal that factors over the factor base. In this case, $\mathfrak{a}$ itself is smooth with $\mathbf{v} = (-1, 1, 0)$.
2. We solve the linear system $A\mathbf{x} = \mathbf{v}$. In this case, the linear system does not have a solution, so we conclude that $\mathfrak{a}$ is not principal.

Notice that this result only holds under the assumption that the factor base used generates Cl_Δ and the ERH, which is required to guarantee that the set of extended relations generate all of Λ'.

The running time of this algorithm is dominated by the cost of producing and verifying a generating system of Λ'. Once that has been computed, requiring expected time $L_\Delta[1/2, \max(2z + 1/(4z), 4z) + o(1)]$, the remaining operations are finding one additional relation, requiring time $L_\Delta[1/2, z + 1/(4z) + o(1)]$, and solving $A\mathbf{x} = \mathbf{v}$, requiring time $L_\Delta[1/2, 3z + o(1)]$. Thus, the overall running time is $L_\Delta[1/2, \max(2z + 1/(4z), 4z) + o(1)]$, and as this is optimized when $z = 1/\sqrt{8}$, we obtain the following theorem.

Theorem 13.24. *The principal ideal problem in a real quadratic order $\mathcal{O}_\Delta$ with $\Delta > 41$ can be solved in expected time $L_\Delta[1/2, \sqrt{2} + o(1)]$ assuming the GRH and ERH.*

If the input ideal $\mathfrak{a}$ is in fact principal and the algorithm returns its distance $\delta(\mathfrak{a})$, then the output can be verified unconditionally by, for example, using infrastructure to compute a compact representation of $\gamma \in \mathbb{Q}(\sqrt{\Delta})$ with

$\log \gamma = \delta(\mathfrak{a})$, verifying that $(\gamma) = \mathfrak{a}$. As in the case of the imaginary quadratic order DLP, the GRH and ERH are required in order to certify that an ideal is not principal, because both are required to ensure that the extended relations produced generate the entire lattice Λ'.

Finally, note that testing whether two ideals are equivalent and solving the DLP in the class group of a real quadratic order can also be done in expected time $L_\Delta[1/2, \sqrt{2} + o(1)]$. Testing whether two ideals are equivalent (i.e., whether there exists $\alpha \in \mathbb{Q}(\sqrt{\Delta})$ such that $\mathfrak{a} = (\alpha)\mathfrak{b}$) amounts to solving the principal ideal problem for $\mathfrak{a}\mathfrak{b}^{-1}$. Solving the DLP can be done as in the imaginary quadratic case, with the exceptions that the extended relation lattice must be computed in order to certify that no solution exists and the relative generator α such that $\mathfrak{a} = (\alpha)\mathfrak{g}^x$ is also required to prove that the computed value of x is correct. Given x, we can compute α by solving the principal ideal problem for $\mathfrak{a}\mathfrak{g}^{-x}$.

13.6 Complexity

We have now seen algorithms with the best known complexities for computing class groups and regulators of quadratic orders, all of which are subexponential in $\log|\Delta|$. It is also interesting to consider to which computational complexity class each of these problems belongs. The index-calculus algorithms described above allow us to give some results in this direction.

As we know of no algorithms that solve these problems in deterministic polynomial time, it is not known whether they belong to $\mathcal{P}$, the class of problems that can be solved in polynomial time. The complexity class $\mathcal{NP}$ denotes the set of decision problems (i.e., problems whose output is either "yes" or "no") that have short proofs that the answer is "yes." In other words, if a problem is in $\mathcal{NP}$, then there exists a short (polynomial sized) certificate that a verifier can process in polynomial time in order to verify that the output of the decision problem is "yes." Certainly, it may be difficult to compute such a certificate, but once it is found, it can be verified efficiently. Similarly, the class co-$\mathcal{NP}$ denotes the set of decision problems that admit short proofs that the answer is "no." Formal definitions for these complexity classes can be found in standard textbooks on the subject.[66]

A major open problem in theoretical computer science is whether $\mathcal{P} = \mathcal{NP}$ (i.e., whether there exist polynomial-time algorithms that can solve every problem in $\mathcal{NP}$). Any results on classifying problems into complexity classes furthers our understanding as to how difficult a problem actually is.

The problems of computing the class group and regulator have decision problem versions and thus can be considered in this paradigm. For example, the decision problem version of computing the regulator is to determine whether or not a given approximation of a real number is an approximation of the regulator. If a short certificate, of polynomial length in $\log|\Delta|$, exists that a verifier can use to certify that a given floating point number is an approximation of the regulator in polynomial time, then the problem of computing R_Δ would be considered to be in $\mathcal{NP}$. If a short proof exists that a given floating point number is not an approximation of the regulator, then the problem would be in co-$\mathcal{NP}$.

It can be shown that the DLP in the ideal class group, computing h_Δ, Cl_Δ, and R_Δ and solving the principal ideal problem belong to $\mathcal{NP} \cap$ co-$\mathcal{NP}$ under the assumptions of the GRH and ERH.[67] The idea is to devise short certificates making use of the index-calculus methods described above and compact representations of elements in $\mathbb{Q}(\sqrt{\Delta})$.

We will describe a slightly modified version of the proof of Buchmann and Williams[68] showing that the problems of computing the class number and regulator of a real quadratic order are both in $\mathcal{NP} \cap$ co-$\mathcal{NP}$. Their method is a generalization of McCurley's proof[69] that computing the class number of an imaginary quadratic order is in the same complexity class, so the techniques we describe here also apply in that context.

The idea of the proof is to submit as a certificate certain components from the index-calculus algorithm described above to a verifier to check their correctness. In particular, the certificate consists of the following:

1. a factor base $FB = \{\mathfrak{p}_1, \ldots, \mathfrak{p}_k\}$ with $N(\mathfrak{p}_k) < 12 \log^2 |\Delta|$,
2. for each prime ideal $\mathfrak{p} \notin FB$ with $N(\mathfrak{p}) \le 12 \log^2 |\Delta|$, a vector $\mathbf{v} \in \mathbb{Z}^k$ and compact representation of $\gamma \in \mathbb{Q}(\sqrt{\Delta})$ such that

$$\mathfrak{p} \prod_{i=1}^{k} \mathfrak{p}_i^{v_i} = (\gamma) \,,$$

3. an HNF basis of the relation lattice Λ consisting of relations $\mathbf{v}_1, \ldots, \mathbf{v}_k \in \mathbb{Z}^k$,
4. compact representations of $\gamma_1, \ldots, \gamma_k$ such that

$$FB^{\mathbf{v}_i} = (\gamma_i) \,,$$

5. a compact representation of a unit $\epsilon \in \mathcal{O}_\Delta$,
6. an integer Q such that the value h^* computed using Bach's method for approximating $L(1, \chi)$ and (10.19) or (10.20) satisfies $h^* \le h_\Delta R_\Delta \le 2h^*$ assuming the ERH.

The verification proceeds as follows. The first step is to verify that the factor base presented generates Cl_Δ. The verifier checks that the ideals presented in the factor base are all prime ideals by proving that their norms are prime, which can be done in polynomial time using the algorithm of Agrawal, Kayal, and Saxena.[70] Alternatively, short proofs of the primality of each of these norms could be submitted as part of the certificate. Next, the verifier checks that

$$\mathfrak{p} \prod_{i=1}^{k} \mathfrak{p}_i^{v_i} = (\gamma) \tag{13.3}$$

holds for each prime ideal $\mathfrak{p} \notin FB$ with $N(\mathfrak{p}) \le 12 \log^2 |\Delta|$. If so, then, assuming the GRH, the factor base provided does generate Cl_Δ, because each of the prime ideals less than Bach's bound can be factored over the factor base. This step can also be done in polynomial time, because the number of prime ideals to check is polynomial in $\log |\Delta|$ and because the γ are given in compact representation. In particular, the left-hand side of (13.3) can be evaluated using ideal arithmetic in polynomial time. Given a compact representation of γ, we can compute a $\mathbb{Z}$-basis representation of (γ) in polynomial time using Proposition 12.2 and the subsequent remarks, and this is simply compared to the ideal resulting from evaluating the left-hand side of (13.3).

The next step is to verify that the HNF basis provided is a basis of a sublattice of Λ. First, each of the relation vectors $\mathbf{v}_i$ is verified by checking that

$$FB^{\mathbf{v}_i} = (\gamma_i) \, .$$

This can be done in polynomial time because the number of relations is equal to the size of the factor base, which is polynomial in $\log |\Delta|$, and as in the previous step, each test can be done in polynomial time because the γ_i are given in compact representation. As long as the relations are all valid, we know that they form a sublattice $\Lambda' \subseteq \Lambda$, and because they are given in HNF, Λ' has full rank and $\det \Lambda'$ is a multiple of h_Δ.

It remains to verify that ϵ is a unit with $\epsilon > 1$ (i.e., a positive power of the fundamental unit) and to prove that $h = \det \Lambda' = h_\Delta$ and $R = \log \epsilon = R_\Delta$. As ϵ is given in compact representation, we can verify that it is a unit in polynomial time by computing its norm as described in §12.3 and verifying that it is ± 1. We can evaluate $h = \det \Lambda'$ in polynomial time because we are given an HNF basis of Λ', and the matrix with the basis vectors as columns is upper triangular. We can also approximate $R = \log \epsilon$ in polynomial time because ϵ is in compact representation. If $R > 0$, then we know that $\epsilon > 1$. Next, we verify, using (10.2), that Bach's method with the given bound Q yields an approximation of $L(1, \chi)$ accurate to within $\sqrt{2}$ and compute h^* using (10.19) or (10.20) such that $h^* < h_\Delta R_\Delta < 2h^*$. Finally, to prove that $h = h_\Delta$ and $R = R_\Delta$, we have only to check that $hR < 2h^*$. Because Λ' is a full-rank sublattice and ϵ is a positive power of the fundamental unit, $h = c_1 h_\Delta$ and $R = c_2 R_\Delta$ for $c_1, c_2 \in \mathbb{Z}_{\ge 1}$, so this inequality will only be satisfied if $h = h_\Delta$ and $R = R_\Delta$.

The total length of the contents of the certificate is polynomial in $\log |\Delta|$, because all of the elements of $\mathbb{Q}(\sqrt{\Delta})$ are given in compact representation, and the total number of items is polynomial in $\log |\Delta|$. Therefore, we have exhibited a short certificate that provides a proof of the class number and regulator of a quadratic order. This implies that the problems of computing these invariants are in $\mathcal{NP}$. We can use almost the same proof to show that these problems are also in co-$\mathcal{NP}$. To prove that the given values h and R are not equal to h_Δ and R_Δ, respectively, we simply check the certificate described above, thereby obtaining h_Δ and R_Δ, and compare these to the submitted values h and R. Thus, the problems of computing the class number and regulator are in $\mathcal{NP} \cap \text{co-}\mathcal{NP}$. However, this result is conditional on the GRH and ERH. Once again, the GRH is required to ensure that the number of prime ideals needed to generate Cl_Δ is polynomial in $\log|\Delta|$, and the ERH is required to guarantee that $h^* < h_\Delta R_\Delta < 2h^*$.

The proof described above makes use of short proofs of ideal principality, namely verifying that an element $\gamma \in \mathbb{Q}(\sqrt{\Delta})$ given in compact representation is the generator of the input ideal. Thus, principal ideal testing is in $\mathcal{NP}$. Similarly, the DLP is in $\mathcal{NP}$. The certificate consists of an integer x and a compact representation of $\gamma \in \mathbb{Q}(\sqrt{\Delta})$. The certificate is verified by checking that $\mathfrak{g}^x \mathfrak{a}^{-1} = (\gamma)$, implying that $\mathfrak{g}^x \sim \mathfrak{a}$. Note that these results are unconditional; that is, the assumption of GRH and ERH is not required.

Proving that the principal ideal problem and the DLP are in co-$\mathcal{NP}$ requires two components in the certificates. In both cases, we require the certificate proving the correct values of h_Δ and R_Δ. Given that, we can prove that an ideal $\mathfrak{a}$ is not principal by exhibiting a vector $\mathbf{v} \in \mathbb{Z}^k$ and a compact representation of $\gamma \in \mathbb{Q}(\sqrt{\Delta})$ such that

$$\mathfrak{a} = (\gamma) \prod_{i=1}^{k} \mathfrak{p}_k^{v_i} \ .$$

As described above, this can be verified in polynomial time. Then because the proof that h_Δ and R_Δ are correct gives us a basis of Λ, we can prove that $\mathfrak{a}$ is not principal by showing that the linear system $H\mathbf{x} = \mathbf{v}$ has no solution, where H is an HNF matrix whose columns are the basis of Λ. The fact that the columns of H form a basis of Λ implies that all principal ideals can be factored over the factor base via a linear combination of the basis vectors; so the fact that the linear system has no solution implies that $\mathfrak{a}$ is not principal. This can be done in polynomial time because H is in HNF, and, hence, upper triangular. The proof that the DLP is in co-$\mathcal{NP}$ is similar and follows Vollmer's algorithm described above. Note that these results are dependent on the GRH and ERH, because these are needed for the certification of h_Δ and R_Δ.

It is interesting to note that we do not know of a short proof for R_Δ that does not also prove h_Δ. We can give a compact representation of any element in $\mathbb{Q}(\sqrt{\Delta})$ with small norm and check in polynomial time whether it is a unit, but we currently require the class number in order to prove fundamentality by using an approximation of $L(1, \chi)$ and the analytic class number formula to determine that the product of the class number and regulator is correct.

Note that because knowledge of the regulator R_Δ is sufficient to find solutions to the Pell equation, this also shows that the problem of solving the Pell equation is in $\mathcal{NP} \cap \text{co-}\mathcal{NP}$.

13.7 Practical Improvements

13.7.1 Improvements to the Random Exponents Method

There are a number of ways to improve the algorithms described above. The first results discussing fit-for-implementation versions of these algorithms in the imaginary case are due to Buchmann and Düllmann.[71] Their first improvement was to simplify the relation generation process. Rather than selecting random exponent vectors from $\mathbb{Z}^k$, the idea was to force the majority of the coefficients to be zero and to select the remaining coefficients randomly using a constant bound such as 30 as opposed to $|\Delta|$. The result is that the time required to find relations is significantly faster, as the number of ideal multiplications required to generate a relation will be reduced. This can be improved further by precomputing a few tables of factor base elements raised to small powers up to the exponent bound. A second benefit of this approach is that the resulting relation matrices are sparse, because the number of non-zero elements per relation is small. This reduces the amount of memory required to store the relation matrix and permits the use of special-purpose linear algebra algorithms that can exploit the sparseness of the matrix. For example, structured Gaussian elimination[72] can be used to reduce the relation matrix to a much smaller dense matrix on which standard HNF algorithms can be applied.

A second observation by Buchmann and Düllman was that in practice it is often unnecessary to force the relation matrix to be strictly diagonally dominant. In addition, the total number of relations required to generate Λ tends to be significantly smaller than that predicted by the analysis; in particular, generating k random relations tends to work reasonably well in practice, and if not, only a small number of additional relations are required. If, during the process of computing the HNF, it is discovered that the relation matrix is in fact singular, it is possible to compute the rank profile of the row space of the matrix, a list of row indices of the rows that are linearly independent. Given this information, additional relations can be generated with non-zero coefficients in the locations of the linearly dependent rows, by choosing the random vectors to have non-zero coefficients in these locations, in an effort to remove the dependencies. If the relation matrix is non-singular but the columns do not generate Λ, Buchmann and Düllman suggested computing one additional relation and computing the HNF and determinant again. Note that this additional relation can be appended to the previous HNF matrix, so computing the new HNF is easy as the matrix is already in triangular form.

Using these ideas, Buchmann and Düllman were able to compute class groups and solve instances of the DLP for several imaginary quadratic fields. The largest of these has the 40-digit discriminant $-4F_7$, where $F_7 = 2^{128} + 1$ denotes the seventh Fermat number. To the best of our knowledge, this was the first implementation of these algorithms.

Buchmann and Düllman's ideas were adapted to the real case by Cohen, Diaz y Diaz, and Olivier.[73] They suggested that computing an initial $k + 10$

random extended relations was usually enough to generate Λ' in practice, and if not, generating a small number of additional relations and recomputing the HNF would produce a complete generating system after very few iterations.

Cohen et al. also observed that the subexponential-sized factor bases are much larger than necessary for practical use and suggested the use of much smaller factor bases. Using a smaller factor base decreases the probability of finding relations because the chance that a random ideal is smooth is smaller if the factor base is smaller, but this is often offset by the fact that the linear algebra will be faster. Thus, there is some opportunity to fine-tune the factor base size to obtain a reduced overall running time in practice. However, in order to certify that the output is correct under the assumption of the GRH, it is necessary to show that all prime ideals $\mathfrak{p} \notin FB$ but with $N(\mathfrak{p}) < 12 \log^2 |\Delta|$ (or $6 \log^2 |\Delta|$ if Δ is known to be fundamental) can be factored over the factor base, implying that their equivalence classes belong to the subgroup of Cl_Δ generated by the factor base. This can be done for each such $\mathfrak{p}$ by using the method described above to factor an arbitrary ideal over the factor base.

Note that this observation also allows an implementer to choose the factor base size based on the amount of available memory. Even if the largest factor base size that one can use in this context does not generate Cl_Δ under the GRH, one can perform the computation and verify that the smaller factor base does in fact generate Cl_Δ using this idea.

Finally, Cohen et al. suggested using Lenstra's distance $\frac{1}{2} \log |\gamma/\overline{\gamma}|$ for the principal ideal generators γ of each relation instead of Shanks' distance $\log |\gamma|$. Notice that if γ is a unit, then Lenstra's distance will yield the same regulator multiple as Shanks' distance. The advantage of using Lenstra's distance is that it is not necessary to keep track of rational factors of γ that arise from ideal inversion. In particular, whenever an exponent in a relation is negative, the inverse of the corresponding prime ideal $\mathfrak{p}$ is involved, and as the fractional ideal $\mathfrak{p}^{-1} = (1/p)\overline{\mathfrak{p}}$, one must account for the rational factors of $1/p$ when computing γ. However, these rational factors are eliminated when using Lenstra's distance, because $\gamma/\overline{\gamma} = (a + b\sqrt{\Delta})/(a - b\sqrt{\Delta})$ when $\gamma = (a + b\sqrt{\Delta})/c$.

Cohen, Diaz y Diaz, and Olivier described the results of an implementation of these algorithms incorporating the improvements described above. To the best of our knowledge, this was the first implementation of an index-calculus algorithm for computing invariants of real quadratic fields. Results and running time data are given for six examples, including the 40-digit discriminant $10^{40} + 1$.

13.7.2 The Large Prime Variation

The large prime variation has proved to be a useful practical improvement for index-calculus algorithms in other settings, including integer factorization and discrete logarithm computation in finite fields. Buchmann and Düllmann[74] generalized this technique to the quadratic field-based index-calculus algorithms described above.

The idea behind this variation is that many of the randomly produced reduced ideals that we attempt to factor over the factor base almost factor completely except for one large prime factor lying outside the factor base. If we find two such factorizations with the same large prime factor, then they can be combined in such a way that the large primes cancel, yielding a relation. In practice, many of these partial factorizations are found, and the time spent generating relations can be decreased considerably.

Suppose that we compute a reduced ideal $\mathfrak{b}$ such that $\mathfrak{a}(\gamma) = \mathfrak{b}$ with $\mathfrak{a} = FB^{\mathbf{v}}$ as before. In addition, suppose that $\mathfrak{b}$ factors over FB except for one additional prime ideal factor, that is, that

$$\mathfrak{b} = \mathfrak{p}^s FB^{\mathbf{w}} ,$$

with $s = \pm 1$. Then

$$(\gamma) = \mathfrak{p}^s FB^{\mathbf{w} - \mathbf{v}}$$

and the vector $\mathbf{e} = \mathbf{w} - \mathbf{v}$ almost is a relation, except for the single large prime factor $\mathfrak{p}^s$. We call the tuple $(\mathfrak{p}, s, \mathbf{e}, \log |\gamma|)$ a *partial relation*, in contrast to the *full relations* described above that correspond to principal ideals that factor completely over the factor base.

Suppose now that we have two partial relations $(\mathfrak{p}, s_1, \mathbf{e}_1, \log |\gamma_1|)$ and $(\mathfrak{p}, s_2, \mathbf{e}_2, \log |\gamma_2|)$ that have the same large prime ideal factor $\mathfrak{p}$, but possibly with different signs s_i. In order to form a full relation, we need to find some combination of these two partial relations that yields a principal ideal which factors completely over the factor base. If $s_1 \neq s_2$ (i.e., $s_1 + s_2 = 0$), then the two almost smooth principal ideals can be multiplied, yielding

$$(\gamma_1 \gamma_2) = \mathfrak{p}^{s_1} FB^{\mathbf{e}_1} \mathfrak{p}^{s_2} FB^{\mathbf{e}_2} = \mathfrak{p}^{s_1 + s_2} FB^{\mathbf{e}_1 + \mathbf{e}_2} = FB^{\mathbf{e}_1 + \mathbf{e}_2} ,$$

and $(\mathbf{e}_1 + \mathbf{e}_2, \log |\gamma_1| + \log |\gamma_2|)$ is a full relation. On the other hand, if $s_1 = s_2$ (i.e., $s_1 - s_2 = 0$), then

$$(\gamma_1 \gamma_2^{-1}) = \mathfrak{p}^{s_1} FB^{\mathbf{e}_1} \mathfrak{p}^{-s_2} FB^{-\mathbf{e}_2} = \mathfrak{p}^{s_1 - s_2} FB^{\mathbf{e}_1 - \mathbf{e}_2} = FB^{\mathbf{e}_1 - \mathbf{e}_2} ,$$

and $(\mathbf{e}_1 - \mathbf{e}_2, \log |\gamma_1| - \log |\gamma_2|)$ is a full relation.

To apply the large prime variation to relation generation, we select a bound LB for the norms of the large prime ideals which we will accept. Then, in the course of searching for relations, we save any partial relations that occur for with the large prime factor $\mathfrak{p}$ satisfies $N(\mathfrak{p}) \leq LB$. At certain intervals during the relation generation process, we sort the list containing the partial relations according to the large primes $\mathfrak{p}_i$. Now, suppose we have found l partial relations containing the same large prime ideal $\mathfrak{p}_i$. Note that these l partial relations are easy to find, since the list is sorted. We can compute a full relation from each distinct pair of these partial relations, yielding $\binom{l}{2}$ new full relations.

Since sorting the list of partial relations is relatively time-consuming, we only collect the full relations at a few points during the relation generation procedure. The following method works well in practice.[75] Assume that we need to generate n relations in total. As soon as we have generated $n' = 0.1n$ full relations (10% of n), we sort the list of partial relations and count the number p of new full relations that could be generated. We do not actually compute these new full relations; rather we note the fact that they can be generated if needed. We repeat this step each time we have $n' = 0.2n$, $n' = 0.3n$, etc.... until $n' + p \geq n$. At this point, we sort the list of partial relations one last time and explicitly compute the full relations produced by combinations of the partial relations.

13.7.3 Parallelism

Buchmann and Düllman[76] observed that the relation generation stage can be parallelized using master-slave parallelism, in which a number of "slave" computers compute relations and send them to a single "master" computer responsible for coordinating the entire computation . When using the random exponents strategy, each slave independently computes relations and sends them to the master. As long as the sequences of random numbers used are different for each slave, obtained by, for example, ensuring that the pseudo-random number generator used by each slave receives a different seed, the relations produced are likely to be distinct. The result is that the relation generation can be sped up by a factor of n, where n slave nodes are used. By using large primes and parallel relation generation, Buchmann and Düllman were able to compute Cl_Δ for an imaginary quadratic field with the 55-decimal digit discriminant $\Delta = -4(10^{54} + 1)$.

13.7.4 Computing Relations Using Sieving

As with factoring algorithms, an important breakthrough in improving efficiency was the realization that sieving could be used to simultaneously test a set of equivalent ideals for smoothness. Using the relation generation strategies described above requires that each candidate reduced ideal be factored individually using trial division or some other method such as the elliptic curve method (ECM)[77]. However, as shown by Jacobson[78], it is possible to use a sieve procedure similar to the multiple polynomial quadratic sieve[79] (MPQS) integer factorization algorithm to find relations in both the imaginary and real quadratic cases.

The main idea is due to Paulus,[80] who observed that in order to find a relation given a reduced ideal $\mathfrak{b}$ equivalent to some ideal that is smooth with respect to the factor base, it is enough to factor any ideal equivalent to $\mathfrak{b}$. In addition, he observed that the norms of ideals equivalent to $\mathfrak{b}$ are equal to values of a bivariate quadratic polynomial whose coefficients correspond to the coefficients of $\mathfrak{b}$. This polynomial can be sieved using the norms of the prime ideals in the factor base to identify norms of ideals equivalent to $\mathfrak{b}$ that are smooth. In addition, given the values of the variables corresponding to the smooth norm, it is possible to explicitly generate the smooth ideal itself and factor it.

Paulus' idea was improved by Jacobson[81] by extending it to model the successful MPQS factoring algorithm and compute the extended relations required for the algorithms set in real quadratic orders. We begin, as before, by computing a random $\mathcal{O}_\Delta$-ideal

$$\mathfrak{a} = (d) \prod_{i=1}^{k_0} \mathfrak{p}_i^{v_i} = \left[\frac{Q}{r}, \frac{P + \sqrt{D}}{r} \right]$$

with k_0 small and $v_i \in \{0, 1, -1\}$, where by d we denote the denominator of the (possibly) fractional ideal $\prod_{i=1}^{k_0} \mathfrak{p}_i^{v_i}$. Note that this is similar to Düllmann's strategy, except the exponents are chosen from an even more restrictive set. As suggested by Paulus, rather than reducing $\mathfrak{a}$ and testing it for smoothness, we use a sieve to test many ideals equivalent to $\mathfrak{a}$ for smoothness. Any ideal equivalent to $\mathfrak{a}$ that factors over the factor base yields a relation. We can sieve over many ideals $\mathfrak{a}$, thus obtaining a multiple polynomial-type algorithm.

The following lemma gives us the basis for setting up the sieve.

Lemma 13.25. *If* $\gamma \in \mathfrak{a} = [Q/r, (P + \sqrt{D})/r]$, *so that* $\gamma = (Q/r)x + \frac{P+\sqrt{D}}{r}y$ *for* $x, y \in \mathbb{Z}$, *then there exists an ideal* $\mathfrak{b}$ *such that* $(\gamma) = \mathfrak{ab}$ *and*

$$N(\mathfrak{b}) = (Q/r)x^2 + (2P/r)xy + \frac{P^2 - D}{rQ}y^2 \ .$$

Proof. Put $R = (P^2 - D)/(rQ)$ and note that since we get $\gamma \in \mathfrak{a}$, we have $(\gamma) = \mathfrak{ab}$ for some $\mathcal{O}_\Delta$-ideal $\mathfrak{b}$. We also have

$$
\begin{aligned}
N((\gamma)) = |\gamma\bar{\gamma}| &= \left| \left(\frac{Q}{r}x + \frac{P + \sqrt{D}}{r}y \right) \left(\frac{Q}{r}x + \frac{P - \sqrt{D}}{r}y \right) \right| \\
&= \left| (Q/r) \left((Q/r)x^2 + (2P/r)xy + Ry^2 \right) \right| \\
&= N(\mathfrak{a})N(\mathfrak{b}) \ .
\end{aligned}
$$

Because ideal norm is multiplicative, we get an ideal $\mathfrak{b}$ with norm $N(\mathfrak{b}) = (Q/r)x^2 + (2P/r)xy + Ry^2$. □

Using the lemma, if we can find $x, y \in \mathbb{Z}$ such that $N(\mathfrak{b}) = f(x, y)$ factors over the norms of the prime ideals in the factor base (i.e., $\mathfrak{b}$ factors over FB) then we get a relation. Note that if an extended relation is required, the value of γ is obtained immediately, which, after adjusting for negative powers in the factorizations of $\mathfrak{a}$ and $\mathfrak{b}$, will be the generator of a smooth principal ideal. Thus, the problems of finding both relations and extended relations are reduced to finding smooth values of quadratic polynomials, which can be done with a sieve. In practice, given a *sieve radius* $M \in \mathbb{Z}$, we set $y = 1$ and search for values of the quadratic polynomial $f(x, 1) = (Q/r)x^2 + (2P/r)X + R$ with

$-M \le x \le M$ that factor completely over the norms of the prime ideals in the factor base. Sieving is possible due to the observation that if $p \mid f(x, 1)$, then $p \mid f(x + mp, 1)$ for all $m \in \mathbb{Z}$, allowing one to determine rapidly all values of x for which the quadratic polynomial is divisible by a prime p, given its roots modulo p.

We can also make use of the ideas of Silverman's MPQS in order to maximize the chances of finding relations, by forcing the values of $f(x, y)$ to be as small as possible, given bounds on x and y. In particular, for a given sieve radius M, ensuring that $N(\mathfrak{a}) \approx (\sqrt{|\Delta|/2})/M$ will achieve this goal when we take $y = 1$. Also, as with the MPQS factoring algorithm, this method is trivial to parallelize; each processor simply performs sieving on its own unique set of ideals $\mathfrak{a}$.

The sieving idea can be extended to factor an ideal $\mathfrak{a}$ over the factor base. For a random $\mathbf{v} \in \mathbb{Z}^k$, compute $\mathfrak{c} = [Q/r, (P + \sqrt{D})/r]$ with $\mathfrak{c} \sim \mathfrak{a}FB^{\mathbf{v}}$. For $\gamma = (Q/r)x + ((P + \sqrt{D})/r)y \in \mathfrak{c}$, there exists an ideal $\mathfrak{b}$ such that $(\gamma) = \mathfrak{c}\mathfrak{b}$ and $N(\mathfrak{b}) = (Q/r)x^2 + (2P/r)xy + Ry^2 = f(x, y)$ as above. Thus, $(\gamma) = \mathfrak{c}\mathfrak{b}$ implies that $\mathfrak{a} \sim \mathfrak{b}^{-1}FB^{-\mathbf{v}}$, and the problem is reduced to finding smooth values of $f(x, y)$. As above, the idea of MPQS factoring can be applied, also in parallel.

13.7.5 Self-initialization

After sieving with one ideal, if additional relations are required, we have to start over by computing a new ideal via a random power product of factor base elements. In addition, we have to compute the roots of the sieve polynomial corresponding to the new ideal modulo the norm of each prime ideal in the factor base in order to initialize the sieving procedure. Both of these tasks are significantly more expensive than performing the sieving itself.

For integer factorization, Alford and Pomerance[82] addressed this problem by extending the MPQS algorithm to incorporate a technique called self-initialization. Their method was extended to computations in quadratic fields by Jacobson.[83] The idea behind self-initialization is to use a special set of sieve polynomials that enables fast switching from one sieve polynomial to the next. Select $S = \{\mathfrak{q}_1, \ldots, \mathfrak{q}_t\} \subset FB$ such that

$$\prod_{j=1}^{t} N(\mathfrak{q}_j) = \prod_{j=1}^{t} q_j \approx \frac{\sqrt{|\Delta|/2}}{M}.$$

As before, the bound on the norm ensures that the values of the sieve polynomial will be as small as possible for values of x in the interval $[-M, M]$. The $\mathcal{O}_\Delta$-ideals we use for sieving are computed as $\mathfrak{a} = (d)S^{\mathbf{v}}$, where $\mathbf{v} \in \{-1, 1\}^t$ and d denotes the denomiantor of the fractional ideal $S^{\mathbf{v}}$. Note that there are 2^{t-1} possibilities for $\mathbf{v}$ if we only count one of $\mathfrak{a}$ and $\mathfrak{a}^{-1}$. For each of these different possible ideals, the norm $Q/r = N(\mathfrak{a}) = \prod_{j=1}^{t} q_j$ is fixed, but there are 2^{t-1} different possible P coefficients with $P^2 \equiv D \pmod{rQ}$.

The point behind self-initialization is to avoid explicitly computing each of the 2^{t-1} ideals $S^{\mathbf{v}}$ via ideal multiplication. We will instead construct the 2^{t-1} solutions P_i of the congruence

$$P_i^2 \equiv D \pmod{rQ} \tag{13.4}$$

corresponding to a given value of $Q/r = q_1 \cdots q_t$. If $\Delta \equiv 0 \pmod 4$, i.e., $r = 1$, the solutions P_i can be constructed directly using the Chinese remainder theorem modulo the q_j. Given integers B_j for $1 \le j \le t$ such that

$$\begin{cases} B_j \equiv 0 \pmod{q_k} & \text{if } k \ne j \\ B_j^2 \equiv D \pmod{q_j} & \text{otherwise}, \end{cases}$$

the Chinese Remainder Theorem tells us that any combination of the form $B = \pm B_1 \pm B_2 \pm \cdots \pm B_{t-1} + B_t$ is a solution of (13.4). We fix the sign of B_t in order to obtain only one of B and $-B$, and thus one of $\mathfrak{a}$ or $\mathfrak{a}^{-1}$. The values of P_i correspond to the least positive remainders of $B \bmod Q$. As shown by Alford and Pomerance, the B_j are given by

$$B_j = \left(\frac{Q}{q_j}\right) \left(\left(\frac{Q}{q_j}\right)^{-1} t_{q_j} \pmod{q_j} \right) \pmod Q,$$

where t_{q_j} is the solution of $T^2 \equiv D \pmod{q_j}$ yielding the smallest least positive residue of $(Q/q_j)^{-1} t_{q_j} \pmod{q_j}$. If $\Delta \equiv 1 \pmod 4$ (i.e., $r = 2$) we follow the same procedure, and at the end, we take the least positive remainder of $B + Q \pmod{2Q}$.

It remains to be seen how to generate efficiently each of the 2^{t-1} solutions P_i of (13.4). Recall that each P_i corresponds to a combination of the t values of B_j, where each B_j is either added or subtracted. We associate a binary vector $\mathbf{u} \in \{-1, 1\}^t$ to each of the P_i, where $u_j = 1$ if B_j is added and $u_j = -1$ if B_j is subtracted. Take $P_1 = B_1 + B_2 + \cdots + B_t$ so that $\mathbf{u}_1 = (1, 1, \ldots, 1)$. Starting with $\mathbf{u}_1$, we order the 2^{t-1} possible vectors $\mathbf{u}_i$ in a Gray code:

$$\mathbf{u}_{i+1} = \mathbf{u}_i + \mathbf{n}_i.$$

Here, entry ν of $\mathbf{n}_i$ is $2(-1)^{\lceil i/2^\nu \rceil}$, where ν is the unique positive integer such that $2^\nu \parallel 2i$, and all other entries are zero. Note that any two consecutive vectors $\mathbf{u}_i$ and $\mathbf{u}_{i+1}$ in this ordering differ in only a single entry. If entry ν of $\mathbf{u}_i$ is -1, we add 2 to it in order to obtain 1, and if it is 1, we subtract 2. This ordering of the $\mathbf{u}_i$ yields a similar ordering of the P_i given by

$$P_{i+1} = P_i + 2B_\nu(-1)^{\lceil i/2^\nu \rceil} \quad (2^\nu \parallel 2i). \tag{13.5}$$

If P_i was formed by subtracting B_ν, then we add $2B_\nu$ in order to obtain P_{i+1}, and if P_i was formed by adding B_ν, we subtract $2B_\nu$.

The roots of the sieving polynomials $F_i(X) = (Q/r)X^2 + (2P_i/r)X + R_i$, where $R_i = (P_i^2 - D)/(rQ)$, can also be computed iteratively. For some $p_j =$

$N(\mathfrak{p}_j)$, where $\mathfrak{p}_j \in FB$ and $p_j \nmid Q$, let $r_{1,j}$ be a root of $F_1(X)$ (mod p_j). If $p_j \nmid \Delta$, the two roots of $F_1(X)$ are given by

$$- \left(\frac{2P_1}{r} \right) \left(\frac{2Q}{r} \right)^{-1} \pm \sqrt{\Delta} \left(\frac{2Q}{r} \right)^{-1} \pmod{p_j} ,$$

and if $p_j \mid \Delta$, the single root is given by $-(2P_1/r)(2Q/r)^{-1}$ (mod p_j). By (13.5), for any P_i and P_{i+1} we have

$$\frac{2P_{i+1}}{r} - 2\frac{2P_i}{r} = \left(\frac{2}{r} \right) 2B_\nu(-1)^{\lceil i/2^\nu \rceil} ,$$

and it follows that

$$r_{i+1,j} \equiv r_{i,j} - 2B_\nu Q^{-1}(-1)^{\lceil i/2^\nu \rceil} \pmod{p_j} . \tag{13.6}$$

We have to compute the root of $F_i(X)$ (mod q_j) for each $q_j \mid Q$ separately, using the fact that these roots are given by $-R_i(2P_i/r)^{-1}$ (mod q_j). However, only t of these special roots have to be computed for each $F_i(X)$; all the others can be computed using (13.6).

To use these polynomials for relation generation, we also need to know the factorizations of the ideals $\mathfrak{a}_i = (Q, P_i)$ over the factor base as $\mathfrak{a} = (d)FB^{\mathbf{w}_i}$. The non-zero coefficients of $\mathbf{w}_i$ will always correspond to the same t ideals in the set S; the only question is to determine their signs according to whether $P_i \equiv t_j$ (mod q_j) for each of the prime ideals $\mathfrak{q}_j = (q_j, t_j) \in S$. Notice that once the correct signs are determined for the factorization of $\mathfrak{a}_1$, these can be iteratively updated using Grey-code-based ideas similar to the methods described above.

If more than 2^{t-1} polynomials are required to generate a sufficient number of relations, we simply choose another set S of t ideals in the factor base, from which we can generate another 2^{t-1} ideals. Any set of t ideals whose product yields an ideal $\mathfrak{a}$ with $N(\mathfrak{a}) \approx (\sqrt{|\Delta|/2})/M$ will suffice for this purpose, so in general there are plenty of possibilities for S, and the probability of running out of sieve polynomials is negligible.

In practice, the time to compute the P_i and the roots of $F_i(X)$ (mod p_j) from (13.5) and (13.6) can be shortened considerably by storing a few pre-computed values for each different value of Q. For example, the values of $2B_i$ and Q^{-1} (mod p_j) only depend on $Q = q_1 \ldots q_t$, and can be computed once when the set S is chosen. In addition, the values $\sqrt{\Delta}$ (mod p_j) can be computed once for the entire computation, as these are independent of the set S. With these precomputed values, only one addition or subtraction is required to compute P_{i+1} given P_i, and only one subtraction and one multiplication modulo p_j for each of the roots $r_{i+i,j}$ given $r_{i,j}$, considerably fewer operations than would be required using ideal arithmetic to generate each sieving polynomial and computing the roots directly.

13.8 Computational Results

The introduction of sieving for relation generation has resulted in a significant increase in speed. For example, Jacobson[84] reports that computing the
class group of $\mathcal{O}_\Delta$ with $\Delta = -4(10^{39} + 1)$ requires over 4 hours on a 296-
MHz Sun UltraSPARC-II using Buchmann and Düllman's improved version
of Hafner and McCurley's algorithm, but only about 2 minutes using MPQS-
style sieving. Using self-initialization improves this to less than half a minute.
In addition, the class group for the 55-decimal digit $-4(10^{54} + 1)$ was computed in just over 15 minutes on this machine, whereas a parallel computation
using 14 processors lasting more than 10 days was required using random exponents. Similarly, the class group and regulator of $\mathcal{O}_\Delta$ with $\Delta = 10^{40} + 1$
required over 8 hours on this machine using the improvements of Cohen et
al. described above, but only about 48 seconds using MPQS and 22 seconds
using self-initialization.

The most recent computational results for solving DLPs and computing
invariants of quadratic orders make use of self-initialized sieving and the large
prime variant. Using a single computer, the following results have been obtained by Jacobson.[85] The class group of an imaginary quadratic order with
discriminant $|\Delta| \approx 10^{80}$ was computed on a 296-MHz UltraSPARC-II in about
5.5 days. Solutions to instances of the DLP were each computed in about 4.5
hours using Buchmann and Düllman's method of finding the images of the
equivalence classes of $\mathfrak{a}$ and $\mathfrak{g}$ in the canonical representation of the class group
as a direct product of cyclic subgroups.

The largest instance of the principal ideal problem solved to date[86] is for
a discriminant $\Delta \approx 10^{66}$. This computation, also on a 296-MHz UltraSPARC-
II, required 1 day to compute R_Δ plus 4.5 hours per instance of the principal
ideal problem. The largest class number and regulator computed sequentially
using the methods described above[87] are for the 80-digit discriminant

$$\Delta = 12779403100260586715025492824657916044067403863724697039719773038860596565553681$$

for which, assuming the GHR and ERH, we have

$$R_\Delta = 1828710892199575366719923026577142676945.486446669 \,,$$

$$h_\Delta = 1 \,.$$

On a 296-MHz UltraSPARC-II, this computation required 3.45 days of computing time.

It is interesting to note that there is some disparity in the performance of
the index-calculus algorithm. For example, the next largest discriminant for
which the class number and regulator was computed using this algorithm[88] is
the 72-digit

$$\Delta = 133007243922787512412600341028518035429251391005992761399935498154029253 \,,$$

for which, assuming the GHR and ERH, we have

$$R_\Delta = 66252913306616520534293587275455606.557249020\,,$$
$$h_\Delta = 4\,,$$
$$Cl_\Delta = C(2) \times C(2)\,.$$

Even though this discriminant is smaller than the previous one presented, over 18 days were required on the same computer to compute h_Δ and R_Δ. The reason for this disparity lies in the fact that it is harder to find relations for quadratic orders that have few prime ideals of small norm in the factor base, because it is more probable to find smooth integers (norms of ideals in our case) with small prime factors. When using the index-calculus algorithm to compute the regulator and class number, the 80-digit discriminant has all primes less than 239 in the factor base, whereas the smallest prime in the factor base for the 72-digit discriminant is 347. Thus, we would expect finding relations to be more difficult in the latter case.[89]

These two examples were specially constructed for different applications. The 80-digit example was constructed so that the ratio between the fundamental solutions to Pell's equation for $D = \Delta/r^2$ and $D - 1$ was forced to be large. In particular, if we define $\rho(D) = \log t_1 / \log t_0$, where (t_1, u_1) is the fundamental solution of Pell's equation for D and (t_0, u_0) is the fundamental solution for $D - 1$, then, for this example,

$$\rho(D) \approx 3985513948589296828176189144643791044 88.73182644883\,.$$

In contrast, the example of $D = 1621$ mentioned in the Preface only has $\rho(D) = 34.35$. Jacobson and Williams[90] proved that $\rho(D)$ can be arbitrarily large by exhibiting an infinite family of D for which $\rho(D) \gg D^{1/6}/\log D$. However, if the ERH is assumed, then we would expect that there exists an infinitude of values of D for which $\rho(D) \gg \sqrt{D}\log\log D/\log D$. The 80-digit example and others provided by Jacobson and Williams provide evidence that this should indeed be the case. These examples were found, in part, by maximizing the first few terms in the Euler product expansion of $L(1, \chi)$. The construction used to find the 80-digit example forced the Kronecker symbol $(\Delta/p) = 1$ for all primes $p \leq 239$.

In contrast, the 72-digit example was constructed in such a way that the asymptotic density of prime values of the quadratic polynomial $f_A(x) = x^2 + x + A$ with $\Delta = 1 - 4A$ is high. The construction used minimizes the first few terms of the Euler product expansion of $L(1, \chi)$—in this case, by forcing $(\Delta/p) = -1$ for all primes $p \leq 337$. A conjecture of Hardy and Littlewood,[91] called Conjecture F, implies that $P_A(n)$, the number of prime values assumed by $f_A(x)$ for $0 \leq x \leq n$, satisfies

$$P_A(n) \sim C(\Delta) L_A(n)\,,$$

where

$$L_A(n) = 2 \int_0^n \frac{dx}{\log f_A(x)}$$

and

$$C(\Delta) = \prod_{p \geq 3} 1 - \frac{(\Delta/p)}{p-1} \, ,$$

indicating that values of Δ for which $C(\Delta)$ is large should yield polynomials $f_A(x)$ that have large asymptotic densities of prime values. Jacobson and Williams[92] described how to approximate $C(\Delta)$ efficiently and accurately given h_Δ and R_Δ, together with constructions that force $(\Delta/p) = -1$ for several small primes p, thereby maximizing the first terms of $C(\Delta)$. The 72-digit example mentioned here was produced in this manner and has $C(\Delta) = 5.65726388$, the largest known $C(\Delta)$ value to date. Thus, assuming Conjecture F, the ERH, and the GRH, the latter two being required for the correctness of h_Δ and R_Δ, and hence for the approximation of $C(\Delta)$, the polynomial $x^2 + x + A$ for

$A = -33251810980696878103150085257129508857312847751498190349983$
$\quad 874538507313$

has the largest asymptotic density of prime values for any polynomial of this type currently known.

Using parallel computations, it is possible to compute h_Δ and R_Δ for even larger Δ. For example, Hühnlein, Jacobson, and Weber[93] reported the computation of the class number and class group of an imaginary quadratic field whose discriminant $|\Delta| \approx 10^{90}$. Using a cluster of 16, 550-MHz Pentium III computers, this computation required 3 days to compute Cl_Δ plus 3 minutes per instance of the DLP. Using the same computers, Jacobson, Scheidler, and Williams[94] reported the computation of the class number, class group, and regulator of a real quadratic field $\mathbb{Q}(\sqrt{\Delta})$ with a 90-decimal-digit discriminant. In particular, for

$\Delta = 2152246981037284004104837712406016716686342009150185060 46263$
$\quad 9189777165915901265583086318 04$

we get, under the GRH and ERH,

$R_\Delta = 13141178379338133605434507674050601151666 86144.03321787 \, ,$

$$h_\Delta = 1 \, .$$

This computation took a total of 10 days, or approximately 5.2 months of CPU time, using the same cluster. In addition, the 101-decimal-digit discriminant

$\Delta = 1302219410219035041031908532979320512731946413288477616336 15$
$\quad 783665713790925835602630873971846 69099836$

was shown, under the GRH and ERH, to have

$$R_\Delta = 31780254623174755539291764915494863617276316347826 0.945231457,$$
$$h_\Delta = 1.$$

This computation took 87 days of real time using the cluster, approximately 3.8 years of CPU time.

The latter result is the first computation of a regulator corresponding to a discriminant of 100 or more decimal digits and the largest regulator found to date. Recall that, given the regulator, it is possible to compute a compact representation of the fundamental unit and, hence, the fundamental solution of Pell's equation. However, once again, it must be emphasized that the correctness of this result, as well as the other class number and class group computations mentioned above, is dependent on the truth of the GRH and ERH. On the other hand, the discrete logarithms and solutions to the principal ideal problem mentioned above are unconditionally correct.

13.9 Open Problems and Further Improvements

There remain numerous potential improvements to the subexponential method that have yet to be explored. For relation generation, one possibility is the use of double large prime relations. This strategy, which has been employed in other settings including integer factorization, and discrete logarithm computation in finite fields and hyperelliptic curves, involves making use of principal ideals that factor completely over the factor base except for one or two additional large prime factors. It is possible to find collections of such double large prime relations for which, when combined by multiplying or dividing, all the large prime factors cancel, resulting in a relation. This process of finding combinations of double large prime relations is slightly more complicated than in the integer factorization context, as the large prime factors must completely cancel as opposed to combining into even multiplicities that become zero when reduced modulo 2. Nevertheless, similar techniques can be used, as illustrated by Gaudry et al. for the low-genus hyperelliptic curve DLP.[95] These ideas should carry over easily to the index-calculus algorithms for quadratic orders described above and will likely offer improvements in practice for sufficiently large discriminants.

The linear algebra algorithms used to produce the examples described above can also be improved significantly. For example, the algorithms of Giesbrecht, Jacobson, and Storjohann referred to in the analyses have yet to be tried in practice. These algorithms for solving linear systems over the integers and computing the HNF and kernel basis of an integer matrix should work very well in practice, allowing class groups and regulators for significantly larger discriminants of quadratic orders to be computed. The algorithm for solving linear systems, in particular, should be very effective in conjunction with Vollmer's algorithm for computing discrete logarithms, but this also has not yet been attempted. One drawback of these methods is that they do not take advantage of the sparseness of the relation matrices that arise in practice, especially using sieving. Efficient algorithms that solve linear systems, compute a basis of the kernel, compute the determinant, and compute the SNF of a large sparse integer matrix exist[96] that all have subcubic complexity in the dimensions of the input matrices and may offer an even greater improvement in practice. To the best of our knowledge, there is no known HNF algorithm with subcubic complexity that exploits sparseness.

In addition to these practical issues, there are two important open theoretical problems that have yet to be solved. The first is whether it is possible to find an algorithm that breaks the $L_\Delta[1/2, \alpha]$ complexity barrier. In the context of integer factorization and computing discrete logarithms in finite fields $\mathbb{F}_p$, this has been achieved by the number field sieve, for which a heuristic analysis suggests a running time of $L_n[1/3, \alpha]$ for factoring n and $L_q[1/3, \alpha]$ for solving discrete logarithms in the finite field $\mathbb{F}_q$. Given the many similarities between factoring algorithms and algorithms for computing class groups and regulators and the fact that most of the advances in factoring have successfully been translated to the quadratic order setting, it is tempting to believe that a number field sieve analogue exists. At the moment, we have no idea how to do this, and in fact, Bauer and Hamdy[97] provided some evidence that it may not be possible.

The other important theoretical problem is whether the dependence on the GRH and ERH of the running time analysis and correctness can be removed. This appears to be a very difficult problem, and we currently know of no Las Vegas type algorithm that removes this dependence on unproven hypotheses and retains subexponential complexity. However, it is possible to certify unconditionally the regulators and the output to the principal ideal problem produced by the index-calculus algorithms described above, resulting in algorithms whose expected runtimes are faster than the deterministic algorithms presented in Chapters 7 and 10. The problems of unconditionally verifying the regulator and class number of a real quadratic order are discussed in Chapter 15 and that of unconditionally verifying the output to the principal ideal problem is discussed in Chapter 16.

Notes and References

[1] A *Las Vegas* algorithm is a probabilistic algorithm for which the output is correct but the running time is probabilistic. This is in contrast to *Monte Carlo* algorithms, which have deterministic running time but can only guarantee the correctness of their output up to some small probability of error.

[2] This idea seems to have originated with Kraitchik [Kra22], pp. 120–123, who used it to solve the discrete logarithm problem in the group of residues modulo a prime p.

[3] See [CP05] or [Coh93] for a survey of factoring algorithms.

[4] [LL93], [Gor93], and [Sch00].

[5] [LL90].

[6] [Sey87].

[7] [HM89].

[8] [Buc90].

[9] See [BD91b], [CDyDO93], [BD92], and [Abe94] for some significant examples of improvements to Buchmann's method.

[10] [Vol00] and [Vol02].

[11] [Jac99b] and [Jac00].

[12] [BV07].

[13] [McC89].

[14] [Vol00].

[15] See [Jac98], [JRW06], and [Ram06] for descriptions of numerical experiments supporting this phenomenon.

[16] [HM89].

[17] See [BV07], Lemma 11.4.6.

[18] See, for example, [GJS01] or [MS99].

[19] Seysen's result on ideal smoothness for imaginary quadratic orders appears in [Sey87]. Buchmann and Hollinger [BH96] generalized this to real quadratic fields, and Abel (née Hollinger) extended the result to real quadratic orders [Abe94], Proposition 7.2.2. It is interesting to note that no analogous result on smooth ideals has been proved for non-quadratic number fields. Thus, the analysis of index-calculus algorithms for arbitrary degree number fields requires the assumption of a smoothness result of this type.

[20] See [BV07], Proposition 11.4.3, for the statement and proof of this proposition in the context of quadratic fields.

[21] Adapted from [BV07], Lemma 11.4.6.

[22] [Sey87].

[23] See, for example, [BV07], Lemma A.5.5.

[24] Adapted from [BV07], Proposition 11.4.9.

[25] [BV07], Lemma 11.4.11.

[26] Adapted from [BV07], Proposition 11.4.13.

[27] [GJS01].

[28] [PZ79].

[29] [PZ85].

[30] [HM89].

[31] [Bac90].

[32] [GJS01].

[33] See [BV07], p. 235.

[34] [Sto00].

[35] Adapted from [BV07], Corollary 11.4.16.

[36] See [BV07], p. 268. Here, Buchmann and Vollmer argue that complexity

$$L_\Delta[1/2, 3\sqrt{2}/4 + o(1)] = L_\Delta[1/2, 1.061 + o(1)]$$

should be possible. Our improvement comes from combining their observations with an asymptotically faster HNF algorithm due to Storjohann and Labahn [SL96b].

[37] In fact, Vollmer stated in [Vol00] that the imaginary quadratic order DLP could be solved in expected time $L_\Delta[1/2, 3\sqrt{2}/4 + o(1)]$, but subsequently acknowledged in [Vol02], Corrigendum, that his proof was not valid because it did not differentiate between the costs of finding these "dense" and "sparse" relations.

[38] [SL96b].

[39] [BD91a].

[40] See Exercise 3 on p. 292 of [Knu98], and the solution on p. 630 for an explicit version of the generalized Chinese Remainder Theorem.

[41] [Jac00].

[42] [Buc90].

[43] [Abe94].

[44] [Vol02].

[45] See [BV07] and [Vol02].

[46] This result originally appeared as [Buc90], Theorem 2.1. See [BV07], Proposition 11.5.2, for a complete proof.

[47] [CDyDO93].

[48] [Mau00].

[49] See [BV07], Section 11.5.4, for a precise description of this method.

[50] Adapted from [BV07], Lemma 11.5.17.

[51] [BV07], Corollary 11.5.21.

[52] Adapted from [BV07], Proposition 11.5.23.

[53] [BV07], Lemma 11.5.24.

[54] Adapted from [BV07], Proposition 11.5.26.

[55] [Vol02] and [BV07], Algorithm 11.15.

[56] See [Vol02] and [BV07], Proposition 11.5.28.

[57] See, for example, [GJS01].

[58] Adapted from [BV07], Proposition 11.5.29.

[59] See [Mau00], Theorem 12.1.5.

[60] See [BV07], p. 235.

[61] [Sto00].

[62] Adapted from [BV07], Theorem 11.5.30.

[63] [Vol02].

[64] Adapted from [Vol02], Theorem 1.

[65] [Mau00].

[66] See, for example, [HMU07].

[67] See [McC89], [BW89b], and [BW91].

[68] [BW89b].

[69] [McC89].

[70] [AKS04].

[71] [BD91b] and [BD91a].

[72] As described in, for example, [BD91b].

[73] [CDyDO93].

[74] [BD92].

[75] [Jac99b].

[76] [BD92].

[77] [Len87].

[78] [Jac99a].

[79] [Sil87].

[80] [Pau96a].

[81] See [Jac99b] and [Jac99a].

[82] [AP95].

[83] [Jac99b].

[84] See Tables 5.12, 5.33, and 5.34 of [Jac99b] for additional examples.

[85] [Jac00].

[86] [Jac00].

[87] [JW00].

[88] [JW03].

[89] This phenomenon is explored in more detail by Jacobson in [Jac99b].

[90] [JW00].

[91] [HL23].

[92] [JW03].

[93] [HJW03].

[94] [JSW01].

[95] [GTTD07].

[96] [EGG$^+$06] and [EGG$^+$07].

[97] [BH03].

14

Applications to Cryptography

14.1 Introduction

We now live in a world where the security and integrity of our information and communications is not guaranteed, where the number and sophistication of attacks on these systems are increasing rapidly, and where the impact of those attacks can be measured in billions of dollars and in the loss of reputation and personal integrity. Thus, it is essential today, more than ever, to develop techniques that will protect our communications. One essential component of any installation in which secure communication is needed is cryptography. Its use goes back to Julius Caesar, or perhaps even earlier.[1]

Briefly put, *cryptography* is the study and development of techniques for rendering information unintelligible to all but intended recipients of that information.[2] If a sender, Alice, and a receiver, Bob, of a message wish to communicate over an insecure channel (e.g., mobile phone, internet) and want to ensure that no other unauthorized party can read their transmission, they will make use of a particular *cryptosystem*. A conventional cryptosystem can be thought of as a large collection of transformations (*ciphers*), any one of which will render the original message (*plaintext*) to unintelligible *ciphertext*, but in order for the receiver to read the message, Bob must know which particular transformation was used by Alice. The information that identifies the transformation used by the sender is called the *key*. It is important to point out that if an eavesdropper (Eve) acquires some message and its encrypted equivalent, she should not be able to extract the key from this information. Nor should the system be vulnerable to an adaptive attack; such attacks make use of information previously acquired to obtain new information from the sender and so on until the system is broken.

Cryptanalysis is the process by which Eve, on receiving some ciphertext, determines the original message without prior knowledge of the key. When this is successful, we say that Eve has broken the system. *Cryptology* is the study of both cryptography and cryptanalysis. With regard to cryptology, we should always bear in mind a famous quote of Poe[3]:

Few persons can be made to believe that it is not quite an easy thing to invent a method of secret writing which shall baffle investigation. Yet it may be roundly asserted that human ingenuity cannot concoct a cipher which human ingenuity cannot resolve.

To maintain security, it is vital that the key be known to only the sender and receiver of the messages. Of course, this means that at some point the key must be communicated between the sender and receiver in a very secure manner. In one-key or symmetric cryptosystems, this must be done over a different and more secure transmission channel than that used for the transformed (encrypted) messages. As separate communication channels are expensive and often inconvenient to use, one important objective of modern cryptography has been to try to eliminate them altogether. An extremely important event of the mid-1970s was the landmark contribution of Diffie and Hellman in their paper New Directions in Cryptography.[4] In this work they introduced several important concepts, such as the one-way function and a kind of scheme now referred to as a *Diffie-Hellman key exchange protocol*. In such a scheme, the transmitter and the receiver exchange information over a public channel that they can then assemble into a common communication key. An eavesdropper, however, does not acquire sufficient information to construct this key. They also suggested, but provided no example, another method of avoiding the use of a separate key channel by using a *public-key cryptosystem*. In such a system, each participant has two keys: a private one and a public one. The idea is that knowledge of the public key should not reveal anything about the private key. Thus, anyone who wants to send a secure message to one of the participants uses that individual's public key, available in an easily accessible directory, for example, to encrypt the message; as only this same individual knows his private key, he can use this to decrypt the enciphered message, but no one else can. The first example of a public-key cryptosystem was discovered a year later and is now well known as the RSA system after its inventors: Rivest, Shamir, and Adelman.[5] The notion of a digital signature quickly followed and versions of these two protocols in particular have been adopted in a variety of standards and are in everyday use.[6]

In this short description of the development of public-key cryptography, it should be noted that all of the important concepts mentioned above had been discovered earlier by workers at the Communications-Electronics Security Group (CESG), a division of GCHQ in Cheltenham, U.K., but as this information was classified, credit today is generally given to the individuals mentioned above.[7] We also emphasize that Poe's dictum seems to be true for these modern techniques, but the difference is that the amount of time that is required to break these systems is very long, and even when this happens, all one has to do to frustrate the cryptanalyst is make the key larger.

The designer, say Alice, of an RSA cryptosystem selects at random two large (1024 bits is now being widely advocated) primes p and q and calculates $n = pq$. She also selects at random a value of e ($< n$) such that $(e, \phi(n)) = 1$

and finds d such that
$$de \equiv 1 \pmod{\phi(n)}$$
and $0 < d < n$. For this scheme, Alice's public key is $\{e, n\}$ and her private key is d.

If Bob wishes to send a secure message M $(< n)$ such that $(M, n) = 1$ to Alice, he sends
$$C \equiv M^e \pmod{n} ,$$
where $0 < C < n$. Alice can recover M from C by calculating
$$C^d \equiv M^{ed} \equiv M^{1+k\phi(n)} \equiv M \pmod{n} .$$

Since $M < n$, it can now be determined easily.

It is important to stress that great care much be exercised in actually implementing this encryption technique,[8] but if properly done the scheme resists attack very well.[9] This is still the most widely used public-key encryption method, although new methods, for example based on the arithmetic of elliptic curves,[10] which make use of smaller key sizes, especially applicable to small, hand-held devices, are becoming popular. It is clear that if a cryptanalyst succeeds in solving the difficult problem of factoring[11] n, then she has broken the system. However, it has never been shown that if she can break the system, then she can factor n; that is, we do not know whether breaking RSA and factoring the modulus n are of equivalent difficulty[12].

14.2 The Pell Equation in a Public-Key Cryptosystem

In 1979, Rabin[13] succeeded in developing a method for encrypting which is demonstrably as difficult to break as it is to factor a modulus n. However, Rabin's scheme has several problems, and he only advocated its use as a technique for producing digital signatures. In what follows, we will describe an RSA-like technique[14] for encrypting M which does not have the problems of Rabin's method and is as difficult to break as factoring n.

Let[15] $t, u \in \mathbb{Z}$ and $t^2 - Du^2 = 1$. We will now produce some properties of the T_n and U_n introduced in §1.4. If we let $i, j \in \mathbb{Z}$, it is easy to establish from
$$T_n + \sqrt{D}U_n = (t + \sqrt{D}u)^n$$
that
$$T_{i+j} = T_i T_j + DU_i U_j, \quad U_{i+j} = T_i U_j + T_j U_i , \tag{14.1}$$
and
$$T_{i+j} = 2T_i T_j - T_{j-i}, \quad U_{i+j} = 2T_i U_j - U_{j-i} . \tag{14.2}$$
Hence, if $i = j$, we get
$$T_{2i} = T_i^2 + DU_i^2 = 2T_i^2 - 1, \quad U_{2i} = 2T_i U_i .$$

Since $T_1 = t$, we see by the first formula of (14.1) that T_i can be expressed as a polynomial in $\mathbb{Z}[t]$, which we will sometimes denote by $T_i(t)$. Also, we deduce from

$$(t + \sqrt{D}u)^{ij} = (T_j + \sqrt{D}U_j)^i$$

that

$$T_i(T_j(t)) = T_{ij}(t) . \tag{14.3}$$

If, for some values of k and $n \in \mathbb{Z}$, we wish to compute $T_k(t) \pmod{n}$, we first write the binary expansion of k as

$$k = b_0 2^l + b_1 2^{l-1} + \cdots + b_l ,$$

where $b_0 = 1, b_1, \ldots, b_l \in \{0, 1\}$ and $l = \lfloor \log_2 k \rfloor$. Put $P_1 \equiv (T_2, T_1) \pmod{n}$. Here, we use the notation $(A_1, B_1) \equiv (A_2, B_2) \pmod{n}$ to denote that $A_1 \equiv A_2 \pmod{n}$ and $B_1 \equiv B_2 \pmod{n}$. If

$$P_j \equiv (T_{s+1}, T_s) \pmod{n} ,$$

we define

$$P_{j+1} \equiv \begin{cases} (2T_s T_{s-1} - T_1, 2T_s^2 - 1) \pmod{n} & \text{when } b_{s+1} = 0 \\ (2T_{s+1}^2 - 1, 2T_s T_{s+1} - T_1) \pmod{n} & \text{when } b_{s+1} = 1. \end{cases}$$

By the first formula of (14.2), we see that

$$P_{j+1} \equiv (T_{2s+b_{s+1}+1}, T_{2s+b_{s+1}}) \pmod{n} .$$

By the same reasoning as that used at the beginning of §11.3, it follows that

$$P_l \equiv (T_{k+1}, T_k) \pmod{n} .$$

Also, the amount of work needed to compute $P_l \pmod{n}$ is not much more than that needed to compute $a^k \pmod{n}$.

Let p and q be distinct odd primes and put $\eta_p \equiv p \pmod{4}$ and $\eta_q \equiv q \pmod{4}$, where $\eta_p, \eta_q \in \{1, -1\}$. Find a non-square integer $D > 0$ such that

$$(D/p) = -\eta_p \quad \text{and} \quad (D/q) = -\eta_q .$$

Since there are $(p-1)/2$ values of $D \pmod{p}$ satisfying the first of these conditions and $(q-1)/2$ values of $D \pmod{q}$ satisfying the second, there must, by the Chinese Remainder Theorem, be $(q-1)(p-1)/4$ values of $D \pmod{n}$, where $n = pq$, satisfying these conditions. Thus, under the reasonable assumption that such values of D are randomly distributed, we would expect to find a rather small value of D by trial, and this is what happens in practice.[16]

We now put $n = pq$ and $m = (p - \eta_p)(q - \eta_q)/4$. We also observe that m is odd and $(D, n) = 1$. We are now able to prove the following useful theorem.

Theorem 14.1. *Let $D, n,$ and m be defined as above and let $\gamma = a + b\sqrt{D}$, where $a, b \in \mathbb{Z}$ and the Jacobi symbol $(N(\gamma)/n) = 1$. If $\alpha = \gamma/\overline{\gamma}$, then*

$$\alpha^{2k} \equiv \pm\alpha \pmod{n} \;,$$

where $k \equiv (m+1)/2 \pmod{m}$.

Proof. It is well known that

$$\gamma^p \equiv \begin{cases} \gamma \pmod{p} & \text{where } \eta_p = 1 \\ \overline{\gamma} \pmod{p} & \text{where } \eta_p = -1 \;; \end{cases}$$

thus,

$$\gamma^{p-\eta_p} \equiv N^{(1-\eta_p)/2},$$

where $N = N(\gamma)$. Since $N\alpha = \gamma^2$, we get

$$\alpha^{(p-\eta_p)/2} \equiv N^{\frac{p-1}{2}} \equiv (N/p) \pmod{p} \;.$$

Similarly,

$$\alpha^{(q-\eta_q)/2} \equiv (N/q) \pmod{q} \;.$$

Thus, $\alpha^m \equiv (N/p) \pmod{p}$ and $\alpha^m \equiv (N/q) \pmod{q}$. Since $(N/pq) = 1$, we have $(N/p) = (N/q)$ and

$$\alpha^m \equiv \pm 1 \pmod{n} \;.$$

Since $k = tm + (m+1)/2$, we get

$$\alpha^{2k} \equiv \alpha^{2tm}\alpha^{m+1} \equiv \alpha^{m+1} \equiv \pm\alpha \pmod{n} \;,$$

as required. $\square$

We can now use this result as the basis of a two-key cryptosystem. Alice first selects $p, q,$ and a value for $S \in \mathbb{Z}$ such that the Jacobi symbol $((S^2 - D)/n) = -1$. As there are close to $\phi(n)/2$ such values of S, she can usually find a small value for S by trial. Alice also selects a value for e such that $(e, m) = 1$ and makes her key $\{n, e, S, D\}$ public. She also solves

$$de \equiv (m+1)/2 \pmod{m}$$

for d and keeps d as her secret key. As we have noted that S and D are usually small, Alice's public key will not be much larger than a typical RSA key.

Let M be a message that Bob wishes to communicate to Alice. He computes $j_1 = ((M^2 - D)/n)$. As the chance that $j_1 = 0$ is very remote, we may assume that $|j_1| = 1$. If $j_1 = 1$, Bob puts

$$t \equiv (M^2 + D)/(M^2 - D) \pmod{n} \;,$$
$$u \equiv 2M/(M^2 - D) \pmod{n} \;;$$

if $j_1 = -1$, Bob puts

$$t \equiv \left((M^2 + D)(S^2 + D) + 4DMS\right)/((M^2 - D)(S^2 - D)) \pmod{n} \ ,$$
$$u \equiv \left(2S(M^2 + D) + 2M(S^2 + D)\right)/((M^2 - D)(S^2 - D)) \pmod{n} \ .$$

He also selects $j_2 \equiv t \pmod 2$, where $j_2 \in \{0,1\}$. In other words, if $\mu = M + \sqrt{D}, \nu = S + \sqrt{D}$, then

$$\alpha = t + u\sqrt{D} \equiv \gamma/\bar{\gamma} \pmod{n} \ ,$$

where

$$\gamma = \mu\nu^{(1-j_1)/2}$$

and $(N(\gamma)/n) = 1$. Also,

$$t^2 - Du^2 \equiv 1 \pmod{n}$$

for these values of t and u. Throughout the discussion that follows, we will assume that $(u,n) = 1$, a restriction[17] similar to $(M,n) = 1$ for the RSA scheme.

For this value of t, Bob computes the pair $(T_{l+1}, T_l) \pmod{n}$ by the technique described earlier. As $T_1 = t$ and $U_1 = u$, by the first formula in (14.1), we get

$$DU_l = T_{l+1} - tT_l \ .$$

By Theorem 14.1 we must have

$$T_{2ed} \equiv \sigma t \pmod{n} \ ,$$
$$U_{2ed} \equiv \sigma u \pmod{n} \ ,$$

where $\sigma \in \{-1.1\}$. Also, if p or q divides $U_l T_l$, then $p \mid U_{2e}$ and therefore $p \mid U_{2ed}$, but this means that $p \mid u$, which is impossible. Hence, $(n, U_l T_l) = 1$. Thus, since $(n, uDU_l) = 1$, Bob can compute

$$E \equiv T_l/E_l \equiv DuT_l(T_{l+1} - tT_l)^{-1} \pmod{n} \ ,$$

with $0 < E < n$. Bob sends $E(M) = \{E, j_1, j_2\}$ to Alice.

To recover M from $E(M)$, Alice computes T_{2l} and $U_{2l} \pmod{n}$ by

$$T_{2l} \equiv T_l^2 + DU_l^2 \equiv \frac{T_l^2 + DU_l^2}{T_l^2 - DU_l^2} \equiv \frac{E^2 + D}{E^2 - D} \pmod{n}$$

$$U_{2l} \equiv \frac{2T_l U_l}{T_l^2 - DU_l^2} \equiv \frac{2E}{E^2 - D} \pmod{n} \ .$$

She next computes

$$T_d(T_{2e}) \equiv T_{2de}(t) \pmod{n} \text{ and } T_{d+1}(T_{2e}) \equiv T_{2de+2e}(t) \pmod{n}$$

by the technique mentioned earlier. By (14.1),

$$DU_{2e}U_{2ed} = T_{2ed+2e} - T_{2e}T_{2ed} ;$$

thus, Alice can compute T_{2de} (mod n) and $DU_{2e}U_{2ed}$ (mod n). Since, by Theorem 14.1, $T_{2de} \equiv \sigma t$ (mod n) and Alice knows $j_2 \equiv t$ (mod 2), she can find σ and therefore determine t (mod n). Since

$$u \equiv \sigma U_{2ed} \equiv \sigma(T_{2ed+2e} - T_{2e}T_{2ed})/(DU_{2e}) \pmod{n} ,$$

she can also find u.

She now has $\alpha \equiv t + u\sqrt{D}$ (mod n). On putting

$$\alpha' = \begin{cases} \alpha & \text{when } j_1 = 1 \\ \dfrac{\alpha(S - \sqrt{D})}{S + \sqrt{D}} & \text{when } j_1 = -1 , \end{cases}$$

she has

$$\alpha' \equiv (M + \sqrt{D})/(M - \sqrt{D}) \pmod{n}$$

and

$$M \equiv (\alpha' + 1)\sqrt{D}/(\alpha' - 1) \pmod{n} .$$

Of course, if anyone finds p or q, he can immediately compute M and d and break the system. Suppose, on the other hand, that a cryptanalyst has found an algorithm which can be used to break the system. We now show that this algorithm can be used to factor n. We select by trial some X such that $((X^2 - D)/n) = -1$. We encrypt X by putting $\gamma = X + \sqrt{D}$ and creating the "ciphertext" $E(X) = (E, 1, j_2)$. The false value of $j_1 = 1$ is selected on purpose. The cryptanalyst uses his algorithm to decrypt $E(X)$ and finds the corresponding plaintext Y. However, Y will not be the same as X, because of the false value for j_1. Indeed, it can be shown that $(X - Y, n) = p$ or q and, therefore, the cryptanalyst's algorithm can be used to factor n.

Results from many different areas of mathematics have been applied to the development of cryptographic systems. One reason for this is that it is always sound cryptographic practice to have access to as many different systems as possible; this ensures that the sender has a choice of possible schemes, a very useful feature if one or more of them is compromised. The above encryption technique[18] appeared in 1984 and represents an early instance of the application of algebraic number theory to produce cryptosystems. Since that time, several other secure communication algorithms have been developed that involve the properties of algebraic number fields. Most often these make use of quadratic fields, as these structures possess many of the complicating features that make them much more difficult, particularly from the perspective of a cryptanalyst, to deal with than the field of rational numbers. Furthermore, as we have seen, conducting arithmetic in them is relatively simple and efficient compared to the same operations in number fields of higher degree. In the remainder of this chapter, we will present some other cryptosystems that make use of various properties of quadratic fields.

14.3 Cryptography in Imaginary Quadratic Fields

The class group of an imaginary quadratic field[19] was first proposed for use in cryptographic protocols in 1988 by Buchmann and Williams.[20] Although currently not used in practical applications, there are a number of factors which make these protocols interesting to study. The first is that, unlike most widely used public-key cryptographic protocols, the security of quadratic field-based protocols does not rely on the presumed difficulty of integer factorization, computing discrete logarithms in finite fields, or the elliptic curve discrete logarithm problem. As pointed out in Chapter 7, it is known that integer factorization reduces to the problem of computing the class number h_Δ, implying that computing these invariants (and likely solving the associated discrete logarithm problems) is at least as hard as integer factorization. The current belief, due to the fact that no number field sieve analogue with complexity $L_\Delta(1/3, \alpha)$ has been discovered for computational problems in quadratic fields, is that these problems are likely harder than integer factorization. Although the elliptic curve discrete logarithm problem is currently believed to be harder (the best known general-purpose algorithms have exponential complexity), it is still important to have alternative cryptosystems whose security is unrelated to those currently being used. To the best of our knowledge, there are no reductions between the elliptic curve discrete logarithm problem and discrete logarithm problems in quadratic fields.

Cryptography in imaginary quadratic fields has been described elsewhere in some detail.[21] Hence, we will only present here some of the main highlights and refer the interested reader to these sources for further details.

Buchmann and Williams' original contribution was a version of Diffie-Hellman key exchange set in the ideal class group of an imaginary quadratic field. The participants, Alice and Bob, agree on a publicly-available discriminant $\Delta < 0$ and a reduced $\mathcal{O}_\Delta$-ideal $\mathfrak{g}$. Alice selects a random integer x and sends the reduced ideal equivalent to $\mathfrak{g}^x$ to Bob, who selects a random integer y and sends $\mathfrak{g}^y$ to Alice. Then, Alice and Bob can both compute the same reduced ideal $\mathfrak{K} \sim \mathfrak{g}^{xy}$. Notice that an eavesdropper, who only has Δ, $\mathfrak{g}$, $\mathfrak{g}^x$, and $\mathfrak{g}^y$, has to solve the Diffie-Hellman problem in the class group in order to obtain $\mathfrak{K}$. One method to do this is to solve an instance of the discrete logarithm problem in the class group to obtain either x or y.

In general, it is straightforward to use any finite abelian group in this protocol; however, any such group should satisfy the following requirements if it is to be used for cryptography:

1. efficient, unique representation of group elements,
2. efficient group operation,
3. exponentiation is a one-way function (e.g. the discrete logarithm problem should be computationally infeasible).

We have seen in previous chapters that all three of these requirements are satisfied for class groups of imaginary quadratic fields. Group elements (ideal equivalence classes) are represented by reduced ideals, essentially an ordered pair of integers, each of which is bounded in absolute value by $\sqrt{\Delta}$. Ideal arithmetic, especially using NUCOMP, is not as efficient as, say, adding points on an elliptic curve, but it is nevertheless polynomial in $\log|\Delta|$ and fairly efficient in practice. Finally, in the previous chapter, we saw that the best known algorithms for solving the discrete logarithm problem in the class group are of subexponential complexity in $\log|\Delta|$. Thus, class groups of imaginary quadratic fields are, at least in principle, suitable for cryptographic applications.

More generally, the usual problems related to finite abelian groups are believed to be hard in the class group of an imaginary quadratic order. The following are the main problems:

- IQ-DHP (computational Diffie-Hellman problem): Given ideals $\mathfrak{g}$, $\mathfrak{a}$, and $\mathfrak{b}$ of an imaginary quadratic order with $\mathfrak{a} \sim \mathfrak{g}^a$ and $\mathfrak{b} \sim \mathfrak{g}^b$ for some unknown integers a and b, compute an ideal equivalent to $\mathfrak{g}^{ab}$.
- IQ-DLP (discrete logarithm problem): Given ideals $\mathfrak{g}$ and $\mathfrak{a}$ of an imaginary quadratic order, find the smallest positive integer x such that $\mathfrak{a} \sim \mathfrak{g}^x$ or decide that no such x exists.
- IQ-OP (order problem): Given an ideal $\mathfrak{a}$ of an imaginary quadratic order, compute the order of $[\mathfrak{a}] \in Cl_\Delta$.
- IQ-RP (root problem): Given an ideal $\mathfrak{a}$ of an imaginary quadratic order and an integer $x > 1$, compute $\mathfrak{g}$ such that $\mathfrak{a} \sim \mathfrak{g}^x$ or decide that no such $\mathfrak{g}$ exists.

The following reductions between these problems are known.[22] The notation $A \leq_P B$ indicates that there exists a polynomial-time reduction from problem A to problem B.

- IQ-RP $\leq_P$ IQ-OP
- IQ-OP $\leq_P$ IQ-DLP
- IQ-DHP $\leq_P$ IQ-DLP

The existence of these reductions indicates that the discrete logarithm problem is at least as hard as the root problem and the order problem, as any algorithm for computing discrete logarithms can also be used to solve the the other two problems by applying the polynomial-time reductions. In fact, there is some evidence that the root and order problems may be easier in practice. Sutherland[23] has discovered an algorithm for solving the order problem in a generic group of order N that runs in time $O((N/\log\log N)^{1/2})$ in the worst case (N is prime). In contrast, Shoup[24] proved that any algorithm for solving the discrete logarithm problem in a generic group has complexity $\gg (N^{1/2})$, so the order problem is in fact easier in a generic group. In class groups of imaginary quadratic fields, Buchmann and Vollmer have shown[25] that a modification of the index-calculus algorithm for computing the class group described in Chapter 13 can solve the IQ-RP in expected time $L_\Delta[1/2, 1+o(1)]$. Thus, according to our current knowledge, the root problem is easier than both the order and discrete logarithm problems in class groups of imaginary quadratic orders.

On the other hand, as shown in Chapter 7, we also have that the IFP $\leq_P$ IQ-OP, where by IFP we denote the integer factorization problem. In addition, the IQ-RP and IFP are equivalent[26] if $x = 2$ (i.e., computing square roots). This suggests that computing element orders, the class number, and even discrete logarithms and roots are all likely intractable, so cryptosystems whose security relies on any of these can, and have been, proposed.

There is one main difficulty in adapting generic public-key cryptosystems to imaginary quadratic fields. Many of these systems, especially digital signature protocols such as the digital signature algorithm[27] (DSA), require knowledge of the order of the group. As we have seen in Chapters 10 and 13, computing the class number is essentially as hard in practice, given our current knowledge, as computing discrete logarithms, so cryptographic protocols in class groups cannot, in general, make use of the group order. In addition, even when the class number is not explicitly required for the protocol itself, it is desirable to compute it to verify that the group is not cryptographically weak in the sense that it admits special-purpose algorithms that can solve the discrete logarithm problem significantly faster. For example, if the class number is smooth, the Pohlig-Hellman algorithm[28] can be used to solve the discrete logarithm problem by computing discrete logarithms in the small order p subgroups corresponding to each prime $p \mid h_\Delta$ and combining these via the Chinese Remainder Theorem.

Fortunately, at least heuristically, a random class group is cryptographically suitable with high probability. The following summarizes results from previous chapters supporting this conclusion:

1. On average, $h_\Delta \approx \sqrt{|\Delta|}$ by Siegel's theorem (see Chapter 9, (9.21)).
2. Assuming the extended Riemann hypothesis (ERH)

$$h_\Delta \gg \sqrt{|\Delta|}/(\log\log|\Delta|)$$

   due to a result of Littlewood (see Chapter 9, (9.23)).
3. The class number h_Δ is odd if and only if $|\Delta|$ is a prime congruent to 3 modulo 4 (see §7.3).
4. If h_Δ is odd, Cl_Δ is cyclic with probability > 0.97 according to heuristics of Cohen-Lenstra (Conjecture 7.7), and these, while still conjectural, are supported by extensive numerical data.

In addition, Hamdy and Möller[29] showed that h_Δ behaves asymptotically like an arbitrary integer with respect to smoothness assuming the Cohen-Lenstra heuristics. Thus, with high probability, simply taking $|\Delta| \equiv 3 \pmod 4$ prime at random yields a cryptographically suitable imaginary quadratic field. It is not known how to construct a cryptographically weak imaginary quadratic field of large discriminant; for example, one whose class number is smooth.

14.3.1 Cryptographic Protocols

Diffie-Hellman key exchange and El Gamal encryption[30] using the ideal class group Cl_Δ are straightforward. Versions that are provably secure against chosen-ciphertext attacks, including a generalization of the Diffie-Hellman integrated encryption scheme,[31] are also straightforward, and the security proofs carry over directly.

Signature schemes are more difficult, because computing h_Δ is believed to be computationally infeasible and most group-based signature schemes require knowing the order of the group (e.g. to reduce the sizes of exponents in DSA). Nevertheless, the following have been proposed for imaginary quadratic fields:

- IQ-RDSA[32]: a variant of the NIST standard Digital Signature Algorithm (DSA) whose security is based on the intractability of the IQ-RP. This protocol was broken by Fouque and Poupard[33] in 2003.
- IQ-DSA[34]: a variant of Schnorr signatures due to Poupard and Stern[35] that does not require the group order h_Δ. This is secure against existential forgery in the random oracle model[36] using an adaptive chosen message attack assuming the intractability of IQ-DLP. It requires significantly larger exponents and has larger signatures than DSA because the exponents cannot be reduced modulo the unknown group order.
- IQ-GQ[37]: an analogue of the Guillou-Quisquater signature scheme[38] whose security is based on the intractability of the IQ-RP. IQ-GQ is also secure in the random oracle model against existential forgery using an adaptive chosen message attack.

14.3.2 Efficiency

As shown in Chapter 13, the IQ-DLP in Cl_Δ can be solved in time

$$L_\Delta[1/2, \sqrt{2} + o(1)]$$

using Vollmer's algorithm, where

$$L_\Delta[a, b] = \exp\left(b(\log|\Delta|)^a (\log\log|\Delta|)^{1-a}\right) .$$

In practice, an analogue of the self-initializing quadratic sieve factoring algorithm is used, which, according to a numerical investigation of Hamdy,[39] likely runs in time $L_\Delta[1/2, 1 + o(1)]$. Thus, the IQ-DLP can be solved in subexponential time, but not as fast as the number field sieve (NFS) for factoring or computing discrete logarithms in a finite field. The discriminants for imaginary quadratic field-based protocols required to provide the same level of security as RSA will therefore be smaller than the corresponding RSA moduli. For example, Hamdy estimates that a 795-bit Δ provides roughly the same security as a 1024-bit RSA. The discrepancy in sizes increases with the security level; Hamdy's estimates suggest that a 5704-bit discriminant provides the same level of security as a 15360-bit RSA. As mentioned in Chapter 13, it is unknown whether an NFS analogue exists for the IQ-DLP.

Because inversion of ideal classes is almost free (simply negate the P coefficient), standard improvements to binary exponentiation such as non-adjacent form and window-based extensions can be applied directly to speed the required ideal exponentiations in these protocols. Hamdy[40] gives some performance data on optimized implementations of IQ-DSA and IQ-GQ. From these data, it is clear that these protocols do not offer any improvements in efficiency over more established schemes such as DSA and RSA. However, the times listed are within a factor of at most six of these faster protocols, and perhaps with a concentrated effort, the efficiency of cryptographic protocols based on class groups of imaginary quadratic fields can be improved to the point that they will be competitive with other protocols.

14.4 Cryptography in Real Quadratic Fields

In 1989, Buchmann and Williams[41] described how to perform an analogue of Diffie-Hellman key exchange in the infrastructure of the principal class of a real quadratic field. This was especially noteworthy, as this key exchange protocol was the first based on arithmetic in a structure that is not a group.

Recall that ideal representation and arithmetic are similar to the imaginary case, except for the following main differences (see Chapter 7):

- By the Cohen-Lenstra heuristics, we expect that h_Δ is usually small. As $h_\Delta R_\Delta \approx \sqrt{\Delta}$, we typically have $R_\Delta \approx \sqrt{\Delta}$, so $h_\Delta \approx 1$.
- Ideal reduction is not unique. Each ideal equivalence class contains a finite cycle of reduced ideals that can be traversed using the reduction algorithm (continued fraction expansion of $(P + \sqrt{D})/Q$). The number of ideals in the cycle is roughly R_Δ, and hence, typically of size $\sqrt{\Delta}$.

Thus, DLP-based cryptographic protocols using the ideal class group of a real quadratic field cannot be used directly in general, because there is no efficient method to decide equality of equivalence classes.

One possible solution to this problem is to restrict the use of real quadratic fields in cryptography to special families for which the regulator is guaranteed to be small. In this way, the protocols based on imaginary quadratic fields can be applied almost directly, after adjusting for the fact that there will still be multiple reduced representatives (although the number of these will be small) of each ideal equivalence class. Chapter 6 describes a number of families of discriminants for which the corresponding regulator will be small, and Schielzeth and Pohst[42] described further results in this direction.

Another approach is to instead make use of the fact that equivalence testing is usually difficult and use this (typically, principality testing) as the basis for cryptosystems. In other words, the infrastructure of the principal class, as described in §7.4, is used as the setting for cryptography as opposed to the class group. This was the idea proposed by Buchmann and Williams in their infrastructure-based version of the Diffie-Hellman key exchange protocol.

The main idea behind Buchmann and Williams' key exchange protocol is as follows. Two parties, Alice and Bob, first agree on a public discriminant of a real quadratic field $\mathbb{Q}(\sqrt{\Delta})$ and a reduced principal ideal $\mathfrak{g}$, after which the following steps are executed:

1. Alice selects a random integer a with $0 < a < B$ for some bound B, computes a reduced ideal $\mathfrak{a} = (\theta_a)\mathfrak{g}^a$ with $\theta_a \approx 1$, and sends $\mathfrak{a}$ to Bob.
2. Bob selects a random integer b such that $0 < b < B$, computes a reduced ideal $\mathfrak{b} = (\theta_b)\mathfrak{g}^b$ with $\theta_b \approx 1$, and sends $\mathfrak{b}$ to Alice.
3. Alice computes $\mathfrak{k}_A = (\theta_\alpha)\mathfrak{b}^a$, with $\theta_\alpha \approx 1$.
4. Bob computes $\mathfrak{k}_B = (\theta_\beta)\mathfrak{a}^b$, with $\theta_\beta \approx 1$.

At the end of this protocol, Alice and Bob have ideals $\mathfrak{k}_A$ and $\mathfrak{k}_B$, both of which are equivalent to $\mathfrak{g}^{ab}$. The conditions on the associated relative generators ensure that $\mathfrak{k}_A = (\alpha)\mathfrak{g}^{ab}$ and $\mathfrak{k}_B = (\beta)\mathfrak{g}^{ab}$ with $\alpha, \beta \approx 1$. If sufficiently accurate approximations of these relative generators, or their logarithms, are maintained, it is possible to ensure that $\mathfrak{k}_A = \mathfrak{k}_B$, so that Alice and Bob share the same ideal at the end of the protocol. Buchmann and Williams described a solution that guarantees that Alice and Bob each obtain one of only two possible key ideals in the worst case and described how to resolve this ambiguity by transmitting at most one additional bit.

Although this protocol is clearly similar to standard Diffie-Hellman key exchange set in the class group of an imaginary quadratic field, there are two significant differences. The first is that extra infrastructure operations are required during exponentiation in order to keep the associated relative generators close to 1. After each ideal multiplication, we obtain a reduced ideal equivalent to the product of the inputs, say, $\mathfrak{c} = (\gamma)\mathfrak{ab}$. If NUCOMP as described in §5.4 is used, we expect that the relative generator γ typically satisfies $|\log \gamma| \approx (\log \Delta)/4$ by (5.47). Additional baby steps are required after reduction to find another equivalent ideal $\mathfrak{c}' = (\gamma')\mathfrak{ab}$, with $\log |\gamma| \approx 0$, implying that $\gamma \approx 1$. In comparison to the imaginary case, where only ideal multiplication and reduction are required, this represents additional computational overhead.

The second difference between this protocol and that in the imaginary case is that it is necessary to compute and maintain approximations of the relative generators arising from the ideal arithmetic. This is again additional overhead not required in the imaginary quadratic case. Furthermore, these approximations must be computed to sufficient accuracy in order to guarantee that the participants end up with the same ideal when the protocol terminates.

Since Buchmann and William's original protocol, there have been some efforts to mitigate these complications in an attempt to achieve practical performance comparable to key exchange in imaginary quadratic fields. These improvements focus primarily on the methods used to approximate the relative generators and the analysis of the required precision to ensure successful completion of the protocol. The first significant result is due to Hühnlein and Paulus,[43] who showed that unique key ideals can be obtained if approximations to the logarithms of the relative generators are maintained that are accurate to at least $\log_2(3072\sqrt{\Delta}B^2)$ bits, where B is the bound on the exponents a and b used in the protocol.

The latest results in this direction are due to Jacobson, Scheidler, and Williams,[44] who described how to use NUCOMP for ideal multiplication and (f, p) representations, as described in Chapter 11, to approximate the relative generators. The idea is for Alice and Bob to each compute near-reduced[45] (f, p) representations $(\mathfrak{k}_A, d_A, k_A)$ and $(\mathfrak{k}_B, d_B, k_B)$ of $\mathfrak{g}^{ab}$. Then, as long as the precision p is sufficiently large, $\mathfrak{k}_A$ and $\mathfrak{k}_B$ will both be close to $\mathfrak{g}^{ab}$. In particular, with $p > \log_2(50B^2 \log_2 B)$, a precision bound that is dependent only on the exponent bound B, it can be shown that Alice and Bob each obtain one of five possible key ideals in the worst case and that by communicating an additional five bits, this ambiguity can be resolved.

One advantage to this approach is that the precision requirement is not directly dependent on the discriminant. This becomes particularly advantageous when using an exponent bound B proportional to the number of bits of security offered by a given discriminant. For example, under the assumption that solving the principal ideal problem and the ideal class group discrete logarithm problem requires the same amount of time for discriminants of the same size, Hamdy's estimates suggest that a 795-bit discriminant offers roughly 80 bits of security; that is, the time required to solve the discrete logarithm problem is similar to that for breaking a block cipher with an 80-bit key. Thus, an exponent bound $B = 2^{160}$ can be used without compromising security, as baby-step giant-step methods would be able to solve the principal ideal problem, given the upper bound 2^{160} on the unknown distance, in time 2^{80}. Using these parameters, Hühnlein and Paulus' method would require 730 bits of precision for approximating the relative generators, whereas the method with (f, p) representations requires only 333 bits. In addition to yielding improved practical performance by way of requiring less precision, using (f, p) representations also simplifies the precision analysis itself, as approximations of logarithms are not required.

It is possible to obtain even further improvements by making use of w-near (f, p) representations. By using a value of $w > 0$, it is possible to reduce the extra baby-step operations required by the WNEAR algorithm (Algorithm 11.2) to ensure that the relative generators remain close to w. The idea is based on an observation of Stein as applied to cryptographic key exchange in the infrastructure of a real quadratic function field defined over a finite field.[46] As pointed out in the discussion in §11.3 and in §A.1 of the Appendix, on average, the relative generators γ $(= 1/\mu)$ obtained from NUCOMP tend to satisfy $\log \gamma \approx -(\log \Delta)/4$. Suppose that $\mathfrak{b} = (\theta)\mathfrak{g}$ and that $\log \theta \approx (\log \Delta)/4$. Then the reduced ideal $\mathfrak{c}$ equivalent to $\mathfrak{b}^2$ satisfies $\mathfrak{c} = (\theta')\mathfrak{g}^2$ with $\theta' = \gamma\theta^2$ and $\theta' \approx \theta$. Thus, $\mathfrak{c} = (\theta')\mathfrak{a}^2$, with $\log \theta' \approx (\log \Delta)/4$, and very few (if any) extra baby steps beyond those required for reduction are required to ensure that

θ' is of this size. Carrying this idea further allows one to compute $\mathfrak{d} = (\alpha)\mathfrak{g}^a$ with $\log \alpha \approx (\log \Delta)/4$ using binary exponentiation, again with few additional baby steps beyond those required for reduction expected to ensure that the relative generator is indeed close in size to $(\log \Delta)/4$.

This observation suggests making use of w-near (f, p) representations as described in §11.2. Instead of exponentiating the ideal $\mathfrak{g}$, we exponentiate a w-near representation of $\mathfrak{g}$ using EXP (Algorithm 11.4). As discussed in §11.2 (after the presentation of WMULT), if we take w such that $2^{w-1} < \Delta^{1/4} < 2^w$ (i.e., $w = \lceil (\log_2 \Delta)/4 \rceil$), then we would expect to compute a w-near representation of $\mathfrak{g}^a$ with very few extra baby steps required to reestablish the w-near property after each ideal multiplication during EXP.

We now describe a version of cryptographic key exchange in the infrastructure using w-near (f, p) representations. Alice and Bob first publicly agree on a large discriminant Δ of a real quadratic field, a reduced principal ideal $\mathfrak{g}$ in the maximal order $\mathcal{O}_\Delta$, and a bound $B \in \mathbb{N}$ on the exponents. Let $(\mathfrak{g}_0, d_0, k_0)$ be a w-near (f, p) representation of $\mathfrak{g}$. Using Algorithm 11.2 (WNEAR), we can compute $(\mathfrak{g}_0, d_0, k_0)$ with $f = 1 + 9/8$.

The first result[47] shows the effect on the error term f when truncating the approximation d when using (f, p) representations. This allows us to transmit fewer bits in the key exchange protocol without greatly affecting the required precision.

Lemma 14.2. *Let $(\mathfrak{b}, d, k)$ be a w-near reduced (f, p) representation of some ideal $\mathfrak{a}$. Let $r \in \mathbb{N}$ with $r < p$. Set $d' = 2^r \lceil 2^{-r} d \rceil$. Then $(\mathfrak{b}, d', k)$ is a w-near reduced $(f + 2^r, p)$ representation of $\mathfrak{a}$.*

The proof of this lemma is straightforward using Definition 11.1.

The key exchange protocol consists of each participant computing a w-near representation of $\mathfrak{g}^{ab}$ via two successive exponentiations. Theorem 11.4 states that if we select the precision p sufficiently large that at the end of these exponentiations we can ensure that the error $f < 2^{p-4}$ in both cases, then the ideal computed by each participant is one of at most five that are easily computed by the other. The next result[48] shows what precision p is required to ensure that this happens.

Theorem 14.3. *Let $(\mathfrak{g}_0, d_0, k_0)$ be a w-near $(1 + 9/8, p)$ representation of a reduced principal ideal $\mathfrak{g}$ and let $p, a, b, B \in \mathbb{Z}$ with $B \geq 14$, $0 < a, b \leq B$, and $2^p \geq 66B^2 \max\{16, \log_2 B\}$. Set $r = \lfloor \log_2 B \rfloor$ and*

$$(\mathfrak{a}, d_a, k_a) = \mathrm{EXP}((\mathfrak{g}_0, d_0, k_0), a, w, p),$$
$$(\mathfrak{k}, d, k) = \mathrm{EXP}((\mathfrak{a}, 2^r \lceil 2^{-r} d_a \rceil, k_a), b, w, p).$$

Then $(\mathfrak{k}, d, k)$ is a w-near (f, p) representation of $\mathfrak{g}^{ab}$ with $f < 2^{p-4}$.

Proof. Set $h = \max\{16, \log_2 B\}$. As $a \leq B$ and

$$h(3.43(1 + 9/8) + 10.72)B < (18.0088)Bh < 2^p.$$

Theorem 11.8 implies that $(\mathfrak{a}, d_a, k_a)$ is a w-near (g, p) representation of $\mathfrak{g}^a$ with $g < 18.0088a < 18.0088B$ and $hg < 2^p$. By Lemma 14.2, $(\mathfrak{a}, 2^r \lceil 2^{-r} d_a \rceil,$ $k_a)$ is a w-near $(g + 2^r, p)$ representation of $\mathfrak{g}^a$, where $g + 2^r \leq g + B <$ $19.0088B$. As $B \geq 14$, we have $0.7998B > 10.72$, so

$$h(3.43(19.0088B) + 10.72)b < h(65.2002B + 10.72)B < 66B^2 h \leq 2^p .$$

Thus, by Theorem 11.8, $(\mathfrak{k}, d, k)$ is a w-near (f, p) representation of $\mathfrak{g}^{ab}$ with $f < 2^p/h \leq 2^{p-4}$. $\qquad\square$

Corollary 14.3 and Theorem 11.4 show how to use w-near representations and EXP to construct a key exchange protocol, as well as the precision p required to ensure that at the end, the approximations of the relative generators involved are sufficiently accurate that the ideal Alice obtains is one of at most five that Bob can compute easily. In particular, the following theorem[49] is a straightforward consequence of these two results.

Theorem 14.4. *Let* $\mathfrak{g}_0, d_0, k_0, r, p, a, b,$ *and* B *be as in Theorem 14.3 and set*

$$(\mathfrak{a}, d_a, k_a) = \mathrm{EXP}((\mathfrak{g}_0, d_0, k_0), a, w, p) ,$$
$$(\mathfrak{b}, d_b, k_b) = \mathrm{EXP}((\mathfrak{g}_0, d_0, k_0), b, w, p) ,$$
$$(\mathfrak{k}, d, k) = \mathrm{EXP}((\mathfrak{a}, 2^r \lceil 2^{-r} d_a \rceil, k_a), b, w, p) ,$$
$$(\mathfrak{m}, e, h) = \mathrm{EXP}((\mathfrak{b}, 2^r \lceil 2^{-r} d_b \rceil, k_b), a, w, p) .$$

Then $(\mathfrak{k}, d, k)$ *and* $(\mathfrak{m}, e, h)$ *are* w-near (f, p) *representations of* $\mathfrak{g}^{ab}$ *with* $f <$ 2^{p-4} *and* $\mathfrak{k} \in \{\rho^{-2}(\mathfrak{m}), \rho^{-1}(\mathfrak{m}), \mathfrak{m}, \rho(\mathfrak{m}), \rho^2(\mathfrak{m})\}$.

Theorem 14.4 immediately suggests the following protocol.

Protocol 14.1: Cryptographic Key Exchange

Parameters:

- discriminant Δ of a real quadratic field,
- exponent bound $B \geq 14$,
- precision $p \geq \log_2(66B^2 \max\{16, \log_2 B\})$,
- $w = \lceil (\log_2 \Delta)/4 \rceil$,
- $(\mathfrak{g}_0, d_0, k_0)$, w-near $(1 + 9/8, p)$ representation of a reduced principal ideal $\mathfrak{g}$
- $r = \lfloor \log_2 B \rfloor$

Alice
1:    secretly generates $a \in \mathbb{N}$, $a \leq B$;
2:    computes $(\mathfrak{a}, d_a, k_a) = \mathrm{EXP}((\mathfrak{g}_0, d_0, k_0), a, w, p)$;
3:    sends $(\mathfrak{a}, \lceil 2^{-r} d_a \rceil, k_a)$ to Bob.

Bob
1:    secretly generates $b \in \mathbb{N}$, $b \leq B$;
2:    computes $(\mathfrak{b}, d_b, k_b) = \mathrm{EXP}((\mathfrak{g}_0, d_0, k_0), b, w, p)$;

3:      sends $(\mathfrak{b}, \lceil 2^{-r}d_b \rceil, k_b)$ to Alice.
4: Alice computes $(\mathfrak{k}, d, k) = \mathrm{EXP}((\mathfrak{b}, 2^r \lceil 2^{-r}d_b \rceil, k_b), a, w, p)$.
5: Bob computes $(\mathfrak{m}, e, h) = \mathrm{EXP}((\mathfrak{a}, 2^r \lceil 2^{-r}d_a \rceil, k_a), b, w, p)$.

Note that Alice transmits roughly $\log_2 \Delta + \log_2 B \log_2 \log_2 B$ bits: the co-efficients of the ideal $\mathfrak{a}$ are of approximate size $\log_2(\sqrt{\Delta})$, $|k_a|$ tends to be very small, and $2^{-r}d_a \approx 2^{p-r} \approx 66B \log_2 B$; similarly for Bob. This is an improvement of approximately $\log_2(B)$ bits over the original version of this protocol.[50] Notice also that the required precision is larger than that in the first improvement to the original protocol, but this is offset by the improvement offered by using w-near representations to eliminate most of the extra baby steps required to maintain small relative generators.

Even though Theorem 14.4 only guarantees that

$$\mathfrak{k} \in \{\rho^{-2}(\mathfrak{m}), \rho^{-1}(\mathfrak{m}), \mathfrak{m}, \rho(\mathfrak{m}), \rho^2(\mathfrak{m})\} \,,$$

Theorem 11.6 and the subsequent remarks suggest that we should expect to have $\mathfrak{k} = \mathfrak{m}$ with high probability. Indeed, since $f < 2^p / \log B$, we see that

$$\frac{f}{2^p - f} < \frac{1}{\log_2 B - 1} \,.$$

Thus, the probability that $\mathfrak{k} = \mathfrak{m}$ is about

$$1 - \frac{8}{(\log_2 B)^2} \,.$$

In fact, it is usually better than this, as Table 14.1 indicates. However, if Alice and Bob have doubts about whether they computed the same ideal, they could simply derive a shared symmetric key K from $\mathfrak{k}$ and verify that they have the same key K by, for example, encrypting and decrypting a challenge ciphertext. Alternatively, they can choose $\Delta = 4D$ with $D \equiv 3 \pmod 4$ and execute another small protocol that guarantees them a common key ideal.[51]

Table 14.1. Key-Exchange Protocol Mismatches

$\log_2 \Delta$	$\log_2 B$	No. of trials	Mismatches
795	160	12×10^6	32
1384	224	5.5×10^6	7
1732	256	2×10^6	1

14.4.1 Security

Aspects of the security of the general idea underlying the real quadratic fields key exchange protocol have been discussed in some detail.[52] In the following, we summarize the main results, especially those that pertain to the version of key exchange described above.

We note that for the fixed values p and r, a pair of unknown integers a and b, and a given ideal $\mathfrak{g}$, the objects roughly corresponding to a Diffie-Hellman triple here are

$$(\mathfrak{a}, \lceil 2^{-r} d_a \rceil, k_a), \quad (\mathfrak{b}, \lceil 2^{-r} d_b \rceil, k_b), \quad \text{and } \mathfrak{k},$$

where

$$(\mathfrak{a}, d_a, k_a) = \mathrm{EXP}((\mathfrak{g}_0, d_0, k_0), a, w, p),$$
$$(\mathfrak{b}, d_b, k_b) = \mathrm{EXP}((\mathfrak{g}_0, d_0, k_0), b, w, p),$$
$$(\mathfrak{k}, d, k) = \mathrm{EXP}((\mathfrak{a}, 2^r \lceil 2^{-r} d_a \rceil, k_a), b, w, p),$$

because usually $\mathfrak{k} = \mathfrak{m}$. Solving this version of the Diffie-Hellman problem [i.e., computing $\mathfrak{k}$ given $(\mathfrak{a}, \lceil 2^{-r} d_a \rceil, k_a)$ and $(\mathfrak{b}, \lceil 2^{-r} d_b \rceil, k_b)$] breaks the system. Certainly, an attacker can do this if he can deduce a or b from the transmitted information.

If $\mathfrak{g}$ is selected to be a principal ideal, then the discrete logarithm problem in this context, computing a or b, can be solved by solving the infrastructure DLP [i.e., finding a generator α (or a good approximation of $\log |\alpha|$) of the reduced principal ideal $\mathfrak{a}$ or $\mathfrak{b}$]. For example, since $(\mathfrak{a}, \lceil 2^{-r} d_a \rceil, k_a)$ is a w-near (f, p) representation of $\mathfrak{g}^a$, we have

$$\mathfrak{a} = (\theta_a) \mathfrak{g}^a \tag{14.4}$$

for some $\theta_a \in \mathbb{Q}(\sqrt{\Delta})$. Suppose an attacker can find a generator α of $\mathfrak{a}$, or a good approximation of $\log |\alpha|$. Then he could solve (14.4) for a as follows. First, the attacker computes $\gamma \in \mathbb{Q}(\sqrt{\Delta})$ with $\mathfrak{g} = (\gamma)$ by solving the principal ideal problem for γ. Then by (14.4), $|\alpha| = \epsilon_\Delta^m \theta_a \gamma^a$, where $m \in \mathbb{Z}$ and ϵ_Δ is the fundamental unit of $\mathbb{Q}(\sqrt{\Delta})$, so

$$\log \alpha = m R_\Delta + \log \theta_a + a \log \gamma,$$

where $R_\Delta = \log \epsilon_\Delta$ is the regulator of $\mathbb{Q}(\sqrt{\Delta})$. Now, $\log \theta_a$ is small by Corollary 11.3.1, and m is small by our choice of the upper bound B on a. So it would not be hard to find a once R_Δ is known, and as seen in Chapter 13, computing R_Δ is closely related to solving the principal ideal problem in practice. In particular, both problems can be solved in expected time $L_\Delta[1/2, \sqrt{2} + o(1)]$ using index-calculus methods.

However, it is possible to attack (14.4) from another point of view. The attacker knows $\lceil 2^{-r} d_a \rceil$ and k_a. Note that knowledge of k_a provides him with little information, since k_a is usually small by Corollary 11.3.1 As mentioned right after the proof of that corollary, we expect $w - k_a \leq 2$ in 64% of all cases, so many possible values for θ_a could have the same k_a value. However, the adversary also knows by (11.4) that $|2^{p-k_a}\theta_a\gamma/d_a - 1| < 2^{-p}f$ for some $f < 19.0088B$ and γ such that $\rho(\mathfrak{a}) = (\gamma)\mathfrak{a}$. Thus, he can use $\hat{\psi} = 2^{p-r-k_a}\gamma/\lceil 2^{-r}d_a \rceil$ as an approximation to $\psi = \theta_a^{-1}$. If we set $\delta = |\hat{\psi}/\psi - 1|$, then it can be shown that $\delta < (1 + 2^{1-r}) \cdot 2^{-p}f < (1 + 2^{-15}) \cdot 2^{-p}f$ when $B \geq 2^{16}$, so $\delta < (1 + 2^{-15}) \cdot 19.0088/(66B\log_2 B)$. We may therefore assume that the attacker knows a good rational approximation $\hat{\psi}$ to ψ as well as an upper bound $\delta < 0.29/(B\log_2 B)$ on the relative error of that approximation.

Let m/n be any convergent of the continued fraction expansion of $\sqrt{\Delta}$. Then for a given n, it is possible to formulate explicitly an infinitude of distinct pairs of rationals $(r(n), s(n))$ such that $r(n), s(n)\sqrt{\Delta} = \Theta(n)$, and if $\psi(n) = r(n) - s(n)\sqrt{\Delta}$, then $|\hat{\psi} - \psi(n)| = O(n^{-1})$, so knowing $\hat{\psi}$ and δ still leaves infinitely many possibilities for ψ.

On the other hand, if the coefficients of ψ were known to lie in a certain range that is not too large, then it might be possible to search for them successfully. If $\psi = r - s\sqrt{\Delta}$, then since $r = (\psi + \overline{\psi})/2$ and $s\sqrt{\Delta} = (\psi - \overline{\psi})/2$, an interval containing r and $s\sqrt{\Delta}$ can be determined from bounds on ψ and $|\overline{\psi}|$. By (14.4), we have

$$N(\mathfrak{g})^a = \psi|\overline{\psi}|N(\mathfrak{a}) . \qquad (14.5)$$

Now, $N(\mathfrak{g})$ is public information, $1 \leq N(\mathfrak{a}) < \sqrt{\Delta}$ since $\mathfrak{a}$ is reduced, and we have the bounds $16/17 < \psi < 16\sqrt{\Delta}/15$ from Lemma 11.3 on ψ. It follows from (14.5) that $|\overline{\psi}| = \Theta(N(\mathfrak{g})^a)$ as a grows, so r and $s\sqrt{\Delta}$ are exponentially large in a. Since $a < B$ tends to be quite large, this does not allow for an efficient search for a.

It is important that the discriminant Δ be chosen in such a way that principal ideal testing is as hard as possible. One consideration is that the regulator R_Δ should be large, implying that the set of reduced principal ideals is also large. Although, as described in Chapter 9, Littlewood's bounds on $L(1,\chi)$ and the Cohen-Lenstra heuristics suggest that, under the ERH, $R_\Delta \gg \Delta^{1/2-\epsilon}$ for a large proportion of real quadratic fields, it is unknown how to produce values of Δ for which this is guaranteed. If Δ is selected prime, then the class number h_Δ will be odd and the Cohen-Lenstra heuristics suggest that h_Δ will almost certainly be small. Thus, simply selecting a sufficiently large prime discriminant will suffice with high probability.

The two best available algorithms for solving the principal ideal problem are the index-calculus algorithm described in Chapter 13 and the baby-step giant-step algorithm described in Chapter 7. The complexity of the index-calculus algorithm depends only on the discriminant Δ, so selecting Δ sufficiently large is enough to resist this procedure. As mentioned earlier, if we assume that solving the principal ideal problem and the ideal class group discrete logarithm problem require the same amount of time for discriminants of the same size, Hamdy's estimates suggest that a 795-bit discriminant offers roughly the same level of security as a block cipher with an 80-bit key; that is, the time required to solve the principal ideal problem will be proportional to 2^{80}, the time required to break a block cipher with an 80-bit key. NIST[53] recommends five security levels, providing 80, 112, 128, 192, and 256 bits of security. According to Hamdy's estimates, discriminants of 795, 1384, 1732, 3460, and 5704 bits should provide roughly these levels of security. Further work is required to verify that these estimates do in fact hold for real quadratic fields, but given the similarities between index-calculus algorithms in imaginary and real quadratic fields, these recommendations are likely to be reasonably close to what a more detailed investigation would reveal.

It is also possible to select Δ in such a way that the index-calculus algorithm preforms worse than for a randomly selected discriminant. The idea, as described by Jacobson, Scheidler, and Williams,[54] is to choose Δ such that the Kronecker symbol $(\Delta/p) = -1$ for as many small primes as possible. Then, the factor base constructed for the index-calculus algorithm will not contain any of these small primes, and it will be harder to find smooth principal ideals (relations) because none of these primes will divide the norms of any ideals in $\mathcal{O}_\Delta$. This does indeed appear to adversely affect the performance of the index-calculus algorithm in practice; an example is given at the end of Chapter 13. Such discriminants can be found using special-purpose sieving software or hardware that solves systems of simultaneous linear congruences and can be done easily if the designer of the cryptosystem wishes to invest extra time to make the cryptanalyst's task a little more difficult.

The complexity of the baby-step giant-step algorithm, on the other hand, depends on the distance of the principal ideal itself. In our case, we know that these distances are bounded approximately by the exponent bound B, so this algorithm will take $O(\sqrt{B})$ steps. Notice that if we use a discriminant that we expect will provide b bits of security (i.e., for which solving the principal ideal problem using index calculus will take time approximately 2^b) then using an exponent bound $B = 2^{2b}$ will ensure that attacking the protocol using the baby-step giant-step method would require the same amount of time. Thus, exponent bounds of 2^{160}, 2^{224}, 2^{256}, 2^{384}, and 2^{512} can be used for the 80−, 112−, 128−, 192−, and 256− bit security levels. The fact that the precision required for the (f, p) representations in the protocol described above only depends on B, and not on Δ, indicates that this method of selecting B will also positively affect the performance of the protocol.

The restriction that the base ideal $\mathfrak{g}$ be principal is necessary to ensure that the protocol takes place in the infrastructure of $\mathbb{Q}(\sqrt{\Delta})$. However, the protocol will still work correctly if $\mathfrak{g}$ is not principal. The security considerations are then slightly different, as the adversary would also need to determine the ideal class to which $\mathfrak{g}^{ab}$ belongs by solving a discrete logarithm problem in the ideal class group. This more general version is unlikely to offer any more security because, as seen in Chapter 13, the discrete logarithm problem can be solved in the same expected time as the principal ideal problem. However, it does provide additional flexibility in choosing Δ and alleviates some concerns that we are unlucky and select a field $\mathbb{Q}(\sqrt{\Delta})$ with large class number.

14.4.2 Efficiency

Table 14.2 contains the average CPU time per communication partner for a single application of the key exchange protocol described above. For each discriminant size, the average was taken over 1000 separate runs of the protocol, each using a different randomly selected prime discriminant. The protocol was implemented using the GNU C++ compiler version 4.2 and the C++ computer algebra library NTL,[55] and the computations were performed on a Pentium IV 2.53-GHz computer running Linux. For comparison purposes, key exchange in imaginary quadratic fields and the previous versions of the real quadratic field-based protocol described above were implemented using the same software. In the table, the protocol using imaginary quadratic fields is denoted by IMAG; the first version of the protocol of Jacobson, Scheidler, and Williams[56] that uses (f, p) representations, but regular ideal multiplication is denoted by REAL; the second version[57] uses NUCOMP and WNEAR with $w = 0$ by NEAR; and the new version uses WNEAR described above by WNEAR. We also give the ratio of the average time using WNEAR over the average time using imaginary quadratic fields. The five discriminant sizes are selected to provide 80, 112, 128, 192, and 256 bits of security as described above. For a discriminant providing b bits of security, the exponent bound B is selected to be 2^{2b}, also as described above.

Table 14.2. Average CPU Times (in Seconds) per Key Exchange per Partner

$\log_2 \Delta$	$\log_2 B$	IMAG	REAL	NEAR	WNEAR	WNEAR/IMAG
795	160	0.04	0.38	0.13	0.05	1.25
1384	224	0.11	1.05	0.30	0.14	1.27
1732	256	0.15	1.63	0.43	0.21	1.40
3460	384	0.50	6.34	1.45	0.75	1.50
5704	512	1.32	17.97	3.86	2.04	1.55

It can be seen that the version using WNEAR is the fastest of the three real quadratic field-based versions, and that its performance is close to that of imaginary quadratic fields. In addition, it is clear that significant progress has been made in terms of improving the efficiency of key exchange in the real quadratic case since the introduction of (f, p) representations, and with some more effort, it is possible that real quadratic field-based cryptography will be a viable alternative to its imaginary counterpart. For example, it may be possible to make use of the fact that the baby-step operation (stepping through cycle of equivalent reduced ideals via the continued fraction algorithm) is faster than ideal multiplication, both asymptotically and in practice. Stein has observed[58] that one can improve the efficiency of key exchange in low-genus real quadratic function fields to the point that it compares favourably to the corresponding protocols in imaginary quadratic function fields, by essentially taking advantage of the fact that baby steps in the infrastructure are much less expensive than ideal multiplications. Whether this idea will bear fruit in the real quadratic field case is a topic of further research.

14.4.3 Other Cryptosystems

Most of the literature on real quadratic field-based cryptography deals with the version of Diffie-Hellman key exchange as described above. However, two digital signature schemes have also been proposed, an analogue of El Gamal encryption due to Biehl, Buchmann, and Thiel[59] and a generalization of the Fiat-Shamir signature protocol due to Buchmann, Maurer, and Möller.[60] The (f, p) representation approach has not yet been adapted to either of these protocols, and a detailed investigation of their efficiency remains open.

14.5 Cryptosystems in Non-Maximal Quadratic Orders

Let Δ_1 be a fundamental discriminant, the discriminant of a maximal order $\mathcal{O}_{\Delta_1}$ of the quadratic field $\mathbb{Q}(\sqrt{\Delta_1})$. The *non-maximal order* of conductor $f > 1$ with non-fundamental discriminant $\Delta_f = \Delta_1 f^2$ is the submodule $[1, (\Delta_f + \sqrt{\Delta_f})/2]$ of $\mathcal{O}_{\Delta_1}$ and is denoted by $\mathcal{O}_{\Delta_f}$. As described in §4.5, the fundamental unit ϵ_{Δ_f} of $\mathcal{O}_{\Delta_f}$ is a power of the fundamental unit ϵ_{Δ_1} of $\mathcal{O}_{\Delta_1}$; in particular, $\epsilon_{\Delta_f} = \epsilon_{\Delta_1}^n$, where the positive integer n is called the unit index of ϵ_{Δ_f}. In addition, the class number h_{Δ_f} of $\mathcal{O}_{\Delta_f}$ is a multiple of the class number h_{Δ_1} of $\mathcal{O}_{\Delta_1}$, and by (8.9), we have

$$h_{\Delta_f} = \frac{\alpha f}{n} h_{\Delta_1} \, ,$$

where

$$\alpha = \prod_{p \mid f} \left(1 - \frac{(\Delta_1/p)}{p} \right)$$

and, again, n denotes the unit index of ϵ_{Δ_f} with respect to ϵ_{Δ_1}.

Non-maximal quadratic orders have been proposed for cryptographic applications, based on the fact that exponentiation in Cl_{Δ_f}, the ideal class group of $\mathcal{O}_{\Delta_f}$, is a one-way *trapdoor* function; the factorization of Δ_f serves as trapdoor information, allowing the possessor to invert the function by solving the discrete logarithm problem more efficiently than without it. This has been used to construct cryptosystems whose security is based on the intractability of the integer factorization problem, but with properties beyond those of RSA. The first of these was a version of the El Gamal public-key cryptosystem in the class group Cl_{Δ_f} of a non-maximal imaginary quadratic order due to Hühnlein, Jacobson, Paulus, and Takagi,[61] where knowledge of the conductor is used to improve the speed of decryption. Paulus and Takagi[62] subsequently refined this approach, resulting in a cryptosystem called New Ideal Coset Encryption (NICE) for which decryption can be done in quadratic time in $\log |\Delta_f|$, as opposed to cubic time as in most other public-key cryptosystems, including RSA. An IND-CCA2 secure[63] version of NICE was presented by Buchmann, Sakurai, and Takagi,[64] and NICE has been generalized to non-maximal real quadratic orders by Jacobson, Scheidler, and Weimer.[65] Non-maximal imaginary quadratic orders were also used by Hühnlein, Jacobson, and Weber[66] to construct a non-interactive ID-based cryptosystem, where the trapdoor information is used to compute discrete logarithms in Cl_{Δ_f} efficiently.

The main idea underlying all these cryptosystems is that there exists a surjective homomorphism between Cl_{Δ_f} and Cl_{Δ_1} given by

$$\phi : Cl_{\Delta_f} \to Cl_{\Delta_1} ,$$
$$[\mathfrak{a}] \mapsto [\mathfrak{a}\mathcal{O}_{\Delta_1}]$$

that can be computed in quadratic time if the conductor f is known. Given an ideal $\mathfrak{a} = [Q/r, (P + \sqrt{D_f})/r]$ of $\mathcal{O}_{\Delta_f}$, where $\Delta_f = (2/r)D_f$ and $D_f = f^2 D_1$, we can compute an ideal $\mathfrak{A} = [Q'/r, (P' + \sqrt{D_1})/r] = \mathfrak{a}\mathcal{O}_{\Delta_1}$ using the formulas

$$Q' = Q , \quad P' = P\mu + (Q/r)\lambda \pmod{Q} ,$$

where $1 = \mu f + \lambda(r-1)(Q/r)$; these can be derived using the ideal multiplication method derived in §5.4. Thus, to compute $\phi([\mathfrak{a}])$, we compute $\mathfrak{A}$ as above and reduce it to get a reduced representative of the class $[\mathfrak{A}] \in Cl_{\Delta_1}$. We need to have $\gcd(Q/r, f) = 1$ in order for the above algorithm to work. If $\gcd(Q/r, f) > 1$, we can find an ideal equivalent to $\mathfrak{a}$ whose norm is relatively prime to f and then apply the algorithm. As shown[67] in Theorem 4.37, every ideal equivalence class of an order $\mathcal{O}_{\Delta_f}$ contains an ideal whose norm is relatively prime to f.

For cryptographic applications, we typically choose $|\Delta_1| = p$ and $f = q$ prime, both for simplicity and for security, as $|\Delta_f| = q^2 p$ will be more difficult to factor than $f^2 p$, where f is composite. Then $h_{\Delta_q} = h_{\Delta_1}(q - (\Delta_1/q))/n$. For now, we will restrict our attention to the imaginary case, where the unit index $n = 1$. Then if $|\Delta_1| = p$ and $f = q$ are prime, there are exactly $q - (\Delta_1/q)$ preimages under ϕ for each $[\mathfrak{A}]$ in Cl_{Δ_1}. If we select Δ_1 such that $|\Delta_1|$ is prime and $\gcd(q, h(\Delta_1)) = 1$, then with high probability, Cl_{Δ_q} is cyclic.[68]

Using the algorithm mentioned above, one can evaluate ϕ if the conductor is known. The following theorem illustrates that the converse is also true; any algorithm for computing ϕ can be used to find the conductor. This theorem[69] is presented in the case that $|\Delta_1|$ and the conductor are both prime, but it can be generalized easily.

Theorem 14.5. *Computing ϕ is equivalent to knowing the factorization of Δ_q.*

Proof. If q is known, then we can compute ϕ using the algorithm mentioned above. If there exists an oracle that evaluates ϕ, we can use it to compute q as follows. From the algorithm described above for computing ϕ, it can be seen that if $\phi([\mathfrak{a}]) = [\mathfrak{A}]$ for reduced ideals $\mathfrak{a} = [Q/r, (P + \sqrt{D_f})/r] \in +\mathcal{O}_{\Delta_f}$ and $\mathfrak{A} = [Q'/r, (P' + \sqrt{D_1})/r] \in \mathcal{O}_{\Delta_1}$, we have $P' \equiv Pq^{-1} \pmod{Q}$, or, alternatively, $q \equiv P(P')^{-1} \pmod{Q}$. To find q, we simply obtain $\phi(\mathfrak{p})$ for several prime ideals $\mathfrak{p} \in \mathcal{O}_{\Delta_f}$ and compute q using the Chinese Remainder Theorem. This method will succeed after polynomially many operations in $\log \Delta_q$. $\qquad\square$

14.5.1 NICE

NICE, which stands for New Ideal Coset Encryption, is a public-key cryptosystem set in the class group of a non-maximal imaginary quadratic order $\mathcal{O}_{\Delta_q} = -q^2 p$ for primes p and q. Its security is based on the *hidden kernel problem*: given Δ_q, a generator $[\mathfrak{k}]$ of $\ker(\phi)$, and a random element in the coset $[\mathfrak{m}] \ker(\phi)$, find $[\mathfrak{m}]$. The instances of this problem that arise in NICE can be solved if the homomorphism ϕ can be computed, which, by Theorem 14.5, is equivalent to being able to factor Δ_q. The distinguishing feature of NICE as compared to other factoring-based public-key cryptosystems such as RSA is that decryption is especially fast and can be done in quadratic time as opposed to cubic.

The idea of NICE is to make use of the homomorphism ϕ to decrypt messages efficiently by solving the hidden kernel problem. If the factorization of Δ_q is known only to the decryptor (i.e., the conductor q serves as the private key), then the decryptor can compute ϕ easily, but anyone else cannot without being able to factor Δ_q.

To encrypt a message m, it is first embedded into the Q coefficient of an $\mathcal{O}_{\Delta_q}$-ideal $\mathfrak{m}$, in such a way that $N(\mathfrak{m}) < \Delta_q^{1/4}$. This condition guarantees that $N(\mathfrak{m}) < \sqrt{p}/2$, so that $\phi([\mathfrak{m}])$, a reduced ideal in $\mathcal{O}_p$, will have the same norm as $\mathfrak{m}$. To encrypt $\mathfrak{m}$, a random element in $\ker(\phi)$ is computed by raising a publicly known generator of $\ker(\phi)$ to a random power, yielding the ciphertext ideal $\mathfrak{c} \sim \mathfrak{k}^r \mathfrak{m}$, a reduced $\mathcal{O}_{\Delta_q}$-ideal.

To decrypt $\mathfrak{c}$, the conductor q is used to obtain $\mathfrak{M} = \phi([\mathfrak{c}])$, a reduced ideal in the maximal order $\mathcal{O}_p$. The plaintext can then be recovered from the first coefficient of $\mathfrak{M}$, because we have

$$\phi([\mathfrak{c}]) = \phi([\mathfrak{m}][\mathfrak{k}]^r) = \phi([\mathfrak{m}])\phi([\mathfrak{k}])^r = \phi([\mathfrak{m}])$$

due to the facts that ϕ is a homomorphism and $[\mathfrak{k}] \in \ker(\phi)$. The first coefficient of $\mathfrak{M} = \phi([\mathfrak{m}])$ contains the message because $\phi([\mathfrak{m}])$ is already reduced in $\mathcal{O}_p$ due to the size condition on the first coefficient (note that $N(\mathfrak{m}) = N(\mathfrak{m}\mathcal{O}_{\Delta_1})$).

The operation of NICE is summarized in the following protocol.

Protocol 14.2: NICE

Public key: $\Delta_q = -pq^2$, an $\mathcal{O}_{\Delta_q}$-ideal $\mathfrak{k}$ such that $[\mathfrak{k}]$ generates $\ker(\phi)$
Private key: q
Encryption:
  1: Given a message ideal $\mathfrak{m}$ with $N(\mathfrak{m}) < \sqrt{p}/2$ (embed message in first coefficient) compute the reduced ciphertext ideal $\mathfrak{c} \sim \mathfrak{m}\mathfrak{k}^r$ for random $r \in \mathbb{Z}$.
Decryption:
  1: Compute $\phi([\mathfrak{c}])$ and extract the message from the first coefficient

---

Notice that decryption in NICE only requires one computation of the map ϕ, which can be done in quadratic time in $\log|\Delta_q|$. Encryption, on the other hand, requires exponentiation in Cl_{Δ_q}, which takes cubic time. In comparison with RSA, the most widely used cryptosystem whose security is related to integer factorization, encryption is slower, but decryption is significantly faster.[70]

By Theorem 14.5, the main security consideration is that Δ_q must be chosen so that factoring Δ_q is infeasible. The best known general-purpose factoring algorithm is the number field sieve (NFS), so Δ_q must be sufficiently large to resist factorization by the NFS. However, the elliptic curve method (ECM) is effective at finding prime factors of a given size, and, in general, its complexity depends on the size of the smallest prime factor. Thus, p and q must be sufficiently large to prevent factorization by the ECM as well. There exists a special version of ECM designed to factor numbers of the form pq^2, but this does not appear to yield a significant improvement in practice.[71]

Unfortunately, NICE, as presented above, is vulnerable to a chosen ciphertext attack discovered by Jaulmes and Joux.[72] The following attack allows an adversary to factor Δ_q given two specially chosen ciphertexts:

1. Compute $\mathfrak{m}_1$ and $\mathfrak{m}_2$ with $N(\mathfrak{m}_i) = m_i$ and

$$\sqrt{p/3} < m_i < \sqrt{p} \quad (i = 1, 2) \,.$$

These two ideals are the chosen ciphertexts. The condition on their norms implies that their images under ϕ will likely require just one reduction step to produce a reduced ideal in $\mathcal{O}_{\Delta_1}$.

2. The victim decrypts $\mathfrak{m}_1$ and $\mathfrak{m}_2$, providing the attacker with $m_i' = N(\phi(\mathfrak{m}_i))$. Note that $m_i' = (N_i^2 + p)/m_i \neq m_i$ for some integers N_i (unknown to the attacker), because one reduction step has to be applied to $\mathfrak{m}_i\mathcal{O}_{\Delta_1}$ when computing $\phi([\mathfrak{m}_i])$. Thus, we have $p = m_1 m_1' - N_1^2 = m_2 m_2' - N_2^2$, and if the attacker can find N_1 or N_2, he can factor Δ_q.

3. The attacker computes $k = m_1 m_1' - m_2 m_2'$. By the above relation we have that $k = N_1^2 - N_2^2 = (N_1 - N_2)(N_1 + N_2)$. The attacker attempts to find N_1 and N_2 by factoring $k \approx p$. If successful, then he can compute p, thereby factoring Δ_q.

The NICE-X protocol modifies NICE in order to resist this attack. It can be proved that NICE-X is IND-CCA2 secure in the random oracle model under the assumption that the *smallest kernel-equivalent problem* (SKEP) is intractable. SKEP is a version of the hidden kernel problem that takes into account the norm bound of message ideals in NICE. The additional overhead NICE-X incurs in comparison to NICE is negligible; roughly two hash function applications and one ideal multiplication more than NICE for both encryption and decryption.

The hidden kernel problem as described above has also been used as the basis for a digital signature protocol due to Hühnlein[73] and an undeniable signature scheme due to Biehl, Paulus, and Takagi.[74] A method of distributed RSA key generation due to Biehl and Takagi[75] also makes use of the relationship between the class groups of maximal and non-maximal quadratic orders.

14.5.2 REAL-NICE

A natural question to ask is whether the ideas behind NICE can be generalized to real quadratic orders. Jacobson, Scheidler, and Weimer[76] described a protocol called REAL-NICE that uses ideas inspired by NICE for a public-key cryptosystem set in the class group of a real non-maximal quadratic order that also has quadratic decryption time.

The heart of REAL-NICE is also the homomorphism ϕ, mapping elements of the class group Cl_{Δ_q} of a non-maximal quadratic order to that of the corresponding maximal order, Cl_p. The same results on ϕ hold in the real case, and the same algorithm can be used to evaluate it. The security results of REAL-NICE are thus similar to those of NICE, as decryption involves evaluating ϕ, and this is still equivalent to knowing the factorization of Δ_q. As a result, REAL-NICE also makes use of non-maximal orders with discriminants $\Delta_q = q^2 p$, where p and q are prime. However, the fact that ideal equivalence classes do not have unique reduced representatives requires certain modifications before NICE can be employed in this setting.

The first issue is that decryption in NICE relies on the fact that the reduced ideal obtained by computing $\phi(\mathfrak{c})$ for the ciphertext ideal $\mathfrak{c}$ is unique. In the real setting, this will not be the case; the best we can say is that the output will be one of the approximately R_p reduced ideals equivalent to the ideal $\mathfrak{M}$ whose norm encodes the plaintext. Thus, in order to guarantee that we can efficiently find the correct ideal $\mathfrak{M}$, it is necessary to ensure that R_p be sufficiently small. Fortunately, as we have seen in Chapter 6, there are a number of families of discriminants that satisfy this property. The solution presented by Jacobson, Scheidler, and Weimer is to choose p to be a prime *Schinzel sleeper* (i.e., a prime of the form $a^2 x^2 + 2bx + c$ with $a, b, c, x \in \mathbb{Z}$, $a \neq 0$, and $b^2 - a^2 c$ dividing $4 \gcd(a^2, b)^2$). Recall from §6.3 that if p is a Schinzel sleeper, then $R_p = O(\log p)$, so the number of reduced ideals the decryptor will have to search before finding the plaintext is also $O(\log p)$.

Even when p is chosen in this manner, $\mathfrak{M}$ will still be one of a small set of reduced ideals. Thus, some mechanism must be employed to allow the decryptor to identify which of these ideals is $\mathfrak{M}$. One solution is to encode some redundancy into the message before encrypting; for example, enforcing a fixed bit pattern to occur in the message. The decryptor then computes all the reduced ideals equivalent to $\phi(\mathfrak{c})$, using the baby-step operation ρ described in §5.1, and selects that ideal whose norm contains this bit pattern.

Notice that although decryption is more complicated than in the imaginary case, the extra work required, searching a set of $O(\log p)$ ideals, requires time $O(\log^2 p)$, as each baby step requires time $O(\log p)$. Thus, the overall time required for decryption is still quadratic in the bit length of $\Delta_q = q^2 p$.

The second issue with extending NICE to real non-maximal quadratic orders is that, in NICE, it is necessary that $\ker(\phi)$ be large, so that the coset $\mathfrak{m} \ker(\phi)$ is sufficiently large to hide the message when encrypting. In the real case, recall that

$$| \ker(\phi)| = \frac{h_{\Delta_q}}{h_p} = \frac{q - (p/q)}{n} ,$$

where n is the unit index of the non-maximal order $\mathcal{O}_{\Delta_q}$ with respect to the maximal order $\mathcal{O}_p$. It is possible that n is large, resulting in a very small, perhaps even trivial, kernel.

The solution described by Jacobson, Scheidler, and Weimer is to make use of infrastructure, realizing that in addition to hiding a message ideal within its coset $\mathfrak{m} \ker(\phi)$, we can also hide it within an equivalence class of $\mathfrak{m} \ker(\phi)$ if the regulator R_{Δ_q} is large. In fact, if the unit index, and hence R_{Δ_q}, is sufficiently large, it suffices to encrypt $\mathfrak{m}$ by hiding it in its own equivalence class by, for example, multiplying it by a random principal ideal. Then the resulting ciphertext $\mathfrak{c}$ is still in $\mathfrak{m} \ker(\phi)$, and decryption will work as described above.

To ensure that the unit index n is large, we select p as above and search for a prime conductor q such that $(q - (p/q))$ has a large prime factor. If $(q - (p/q)) = Ld$ for some large prime L, then Weimer[77] proved that $n \geq L$ with high probability. For a particular choice of p and q, this can be verified by checking that $n \nmid d$. Because p is chosen so that $R_p = \log \epsilon_p \in O(\log p)$, it is possible to write down the fundamental unit explicitly as $\epsilon_p = U_1 + V_1 \sqrt{p}$. Then we check that $\epsilon_p^d = U_d + V_d \sqrt{p} \neq \epsilon_{\Delta_q}^k$ (i.e., that $V_d \not\equiv 0 \pmod{q}$). The value of $V_d \bmod q$ can be evaluated efficiently using Lucas functions via a method similar to binary exponentiation, as described in §14.2.[78] Thus, given p, we can find q such that $R_{\Delta_q} \approx q \log p$, and we can make this as large as we like simply by choosing a sufficiently large value of q.

Numerical experiments show that it is not difficult to find suitable parameters for REAL-NICE. However, as expected, the performance is not as good as that of NICE. It is clear that decryption will be somewhat slower, due to the overhead of identifying the correct plaintext ideal in the equivalence class of $\phi(\mathfrak{c})$, but it is, nevertheless, still faster than RSA decryption when using RSA moduli offering the same level of security. Encryption is currently slower than in NICE, but one intriguing aspect of REAL-NICE is that it may be possible to improve this. In particular, the main part of encryption in REAL-NICE involves finding a random principal ideal. The method described by Jacobson, Scheidler, and Weimer makes use of ideal exponentiation. As mentioned above in the context of key exchange in real quadratic fields, it may be possible to replace some ideal multiplications with the faster baby-step operation in order to accomplish the same task faster. Finding such an encryption method that does not compromise the security of the protocol remains an open research problem.

14.5.3 Trapdoor Discrete Logarithm Computation

Suppose that $\mathcal{O}_{\Delta_f}$ is an imaginary non-maximal quadratic order. Hühnlein, Jacobson, and Weber[79] showed that if the factorization of $\Delta_f = f^2 \Delta_1$ is known, the discrete logarithm problem (DLP) in Cl_{Δ_f} can be reduced to two significantly easier DLP instances, one in the class group of the maximal order $\mathcal{O}_{\Delta_1}$ and another in $\ker(\phi)$. The latter can be reduced to the DLP in a finite field.

One consequence of this fact is that it is possible, at least in theory, to set up a non-interactive ID-based cryptosystem using the class group of a non-maximal imaginary quadratic order. In an ID-based cryptosystem, public keys are derived directly from users' identities, so there is no need to authenticate the binding of public keys to user identities, for example, by using certificates.[80] In such a scheme, a trusted key generation centre is required that uses trapdoor information to assign private keys to users based on their fixed public keys. For example, if a cryptosystem is set in the class group of a non-maximal order $\mathcal{O}_{\Delta_f}$, a key generation centre which knows the factorization of Δ_f could use the trapdoor reduction mentioned above to compute discrete logarithms in Cl_{Δ_f} and assign these as private keys to be used with the Diffie-Hellman integrated encryption scheme or the IQ-DSA digital signature scheme. In this case, none of the users of the cryptosystem would know the factorization of Δ_f; they would simply use the class group as described in §14.3.

Another consequence is that if the class group Cl_{Δ_f} is to be used for standard cryptographic protocols, then the secret factorization of Δ_q could be kept by an escrow agent and used to compute discrete logarithms upon request. Once again, the users of the system would simply use cryptographic protocols in the class group as described in §14.3. If it is desirable to have assurance that such an escrow system is not being used, then it would be necessary to provide a proof that the discriminant Δ being used is fundamental; for example, that $|\Delta|$ is prime.

Computing discrete logarithms in Cl_{Δ_f} is done as follows.[81] Suppose that we wish to solve $[\mathfrak{g}]^x = [\mathfrak{a}]$ in Cl_{Δ_f} for x.

1. Compute x_1 such that $\phi([\mathfrak{g}])^{x_1} = \phi([\mathfrak{a}])$ by solving the DLP in Cl_{Δ_1}. Because ϕ is a homomorphism, we know that $x = x_2 h + x_1$, where h is the order of $\phi([\mathfrak{g}]) \in Cl_{\Delta_1}$.
2. Compute $\mathfrak{G} \sim \mathfrak{g}^h$, $\mathfrak{A} \sim \mathfrak{a}\mathfrak{g}^{-x_1}$. Notice that both of $[\mathfrak{G}]$ and $[\mathfrak{A}]$ are in $\ker(\phi)$.
3. Compute x_2 such that $\mathfrak{G}^{x_2} \sim \mathfrak{A}$. We now have $x = x_2 h + x_1$, because

$$\left(\mathfrak{g}^h\right)^{x_2} \sim \mathfrak{a}\mathfrak{g}^{-x_1} \implies \mathfrak{g}^{x_2 h} \sim \mathfrak{g}^x \mathfrak{g}^{-x_1} \implies x = x_2 h + x_1 \,.$$

We therefore need to solve two problems in order to compute x :

- compute a discrete logarithm in Cl_{Δ_1}
- compute a discrete logarithm in $\ker(\phi)$.

Both of these are easier than solving the DLP directly in Cl_{Δ_f}. As shown in Chapter 13, the DLP in Cl_{Δ_1} can be solved in expected time $L_{\Delta_1}[1/2, \sqrt{2} + o(1)]$. As $\Delta_1 < \Delta_f$; this is faster than solving the DLP in Cl_{Δ_f}.

To compute discrete logarithms in $\ker(\phi)$, we first note the existence of the following surjective homomorphism[82]:

$$\psi : (\mathcal{O}_{\Delta_1}/f\mathcal{O}_{\Delta_1})^* \to \ker(\phi) \,,$$
$$[\alpha] \mapsto [\alpha \mathcal{O}_{\Delta_1} \cap \mathcal{O}_{\Delta_f}] \,.$$

Hühnlein et al. showed how to compute preimages of ψ given elements in $\ker(\phi)$, allowing the DLP in $\ker(\phi)$ to be solved by solving it in $(\mathcal{O}_{\Delta_1}/f\mathcal{O}_{\Delta_1})^*$ and reducing the result modulo $|\ker(\phi)|$. Assuming $f = p_1^{e_1} p_2^{e_2} \cdots p_k^{e_k}$, where $p_1, p_2, \ldots, p_k$ are primes, the DLP in $(\mathcal{O}_{\Delta_1}/f\mathcal{O}_{\Delta_1})^*$ reduces to DLPs in $(\mathcal{O}_{\Delta_1}/p_i \mathcal{O}_{\Delta_1})^*$ via the Chinese Remainder Theorem and the Pohlig-Hellman algorithm. The group $(\mathcal{O}_{\Delta_1}/p_i \mathcal{O}_{\Delta_1})^*$ is isomorphic to

$$\begin{cases} \mathbb{F}_{p_i}^* \otimes \mathbb{F}_{p_i}^* & \text{if } (\Delta_1/p_i) \in \{0,1\} \\ \mathbb{F}_{p_i^2}^* & \text{if } (\Delta_1/p_i) = -1 \,, \end{cases}$$

where $\mathbb{F}_q$ denotes the finite field of q elements and, as usual, $\mathbb{F}_q^* = \mathbb{F}_q \setminus \{0\}$ is the multiplicative group of units in $\mathbb{F}_q$. Let $[\gamma] = [x + y\omega] \in (\mathcal{O}_{\Delta_1}/p_i \mathcal{O}_{\Delta_1})^*$. Then

$$[\gamma] \mapsto \begin{cases} (x + y\omega, x + y\overline{\omega}) \pmod{p_i} & \text{if } (\Delta_1/p_i) \in \{0,1\} \\ x + y\omega \pmod{p_i} & \text{if } (\Delta_1/p_i) = -1 \,. \end{cases}$$

Thus, to solve the DLP in $\ker(\phi)$ we only need to compute DLPs in $\mathbb{F}_{p_i}^*$ or $\mathbb{F}_{p_i^2}^*$. The number field sieve can be used to do this in expected time $L_{p_i}[1/3, \alpha + o(1)]$ for some constant α, which, once again, is significantly faster than computing the DLP in Cl_{Δ_f} directly.

These ideas can be applied almost directly to compute discrete logarithms in Cl_{Δ_f} when $\mathcal{O}_{\Delta_f}$ is real. It may even be possible to use a similar method to reduce the principal ideal problem in $\mathcal{O}_{\Delta_f}$ to instances of the principal ideal problem in $\mathcal{O}_{\Delta_1}$ and a finite field; this is a subject of ongoing research.

In order to use $\mathcal{O}_{\Delta_q}$ with $\Delta_q = -pq^2$, p, and q prime for an ID-based or key escrow system, we require[83] (at least) the following restrictions on p and q :

- $\Delta_q > 2^{576}$ (to prevent NFS factorization of Δ_q),
- $p, q > 2^{222}$ (to prevent ECM factorization of Δ_q),
- $p < 2^{300}$ (to ensure that the DLP in Cl_p is feasible),
- $q < 2^{300}$ (to ensure that the DLP in $\ker(\phi)$, essentially in $\mathbb{F}_q^*$, is feasible).

Using $p \approx 2^{304}$ and $q \approx 2^{305}$, Hühnlein, Jacobson, and Weber showed that the reduction described above can be used to compute a discrete logarithm in 3 days using a cluster of 16, 500-Mhz PIII processors running Linux.

These results show that the method is really not suitable for a practical deployment of an ID-based cryptosystem, especially since bilinear maps[84] offer much more practical solutions. Nevertheless, non-maximal imaginary quadratic orders should be viewed as an interesting alternative setting, especially for key escrow systems, where the escrow authority may be able to invest more time and resources in retrieving private keys when required. In addition, the lesson that only maximal orders should be used if key escrow is to be avoided should be emphasized.

We hope that the reader now has an appreciation of the various applications of quadratic fields and the Pell equation to public-key cryptography. Although none of the cryptosystems described above is currently used in practice, it is possible that this will change if some of the more widely used systems such as RSA or elliptic curve cryptography are broken or if the ongoing work on improving ideal arithmetic continues to progress. In any event, the existence of these applications certainly provides extra motivation for further investigations into efficient algorithms for computational problems in quadratic fields and, perhaps, another entry point for introducing more people to this fascinating subject.

Notes and References

[1] For a discussion of the lengthy history of this subject, see [Kah96] and [Mol05].

[2] Over the past 30 years, cryptography has become a subject of intense research activity. To get some idea of the breadth and depth of this work, the reader is advised to consult [MvOV96] and [Sch96]. Furthermore, many textbooks on the subject of cryptography have also appeared, the most popular of which seem to be [Mol07], [Sti05], and [TW06].

[3] [Poe41].

[4] [DH76].

[5] [RSA78].

[6] For a history of these developments, see [Sin99] and [Lev02].

[7] See [Sin99], pp. 279–292, and [Ell87].

[8] See, for example, [BJN00] and [FOPS04].

[9] See [Bon99].

[10] Elliptic curves cryptography was independently invented by Koblitz [Kob87] and Miller [Mil86]. See [HMV04] for information on current practical recommendations and protocols.

[11] Although much progress has been made on this problem in the last 30 or more years, it still seems to be very difficult to solve. The best general-purpose technique for factoring an integer is still the number field sieve technique, which is discussed in [LL93] and [Pom94]. More recent developments can be found in [Kle06]. So far, the largest (difficult) number that has been factored by these techniques is RSA200, a specially crafted 200-digit integer which was factored in 2005 [BBFK05]. At this writing it seems that factoring a 300-digit number is an impossible dream, but then factoring a 200-digit number was also an impossible dream 20 years ago.

[12] In fact, it appears that these problems may not be equivalent. See [BV99].

[13] [Rab79].

[14] See [Wil85d], [Wil85b], and [Sal90], pp. 159–166. A more recent analysis of this idea can be found in [Mül06]. This work developed from a technique in which $n = pq$, where $p \equiv 3 \pmod 8$ and $q \equiv 7 \pmod 8$ in [Wil80a].

[15] In this discussion, it is not necessary that t, u be a fundamental solution of the Pell equation.

[16] In fact, there exists a constant c such that $D < c(\log n)^2$. This is guaranteed by the effective version of the Chebotarev density theorem; however, the truth of this result requires the assumption of a generalized Riemann hypothesis. See [LO77]. Also, by using [Oes79], it is possible to compute an explicit value of c. For a charming introduction to Chebotarev and his important theorem, the reader is urged to consult [SL96a].

[17] In [Mül06], a version of this process is given in which this assumption need not be made, but then a different (and more lengthy) process must be used to compute T_l, U_l and T_{2ed}, U_{2ed}. See [Sal90], p. 161.

[18] [Wil85d] and [Wil85c].

[19] The earliest application of quadratic number fields seems to be [OSS84], a digital signature scheme whose security is related to solving norm equations in imaginary quadratic fields.

[20] [BW88a].

[21] See, for example, the survey articles of Buchmann, Hamdy, Takagi, and Vollmer [BTV04, BH01], Hamdy's thesis [Ham02], and the recent book of Buchmann and Vollmer [BV07].

[22] See [BV07] for discussion of how these reductions are constructed.

[23] [Sut07].

[24] [Sho97b].

[25] [BV07], p. 276.

[26] See [BS96c] [BS97].

[27] [Nat00].

[28] [PH78].

[29] [HM00a].

[30] [Gam85].

[31] [ABR01].

[32] [BBHM02].

[33] [FP03].

[34] [BH01].

[35] [PS98].

[36] The random oracle model is an assumption that certain functions used in a protocol are random functions (mapping of inputs to outputs is fixed but random) and is in many cases required to prove strong notions of security such as resistance to existential forgery under an adaptive chosen message attack. Although proofs of security that rely on the random oracle model do not carry over to an actual implementation of the protocol, because the required random functions are typically approximated by hash functions, it is believed that these protocols nevertheless offer more security than protocols without any associated proof. For more on the random oracle model, see [Gol04].

[37] [BH01].

[38] [GQ88].

[39] [Ham02].

[40] [Ham02].

[41] The original idea appeared in [BW89a]. Subsequently, Scheidler, Buchmann, and Williams published a more detailed version in [SBW94].

[42] [SP05].

[43] [HP00].

[44] [JSW01] describes the first version of (f, p) representations as applied to cryptographic key exchange in the infrastructure. [JSW06b] extends these ideas to incorporate the parameter k of Definition 11.1 and NUCOMP for ideal multiplication.

[45] The notion of near-reduced, as described in [JSW06b], is equivalent to w-near, as presented in §11.2, with $w = 0$.

[46] [JSS07a].

[47] Adapted from [JSW06b], Lemma 7.1.

[48] Adapted from [JSW06b], Corollary 7.1.

[49] Adapted from [JSW06b], Theorem 7.1.

[50] [JSW01].

[51] See [JSW06b], Protocol 8.1, for details.

[52] See, for example, [SBW94] and [JSW01].

[53] [Nat07].

[54] [JSW01].

[55] [Sho01].

[56] [JSW01].

[57] [JSW06b].

[58] This observation, substituting certain ideal multiplications with baby steps during exponentiation, is described by Jacobson, Scheidler, and Stein in [JSS07a], together with numerical experiments demonstrating that significant performance improvements can indeed be realized.

[59] [BBT95].

[60] [BMM00].

[61] [HJPT98].

[62] [PT00].

[63] IND-CCA2 stands for indistinguishability under an adaptive chosen ciphertext attack. Informally, this means that when presented with a ciphertext that is known to be the encryption of one of two possible plaintexts, an adversary with polynomially bounded time and computational resources is unable to gain any significant advantage in determining which of the two plaintexts was encrypted. In addition, the adversary is unable to do so even when mounting an adaptive chosen ciphertext attack, where he or she can obtain the decryptions of any ciphertexts (except the one in question) adaptively during the course of the attack. This is considered to be one of the strongest notions of security for a public-key cryptosystem and is, in general, expected of any cryptosystem to be used in practice. For a formal definition, and more information about "provable security" in general, see [Gol04], and, for a more critical examination, see [KM07].

[64] [BST02].

[65] [JSW08].

[66] [HJW03].

[67] Also, see [HJPT98] for a constructive proof.

[68] See [HJW03].

[69] [PT00], Theorem 1.

[70] See [PT00] for some running times in support of this claim.

[71] This method is described in [PO96], and its practical performance is investigated in [ET02].

[72] [JJ00].

[73] [Hüh01].

[74] [BPT04].

[75] [BT02].

[76] This protocol, published in [JSW08], is also described in Weimer's master's thesis [Wei04].

[77] [Wei04], Theorem 5.8.

[78] See also [Wil98], Ch. 4, pp. 69–95.

[79] [HJW03].

[80] The binding between public keys and user identities is in some sense the Achilles heel of public-key cryptography. If this binding is not guaranteed, there is nothing to prevent an active adversary from impersonating a legitimate user by replacing the user's public key with one of his or her choosing. Public-key infrastructures, in which a trusted authority issues unforgeable certificates guaranteeing that a given public key belongs to a particular user, is the most widely deployed solution to this problem. ID-based cryptography offers an interesting alternative. For more information on early work on these topics, see [Gag03].

[81]The method here is described in detail by Hühnlein, Jacobson, and Weber in [HJW03]. It is interesting to note that this approach is essentially a special case of the more general setting of computations in ray class groups described by Cohen, Diaz y Diaz, and Olivier in [CDyDO98], where the modulus is an integer.

[82]For more details and proofs of these results, see [HJW03].

[83]These recommendations are based on somewhat out dated estimates from [HJW03], but nevertheless illustrate the difficulties of using such a system in practice.

[84]The Weil or Tate pairing on an elliptic curve is an example of a bilinear map that can be used to construct practical ID-based cryptosystems. See, for example, the seminal paper by Boneh and Franklin [BF03].

15

Unconditional Verification of the Regulator and the Class Number

15.1 Introduction

We have seen that if ϵ_Δ is the fundamental unit of $\mathcal{O} = [1, \sqrt{D}]$, then

$$t + u\sqrt{D} = \begin{cases} \epsilon_\Delta & \text{when } N(\epsilon_\Delta) = 1 \\ \epsilon_\Delta^2 & \text{when } N(\epsilon_\Delta) = -1 \,, \end{cases}$$

where t, u is the fundamental solution of the Pell equation. We have also seen in Chapter 12 that if we have a sufficiently accurate approximation R_Δ' to $R_\Delta = \log \epsilon_\Delta$, then we can compute a compact representation of ϵ_Δ, from which it is a simple matter to determine certain properties of t and u. This is of particular importance when ϵ_Δ is very large, which, as we have pointed out, is often the case when Δ is large.

In Chapter 13 we presented a fast technique for computing both R_Δ' and h_Δ, the latter being the class number of the quadratic order $\mathcal{O}$ of discriminant Δ. However, the correctness of these values is contingent on unproved hypotheses. In this chapter, we discuss techniques[1] for verifying the values of R_Δ' and h_Δ that the index-calculus method provides. This process is in certain respects very similar to the techniques described in Chapter 10. As much of the work discussed here is intended for computer implementation, it is more convenient to make use of $\mathcal{R}_\Delta = \log_2 \epsilon_\Delta$ instead of R_Δ. This will cause no real problem because

$$R_\Delta = (\log 2)\mathcal{R}_\Delta$$

and $\log 2 = 0.6931471805\ldots$ can be easily computed to great precision. We remind the reader that the most we can be sure of on applying the techniques of Chapter 13 is a value for R_Δ', which is close to an integral multiple of the actual R_Δ, and a value for h_Δ, which is a divisor of the actual h_Δ. We will assume here, then, that we have been presented with a value of $\mathcal{R}_\Delta'$ that is sufficiently close to some integral multiple of $\mathcal{R}_\Delta$ (i.e., a value of $\mathcal{R}_\Delta' \in \mathbb{Q}$ for which it should be true that

$$|\mathcal{R}'_\Delta - c\mathcal{R}_\Delta| < 1 \qquad (15.1)$$

for some $c \in \mathbb{Z}^{\geq 0}$). The goal of our verification technique will be to prove (15.1) and subsequently establish that $c = 1$. As this problem is very easily solved when $\epsilon_\Delta < \Delta^{3/2}$, we will also assume that $\epsilon_\Delta > \Delta^{3/2}$.

We first examine the problem of verifying that (15.1) holds. Certainly, if (15.1) does hold, then on putting $x = \lfloor \mathcal{R}'_\Delta \rfloor - 2$, $\mathfrak{a}_1 = \mathcal{O}$, $\mathfrak{a}_i = \mathfrak{a}[x]$, $\theta_j = \eta = \epsilon^c_\Delta$, $a = 1$, and $b = 4$ in Theorem 11.12, we must have

$$\mathfrak{a}_j \in \{\mathfrak{a}_i, \mathfrak{a}_{i+1}, \dots, \mathfrak{a}_{i+8}\} \qquad (15.2)$$

because $\mathfrak{a}_j = \theta_j \mathfrak{a}_1 = \eta \mathfrak{a}_1 = \mathfrak{a}_1$. Thus, we can use the AX algorithm to compute an (f, p) representation $(\mathfrak{a}_i, d_i, k_i)$ of $\mathfrak{a}_1$ ($f < 2^{p-4}$) and then check to see whether $\mathfrak{a}_1$ $(= \mathfrak{a}_j)$ is an ideal in $\{\mathfrak{a}_i, \mathfrak{a}_{i+1}, \dots, \mathfrak{a}_{i+8}\}$. However, if (15.2) holds, it may not be the case that $|\mathcal{R}'_\Delta - c\mathcal{R}_\Delta| < 1$ for the value of $\mathcal{R}'_\Delta$ that we have been given. Nevertheless, we can use the FIND algorithm to produce an $(f + 1/4, p)$ representation of $(\mathfrak{a}_j, d_j, k_j)$ of $\mathfrak{a}_1$ and we can use the reasoning following Algorithm CR in §12.2 to see that if we replace $\mathcal{R}'_\Delta$ by

$$\mathcal{R}'_\Delta = k_j - p + \frac{s}{2},$$

where

$$2^s < d_j^2 \leq 2^{s+1},$$

then we can be sure that (15.1) is true for this value of $\mathcal{R}'_\Delta$.

Thus, the process of verifying (15.1) will execute in $O(\log \mathcal{R}'_\Delta \log \Delta)$ elementary operations. The difficult problem is proving that $c = 1$. We will approach this in two stages. In the first stage, we will verify that $\mathcal{R}_\Delta > K$ for some preassigned value of K. In the second, we will show that (15.1) cannot hold for any integer c such that

$$1 < c < \frac{\mathcal{R}'_\Delta}{K} + 1.$$

Since it is clear that c cannot exceed or equal $\mathcal{R}'_\Delta / K + 1$, this means that c must be 1.

15.2 Some Preliminary Results

In this section we establish some results that will be useful in developing our technique of proving that $c = 1$. Throughout this section and the next we will assume that all ideals are $\mathcal{O}$-ideals, $b = \lceil (1/2) \log_2 \Delta \rceil + 1$, and $f < 2^{p-4}$. We will show in the sequel what p should be in order to justify this latter assumption. We now introduce a simple proposition.

Proposition 15.1. *If* $\mathfrak{a}_1 = \mathcal{O}$, $\mathfrak{a}_i = \mathfrak{a}[x]$, *and* $\mathfrak{a}_j = \mathfrak{a}[x + b]$, *then* $j > i$.

Proof. By Lemma 11.3, we know that $\theta_i < (17/16)2^x$ and

$$\theta_{j+1} > \frac{15}{16}2^{x+b} > \frac{15}{8}2^x\sqrt{\Delta}\,.$$

Also, since $\theta_{i+1} = \psi_i\theta_i$ and $\psi_i < \sqrt{\Delta}$, we get $\theta_{i+1} < (17/16)2^x\sqrt{\Delta}$, and therefore

$$\theta_{j+1} > \frac{15\cdot 16}{8\cdot 17}\theta_{i+1} > \theta_{i+1}\,.$$

It follows that $j > i$.   □

In order to prove an important result concerning the baby-step giant-step technique that we will employ in the process of showing that $\mathcal{R}_\Delta > K$, we need to establish two lemmas.

Lemma 15.2. *Let* $x = \lceil c\mathcal{R}_\Delta \rceil$, *where* $c \in \mathbb{Z}^{>0}$, *and let* $(\mathfrak{a}_j, d_j, k_j)$ *and* $(\mathfrak{a}_i, d_i, k_i)$ *be* (f, p) *representations of* $\mathfrak{a}_1 = \mathcal{O}$ *such that* $\mathfrak{a}_j = \mathfrak{a}[x + r]$ *and* $\mathfrak{a}_i = \mathfrak{a}[r]$ *with* $r \in \mathbb{Z}^{>0}$. *Then*

$$\mathfrak{a}[x + r] \in \{\mathfrak{a}_n, \mathfrak{a}_{n+1}, \ldots, \mathfrak{a}_{i+3}\}\,,$$

where $n = \max\{1, i - 2\}$.

Proof. We have $k_j < x + r$, $k_{j+1} \geq x + r$ and $x = \log_2 \eta + \epsilon$ for $\eta = (2^{\mathcal{R}_\Delta})^c$ and some ϵ with $0 < \epsilon < 1$. By Lemma 11.3, we have $\theta_j < (17/16)2^{x+r} = (17/8)2^{\epsilon+r-1}\eta$ and $\theta_{j+1} > (15/16)2^{x+r} = (15/16)2^{\epsilon+r}\eta > 1$ (as $r \geq 1$ and $\eta > 1$).

Put $\theta_m = \eta^{-1}\theta_{j+1}$. Then also $\theta_m > 1$, so we must have $m > 1$, which implies that $m - 1 \geq 1$. Note that by Lemma 11.3, $\theta_{m-1} = \eta^{-1}\theta_j < (17/8)2^{\epsilon+r-1}$ and $\theta_m > (15/16)2^{\epsilon+r}$, and similarly we have $\theta_i < (17/16)2^r$ and $\theta_{i+1} > (15/16)2^r = (15/8)2^{r-1}$.

Now,

$$\theta_m > \frac{15}{17}2^\epsilon\theta_i > \frac{15}{17}\theta_i\,;$$

it follows that

$$\theta_{m+2} > F_3\theta_m = 2\theta_m > \frac{2\cdot 15}{17}\theta_i > \theta_i\,,$$

which implies $m + 1 \geq i$ or $m - 1 \geq i - 2$. Also,

$$\theta_{m-1} < \frac{17}{15}2^\epsilon\theta_{i+1} < 3\theta_{i+1} = F_4\theta_{i+1} < \theta_{i+4}\,,$$

which implies $m - 1 < i + 4$ or $m - 1 \leq i + 3$.

By the definition of θ_m, we have $\mathfrak{a}_{m-1} = (\eta^{-1}\theta_j)\mathfrak{a}_1 = \mathfrak{a}_j$, which, together with our determined bounds on $m - 1$, gives us the required result.   □

Corollary 15.2.1. *Under the conditions of the lemma with the exception now that* $\mathfrak{a}_i = \mathfrak{a}[s]$ *for some* $s \in \mathbb{Z}$ *such that* $s \geq r$, *we have* $\mathfrak{a}[x + r] = \mathfrak{a}_j$, *where* $1 \leq j \leq i + 3$.

Proof. Since $\theta_{i+1} > (15/8)2^{s-1}$, we get

$$\theta_{m-1} < \frac{17}{8}2^{\epsilon+r-1} \le \frac{17}{8}2^{\epsilon+s-1} \le \frac{17}{15}2^{\epsilon}\theta_{i+1} < 3\theta_{i+1} < \theta_{i+4} \ .$$

□

Lemma 15.3. *Let* $r \in \mathbb{Z}^{<0}$, $x = \lceil c\mathcal{R}_\Delta \rceil$ *and* $x+r > 0$. *If we let* $\mathfrak{a}_j = \mathfrak{a}[x+r]$ $(\neq \mathfrak{a}_1)$, $\mathfrak{a}_i = \mathfrak{a}[|r|]$ *and* $\mathfrak{a}_t = \mathfrak{a}[|r|+b]$, *then*

$$\bar{\mathfrak{a}}[x+r] \in \{\mathfrak{a}_n, \mathfrak{a}_{n+1}, \ldots, \mathfrak{a}_t\} \ ,$$

where $n = \max\{2, i-1\}$.

Proof. Again, we have that $x = \log_2 \eta + \epsilon$ for $\eta = (2^{\mathcal{R}_\Delta})^c$ and some ϵ with $0 < \epsilon < 1$. By Lemma 11.3, we have $\theta_j < (17/16)2^{x+r} = (17/16)2^{\epsilon+r}\eta$, from which we can deduce that since $\theta_j|\bar{\theta}_j| = N(\mathfrak{a}_j)$,

$$|\bar{\theta}_j| > \frac{16}{17}2^{|r|-\epsilon}\eta^{-1}N(\mathfrak{a}_j)$$

or

$$\eta|\bar{\theta}_j| > \frac{16}{17}2^{|r|-\epsilon}N(\mathfrak{a}_j) > 1 \ ,$$

as $N(\mathfrak{a}_j) > 1$ and $|r| \ge 1$.

Since $\mathfrak{a}_j$ is reduced and $\mathfrak{a}_j \neq \mathfrak{a}_1$, $\bar{\mathfrak{a}}_j = \mathfrak{a}_m$ for some m such that $m > 1$ and $\theta_m = \eta|\bar{\theta}_j| > 1$. From this it follows that $\eta|\bar{\theta}_{j+1}| = \theta_{m-1}$ because $\bar{\mathfrak{a}}_{j+1} = \mathfrak{a}_{m-1}$. Now, by Lemma 11.3 and Proposition 3.16,

$$\theta_i < \frac{17}{16}2^{|r|} < \left(\frac{17}{16}\right)^2 2^{\epsilon}\eta|\bar{\theta}_j|N(\mathfrak{a}_j)^{-1}$$

$$= \left(\frac{17}{16}\right)^2 2^{\epsilon}\theta_m N(\mathfrak{a}_j)^{-1} < 2\theta_m = F_3\theta_m < \theta_{m+2} \ .$$

So $i < m+2$ and thus $m \ge i-1$ and we already had $m > 1$.

Also, again using Lemma 11.3, as $\theta_{j+1} = \theta_j\psi_j$,

$$\theta_j = \frac{1}{\psi_j}\theta_{j+1} > \frac{1}{\psi_j}\frac{15}{16}2^{x+r} = \frac{1}{\psi_j}\frac{15}{16}2^{r+\epsilon}\eta$$

and, thus,

$$|\bar{\theta}_j| < \frac{16}{15}2^{|r|-\epsilon}\eta^{-1}\psi_j N(\mathfrak{a}_j) \ .$$

Since $\psi_j N(\mathfrak{a}_j) < \sqrt{\Delta}$, we get $\theta_m = \eta|\bar{\theta}_j| < (16/15)2^{|r|-\epsilon}\sqrt{\Delta}$. Now,

$$\theta_{t+1} > \frac{15}{16}2^{|r|+b} > \frac{30}{16}2^{|r|}\sqrt{\Delta} > \frac{30}{16}\cdot\frac{15}{16}2^{\epsilon}\theta_m > \theta_m$$

and $t+1 > m$; hence, $m \le t$.

□

Let

$$\mathcal{L} = \{\mathfrak{a}_1, \mathfrak{a}_2, \ldots, \mathfrak{a}_t\},$$

where $\mathfrak{a}_t = \mathfrak{a}[s+b]$ for some $s \in \mathbb{Z}^{\geq 0}$. We are now able to prove Theorem 15.4.

Theorem 15.4. *Let $c \in \mathbb{Z}^{>0}$ and $x = \lceil c\mathcal{R}_\Delta \rceil$ and suppose that $x = 2qs - r$ (for $0 < |r| \leq s$, $s > 1$, and $r, s \in \mathbb{Z}$). Then $\mathfrak{a}[2qs]$ or $\bar{\mathfrak{a}}[2qs] \in \mathcal{L}$.*

Proof. Assume that $\mathfrak{a}_l = \mathfrak{a}[s]$ and $\mathfrak{a}_i = \mathfrak{a}[||r||]$. We distinguish between when $r > 0$ and when $r < 0$, but in both cases we have $i < t$ by Proposition 15.1 and the fact that $|r| \leq s$.

CASE 1: $r > 0$:
  We have $\mathfrak{a}_i = \mathfrak{a}[r]$, thus $\mathfrak{a}[x+r] \in \{\mathfrak{a}_{\max\{1,i-2\}}, \ldots, \mathfrak{a}_{i+3}\}$ by Lemma 15.2. Furthermore, $\{\mathfrak{a}_{\max\{1,i-2\}}, \ldots, \mathfrak{a}_{i+3}\} \subseteq \{\mathfrak{a}_1, \mathfrak{a}_2, \ldots, \mathfrak{a}_{t+2}\}$, as $i < t$.
CASE 2: $r < 0$:
  Either $\bar{\mathfrak{a}}[x + r] = \mathfrak{a}_1$ or $\bar{\mathfrak{a}}[x + r] \neq \mathfrak{a}_1$. If $\bar{\mathfrak{a}}[x + r] \neq \mathfrak{a}_1$, then $\bar{\mathfrak{a}}[x + r] \in \{\mathfrak{a}_{\max\{2,i-1\}}, \ldots, \mathfrak{a}_v\}$ for some $v \leq t$ by Lemma 15.3 and the fact that $|r| \leq s$. So, either way, $\bar{\mathfrak{a}}[x + r] \in \{\mathfrak{a}_1, \mathfrak{a}_{\max\{2,i-1\}}, \ldots, \mathfrak{a}_v\} \subseteq \{\mathfrak{a}_1, \mathfrak{a}_2, \ldots, \mathfrak{a}_{t+2}\}$.

Since $x + r = 2qs$, this concludes our proof. $\square$

From Theorem 15.4 we see that if $\mathfrak{a}[2qs]$ and $\bar{\mathfrak{a}}[2qs]$ are not in $\mathcal{L}$ for $q = 1, 2, \ldots, B$, then $\lceil c\mathcal{R}_\Delta \rceil > 2Bs+s > 2Bs$ for all $c \in \mathbb{Z}^{\geq 0}$ (i.e., $\lceil \mathcal{R}_\Delta \rceil > 2Bs$). It follows that if we choose s and B large enough, we can determine a lower bound K for $\mathcal{R}_\Delta$ by verifying that neither $\mathfrak{a}[2qs]$ nor $\bar{\mathfrak{a}}[2qs]$ is in $\mathcal{L}$ for $q = 1, 2, \ldots, B$.
We now derive some results that are useful for proving that $c = 1$.

Theorem 15.5. *Let $x = \lceil \log_2 \theta_j \rceil + \gamma$, where $\gamma \in \{-1, 0, 1\}$. If $\mathfrak{a}_1 = \mathcal{O}$, $\mathfrak{a}_i = \mathfrak{a}[x]$, and $x > 1$, then $\max\{1, j - 3\} \leq i \leq j + 3$.*

Proof. This result follows easily on putting $b = 1$ and $a = -2$ in Theorem 11.12. $\square$

Corollary 15.5.1. *If η $(> \Delta^{3/2})$ is any unit of $\mathcal{O}$ and $x = \lceil \log_2 \eta \rceil + \gamma \geq 2$ $(\gamma \in \{-1, 0, 1\})$, we must have $\mathfrak{a}[x] \in \mathcal{S}$, where*

$$\mathcal{S} = \{\bar{\mathfrak{a}}_4, \bar{\mathfrak{a}}_3, \bar{\mathfrak{a}}_2, \mathfrak{a}_1, \mathfrak{a}_2, \mathfrak{a}_3, \mathfrak{a}_4\}.$$

Proof. We know that $\mathfrak{a}_j = \eta\mathfrak{a}_1$, where $\eta = \theta_j$. We also have $\theta_j < \theta_{j-3}(\sqrt{\Delta})^3$; hence, $\theta_{j-3} > 1$ and $j - 3 > 1$ or $j > 4$. Hence, by the theorem,

$$\mathfrak{a}[x] \in \{\mathfrak{a}_{j-3}, \mathfrak{a}_{j-2}, \mathfrak{a}_{j-1}, \mathfrak{a}_j, \mathfrak{a}_{j+1}, \mathfrak{a}_{j+2}, \mathfrak{a}_{j+3}\}.$$

Furthermore, $\mathfrak{a}_j = \theta_j\mathfrak{a}_1 = \eta\mathfrak{a}_1 = \mathfrak{a}_1$, so $\mathfrak{a}_{j-3} = \bar{\mathfrak{a}}_4$, $\mathfrak{a}_{j-2} = \bar{\mathfrak{a}}_3$, etc. $\square$

Now suppose that $q \in \mathbb{Z}^{>0}$ and $q \mid c$; then $c = kq$ ($k \in \mathbb{Z}^{>0}$). If $x = \lceil \mathcal{R}'_{\Delta}/q \rceil$, $\eta = \epsilon^k_{\Delta}$, and $y = \lceil k \mathcal{R}_{\Delta} \rceil = \lceil \log_2 \eta \rceil$, then since

$$\left| \frac{\mathcal{R}'_{\Delta}}{q} - \log_2 \eta \right| < \frac{1}{q} < 1 \,,$$

we see that $x = \lceil \log_2 \eta \rceil + \gamma$, where $\gamma \in \{-1, 0, 1\}$. Hence, if $\mathfrak{a}[x] \notin \mathcal{S}$, then $q \nmid c$. However, what can we say if we find that $\mathfrak{a}[x] \in \mathcal{S}$? It is not immediately clear that $q \mid c$ in this case because if $q \approx \mathcal{R}_{\Delta}$, we might find that x is sufficiently small that $\mathfrak{a}[x] \in \mathcal{S}$ even though this value of q would not divide c.

In order to deal with this problem, we first point out that if $x = y + \gamma$, where $x, y \in \mathbb{Z}$, $\gamma \in \{-1, 0, 1\}$, then $\lceil x/q \rceil = \lceil y/q \rceil + \gamma'$, where $\gamma' \in \{-1, 0, 1\}$, for any $q \geq 1$. We are now able to prove a useful result which puts an upper bound of those values of q such that if $\mathfrak{a}[x] \in \mathcal{S}$, then $q \mid c$.

Theorem 15.6. *Let* $\eta = \epsilon^m_{\Delta}$ *(*$m \in \mathbb{Z}^{>0}$*) and* $x = \lceil (\log_2 \eta + \gamma)/q \rceil$*, where* $\gamma \in \{-1, 0, 1\}$*. If* $q \in \mathbb{Z}^{>0}$ *and* $\mathcal{R}_{\Delta} > q(2 \log_2 \Delta + \log_2 17/4)$*, we must have* $q \mid m$ *when* $\mathfrak{a}[x] \in \mathcal{S}$*.*

Proof. Let $\mathfrak{a}_j = \mathfrak{a}[x]$. If we do have $\mathfrak{a}_j = \mathfrak{a}_i$ or $\mathfrak{a}_j = \bar{\mathfrak{a}}_i$ with $i \leq 4$, then either $\theta_j/\theta_i = \epsilon^k_{\Delta}$ or $\theta_j/|\bar{\theta}_i| = \epsilon^k_{\Delta}$ for some $i \leq 4$ and $k \in \mathbb{Z}$. Using Lemma 11.3, we can deduce that

$$\theta_j < \frac{17}{16} 2^{\lceil (\log \eta + \gamma)/q \rceil} = \frac{17}{16} 2^{\lceil \log \eta/q \rceil + \gamma'} = \frac{17}{16} \epsilon^{m/q}_{\Delta} 2^{\epsilon + \gamma'} \qquad (15.3)$$

and

$$\theta_{j+1} > \frac{15}{16} 2^{\lceil (\log \eta + \gamma)/q \rceil} = \frac{15}{16} 2^{\lceil \log \eta/q \rceil + \gamma'} = \frac{15}{16} \epsilon^{m/q}_{\Delta} 2^{\epsilon + \gamma'} \,;$$

hence,

$$\theta_j > \frac{1}{\psi_j} \frac{15}{16} \epsilon^{m/q}_{\Delta} 2^{\epsilon + \gamma'} \qquad (15.4)$$

for some $\gamma' \in \{-1, 0, 1\}$ and some ϵ with $0 < \epsilon < 1$.

Now, put $\nu = \theta_i$ if $\mathfrak{a}_j = \mathfrak{a}_i$ or $\nu = |\bar{\theta}_i|$ if $\mathfrak{a}_j = \bar{\mathfrak{a}}_i$, so that $\theta_j = \nu \epsilon^k_{\Delta}$. If $\nu = \theta_i$, then $1 \leq \nu \leq \theta_4 < (\sqrt{\Delta})^3$, as $i \leq 4$. If $\nu = |\bar{\theta}_i|$, then since $\theta_i|\bar{\theta}_i| = N(\mathfrak{a}_i)$ and $i \leq 4$, we get

$$\left(\sqrt{\Delta} \right)^{-3} < \frac{N(\mathfrak{a}_i)}{\theta_i} \, (= \nu) < \sqrt{\Delta} \,.$$

Hence, $-(3/2) \log_2 \Delta < \log_2 \nu < (3/2) \log_2 \Delta$.

Using (15.3), we obtain

$$\nu \epsilon^k_{\Delta} = \theta_j < \frac{17}{16} 4 \epsilon^{m/q}_{\Delta} \,,$$

which means that

$$\log_2 \nu + k \mathcal{R}_{\Delta} < \log_2 \frac{17}{4} + \frac{m}{q} \mathcal{R}_{\Delta} \,.$$

Using (15.4), we get

$$\nu \epsilon_\Delta^k = \theta_j > \frac{1}{2}\frac{15}{16}\eta_0^{m/q}\frac{1}{\sqrt{\Delta}} = \frac{15}{32}\epsilon_\Delta^{m/q}\frac{1}{\sqrt{\Delta}}$$

and

$$\log_2 \nu + k\mathcal{R}_\Delta > -\log_2\sqrt{\Delta} + \log_2\frac{15}{32} + \frac{m}{q}\mathcal{R}_\Delta .$$

Hence,

$$-\log_2\nu - \log_2\sqrt{\Delta} + \log_2\frac{15}{32} < \left(k - \frac{m}{q}\right)\mathcal{R}_\Delta < -\log_2\nu + \log_2\frac{17}{4} ,$$

$$\left|\left(k - \frac{m}{q}\right)\mathcal{R}_\Delta\right| < 2\log_2\Delta + \log_2\frac{17}{4} ,$$

and

$$|kq - m|\mathcal{R}_\Delta < q\left(2\log_2\Delta + \log_2\frac{17}{4}\right) .$$

Thus, when $\mathcal{R}_\Delta > q(2\log_2\Delta + \log_2(17/4))$, we must have $kq - m = 0$ and $q \mid m$.   □

From this result it follows that we can verify that $c = 1$ by determining that

$$\mathfrak{a}[x_q] \notin \mathcal{S} ,$$

for all primes $q < \mathcal{R}'_\Delta/K + 1$. Here, $x_q = \lceil \mathcal{R}'_\Delta/q \rceil$ and we assume that $\mathcal{R}_\Delta > q(2\log_2\Delta + \log(17/4))$.

15.3 The Algorithm and Some Implementation Issues

In order to verify our value for $\mathcal{R}'_\Delta$, we must compute $\mathcal{L}$, a process requiring the computation of $t + 2$ baby steps. Since $\mathfrak{a}_t = \mathfrak{a}[s + b]$, we see by taking logarithms to base 2 of the result of Theorem 3.17 that we would expect that

$$t \approx \frac{s + b}{1.7} . \tag{15.5}$$

Thus, we require $O(s)$ elementary operations to compute $\mathcal{L}$. We next have to compute the giant steps $\mathfrak{a}[2qs]$ for $q = 1, 2, \ldots, B$. Since $\mathfrak{a}[2(q+1)s] = \mathfrak{a}[2qs + 2s]$, we need to perform algorithm ADDXY once for each value of q, which means that in order to show that $\mathcal{R}_\Delta > K$, we must execute $O(s) + O(B\log\Delta)$ elementary operations, where $2Bs > K$. If we equate B and s, the complexity of this algorithm is $O(K^{1/2} + K^{1/2+\epsilon}) = O(K^{1/2+\epsilon})$.

In order to verify that $c = 1$, we must execute algorithm AX approximately $M/\log M$ times, where $M = \mathcal{R}'_\Delta/K + 1$. The cost of this part of the process is

$$O\left(\left(\frac{M}{\log M}\right)\log \mathcal{R}'_\Delta \log \Delta\right) = O\left(M^{1+\epsilon}\right)$$

elementary operations. To make the complexities of the two main components of the algorithm roughly equal, we need $\sqrt{K} = \mathcal{R}'_\Delta/K$ (ignoring log factors), which means that we should select $K = \mathcal{R}'_\Delta{}^{2/3}$. On doing this, we see that the complexity of the algorithm is $O(\mathcal{R}'_\Delta{}^{1/3+\epsilon})$. Under the general Riemann hypothesis (GRH), we expect that $\mathcal{R}_\Delta$ is close to $\mathcal{R}'_\Delta$; indeed, since we have an explicit upper bound on $\mathcal{R}_\Delta$ from (9.16) and the analytic class number formula, we would certainly not execute this process if $\mathcal{R}'_\Delta$ exceeds this bound by more than 1. Thus, we have $O(\mathcal{R}'_\Delta{}^{1/3}) = O(\Delta^{1/6+\epsilon})$. Of course, if the second stage of the algorithm is to give a correct answer, we must guarantee that

$$\mathcal{R}_\Delta > M\left(2\log_2 \Delta + \log\frac{17}{4}\right).$$

Since $\mathcal{R}_\Delta > K$, this will be the case if

$$K > M\left(2\log \Delta + \log_2\frac{17}{4}\right).$$

Now, $K = \mathcal{R}'_\Delta{}^{2/3}$ and $M = \mathcal{R}'_\Delta{}^{1/3} + 1$; thus, this will certainly happen if

$$\mathcal{R}'_\Delta > 216\left(\log \Delta\right)^3.$$

If $\mathcal{R}'_\Delta$ is less than this bound, then $\mathcal{R}_\Delta$ is small and $\mathcal{R}'_\Delta$ can be easily and quickly verified by other methods, such as the continued fraction technique.

Notice that this algorithm for verifying that $c = 1$ is completely deterministic. If we include the running time of the subexponential algorithm for computing $\mathcal{R}'_\Delta$, then we obtain a Las Vegas algorithm for computing R_Δ given the discriminant Δ.

Theorem 15.7. *The regulator of a real quadratic order $\mathcal{O}_\Delta$ can be computed in expected time $O(\Delta^{1/6+\epsilon})$ under the extended Riemann hypothesis (ERH) and GRH. The output is unconditionally correct.*

This result follows from the fact that the subexponential algorithm will compute $\mathcal{R}'_\Delta$ in expected subexponential time in $\log \Delta$, and that, assuming the ERH, we have $c = 1$. Although the algorithm is not deterministic, it is the fastest known algorithm for computing R_Δ unconditionally in practice, as we will see below.

The basic algorithm is quite simple, but in order to get the best performance from it on a computer, there are a number of issues that should be discussed.[2]

In practice, we only have a limited amount of storage space available to store the baby-step list $\mathcal{L}$. Since we have a fixed set of ideals to compute for the baby-step list (all ideals up to $\mathfrak{a}[s + b]$ and two beyond), we have to somehow limit the number of ideals that we are going to store, which we do by introducing gaps in the baby-step list. Note that (15.5) shows us there are roughly $u = (s + b)/1.7$ ideals in $\mathcal{L}$.

Assume that $\mathfrak{a}_t = \mathfrak{a}[s + b]$. Furthermore, assume that we have space for storing $N + 4$ ideals and that our approximation t' of t (the number of ideals below $\mathfrak{a}[s + b]$) is such that $t < 1.05t'$.[3] When $1.05t' \leq N$, we should have enough space to store the entire baby-step list $\mathcal{L}$. However, as soon as $1.05t' > N$, we need to leave out part of the baby-step list. Let $l = \lceil 1.05/N \rceil$. Instead of storing the entire list $\mathcal{L}$, we store the sublist

$$\mathcal{L}' = \{\mathfrak{a}_1, \mathfrak{a}_l, \mathfrak{a}_{2l}, \ldots, \mathfrak{a}_t = \mathfrak{a}[s + b], \mathfrak{a}_{t+1}, \mathfrak{a}_{t+2}\},$$

which contains every lth ideal before ideal $\mathfrak{a}_t$ and the ideals $\mathfrak{a}_1, \mathfrak{a}_t, \mathfrak{a}_{t+1}$, and $\mathfrak{a}_{t+2}$. Because of the way in which we determined l, we know that we have enough space to store these ideals. It is now fairly easy to show that if we have an ideal $\mathfrak{a}_j$ that is in $\mathcal{L}$, then at least one of the ideals in the list

$$\mathcal{N} := \{\mathfrak{a}_j, \mathfrak{a}_{j+1}, \ldots, \mathfrak{a}_{j+l-1}\}$$

has to be in $\mathcal{L}'$, so that we can replace matching of the ideal $\mathfrak{a}_j$ with the ideals in the list $\mathcal{L}$ with an iteration that tries to match the ideals in $\mathcal{N}$ with the ideals in $\mathcal{L}'$.

We next present an improvement, first introduced by Jacobson et al.,[4] to the algorithm for determining that $c = 1$. In our presentation below, the original method is discussed in greater detail and has been modified to work with (f, p) representations. While trying to determine the multiplier c, we repeatedly have to compute an ideal $\mathfrak{a}[\lceil \mathcal{R}'_\Delta/q_i \rceil]$ and determine whether this ideal is in the list $\mathcal{S}$. We start this procedure with the largest prime and proceed with the primes sorted in a decreasing order, so that the values $\mathcal{R}'_\Delta/q_j$ increase during the algorithm. Instead of computing the ideal $\mathfrak{a}[\lceil \mathcal{R}'_\Delta/q_i \rceil]$ from scratch for every prime q_i, we can now use the fact that we have already computed the ideal $\mathfrak{a}[\lceil \mathcal{R}'_\Delta/q_{i-1} \rceil]$ in the preceding step (where $q_i < q_{i-1}$), and this lies close to the new ideal $\mathfrak{a}[\lceil \mathcal{R}'_\Delta/q_i \rceil]$. Because we already have this ideal, we can put $\delta' := \lceil \mathcal{R}'_\Delta/q_{i-1} \rceil$, determine the difference $\delta := \lceil \mathcal{R}'_\Delta/q_i \rceil - \delta'$, and use the infrastructure to take a giant step from $\mathfrak{a}[\lceil \mathcal{R}'_\Delta/q_{i-1} \rceil]$ to $\mathfrak{a}[\lceil \mathcal{R}_\Delta/q_i \rceil]$ using the ideal $\mathfrak{a}[\delta]$ and the algorithm ADDXY.

To make the process of computing the ideals $\mathfrak{a}[\delta]$ more efficient, we can further optimize the algorithm by approximating these ideals using products of precomputed ideals. We first precompute all ideals $\mathfrak{a}_{t_0}, \mathfrak{a}_{t_1}, \ldots, \mathfrak{a}_{t_m}$, where $\mathfrak{a}_{t_i} = \mathfrak{a}[\delta_{t_i}]$, $\delta_{t_i} = 2^i s'$, $s' = s - 1$ (here s is the s in Theorem 15.4), and m depends on which ideals we expect to use. Now, after computing the ideal $\mathfrak{a}[\delta']$, where $\delta' \approx \lceil \mathcal{R}'_\Delta/q_{i-1} \rceil$, we compute δ and put $\rho := \lfloor \delta/s' \rfloor + 1$, so that $\delta < \rho s' \leq \delta + s'$. We can use the binary expansion $b_r 2^r + b_{r-1} 2^{r-1} + \cdots + b_0$ of ρ to get

$$\delta \approx \rho s' = \rho \delta_{t_0} = 2^r \delta_{t_0} + b_{r-1} 2^{r-1} \delta_{t_0} + \cdots + b_0 \delta_{t_0}$$
$$= b_r \delta_{t_r} + b_{r-1} \delta_{t_{r-1}} + \cdots + b_0 \delta_{t_0}.$$

So if we have $\mathfrak{a}[\delta']$, where $\delta' \approx \lceil \mathcal{R}'_\Delta / q_{i-1} \rceil$, we can find an ideal near $\mathfrak{a}[\lceil \mathcal{R}'_\Delta / q_i \rceil]$ by computing the ideal $\mathfrak{a}[\delta' + \rho s']$. Here, we can easily compute ideal $\mathfrak{a}[\rho s']$ by applying ADDXY to those ideals $\mathfrak{a}_{t_i}$ for which $b_i = 1$, after which we can then use $\mathfrak{a}[\rho s']$ and $\mathfrak{a}[\delta']$ to compute efficiently the ideal $\mathfrak{a}[\delta' + \rho s']$.

Because we actually add $\rho s'$ instead of δ to δ', we do not in general compute the ideal $\mathfrak{a}[\lceil \mathcal{R}'_\Delta / q_i \rceil]$ but obtain an ideal that is close by. The relevance of the next theorem is that it tells us that we can use the technique described above to compute rapidly an ideal near $\mathfrak{a}[\lceil \mathcal{R}'_\Delta / q \rceil]$ for each prime q and then determine that the q does not divide the multiplier c for which $|\mathcal{R}'_\Delta - c\mathcal{R}_\Delta| < 1$ when the ideal that we compute does not end up in the list mentioned in the theorem.

Theorem 15.8. *Suppose that c is a positive integer and $|\mathcal{R}'_\Delta - c\mathcal{R}_\Delta| < 1$. Let s (> 1) and t be defined as in Theorem 15.4 and suppose that $0 < \delta' < \lceil \mathcal{R}'_\Delta / q \rceil + s$, where q is a positive integer. Then if $q \mid c$, we must have that $\mathfrak{a}[\delta' + \rho s']$ is an element of $\{\bar{\mathfrak{a}}_4, \bar{\mathfrak{a}}_3, \bar{\mathfrak{a}}_2, \mathfrak{a}_1, \mathfrak{a}_2, \ldots, \mathfrak{a}_{t+2}\}$, where $\rho = \lfloor \delta / s' \rfloor + 1 > -1$, $s' = s - 1$, and $\delta = \lceil \mathcal{R}'_\Delta / q \rceil - \delta'$.*

Proof. Since $\rho = \lfloor \delta / s' \rfloor + 1$, we have $\delta < \rho s' < \delta + s$. Also,

$$\begin{aligned} \rho > \delta / s' &= \lceil \mathcal{R}'_\Delta / q \rceil / s' - \delta' / s' \\ &> \lceil \mathcal{R}'_\Delta / q \rceil / s' - (\lceil \mathcal{R}'_\Delta / q \rceil + s') / s' \\ &= -s' / s' = -1 \,. \end{aligned}$$

Now, $\delta + \delta' < \rho s' + \delta' < \delta + \delta' + s$ implies that $\lceil \mathcal{R}'_\Delta / q \rceil < \rho s' + \delta' < \lceil \mathcal{R}'_\Delta / q \rceil + s$. Since $\lceil \mathcal{R}'_\Delta / q \rceil = \lceil c\mathcal{R}_\Delta / q \rceil + \gamma$, where $\gamma \in \{-1, 0, 1\}$, we have $\lceil c\mathcal{R}_\Delta / q \rceil \le \rho s' + \delta' \le \lceil c\mathcal{R}_\Delta / q \rceil + s$. If $q \mid c$, then $\lceil k\mathcal{R}_\Delta \rceil \le \rho s' + \delta' \le \lceil k\mathcal{R}_\Delta \rceil + s$, where $k = c/q$ is an integer.

By Corollary 15.5.1, we have $\mathfrak{a}[\lceil k\mathcal{R}_\Delta \rceil] \in \{\bar{\mathfrak{a}}_4, \bar{\mathfrak{a}}_3, \bar{\mathfrak{a}}_2, \mathfrak{a}_1, \mathfrak{a}_2, \mathfrak{a}_3, \mathfrak{a}_4\}$. If $\mathfrak{a}_i = \mathfrak{a}[s]$, $\rho s' + \delta' = \lceil k\mathcal{R}_\Delta \rceil + r$, and $r \ge 1$, then $r \le s$, and for $\mathfrak{a}_w = \mathfrak{a}[\lceil k\mathcal{R}_\Delta \rceil + r]$, we have $w < i + 3 \le t + 2$ by Corollary 15.2.1 and Lemma 15.2. It follows that $\mathfrak{a}[\rho s' + \delta'] \in \{\bar{\mathfrak{a}}_4, \bar{\mathfrak{a}}_3, \bar{\mathfrak{a}}_2, \mathfrak{a}_1, \mathfrak{a}_2, \ldots, \mathfrak{a}_{t+2}\}$. $\square$

In order to use the theorem, we first order all the k possible primes ($\le M$) such that q_1 is the largest such prime and $q_i > q_{i+1}$ ($i = 1, 2, \ldots, k-1$). Put $\delta'_1 = \lceil R_\Delta' / q_1 \rceil$ and define

$$\begin{aligned} \delta_i &= \lceil R_\Delta' / q_i \rceil - \delta'_i \,, \\ \rho_i &= \lceil \delta_i / s' \rceil + 1 \,, \\ \delta_{i+1} &= \delta_i + \rho_i s' \end{aligned}$$

for $i = 1, 2, \ldots, k-1$. We certainly have $0 < \delta'_1 < \lceil R_\Delta' / q_1 \rceil + s$. Suppose that $0 < \delta_i < \lceil R_\Delta' / q_i \rceil + s$; from the proof of the theorem, we must have

$$\delta'_{i+1} < \lceil R_\Delta / q_i \rceil + s \le \lceil R_\Delta / q_{i+1} \rceil + s \,.$$

Thus, by induction, $\delta_i' < \lceil R_\Delta/q_i \rceil + s$ for $1 \leq i \leq k$. It follows that in order to show that $c = 1$, all we need do is check that

$$\mathfrak{a}[\delta_i] \notin \{\bar{\mathfrak{a}}_4, \bar{\mathfrak{a}}_3, \bar{\mathfrak{a}}_2, \mathfrak{a}_1, \mathfrak{a}_2, \ldots, \mathfrak{a}_{t+2}\} \ .$$

Also, as mentioned earlier, $\mathfrak{a}[\delta_i]$ is easily computed from the previously determined $\mathfrak{a}[\delta_{i-1}]$ and $\mathfrak{a}[\rho_{i-1}s']$. We begin this process by using AX to compute $\mathfrak{a}[\delta_1]$.

Because the computations required for the verification procedure can become very time-consuming, it is useful to have the ability to run the algorithm in parallel. Being able to do this allows us to share the workload among multiple machines and/or processors and thereby to handle the verification for orders $\mathcal{O}$ of discriminant Δ, where the value for Δ is larger than the values that we can handle with a non-parallelized version of the algorithm.

Dividing the workload for various parts of the algorithm can be done in a relatively simple way, as all parts basically work with intervals that can simply be divided into smaller, equally sized intervals. For example, while computing the baby-step list, we compute $\mathfrak{a}_1$ ($= \mathcal{O}$), $\mathfrak{a}_2, \mathfrak{a}_3, \ldots$ and determine when we are done by checking whether we encounter the ideal $\mathfrak{a}[s + b]$ computed by AX. To divide this interval between 1 and $s + b$ into y smaller intervals, we introduce lower and upper bounds $(i-1)(s+b)/y$ and $i(s+b)/y$ for each new interval $i = 1, 2, \ldots, y$ and compute the corresponding ideals with distances close to these bounds using AX. If we work with gaps in the baby-step lists, we also need to ensure that the number of omitted ideals is not larger than l ideals by storing the ideal corresponding to the lower bound in the list $\mathcal{L}'$. Because this may introduce additional ideals to store, we need to adjust l accordingly before initiating the computation. Similar techniques can be used for the other parts of the algorithm, for which identifying the intervals is more straightforward.

The part of the verification algorithm that computes the multiplier is easy to parallelize, as each processor simply works on a different interval of primes. In our implementation, we employed a heuristic method to balance roughly the time spent by each processor by creating a large number of subintervals and supplying processors with a new interval only after they finished checking the previous interval obtained. Because an interval containing small primes takes longer to process than one containing the same number of large primes, the intervals with small primes are distributed first.

The baby-step giant-step part of the verification algorithm is unfortunately difficult to parallelize optimally unless all processors have access to the same memory. In that case, a straightforward parallelization in which each processor computes a subset of the baby steps and a subset of the giant steps would work. This approach will most likely fail in a message passing model because the communication overhead would be too high. For example, in order to determine whether a giant-step ideal is in the list of baby steps, a machine would have to query every other machine to determine whether the ideal is in that machine's piece of the baby-step list. As a result, we opted for a sub-optimal solution in which each machine computes an identical copy of the baby-step list. The disadvantage is that the number of baby steps is confined to that which a single machine can store, but the advantage is that there is no interprocessor communication other than the coordinating processor sending initialization information to the other processors at the beginning of the computation.[5]

It is, of course, necessary to have $f < 2^{p-4}$ throughout this process. This can be done by using the following bounds on p.

- For the baby-step giant-step algorithm, if $s > 16$ and $B > \max\{16, \log s\}$, then $f < 2^{p-4}$ for all (f, p) representations computed during the algorithm if p is chosen such that $2^p \geq 221B^2 s$.

- For verifying that $c = 1$ we first assume that $\mathcal{R}'_\Delta > 10^6$, $\mathcal{R}'_\Delta{}^{1/2} < K < \mathcal{R}'_\Delta{}^{5/6}$ and $\max\{16, K^{2/5}\} < s < K^{3/5}$, which does not restrict the choice for the values of the variables too much when $\mathcal{R}'_\Delta$ is large. It turns out that, under these conditions, $f < 2^{p-4}$ for all (f, p) representations that are computed during the determination algorithm if $2^p > 19\mathcal{R}'_\Delta \log \mathcal{R}'_\Delta$.

- If the initial approximation $\mathcal{R}'_\Delta$ needs to be refined, as mentioned at the beginning of §15.2, we assume that $\mathcal{R}'_\Delta > 10^6$ and require that $2^p > 21\mathcal{R}'_\Delta \log \mathcal{R}'_\Delta$ in order to ensure that AX will produce an (f, p) representation of $\mathfrak{a}[\lceil \mathcal{R}'_\Delta \rceil - 2]$ with $f < 2^{p-4}$.

The proofs[6] of these results can be derived by an analysis similar to that employed in Chapter 11.

To get some idea of how effective this verification is, we first programmed it to run on a machine with two Intel P4 Xeon 2.4-GHz processors and 2 GB of RAM. The resulting runtimes are listed in Table 15.1, where we have incorporated the time needed for the subexponential algorithm in the time needed for the $\Delta^{1/6+\epsilon}$ algorithm, as the latter depends on the former for its input. In our examples, Δ was selected to be a fundamental discriminant.

The data in Table 15.1 show that even though Lenstra's $\Delta^{1/5+\epsilon}$ algorithm is initially faster, the $\Delta^{1/6}$ algorithm becomes significantly faster as soon as $\Delta \approx 10^{20}$ or larger. In fact, the only reason why the $\Delta^{1/5+\epsilon}$ has this initial advantage is because of the standard overhead of the subexponential algorithm, which for $\Delta \approx 10^{15}$ can be seen to be the most costly part of the $\Delta^{1/6+\epsilon}$ algorithm. Furthermore, we can see from the table that even though the $\Delta^{1/6+\epsilon}$ algorithm is significantly faster than the $\Delta^{1/5+\epsilon}$ algorithm for large Δ, it is still very slow compared to the subexponential algorithm.

We also implemented a parallel version of the $\Delta^{1/6+\epsilon}$ algorithm and ran it on 240 processors in a cluster of machines containing two Intel P4 Xeon 2.4-GHz processors and 2 GB of RAM. The timings of the corresponding verifications can be found in Table 15.2, together with a rough approximation of the corresponding regulators.

In particular, we were able to compute unconditionally the regulator for a 65-digit value of Δ; this is far beyond the capabilities of previous unconditional

Table 15.1. Comparison of Runtimes

$\Delta \approx 10^{\cdots}$	Subexponential	$\Delta^{1/6+\epsilon}$	$\Delta^{1/5+\epsilon}$
15	0.29 sec	0.42 sec	0.25 sec
20	0.45 sec	0.93 sec	3.65 sec
25	0.68 sec	3.20 sec	2 min, 20 sec
30	1.44 sec	14.60 sec	44 min, 26 sec
35	2.57 sec	1 min, 27 sec	2 days, 13 hours
40	6.06 sec	6 min, 12 sec	N/A
45	26.27 sec	1 hour, 10 min	N/A
50	1 min, 27 sec	1 day, 9 hours	N/A

Table 15.2. Parallel Version Runtimes

$\Delta \approx 10^{\cdots}$	$R_\Delta \approx \ldots \times 10^{30}$	Subexponential	$\Delta^{1/6+\epsilon}$
62	3.4	1 hour, 14 min	6 days, 3 hours
63	5.2	1 hour, 37 min	8 days, 2 hours
64	8.1	2 hours, 17 min	10 days, 13 hours
65	195.6	2 hours, 5 min	102 days, 7 hours

algorithms. On examining Table 15.2, the runtime for this value of Δ appears to be excessive in light of the complexity discussion at the beginning of this section, which predicts an increase in runtime of approximately a factor 3 over the 64-digit value. However, this is due to some practical issues resulting from the amount of memory available to store $\mathcal{L}'$. For the 65-digit example, the time required for l table look-ups exceeds the time used by the ADDXY operation, which causes the time for all but the baby-step part to increase quadratically.[7]

The precise value of this Δ and its corresponding $\mathcal{R}'_\Delta$ are

$$\Delta = 3928637573454259474975805083515165509211884853083339874356156848 1$$

and

$$\mathcal{R}'_\Delta = 19569629046686552424253842387693615.28 \,.$$

15.4 The Class Number

We now turn to the problem of verifying that the value of h_Δ furnished by the index-calculus method of Chapter 13 is correct. In order to avoid confusion with the correct class number h_Δ, we will denote this value by h'_Δ. Since we know that $h'_\Delta \mid h_\Delta$, it suffices to verify that $h_\Delta/h'_\Delta < 2$. This problem was addressed recently by Booker,[8] and we will briefly discuss his results in this section.

By Corollary 8.35.1, we have[9]

$$\frac{h_\Delta}{h'_\Delta} = \frac{\sqrt{\Delta}}{2h'_\Delta R_\Delta} L(1, \chi_\Delta) . \tag{15.6}$$

By (9.31), we can write this as

$$\frac{h_\Delta}{h'_\Delta} = \frac{1}{h'_\Delta R_\Delta} \sum_{n=1}^{\infty} \chi_\Delta(n) F\left(\frac{n}{\sqrt{\Delta}}\right) , \tag{15.7}$$

where $\chi_\Delta(n) = (\Delta/n)$ and

$$F(x) = \int_x^\infty \left(\frac{1}{x} + \frac{1}{t}\right) e^{-\pi t^2} dt . \tag{15.8}$$

Notice that $F(x)$ is a monotonically decreasing, convex function for $x > 0$.

We now write (15.7) as

$$\left| \frac{h_\Delta}{h'_\Delta} - \frac{1}{h'_\Delta R_\Delta} \sum_{n=1}^{X} \chi_\Delta(n) F\left(\frac{n}{\sqrt{\Delta}}\right) \right| = \frac{1}{h'_\Delta R_\Delta} \left| \sum_{n=X+1}^{\infty} \chi_\Delta(n) F\left(\frac{n}{\sqrt{\Delta}}\right) \right| . \tag{15.9}$$

We can therefore estimate the value of h_Δ/h'_Δ by computing the first X terms of the series in (15.7), provided that the right-hand side of (15.9) is small enough. Our next objective will be to find a value for X for which this will be the case. By partial summation,[10] we can write this as

$$\frac{1}{h'_\Delta R_\Delta} \left| \sum_{n=X+1}^{\infty} \chi_\Delta(n) F\left(\frac{n}{\sqrt{\Delta}}\right) \right|$$

$$= \frac{1}{h'_\Delta R_\Delta} \left| \sum_{n=X+1}^{\infty} S(n) \left[F\left(\frac{n}{\sqrt{\Delta}}\right) - F\left(\frac{n+1}{\sqrt{\Delta}}\right) \right] \right| ,$$

where

$$S(n) = \sum_{j=X+1}^{n} \chi_\Delta(n) . \tag{15.10}$$

For any fixed ϵ and fixed positive integer r, a result of Burgess[11] gives us

$$S(n) = O\left(n^{1-1/r} \Delta^{(r+1)/4r^2 + \epsilon} \right) . \tag{15.11}$$

It is easy to see that $F'(x) = O(x^{-2})$, and since $F(x)$ is a continuous decreasing function, we have

$$\left| F\left(\frac{n}{\sqrt{\Delta}}\right) - F\left(\frac{n+1}{\sqrt{\Delta}}\right) \right| = O\left(\frac{\sqrt{\Delta}}{n^2}\right)$$

by the mean value theorem. Thus, by (15.11),

$$\left| \sum_{n=X+1}^{\infty} \chi_\Delta(n) F\left(\frac{n}{\sqrt{\Delta}}\right) \right| = O\left(\Delta^{1/2+(r+1)/4r^2+\epsilon} \sum_{n=X+1}^{\infty} \frac{1}{n^{1+1/r}}\right)$$

$$= O\left(X^{-1/r}\Delta^{1/2+(r+1)/4r^2+\epsilon}\right) .$$

If, as we are reasonably certain, $h'_\Delta = h_\Delta$, then by Siegel's result (9.22), we get

$$\frac{\left| \displaystyle\sum_{n=X+1}^{\infty} \chi_\Delta(n) F\left(\frac{n}{\sqrt{\Delta}}\right) \right|}{h'_\Delta R_\Delta} = O\left(X^{-1/r}\Delta^{(r+1)/4r^2+\epsilon}\right) . \tag{15.12}$$

This will be small if

$$X \gg \Delta^{1/4+1/4r+2r\epsilon} . \tag{15.13}$$

Since each term of

$$\sum_{n=1}^{X} \chi_\Delta(n) F\left(\frac{n}{\sqrt{\Delta}}\right) \tag{15.14}$$

can be computed to high precision in polynomial (in $\log \Delta$) time and r and ϵ are arbitrary, we can produce an algorithm for calculating a sufficiently accurate value of (15.14) that permits us to assert that $h_\Delta/h'_\Delta < 2$. Furthermore, this algorithm will execute in running time $O(\Delta^{1/4+\epsilon})$ as long as $h'_\Delta = h_\Delta$. The difficulty with this approach to verifying the value of h_Δ is that (15.11) is not effective, which means that we have no way of computing an upper bound on the value of the right-hand side of (15.9).

Booker solved this problem by producing an effective version[12] of Burgess' result, which we give below as Theorem 15.9.

Theorem 15.9. *Let $\Delta > 10^{20}$ be a prime number congruent to 1 (mod 4), $r \in \{2, 3, \ldots, 15\}$, and M and N be integers with $0 < M, N \leq 2\sqrt{\Delta}$. Then*

$$\left| \sum_{M \leq n < M+N} \chi_\Delta(n) \right| \leq \alpha(r)\Delta^{(r+1)/4r^2}(\log \Delta + \beta(r))^{1/2r} N^{1-1/r} , \tag{15.15}$$

where $\alpha(r)$ and $\beta(r)$ are given in Table 15.3.

Table 15.3. Values of $\alpha(r)$ and $\beta(r)$

r	$\alpha(r)$	$\beta(r)$	r	$\alpha(r)$	$\beta(r)$
2	1.8221	8.9077	9	1.4548	0.0085
3	1.8000	5.3948	10	1.4231	-0.4106
4	1.7263	3.6658	11	1.3958	-0.7848
5	1.6526	2.5405	12	1.3721	-1.1232
6	1.5892	1.7059	13	1.3512	-1.4323
7	1.5363	1.0405	14	1.3328	-1.7169
8	1.4921	0.4856	15	1.3164	-1.9808

He next used partial summation to bound the right-hand side of (15.9) as

$$\frac{1}{h'_\Delta R_\Delta}\left|\sum_{n=X+1}^\infty \chi_\Delta(n)F\left(\frac{n}{\sqrt{\Delta}}\right)\right|$$

$$\leq \frac{1}{h'_\Delta R_\Delta}\left(\frac{1}{\sqrt{\Delta}}\sum_{X<n<2\sqrt{\Delta}}\left|S(n)F'\left(\frac{n}{\sqrt{\Delta}}\right)\right|\right.$$

$$\left. + \left|S(2\sqrt{\Delta})\right|F(2) + \left|\sum_{n>2\sqrt{\Delta}}\chi_\Delta(n)F\left(\frac{n}{\sqrt{\Delta}}\right)\right|\right), \quad (15.16)$$

where $S(n)$ is given by (15.10). In this expression it is necessary to bound the three terms on the right.

It is easy to see by integrating by parts that

$$\int_x^\infty t^{-\alpha}e^{-\pi t^2}\,dt < \frac{e^{-\pi x^2}}{2\pi x^{\alpha+1}} \quad (15.17)$$

when $\alpha > -1$ and $x > 0$. Hence,

$$\left|\sum_{n>2\sqrt{\Delta}}\chi_\Delta(n)F\left(\frac{n}{\sqrt{\Delta}}\right)\right| \leq \int_{2\sqrt{\Delta}}^\infty F\left(\frac{t}{\sqrt{\Delta}}\right)dt = \sqrt{\Delta}\int_2^\infty F(x)\,dx.$$

Since $F(x) < (1/\pi x^2)e^{-\pi x^2}$, we get

$$\sqrt{\Delta}\int_2^\infty F(x)\,dx < \frac{\sqrt{\Delta}}{\pi}\int_2^\infty x^{-2}e^{-\pi x^2}\,dx < \frac{\sqrt{\Delta}}{16\pi^2 e^{4\pi}}.$$

As we expect $h'_\Delta R_\Delta$ to be $h_\Delta R_\Delta = \sqrt{\Delta}L(1,\chi_\Delta)/2$, we see that on examining the left-hand side of (9.24) that

$$\frac{1}{h'_\Delta R_\Delta} \left| \sum_{n>2\sqrt{\Delta}} \chi(n) F\left(\frac{n}{\sqrt{\Delta}}\right) \right| < \frac{1}{8\pi^2 e^{4\pi} L(1,\chi_\Delta)} \tag{15.18}$$

should be small, but, of course, this must be checked.

Also, $|S(2\sqrt{\Delta})|F(2)$ can be bounded by using (15.17) and (15.15). The more difficult problem is to select a candidate for X which will permit us to bound the value of

$$\frac{1}{\sqrt{\Delta}} \sum_{X<n<2\sqrt{\Delta}} \left| S(n) F'\left(\frac{n}{\sqrt{\Delta}}\right) \right| \tag{15.19}$$

in such a way that we can conclude that $h_\Delta/h'_\Delta < 2$. For $n < 2\sqrt{\Delta}$, we find from (15.17) that

$$S(n) \leq \alpha(r) \Delta^{(r+1)/4r^2} (\log \Delta + \beta(r))^{1/2r} (n-X)^{1-1/r} . \tag{15.20}$$

We now initiate an interpolation process to find X. We begin by selecting a candidate value of $X = \lfloor \sqrt{\Delta} \rfloor$, say, and estimate the sum (15.19) by using (15.20) and several[13] precomputed upper bounds on $|F'(x)|$ for sample values of x such that $0 < x < 2$. Since F is convex, we can interpolate between them. We can also vary the value of r in (15.20) according to the interval over which we are summing in (15.19). For example, if n is near $2\sqrt{\Delta}$, take $r = 2$, but as n decreases, successive values of $r = 3, 4, \ldots$ work best.

After computing this trial upper bound $T(X)$ on (15.19), we compute $S(X)$, the sum of $T(X)/h'_\Delta R_\Delta$, and our bounds on $S(2\sqrt{\Delta})F(2)/h'_\Delta R_\Delta$ and the term (15.18).[14] If $S(X) < \eta$, some number near 1, say 0.995, we conclude on noting (15.12) that X is too large; if $S(X) > \eta$, we conclude that X is too small. We can now employ a bisection process to obtain the smallest value of X such that $S(X) \leq \eta$. By (15.13), this value of X should be $O(\Delta^{1/4+\epsilon})$, and we use it to compute

$$R(X) = \left| \sum_{n=1}^{X} \chi_\Delta(n) F\left(\frac{n}{\sqrt{\Delta}}\right) \right| .$$

If $R(X) < 2 - \eta$, we know that $h_\Delta/h'_\Delta < 2$ or $h_\Delta = h'_\Delta$.

By using this basic process, together with some clever refinements,[15] Booker was able to verify the class number of $\mathcal{O}$ with fundamental discriminant $\Delta = 10^{31} + 33$ as 43. This required about 95 hours of computation time on a 500-MHz Ultra Sparc II.

Notes and References

[1] These techniques developed from an idea mentioned in [JPW03]. They were reported in [dHJW07] and in much more detail in [dH04].

[2] These matters are discussed much more thoroughly in [dH04].

[3] In some emirical work it was found that t' tends to be somewhat larger than t. This was likely because the value 1.7 is only an approximation.

[4] [JLW95].

[5] Much more information concerning this and the determination of optimal values for the parameters s, B, etc. can be found in §§6.4 and 6.5 of [dH04].

[6] See §6.3 of [dH04]. In fact, some refinement of the arguments presented there would result in a reduction of the constants 221, 19, and 21, inspite of our use here of the less precise technique of Theorem 11.2 as opposed to that of Theorem 3.2 of [dH04].

[7] Furthermore, a significant amount of computation time was added due to synchronization after each part of the verification algorithm and hardware problems in combination with insufficiently fine-grained checkpointing. Under ideal circumstances, we would expect the same computation to take approximately 83 days, given the same memory constraints.

[8] [Boo06].

[9] We are only considering the case of $\Delta > 0$ here, but the method can also be applied to the $\Delta < 0$ case.

[10] See the proof of Theorem 8.24.

[11] [Bur57].

[12] Booker only proves his result for Δ a prime, but his method could be applied to a general Δ with a concomitant increase in the constants α and β.

[13] Booker suggests 10,000 such values geometrically spaced to ensure that there are many more samples near 0. This is because there is a $1/x^2$ singularity there, and as a consequence, these terms will be of greatest importance.

[14] We have written R_Δ here, but this is a convenience because we do not know R_Δ exactly; we would have to use $R_\Delta' - 1$, which we know is a lower bound on R_Δ.

[15] See §4 of [Boo06].

Principal Ideal Testing in $\mathcal{O}$

16.1 Introduction

Let $\mathfrak{i}$ be any given $\mathcal{O}$-ideal. In this chapter we will denote by **P** the problem of determining whether or not $\mathfrak{i}$ is a principal $\mathcal{O}$-ideal. Of course, as we can easily find α and a reduced $\mathcal{O}$-ideal $\mathfrak{b}$ such that $\mathfrak{b} = (\alpha)\mathfrak{i}$, we may consider this problem as applying, instead, to a given reduced $\mathcal{O}$-ideal $\mathfrak{b}$. Also, since a principal ideal is invertible, we may also assume that $\mathfrak{b}$ (or $\mathfrak{i}$) is invertible. An extended version of **P** is the problem of determining, once we know that $\mathfrak{b}$ is principal, a generator β of $\mathfrak{b}$. By our observations in Chapter 12, all we really need to determine β is some $q, b \in \mathbb{Q}$ such that $|\log_2 \beta - b| < q$, where q is small, say $q \leq 10$. Both of these problems can be solved by the index-calculus method described in §13.4, but because we need the truth of unproved hypotheses to be sure our result is correct, particularly when $\mathfrak{b}$ has been declared not principal, this technique is conditional. In this chapter we will develop methods for solving **P** that are either unconditional or are only conditional concerning the runtime; the answer is still mathematically correct (a Las Vegas algorithm). The problem with these processes is that they are unfortunately of exponential complexity.

The simplest unconditional technique for solving **P** is simply to determine the cycle $\mathcal{C}$ of reduced principal $\mathcal{O}$-ideals by computing the period of the continued fraction expansion of ω. The reduced $\mathcal{O}$-ideal $\mathfrak{b}$ is principal if and only if $\mathfrak{b} \in \mathcal{C}$. Also, if $\mathfrak{b} \in \mathcal{C}$, then $\mathfrak{b} = \mathfrak{a}_m$ ($\mathfrak{a}_1 = \mathcal{O}$) for some $1 \leq m \leq l$, where l is the period length of the simple continued fraction (SCF) expansion of ω and therefore $\mathfrak{b} = (\theta_m)$. Since $|\mathcal{C}| = l = O(R_\Delta)$, this algorithm will solve **P** in $O(R_\Delta) = O(\Delta^{1/2+\epsilon})$ elementary operations. We next show how we can use some of our earlier routines to solve **P** in $O(R_\Delta^{1/2+\epsilon})$ elementary operations. The basic idea behind this technique was briefly discussed near the end of Chapter 7; we give here a complete method that is provably exact.

We begin by assuming that $\Delta > 2^{20}$ and that we are given $\mathcal{R}'_\Delta$ such that $|\mathcal{R}'_\Delta - \mathcal{R}_\Delta| < 1$, where, as before, $\mathcal{R}_\Delta = \log_2 \epsilon_\Delta$. We also assume that

$$\sqrt{\mathcal{R}'_\Delta} > 15 + \frac{5}{2} \log_2 \Delta \ (> 65) \,. \tag{16.1}$$

If this is not the case, then $\mathcal{R}_\Delta = O((\log \Delta)^2)$, and we can solve $\mathbf{P}$ by the previous technique in time $O((\log \Delta)^2)$. We present our algorithm as a sequence of three steps.

1. We put $x = \lceil \sqrt{\mathcal{R}'_\Delta} \rceil$ and select p such that

$$2^p > m \max\{16, \log x\} x \,,$$

   where $m = 11.2$. We next use AX to compute an (f, p) representation $(\mathfrak{a}[x], d, k)$ of $\mathfrak{a}_1 \ (= \mathcal{O})$. Since $\mathfrak{a}[x]$ is a reduced ideal, we have $\mathfrak{a}[x] = \mathfrak{a}_n = \theta_n \mathfrak{a}_1$ and by (11.4) and Theorem 11.9, we know that $f < mx$, $f < 2^{p-4}$, and

$$|\log_2 \theta_n - k| < \frac{3}{2} \,. \tag{16.2}$$

   Also, by Corollary 11.3.1, we have

$$\frac{17}{16} x > \log_2 \theta_n > \log_2 \left(\frac{15}{16\sqrt{\Delta}} \right) + x \,. \tag{16.3}$$

   Hence, by (16.1), we get

$$k > \log_2 \left(\frac{15}{16\sqrt{\Delta}} \right) + x - \frac{3}{2} > \frac{3}{4} \log \Delta + 4 \,. \tag{16.4}$$

2. We next compute the SCF expansion of ω to obtain

$$S = \{\mathfrak{a}_1, \mathfrak{a}_2, \mathfrak{a}_3, \ldots, \mathfrak{a}_n\} \,.$$

   Since, by (3.40),

$$n < \frac{\log_2 \theta_n}{\log_2 \tau} + 2 \,,$$

   we get

$$n < \frac{17}{16} \frac{x}{\log_2 \tau} + 2 < 1.6x + 2 \tag{16.5}$$

   by (16.3). Thus, this step can be performed in time $O(\mathcal{R}_\Delta{}^{1/2+\epsilon})$.

3. Put $\mathfrak{c}_1 = \mathfrak{b}$ and use NUCOMP as

$$\mathfrak{c}_i = \text{NUCOMP}(\mathfrak{c}_{i-1}, \mathfrak{a}[x]) \tag{16.6}$$

   to produce a sequence of reduced $\mathcal{O}$-ideals $\mathfrak{c}_1, \mathfrak{c}_2, \mathfrak{c}_3, \ldots$. We now put $\kappa = \lceil (3/4) \log_2 \Delta \rceil$ and

$$j = \left\lceil \frac{\mathcal{R}'_\Delta - k + 3/2}{k - \kappa - 3/2} \right\rceil + 2 \,.$$

   Notice that $k - \kappa - 3/2 > 0$ by (16.4). By using (16.1), (16.2), (16.4) and Corollary 11.3.1, it is a simple matter to show that

$$j < 2\sqrt{\mathcal{R}'_\Delta} \leq 2x \; . \tag{16.7}$$

If, for any $i \leq j$, we get $\mathfrak{c}_i \in S$, then $\mathfrak{b}$ must be a principal $\mathcal{O}$-ideal. If $\mathfrak{c}_i \notin S$ for all i such that $1 \leq i \leq j$, then $\mathfrak{b}$ is not a principal ideal. Since $j = O(R_\Delta^{1/2})$, we see that this step, and hence the entire procedure, can be executed in time $O(R_\Delta^{1/2+\epsilon})$.

We will now explain why this technique works. Note that $\gamma_1 = 1$ and $\mathfrak{c}_i = \gamma_i \mathfrak{b}$, where

$$\gamma_i = \frac{\gamma_{i-1}\theta_n}{\mu_i} \; ,$$

$\mu_i \mathfrak{c}_i = \mathfrak{c}_{i-1}\mathfrak{a}[x]$, and, by (5.45),

$$0 \leq \log_2 \mu_i < \frac{3}{4}\log_2 \Delta \quad (i = 2,3,4,\dots) \; .$$

Suppose $\mathfrak{b}$ is principal; then $\mathfrak{b} = \mathfrak{a}_t$ for some $t \leq l$. Hence, $\mathfrak{c}_i = \gamma_i \theta_t \mathfrak{a}_1$ and $\mathfrak{c}_i$ is reduced and principal for $i = 1, 2, \dots$. Futhermore, if $\mathfrak{b} \notin S$, then $\theta_t > \theta_n$. By (16.3) and (16.1) we have $\mu_i < \theta_n$; thus, $\gamma_i > \gamma_{i-1}$ $(i = 2,3,\dots)$ and there must therefore exist some h such that

$$\gamma_{h-1} < \frac{\epsilon_\Delta}{\theta_t} \text{ and } \gamma_h \geq \frac{\epsilon_\Delta}{\theta_t} \; ;$$

that is,

$$\frac{\epsilon_\Delta}{\theta_t} \leq \gamma_h = \frac{\gamma_{h-1}\theta_n}{\mu_h} < \frac{\epsilon_\Delta \theta_n}{\theta_t}$$

and

$$1 \leq \theta_t \gamma_h \epsilon_\Delta^{-1} < \theta_n \; .$$

Since $\mathfrak{c}_h = \theta_t \gamma_h \mathfrak{a}_1 = \theta_t \gamma_h \epsilon_\Delta^{-1} \mathfrak{a}_1$, we must have $\mathfrak{c}_h \in S$.

We now requre an upper bound on h. Since

$$\log_2 \gamma_i = \log_2 \gamma_{i-1} + \log_2 \theta_n - \log_2 \mu_i \; ,$$

we can easily deduce, by using (16.2), that

$$\log_2 \gamma_i > (i-1)(k - \kappa - 3/2) \; .$$

Also,

$$\log_2 \gamma_{h-1} < \mathcal{R}_\Delta - \log_2 \theta_t < \mathcal{R}_\Delta - \log_2 \theta_n < \mathcal{R}'_\Delta - k + \frac{3}{2} \; ;$$

consequently,

$$\mathcal{R}'_\Delta - k + \frac{3}{2} > (h-2)(k - \kappa - 3/2)$$

and we get

$$h - 2 < \frac{\mathcal{R}'_\Delta - k + 3/2}{k - \kappa - 3/2} \leq j - 2 \; .$$

Unfortunately, when $\mathfrak{b}$ is principal, the above process does not furnish us with any information concerning the size of the generator θ_t of $\mathfrak{b}$. Nevertheless, we can learn something about θ_t by using a similar procedure in which we first ask that p satisfy

$$2^p > 102\mathcal{R}'_\Delta .$$

Since we must find that $\mathfrak{c}_h = \mathfrak{a}_s$ for some s such that $1 \le s \le n$, we can use FPCF (see the Appendix) to obtain an (f_s, p) representation $(\mathfrak{a}_s, d_s, k_s)$ of $\mathfrak{a}_1$, where $f_s < 1 + 2s \le 1 + 2n$. We next use NUMULT instead of NUCOMP and replace (16.6) by

$$(\mathfrak{c}_i, d'_i, k'_i) = \text{NUMULT}((\mathfrak{c}_{i-1}, d'_{i-1}, k'_{i-1}), (\mathfrak{a}[x], d, k), p) ,$$

where we initialize $\mathfrak{c}_1 = \mathfrak{b}$, $d'_1 = 2^{p+1}$, $k'_1 = 0$, and $f'_1 = 1$. Here, $(\mathfrak{c}_i, d'_i, k'_i)$ is an (f'_i, p) representation of $\mathfrak{c}_1 = \mathfrak{b}$.

Since

$$2^p > 102\mathcal{R}'_\Delta ,$$

we get $(m = 11.2)$

$$
\begin{aligned}
2^p &> 2(\sqrt{2}+1)(m+1)(e-1)\mathcal{R}'_\Delta \\
&> 2(\sqrt{2}+1)(e-1)\left[\left(\frac{17}{8}+m\right)\sqrt{R_\Delta'} + mR_\Delta'\right] \\
&> 2(\sqrt{2}+1)(e-1)\left(\frac{17}{8}+mx\right)\sqrt{R_\Delta'} .
\end{aligned}
\tag{16.8}
$$

It certainly follows from (16.8) that

$$2^p > m\max\{16, \log x\}x ,$$

and

$$2^p > (\sqrt{2}+1)(1+2n) \ge (\sqrt{2}+1)f_s \quad \text{(by (16.5) and (16.1))} ; \tag{16.9}$$

hence, by (11.4),

$$|k_s - \log_2 \theta_s| < \frac{3}{2} . \tag{16.10}$$

Also, if we put $f = f_s$, we have

$$f < mx < 2^{p-4}$$

by Theorem 11.9.

Since, $f'_1 = 1$ and

$$f'_{i+1} = \frac{17}{8} + f'_i + f + \frac{f'_i f}{2^p} , \tag{16.11}$$

we get

$$f_i' = \left(\frac{17}{8} + f\right)\left(\frac{c^{i-1} - 1}{c - 1}\right) + c^{i-1}$$

$$< \left(\frac{17}{8} + f\right)\left(\frac{c^i - 1}{c - 1}\right),$$

where $c = 1 + f/2^p < e^{f/2^p}$. Thus,

$$f_i' < \left(\frac{17}{8} + f\right)\left(\frac{e^{fi/2^p} - 1}{f/2^p}\right). \tag{16.12}$$

Now, by (16.11), f_i' is an increasing function of i, and since $f < mx$, we see by (16.8) that

$$2^p > 2mx\sqrt{R_\Delta'} > mxj > fj;$$

consequently, $fj/2^p < 1$. Since for a fixed i, $(e^{xi} - 1)/x$ is an increasing function of x for $x > 0$, we get

$$\frac{e^{fi/2^p} - 1}{f/2^p} < \frac{e^{i/j} - 1}{1/j} \leq j(e - 1)$$

when $i \leq j$. It follows from (16.12), (16.7), and (16.8) that

$$f_i' < \left(\frac{17}{8} + f\right)j(e-1) < 2\sqrt{R_\Delta'}(e-1)\left(\frac{17}{8} + mx\right) < (\sqrt{2} - 1)2^p$$

$(i = 1, 2, 3, \ldots, j)$. Thus, by (11.4),

$$|k_h' - \log \gamma_h| < \frac{3}{2}. \tag{16.13}$$

Since $\mathfrak{c}_h = \mathfrak{a}_s = \theta_s \mathfrak{a}_1$, we get

$$\theta_s = \eta \theta_t \epsilon_\Delta^{-1} \gamma_h, \tag{16.14}$$

where η is some unit of $\mathcal{O}$. By selection of γ_h, we have $\gamma_h \geq \epsilon_\Delta/\theta_t$ and $\gamma_h = \gamma_{h-1}\theta_n/\mu_i \leq \gamma_{i-1}\theta_n < \epsilon_\Delta\theta_n/\theta_t$. Hence, since, by (16.3) and (16.1), $\theta_s \leq \theta_n < \epsilon_\Delta$, we get

$$\epsilon_\Delta^{-1} < \eta < \epsilon_\Delta,$$

and, therefore, $\eta = 1$. By (16.14), we have

$$\log_2 \theta_t + \log_2 \gamma_h = R_\Delta + \log_2 \theta_s,$$

and by (16.10) and (16.13), we get

$$|\log_2 \theta_t - R_\Delta' - k_h' - k_s| < 4.$$

Since we know R_Δ', k_h', and k_s, we can use this result to compute a compact representation of θ_t by the technique in Chapter 12.

16.2 Another Approach to Problem P

In this section we will describe a Las Vegas process for determining whether
a given reduced ideal $\mathfrak{b}$ is (or, more importantly, is not) principal.[1] We will
assume that we have used the method of Chapter 13 to produce the class
number h and the process described in Chapter 15 to compute $\mathcal{R}'_\Delta$. We will
discuss the technique by providing a series of several steps. We assume that
$\mathcal{R}'_\Delta > 11/2 + \log_2(34\sqrt{\Delta}/15)$; if this is not the case, we can solve **P** very easily
in $O(\log \Delta)$ time by simply searching for $\mathfrak{b}$ in $\mathcal{C}$.

1. We put $\mathfrak{a}_1 = \mathfrak{a} = \mathcal{O}$ and quickly check that $\mathfrak{b} \neq \mathfrak{a}_i$ for $1 \leq i \leq 16$. Clearly,
   if $\mathfrak{b}$ is one of these ideals, then $\mathfrak{b}$ is principal. If $\mathfrak{b}$ is not one of these 16
   ideals, then if $\mathfrak{b} = (\beta)$, where $\beta \in \mathcal{O}$ and $\beta > 1$, then $\beta \geq \theta_{17} > F_{17} > 2^{10}$
   and $\log_2 \beta > 10$.
2. We first execute algorithm $\mathrm{EXP}((\mathfrak{b}, 2^{p+1}, -1), h, 0, p)$ of Chapter 11 to find
   a 0-near reduced (f, p) representation $(\mathfrak{c}, d, k)$ of $\mathfrak{b}^h$, where $f < 2^{p-4}$. By
   (11.4), we know that $|\log_2 \phi - k| < 3/2$, where $\mathfrak{c} = \phi \mathfrak{b}^h$.
3. We next make use of the index-calculus algorithm to solve the infrastructure discrete logarithm problem (DLP) for $\mathfrak{c}$ to obtain some $g \in \mathbb{Q}$ such
   that
   $$|\log_2 \gamma - g| < 1,$$
   where $\mathfrak{c} = (\gamma)$ and $1 \leq \gamma < \epsilon_\Delta$. We certainly expect this process to be
   successful because $\mathfrak{b}^h$ must be principal if h is really the class number.
   It is this aspect of our technique that renders it a Las Vegas algorithm
   because we cannot be certain that this step will execute in subexponential
   time.
      If g is small, say $g < 3$, then since $|\log_2 \gamma - g| < 1$, we have $1 < \gamma < 16$.
   By Lemma 3.14, we can verify that $\mathfrak{c}$ is principal by checking that
   $$\mathfrak{c} \in \{\mathfrak{a}_1, \mathfrak{a}_2, \ldots, \mathfrak{a}_7\}.$$
   If this is the case, we can use FIND to compute a $(5/4, p)$ representation
   $(\mathfrak{c}, d'', k'')$ of $\mathfrak{a}$.
4. If $g \geq 3$, we execute algorithm $\mathrm{AX}(\lfloor g \rfloor - 2, p)$ with
   $$2^p > 11.2g \log \max\{16, g\}$$
   to produce $(\mathfrak{d}, d', k')$, a $(\lfloor g \rfloor - 2)$-near (f, p) representation of $\mathfrak{a}$ with $f <$
   2^{p-4} and $\mathfrak{d} = \mathfrak{a}[\lfloor g \rfloor - 2]$.
5. Since $1 < \log_2 \gamma - \lfloor g \rfloor + 2 < 4$, we see by Theorem 11.12 that if g has been
   computed correctly, then
   $$\rho^i(\mathfrak{d}) = \mathfrak{c}$$
   for some $i \in \{0, 1, 2, \ldots, 8\}$. Thus, this step serves to verify the result of
   step 3. Also, in the process of conducting this verification, we can use
   FIND to compute an $(f + 1/4, p)$ representation $(\mathfrak{c}, d'', k'')$ of the ideal $\mathfrak{a}$.
   Thus, if $g < 3$ or $g \geq 3$, by (11.4), we must have

$$|\log_2 \gamma - k''| < 3/2 \,.$$

Therefore,

$$-3 + k'' - k < \log_2 \gamma - \log_2 \phi < 3 + k'' - k \,. \tag{16.15}$$

Before continuting to produce the next steps needed in this process, we should make a few observations. If $\mathfrak{b}$ is principal, then we may assume that $\mathfrak{b} = (\beta)$, where $\beta \in \mathcal{O}$ and $1 \le \beta < \epsilon_\Delta$. Also,

$$\beta^h = \gamma \phi^{-1} \lambda \,,$$

where λ is a positive unit of $\mathcal{O}$. Hence, $\lambda = \epsilon_\Delta^r$ $(r \in \mathbb{Z})$ and

$$h \log_2 \beta = \log_2 \gamma - \log_2 \phi + r \mathcal{R}_\Delta \,. \tag{16.16}$$

We can deduce two useful results from this equation.

Theorem 16.1. *If $\mathcal{R}_\Delta > 9/2 + \log_2(34\sqrt{\Delta}/15)$, then r in (16.16) must satisfy*

$$-1 \le r \le h \,.$$

Proof. By Corollary 11.3.1, we have

$$-\log_2 \frac{34\sqrt{\Delta}}{15} < k < 0 \,.$$

Also, since $|k'' - \log_2 \gamma| < 3/2$, we get

$$\frac{-3}{2} < k'' < \mathcal{R}_\Delta + \frac{3}{2} \,.$$

Since $\log_2 \beta \ge 0$, we have

$$r\mathcal{R}_\Delta \ge \log_2 \phi - \log_2 \gamma > k - k'' - 3$$

by (16.15). Hence,

$$r\mathcal{R}_\Delta > -\log_2 \frac{34\sqrt{\Delta}}{15} - \mathcal{R}_\Delta - \frac{3}{2} - 3$$

and

$$r > -1 - \left(\frac{\log_2 \frac{34\sqrt{\Delta}}{15} + \frac{9}{2}}{\mathcal{R}_\Delta} \right) > -2 \,.$$

Since $\log_2 \beta < \mathcal{R}_\Delta$, we have

$$r\mathcal{R}_\Delta + \log_2 \gamma - \log_2 \phi < h\mathcal{R}_\Delta \,.$$

It follows that

$$r\mathcal{R}_\Delta < h\mathcal{R}_\Delta + 3 - k'' + k < h\mathcal{R}_\Delta + 3 + \frac{3}{2} \,.$$

and

$$r < h + 1 \,.$$

$\square$

If $\mathfrak{b}$ is a principal ideal, then (16.16) must hold for some r such that

$$-1 \leq r \leq h . \tag{16.17}$$

We next look at the problem of establishing whether or not this is the case. To this end, we first define $b(r)$ by

$$b(r) = \left\lceil \frac{r\mathcal{R}'_\Delta + k'' - k}{h} \right\rceil - 6 .$$

It is clear that if (16.16) holds, then $\log_2 \beta$ must be close to $b(r)$ for some r in the range (16.17). The next result establishes a bound on how close this is.

Theorem 16.2. *If (16.16) holds, then*

$$-1 < \log_2 \beta - b(r) < 10 .$$

Proof. We first observe that since $|\mathcal{R}_\Delta - \mathcal{R}'_\Delta| < 1$, we get

$$\frac{-3 + k'' - k + r(\mathcal{R}'_\Delta - 1)}{h} < \log_2 \beta < \frac{3 + k'' - k + r(\mathcal{R}'_\Delta + 1)}{h}$$

by (16.16) and (16.15). Since

$$b(r) = \frac{r\mathcal{R}'_\Delta + k'' - k}{h} + \eta - 6 \quad (0 \leq \eta < 1) ,$$

we get

$$\frac{-3}{h} + b(r) - \eta - \frac{r}{h} + 6 < \log_2 \beta < \frac{3}{h} + b(r) - \eta + \frac{r}{h} + 6 .$$

By Theorem 16.1,

$$\frac{3}{h} - \eta + \frac{r}{h} \leq \frac{3}{h} + \frac{h}{h} = \frac{3}{h} + 1$$

and

$$\frac{-3}{h} - \eta - \frac{r}{h} > \frac{-3}{h} - 1 - 1 ;$$

hence,

$$-1 \leq \frac{-3}{h} + 4 < \log_2 \beta - b(r) < \frac{3}{h} + 7 \leq 10 .$$

$\square$

By step 1, we must have $\log_2 \beta > 10$; hence, $b(r) > 0$. Now, suppose we use AX to compute $\mathfrak{a}[b(r)]$ for $r = -1, 0, 1, \ldots, h$. If $\mathfrak{b}$ is principal, by Theorem 16.2 and Theorem 11.12 we must have $\mathfrak{a}[b(r)] \in S$, where $S = \{\rho^i(\mathfrak{b}) : -16 \leq i \leq 0\}$ for some r in the range (16.17). Thus, since $\mathfrak{a}[b(r)]$ is a principal ideal, $\mathfrak{b}$ is principal if and only if $\mathfrak{a}[b(r)] \in S$ for some r in the range (16.17). We now have our final step.

6. For $r = -1, 0, 1, \ldots, h$, test to determine whether $\mathfrak{a}[b(r)] \in S$. $\mathfrak{b}$ is principal if and only if this happens for some r in the given range. Since, if $\mathfrak{b}$ is principal, $\rho^i(\mathfrak{a}[b(r)]) = \mathfrak{b}$ for some $0 \le i \le 16$, we can use FIND to produce an $(f + 1/4, p)$ representation of $(\mathfrak{b}, g, l)$ of $\mathfrak{a}$ and from this it is easy to compute some $b \in \mathbb{Q}$ such that $|\log_2 \beta - b| < 1$. We can then go on to produce a compact representation of β, if needed.

We can improve the execution of step 6 by determining the value of $b(r+1)$ from that of $b(r)$. We have

$$
\begin{aligned}
b(r+1) &= \left\lceil \frac{(r+1)R_\Delta' + k'' - k}{h} \right\rceil - 6 \\
&= \frac{rR_\Delta' + k'' - k}{h} + \frac{R_\Delta'}{h} + \eta_1 - 6 \quad (0 \le \eta_1 < 1), \\
&= \left\lceil \frac{rR_\Delta' + k'' - k}{h} \right\rceil + \frac{R_\Delta'}{h} + \eta_1 - \eta_2 - 6 \quad (0 \le \eta_2 < 1), \\
&= \left\lceil \frac{rR_\Delta' + k'' - k}{h} \right\rceil - 6 + \left\lceil \frac{R_\Delta'}{h} \right\rceil + \eta_1 - \eta_2 - \eta_3 \quad (0 \le \eta_3 < 1).
\end{aligned}
$$

Hence,

$$
b(r+1) = b(r) + \left\lceil \frac{R_\Delta'}{h} \right\rceil + \eta_1 - \eta_2 - \eta_3,
$$

where

$$
-2 < \eta_1 - \eta_2 - \eta_3 < 1
$$

and $\eta_1 - \eta_2 - \eta_3 \in \mathbb{Z}$. It follows that

$$
b(r+1) = b(r) + \left\lceil \frac{R_\Delta'}{h} \right\rceil + k(r), \tag{16.18}
$$

where $k(r) \in \{0, -1\}$. We can precompute $\mathfrak{a}[\lceil R_\Delta'/h \rceil]$ and $\mathfrak{a}[\lceil R_\Delta'/h \rceil - 1]$ and then we have

$$
\mathfrak{a}[b(r+1)] = \begin{cases} \text{ADDXY}(\mathfrak{a}[b(r)], \mathfrak{a}[\lceil R_\Delta'/h \rceil]) & \text{when } k(r) = 0 \\ \text{ADDXY}(\mathfrak{a}[b(r)], \mathfrak{a}[\lceil R_\Delta'/h \rceil - 1]) & \text{when } k(r) = -1. \end{cases}
$$

The value of $k(r)$ is easily computed from the formula for $b(r+1)$ and (16.18). Clearly, step 6 executes in time complexity $O(h\Delta^\epsilon)$. Thus, if we take into consideration that we must verify R_Δ', a process that requires $O(R_\Delta^{1/3+\epsilon})$ elementary operations, steps 1–6 will execute in expected time complexity

$$
O(R_\Delta^{1/3+\epsilon}) + O(h\Delta^\epsilon).
$$

As we have seen in Chapter 7, the value of h tends to be small, so in most cases this step executes very quickly. In the case where h is large, say $h > \Delta^{1/6}$, an unusual circumstance, the infrastructure method in §16.1 will determine unconditionally whether or not $\mathfrak{b}$ is principal in time complexity $O(R_\Delta^{1/2})$, but in this case, $R_\Delta = O(\Delta^{1/2+\epsilon}/h) = O(\Delta^{1/3+\epsilon})$; hence, the complexity of solving **P** by this method is $O(\Delta^{1/6+\epsilon})$.

In the more usual case of $h < \Delta^{1/6}$, the method of this section will execute in $O(\Delta^{1/6+\epsilon})$ operations provided the subexponential technique required in step 3 succeeds. Thus, we now have a Las Vegas method of solving **P** of complexity $O(\Delta^{1/6+\epsilon})$. Furthermore, once R_Δ' has been computed, the process of determining the principality of any ideal in $\mathcal{O}$ will execute very quickly because, as mentioned earlier, h is likely to be small.

We conclude this section with a problem that was considered by Jacobson and Williams.[2] Let

$$d_1 = 187060083 \,,$$

$$d_3 = 1311942540724389723505929002667880175005208 \,,$$

$$j_1 = 2 \,,$$

$$j_2 = 210404462515563471150485216453334887 \,.$$

It was necessary to show that

$$d_1 x_3^2 - d_3 x_2^2 = \frac{d_3 j_1 - d_1 j_2}{j_2} = c = 880813063496060911643645 \qquad (16.19)$$

has no integer solutions. Since $4 \mid d_3$, it is sufficient to show that all the ideals of norm cd_1 in $\mathcal{O} = [1, \sqrt{D}]$, where $D = d_1 d_3/4$, are not principal. In this case, Δ, the discriminant of $\mathcal{O}$, is the 51-digit number

$$\Delta = d_1 d_3 = 245412080559135221803366130231160886970528733912264 \,.$$

By using the index-calculus algorithm, we found that

$$h = 1024 \text{ and } R' = 6851106675369184895740.24677 \,.$$

Here, R' is an approximation to the regulator R of $\mathcal{O}$. Looking at the prime factors of cd_1, we see

$$cd_1 = \underbrace{5 \cdot 769 \cdot 33809 \cdot 6775714175075849}_{\text{factors of } c} \cdot \underbrace{3 \cdot 7 \cdot 8907623}_{\text{factors of } d_1} \,,$$

and since the prime factors of d_1 ramify in $\mathcal{O}$, we found a total of 16 ideals of norm cd_1. By excluding ideal conjugates, we reduced this to only eight candidates. By invoking the ERH, it was possible to show that (16.19) had no solutions. However, by using the method described here, we are able to show unconditionally that this equation has no solutions. Most (87%) of the time needed to perform this algorithm was required to verify R_Δ'.

16.3 The Equation $X^2 - DY^2 = N$

We now consider the Diophantine equation

$$X^2 - DY^2 = N \quad (N \neq 1), \tag{16.20}$$

an extension of the Pell equation. Notice that if $G^2 \mid (N, D)$, then $G \mid X$; if we put $X' = X/G$, $N' = N/G^2$, and $D' = D/G^2$, then (16.20) reduces to

$$X'^2 - D'Y^2 = N'.$$

Thus, we may assume with no loss of generality that (N, D) is squarefree. Also, if $S = (X, Y)$, then $S^2 \mid N$ and (16.20) becomes

$$X'^2 - DY'^2 = N',$$

where $X' = X/S$, $Y' = Y/S$, $N' = N/S^2$, and $(X', Y') = 1$. Thus, there is no loss of generality in considering only those solutions of (16.20) for which $(X, Y) = 1$. We say that such solutions are *primitive*. For such solutions, we must have $(Y, N) = 1$.

Let $\mathcal{O} = \mathbb{Z}[\sqrt{D}] = [1, \sqrt{D}]$; then $\Delta = 4D$ and $\omega = \sqrt{D}$. We next let X, Y be any primitive solution of (16.20) and consider the principal $\mathcal{O}$-ideal $\mathfrak{a} = (X + Y\sqrt{D})$. By Proposition 12.2, we have

$$\mathfrak{a} = [a, b + \omega], \tag{16.21}$$

where $a = |N| = |N(X + Y\sqrt{D})|$ and $b \equiv XY^{-1} \pmod{a}$. Since $b^2 \equiv D \pmod{a}$, we can find the representation (16.21) without prior knowledge of X and Y by solving

$$Z^2 \equiv D \pmod{N}. \tag{16.22}$$

One of the solutions Z of (16.22) with $0 < Z < |N|$ must be b. We put $a = |N|$ and $b = Z$ in (16.21). Also, since $\mathfrak{a}$ must be principal, it must be invertible, which means by Lemma 4.31 that

$$\left(N, 2Z, \frac{Z^2 - D}{N}\right) = 1.$$

If Z does not satisfy this condition, it must be eliminated as a possible candidate for b.

We may now perform the following steps.

1. Determine whether or not $\mathfrak{a}$ is principal. If $\mathfrak{a}$ is not principal. then there is no solution of (16.20) corresponding to our selected value of Z.
2. If $\mathfrak{a}$ is principal, solve the infrastructure DLP for $\mathfrak{a}$ to produce $g \in \mathbb{Q}$ such that $|\log_2 \gamma - g| < 1$, where $\mathfrak{a} = (\gamma)$, $1 < \gamma < \epsilon_\Delta$. ($\epsilon_\Delta$ is the fundamental unit of $\mathcal{O}$.)

3. From the compact representation of $\gamma = X + Y\sqrt{D}$, we can determine whether $N(\gamma) = N$. If so, we have a solution X, Y of (16.20). If $N(\gamma) = -N$ and $N(\epsilon_\Delta) = -1$, then $N(\gamma\epsilon_\Delta) = N$ and this will also produce a solution of (16.20). Otherwise, there cannot be a solution of (16.20) corresponding to our selection of Z.

Clearly, all solutions of (16.20) can be determined in this way by repeating the process for each distinct S such that $S^2 \mid N$ and each possible candidate for Z that results.

As the quadratic congruence (16.22) is of great importance in this investigation, it is necessary to discuss it in some detail.[3]

Suppose

$$N = 2^\alpha \prod_{i=1}^{k} p_i^{\alpha_k} \quad (\alpha \geq 0, \alpha_i \geq 0) , \tag{16.23}$$

where p_i $(i = 1, 2, \ldots, k)$ are distinct odd primes. We can find all of the solutions of (16.22) by finding all the solutions of

$$Z^2 \equiv D \pmod{2^\alpha} , \tag{16.24}$$

$$Z^2 \equiv D \pmod{p_i^{\alpha_i}} \quad (i = 1, 2, \ldots, k) \tag{16.25}$$

and combining them by use of the Chinese Remainder Theorem. Of course, we must first be able to factor N. As mentioned in Note 11 of Chapter 14, the problem of integer factorization has been studied very intensively during the last 30 years and much progress has been made on it.[4] A complete discussion of this is well beyond the scope of this work; we only mention here that the best general-purpose method for factoring currently known is the general number field sieve (GNFS) technique.[5] The heuristic estimate for the complexity of factoring a general N by this process is $L_N[1/3, (64/9)^{1/3} + o(1)]$.

We observe that if p is any prime such that $p^2 \mid N$ and $p \mid D$, then $p^2 \mid Z$, which means that $p^2 \mid D$, an impossibility by selection of D and N. Thus, if $p^2 \mid N$, then $p \nmid D$. It is well known that there is one solution of (16.24) if $\alpha = 1$; there are two solutions if $\alpha = 2$ and $D \equiv 1 \pmod 4$; and there are four solutions if $\alpha \geq 3$ and $D \equiv 1 \pmod 8$. Otherwise, there are no solutions when $\alpha \geq 2$. Thus, if μ denotes the number of solutions of (16.24), we have $\mu \leq 4$. If p is odd, there are exactly $1 + (D/p)$ solutions[6] of

$$Z^2 \equiv D \pmod{p^\alpha} . \tag{16.26}$$

Thus, the number $\nu(D, N)$ of possible solutions of (16.22) is given by

$$\nu(D, N) = \mu \prod_{i=1}^{k} \left(1 + \left(\frac{D}{p}\right)\right) \leq 2^{\omega(N)+1} ,$$

where $\omega(N)$ denotes the number of distinct prime divisors of N. The behaviour of $\omega(N)$ is quite irregular, but its average value[7] is known to be $\log\log|N|$;

thus, we expect that the usual number of solutions of (16.22) will be of the order of $\log |N|$; this means that in most cases it is only necessary to search for solutions of (16.20) by using only a few values for Z.

In order to find solutions of (16.26), we must first solve

$$Z^2 \equiv D \pmod{p} \tag{16.27}$$

and then use the simple lifting process described in most textbooks on elementary number theory to find the solutions of (16.26) when $\alpha > 1$. Polynomial (in $\log p$) time algorithms[8] are known for solving (16.27) when $(D/p) = 1$ and a quadratic non-residue x of p is known in advance. Practically, the determination of this x is easy because it can usually be found in a few trials. Also, under the extended Riemann hypothesis (ERH), it is possible to show[9] that $x < 2(\log p)^2$. We see, then, that as long as we can factor N, the problem of finding all the solutions of (16.22) is quite tractable.

Thus, we can find all solutions of (16.20) if we can solve the principal ideal problem in $\mathcal{O}$ and exhibit a generator λ for each of the principal ideals which are relevant. However, we will try to abbreviate this process somewhat in the material below. We first consider the case of $D < 0$. Without loss of generality, we will assume that $Y > 0$. The process of solving (16.20) that we will now outline goes back to an old algorithm of Cornacchia.[10] If (16.20) holds, we must have $N > |D| + 1 > 0$ in the non-trivial case of $|X|Y > 1$. We let Z $(0 < Z < N)$ be a solution of (16.22) and note that there must exist some $W \in \mathbb{Z}$ such that

$$X = ZY - NW , \quad (W, Y) = 1 ,$$

if X, Y is a primitive solution of (16.20). Now,

$$0 < 2|X|Y < X^2 + Y^2 \leq X^2 + |D|Y^2 = N ;$$

hence,

$$\left| \frac{W}{Y} - \frac{Z}{N} \right| = \frac{|X|}{NY} < \frac{1}{2Y^2} .$$

By Theorem 3.2, this means that W/Y must be a convergent in the SCF expansion[11] of Z/N; that is, $W/Y = A_i/B_i$ for some $i \leq n$, where $Z/N = [q_0, q_1, q_2, \ldots, q_n]$. Since $(W, Y) = 1$ and $(A_i, B_i) = 1, Y > 0, B_i > 0$, we must have $W = A_i, Y = B_i$ and

$$X = ZB_i - NA_i = (-1)^i R_i$$

by (3.34). It follows, then, that in order to solve (16.20) when $D < 0$, we need only expand Z/N by the Euclidean algorithm and find remainders R_i for which $R_i < \sqrt{N}$. Indeed, it can be shown[12] that if (16.20) has a solution, we need only consider those values of Z satisfying (16.22) for which $N/2 < Z < N$ and the corresponding value of i is such that $R_i < \sqrt{N}$ and $R_{i-1} \geq \sqrt{N}$. Of course, it is necessary to verify that $R_i^2 + |D|B_i^2 = N$. All solutions of (16.20) can be found by this procedure.

We next turn to the more difficult case of $D > 0$. If $\lambda = X + Y\sqrt{D}$ gives a solution of (16.20), then so will $\lambda\eta$, where η is any unit of $\mathcal{O}$ for which $N(\lambda) = 1$. Thus, if (16.20) has a solution, it must have an infinitude of solutions, unlike the case of $D < 0$. A technique for solving (16.20) was given more than two centuries ago by Lagrange,[13] but his procedure is quite inefficient for large values of D. The reason for this is that Langrange's method reduced (16.20) to a similar equation where $|N| < \sqrt{D}$ and then appeals to Theorem 3.3.

Lagrange's technique[14] essentially takes a candidate ideal $\mathfrak{a}$ and finds a reduced ideal $\mathfrak{b} \sim \mathfrak{a}$, where $\mathfrak{b} = \theta\mathfrak{a}$ and $N(\mathfrak{b}) < \sqrt{\Delta}/2$. By Remark 5.14, this can always be done. If $\mathfrak{b} = (\mu)$ is principal, then $\lambda = \mu/\theta$ and we can find μ by searching through the (finite) period of the SCF expansion of $\sqrt{D}$ until we find (or do not) some μ for which $N(\mu)/N(\theta) = N$. However, we know that the continued fraction process is not likely to be fast when D is large.

Thus, for large values of $D > 0$, we are left with the process of solving (16.20), which was described earlier. There remains the problem of characterizing the (possibly) infinite number of solutions of (16.20).

Let T, U be any solution of the Pell equation

$$T^2 - DU^2 = 1 .$$ (16.28)

Following Nagell,[15] we will say that if X_1, Y_1 and X_2, Y_2 are any (they need not be primitive) two solutions of (16.20) such that

$$X_2 + Y_2\sqrt{D} = (X_1 + Y_1\sqrt{D})(T + U\sqrt{D}) ,$$ (16.29)

then they are *associated*. Notice that if $S = (X_1, Y_1)$, then $S = (X_2, Y_2)$. The set of all solutions of (16.20) that are associated with each other forms a *class* of solutions of (16.20).

Theorem 16.3. *If X_1, Y_1 and X_2, Y_2 are solutions of (16.20), they are associated if and only if*

$$N \mid X_1X_2 - DY_1Y_2 \quad and \quad N \mid X_1Y_2 - X_2Y_1 .$$

Proof. Suppose X_1, Y_1 and X_2, Y_2 are associated. By (16.29), we get

$$(X_2 + Y_2\sqrt{D})(X_1 - Y_1\sqrt{D}) = (T + U\sqrt{D})N .$$ (16.30)

The result follows on equating rational and irrational parts. Next, suppose that $N \mid X_1X_2 - DY_1Y_2$ and $N \mid X_1Y_2 - X_2Y_1$. Since

$$X_1^2 - DY_1^2 = X_2^2 - DY_2^2 = N ,$$

we get

$$(X_1^2 - DY_1^2)(X_2^2 - DY_2^2) = N^2 ,$$

which we rewrite as

$$(X_1X_2 - DY_1Y_2)^2 - D(X_1Y_2 - X_2Y_1)^2 = N^2$$

or

$$\left(\frac{X_1X_2 - DY_1Y_2}{N}\right)^2 - D\left(\frac{X_1Y_2 - X_2Y_1}{N}\right)^2 = 1 .$$

It follows that $(X_1X_2 - DY_1Y_2)/N = T$ and $(X_1Y_2 - X_2Y_1)/N = U$, where $T^2 - DU^2 = 1$, and by (16.30), we see that X_1, Y_1 and X_2, Y_2 are associated.

$\square$

Let (t, u) be the fundamental solution of (16.28). If

$$\epsilon = t + u\sqrt{D} ,$$

then

$$\epsilon = \begin{cases} \epsilon_\Delta & \text{when } N(\epsilon_\Delta) = 1 \\ \epsilon_\Delta^2 & \text{when } N(\epsilon_\Delta) = -1 . \end{cases}$$

It follows from our earlier observations that if (16.20) has a solution, we may assume that it has a solution with

$$1 < X + Y\sqrt{D} < \epsilon . \tag{16.31}$$

We call any solution of (16.20) satisfying (16.31) a *fundamental* solution of (16.20). Also, there can only be one such solution in any given class. For if there were two, say X_1, Y_1 and X_2, Y_2, we may assume with no loss of generality that $X_2 + Y_2\sqrt{D} > X_1 + Y_1\sqrt{D}$. Hence, by (16.31),

$$\epsilon > \frac{X_2 + Y_2\sqrt{D}}{X_1 + Y_1\sqrt{D}} = T + U\sqrt{D} = \epsilon^n \quad (n \geq 1) ,$$

which is impossible. If S is the set of all the fundamental solutions of (16.20), then, by our earlier discussion, $|S|$ could not exceed the total number of distinct primitive $\mathcal{O}$-ideals of norm $|N|/S^2$ for each square factor S^2 of N; that is[16]

$$|S| \leq \sum_{S^2 \mid N} \nu\left(D, \frac{N}{S^2}\right) .$$

It follows that we can characterize all solutions of (16.20) by

$$X + Y\sqrt{D} = \nu_i \epsilon^n \quad (n \in \mathbb{Z}; i = 1, 2, \ldots, s) ,$$

where $s = |S|$ and $\nu_i = X_i + Y_i\sqrt{D}$ is a fundamental solution.

We conclude this chapter with a brief discussion of (16.20) for small values of $|N|$. The most interesting of these cases is that of $N = -1$. The Diophantine equation

$$X^2 - DY^2 = -1 \tag{16.32}$$

is of importance because of its relationship to the problem of when $N(\epsilon_\Delta) = -1$. Unlike the case of the Pell equation, (16.32) does not always have a solution for any positive non-square integer D. For example, we see by simple congruence considerations that (16.32) cannot have a solution if a prime p divides D and $p \equiv -1 \pmod 4$. However, even if $D = 2p$, where p is a prime with $p \equiv 1 \pmod 8$, it is possible for (16.32) to not have a solution; for example, consider $D = 194 = 14^2 - 2$. We should mention here that Stevenhagen[17] has produced some very convincing conjectures concerning the density of those values of D for which (16.32) has a solution. The problem of determining a fast technique for recognizing whether or not (16.32) is solvable in integers has attracted a great deal of attention.[18] It was Lagarias[19] who finally developed such a method for doing this when the complete factorization of D is known, together with a quadratic non-residue for each prime which divides D.

Another simple case[20] is that of $N = \pm 2$. Here, we note that if $\nu = x + y\sqrt{D}$ and

$$x^2 - Dy^2 = \pm 2 , \tag{16.33}$$

we may assume that $1 < \nu < \epsilon_\Delta$. Also, $\nu^2/2 \in \mathcal{O}$ and $N(\nu^2/2) = 1$. Hence, $\nu^2/2 = \epsilon_\Delta^n$ and $n > 0$. If $n \geq 2$, then $\nu \geq \sqrt{2}\epsilon_\Delta$, which is impossible; thus, $\epsilon_\Delta = \nu^2/2$. If we have a good approximation to $\log_2 \epsilon_\Delta$, we can find one for $\log_2 \nu$ and use this in conjunction with Algorithm AX and Theorem 11.12 to determine whether or not there exists a principal ideal $\mathfrak{a} = (\nu)$ such that $N(\mathfrak{a}) = 2$. This process will execute very quickly if we know an approximation to $\log_2 \epsilon_\Delta$. This remark also applies to[21]

$$x^2 - Dy^2 = \pm 4 , \tag{16.34}$$

when x and y are odd. For in this case, as mentioned in Chapters 1 and 4, we may assume that $\nu = x + y\sqrt{D}$, $1 < \nu < \epsilon_\Delta$, and we get

$$\epsilon_\Delta = \left(\frac{\nu}{2}\right)^3 .$$

In conclusion, we mention that there are also some scattered results[22] concerning (16.20) for other small values of $|N|$.

Notes and References

[1] This technique is described in some detail in [Sil06].

[2] See [JW02].

[3] See, for example, [Vin54], p. 91ff.

[4] We have had occasion to mention this important problem several times in this book. In addition to the references already given, we also recommend [Len00], [Rie94], [CP05], and [Bre00].

[5] See [LL93].

[6] [Vin54], p. 94.

[7] [CM05], pp. 32–35.

[8] See [BS96b], §§7.1 and 7.2.

[9] [Bac90].

[10] [Cor08].

[11] In our proof of Theorem 3.2 we assumed that ϕ is irrational, but the result also holds when ϕ is rational ([Per57], §13, Satz 11).

[12] See, for example, [Nit95].

[13] [Lag68], pp. 406–495; [Dic19], pp. 360–361.

[14] A good description of this technique can be found in [Chr61], pp. 482–487. This method has been revisited recently by Matthews [Mat00].

[15] [Nag64], pp. 204–212.

[16] More precise values for the number of classes were determined by Stolt in [Sto52]. See also Theorem 2 of [Le95]. From this it can be easily deduced that there can be at most $2^{\omega(n)-1}$ classes of primitive solutions of (16.20).

[17] [Ste95]. Later, Bosma and Stevenhagen [BS96b] extended these observations and acquired much supporting numerical data.

[18] See [BW72] and [Lag80a] for references.

[19] [Lag80a].

[20] The equation is discussed in [Per57], §26.

[21] The problem of determining an *a priori* condition for the solvability of (16.34) when the right-hand side is 4 is called Eisenstein's problem (see [Eis44b]), but it was known previously to Gauss in a somewhat different form ([Gau86], Art. 256, VI). This problem was dealt with in [SW89], but the criterion for solvability derived there is of complexity $O(D^{1/4+\epsilon})$. Other, less efficient criteria can be found in [IKW90] and [Wil90]. Stevenhagen [Ste96b] has also produced conjectures concerning the density of those values of D for which (16.34) is solvable.

[22] See, for example, [MvdPW94] for the case of $N = -3$ and a discussion concerning other values of $|N|$.

17

Conclusion

17.1 A More General Equation

In this section[1] we will investigate the Diophantine equation

$$ax^2 + bxy + cy^2 + dx + ey + f = 0 , \qquad (17.1)$$

where it is required to find integral values of x and y, given $a, b, c, d, e, f \in \mathbb{Z}$. This is the most general form of a quadratic Diophantine equation in two variables and, as such, represents a further generalization of the Pell equation.[2] A method for solving this equation was given over 200 years ago by Lagrange,[3] and this method has not been improved significantly since that time. The reason for this is that Lagrange's method works perfectly well as long as the coefficients in (17.1) do not get very large. However, if we put $H = \max\{|a|, |b|, |c|, |d|, |e|, |f|\}$, it has been shown[4] that there is an infinite collection of equations of the form (17.1) having integer solutions x, y, but none with $\max\{|x|, |y|\} \le 2^{H/5}$. Thus, it is possible for solutions of (17.1) to be very large, even when H is only moderately large. For such cases, Lagrange's method will likely be far too slow to produce the solutions of (17.1). We will show here how our previously developed results can be used to produce a faster method for solving this equation.

If we put $D = b^2 - 4ac$, $E = bd - 2ae$, and $F = d^2 - 4af$, Lagrange realized that (17.1) can be written as

$$DY^2 = (Dy + E)^2 + DF - E^2 , \qquad (17.2)$$

where $Y = 2ax + by + d$. Clearly, if we put $N = E^2 - DF = -4a(ebd + 4acf - ae^2 - fb^2 - cd^2)$, then (17.2) can be written as

$$X^2 - DY^2 = N , \qquad (17.3)$$

where $X = Dy + E$. Thus, if we have any solution X, Y of (17.3) such that there are integers x and y for which

$$X = Dy + E \text{ and } Y = 2ax + by + d \,, \tag{17.4}$$

we get a solution x, y of (17.1).

We have discussed how to solve (17.3) in §16.3. If $D < 0$, this equation has only a finite number of solutions which can be found by making use of the algorithm of Cornacchia. If $D = 0$ or $D > 0$ and D is a perfect integral square, the problem of solving (17.3) reduces to that of factoring N and, once again, there can only be a finite number of solutions of (17.3). Also, if $D > 0$ and $N = 0$, then (17.3) can only have a solution if D is a perfect integral square. In this case, we get an infinitude of solutions of (17.3), but they are very easily characterized.

Thus, the only remaining case is that of $N \neq 0$, $D > 0$, and D not a perfect integral square. Let t, u denote the fundamental solution of the Pell equation

$$T^2 - DU^2 = 1 \,.$$

If

$$X_n + Y_n \sqrt{D} = (X + Y\sqrt{D})(t + u\sqrt{D})^n \quad (n \in \mathbb{Z}) \,, \tag{17.5}$$

where (X, Y) is any solution of (17.3), then (X_n, Y_n) is also a solution of (17.3). Indeed, as we have mentioned in §16.3 and as Lagrange was well aware, there exists a finite set $\mathcal{S}$ made up of ordered pairs (X, Y) of solutions X, Y of (17.3) such that if X', Y' is any solution of (17.3), then $X' = X_n$ and $Y' = Y_n$ for some $n \in \mathbb{Z}$ and some $(X, Y) \in \mathcal{S}$. Thus, after having found $\mathcal{S}$ by the methods of §16.3, the problem reduces to that of identifying for each $(X, Y) \in \mathcal{S}$ those values of n for which

$$\begin{cases} X_n \equiv E \pmod{D} \,, \\ Y_n \equiv b(X_n - E)/D + d \pmod{2a} \,. \end{cases} \tag{17.6}$$

We will now rewrite (17.6) as

$$\begin{cases} X_n \equiv E \pmod{D} \,, \\ DY_n \equiv bX_n - bE + Dd \pmod{2aD} \,. \end{cases} \tag{17.7}$$

Lagrange noted that as there are only a finite number of possible values of

$$(t + u\sqrt{D})^n \pmod{2aD} \,,$$

there must be a least positive integer π for which

$$(t + u\sqrt{D})^\pi \equiv 1 \pmod{2aD} \,.$$

Thus, (17.5) yields values of X_n and Y_n satisfying (17.7) if and only if it does so when n is replaced by $n + \pi$. It follows that in order to test all the solutions of (17.3) produced by (17.5) to see if they satisfy (17.7), it suffices to examine only those for which $0 \leq n \leq \pi - 1$.

Lagrange's method, then, compels us to test up to π values of n to determine those congruence classes of n $(\bmod \pi)$ for which we produce solutions of (17.1) from (17.5). Unfortunately, this could be a very inefficient process when π is large, which is frequently the case when aD is. This difficulty can be overcome by using a technique of Legendre as modified by Dujardin.[5] If we define

$$T_n + U_n\sqrt{D} = (t + u\sqrt{D})^n \quad (n \in \mathbb{Z}) , \qquad (17.8)$$

we see from (17.5) that

$$X_n = XT_n + DYU_n, \quad Y_n = YT_n + XU_n .$$

Since we require that $X_n \equiv E \pmod{D}$, we must have $T_nX \equiv E \pmod{D}$. By (17.8), it is clear that $T_n \equiv t^n \pmod{D}$, and since $t^2 \equiv 1 \pmod{D}$, we get

$$T_n \equiv t^\epsilon \pmod{D}$$

when $n \equiv \epsilon \pmod{2}$, $\epsilon \in \{0,1\}$. Thus, if neither $X \equiv E \pmod{D}$ nor $tX \equiv E \pmod{D}$ holds, then (17.5) will yield no solutions of (17.1).

Suppose that $T_nX \equiv E \pmod{D}$. By (17.7), we also require that

$$dD - bE \equiv (DY - bX)T_n + (DX - bDY)U_n \pmod{2aD} . \qquad (17.9)$$

From (17.4), we can deduce that

$$X - bY = Dy + E - 2abx - b^2y - bd$$
$$= -2a(2cy + e + bx) .$$

Thus, another necessary condition for (17.5) to produce solutions to (17.1) is that

$$2a \mid X - bY . \qquad (17.10)$$

Since $b^2 \equiv D \pmod{2a}$, this means that $2a \mid DY - bX$. We next observe that

$$dD - bE = 2a(eb - 2dc) .$$

Hence, we can now put (17.9) in the form

$$\frac{dD - bE}{2a} \equiv \left(\frac{DY - bX}{2a}\right)T_n + \left(\frac{DX - bDY}{2a}\right)U_n \pmod{D} .$$

By (17.8), we have $U_n \equiv nut^{n-1} \pmod{D}$; hence, (17.9) can be rewritten as

$$\frac{dD - bE}{2a} \equiv \left(\frac{DY - bX}{2a}\right)t^n + nut^{n-1}\left(\frac{DX - bDY}{2a}\right) \pmod{D} . \quad (17.11)$$

Since $t^2 \equiv 1 \pmod{D}$, this becomes a linear congruence in the unknown n.

However, by (17.10) we have $D \mid (DX - bDY)/2a$. Thus, (17.11) can hold for all even n only if

$$D \mid \frac{dD - bE - DY + bX}{2a} \tag{17.12}$$

and for all odd n only if

$$D \mid \frac{dD - bE - DYt + bXt}{2a}. \tag{17.13}$$

Thus, it is no longer necessary to search for all possible values of n up to π. We need only check to see that (17.10) holds. If so and (17.12) holds, then (17.5) produces solutions of (17.1) for any even n, and if (17.13) holds, then (17.5) produces solutions of (17.1) for any odd n. If none of these conditions holds, then (17.5) produces no solutions of (17.1).

The problem of solving (17.1) is therefore very simple once we have determined $\mathcal{S}$. However, even the smallest possible values of X and Y satisfying (17.3) when D is large can be absolutely enormous. Indeed, as we have seen in Chapter 12, it is often not even possible to write them down in standard decimal notation. However, in order to solve (17.1), we need to have the values of X and Y (and t) modulo $2aD$; thus, we need to have a method of finding X and Y that allows for this. Fortunately, this problem is easy to solve because the index-calculus methods can furnish us with an approximation to $\log \lambda$, where $\lambda = X + Y\sqrt{D}$, and this can be used to produce a compact representation of λ. From this, we can then find X and Y modulo $2aD$ by using the process described in §12.3. The value of $t \pmod{2aD}$ can be determined from the compact representation of the fundamental unit of $\mathcal{O} = [1, \sqrt{D}]$.

If R_Δ is the regulator of $\mathcal{O}$, we know that Lagrange's method of solving (17.3) is of complexity $O(R_\Delta)$, whereas the method of §16.3 is, at worst, likely to be of complexity $O(R_\Delta^{1/3+\epsilon})$. Thus, if D is large (or, more precisely, R_Δ is large), we are better off using the methods of Chapter 16 to solve (17.3) than the method of Lagrange.

17.2 Other Generalizations of the Pell Equation

As mentioned in the previous section, (17.1) represents a generalization of the Pell equation, where we retain the conditions of two variables and degree 2. There are, of course, a number of different directions in which we could go in order to produce other versions of the Pell equation. In fact, we could consider almost any Diophantine equation of degree at least 2 to represent a generalization of (1.7), but we will confine our discussion here to equations that are fairly evident extensions of it. Even with this restriction, however, we caution the reader that the literature concerning this topic is vast, and we will barely scratch the surface of it here.

The most obvious generalization of the Pell equation, where we retain the constraint of keeping only two variables, is

$$T^n - DU^n = 1. \tag{17.14}$$

Unlike the Pellian case ($n = 2$), this equation can only have a finite number of solutions for any given D (> 0) when $n > 2$. Indeed, Bennett[6] has shown by very advanced methods that if a, b, and n (≥ 3) are integers with $ab \neq 0$, then the Diophantine equation

$$|ax^n - by^n| = 1$$

has at most one solution in positive integers. There remains, of course, the problem of determining what the solution of (17.14) is or whether it even exists. In the case of even n, this problem can be solved by making use of the following theorem.[7]

Theorem 17.1. *If (17.14), where D is a positive non-square integer, $n = 2m$ ($m \geq 3$), has a solution in positive integers T and U, then*

$$T_1^m + U_1^m \sqrt{D} = \epsilon_\Delta \ or \ \epsilon_\Delta^2 \ ,$$

where ϵ_Δ is the fundamental unit of $\mathcal{O} = [1, \sqrt{D}]$ and $N(\epsilon_\Delta) = 1$. If $N(\epsilon_\Delta) = -1$, there is no solution of (17.14) with $U \neq 0$.

There is a similar result[8] for $m = 2$.

One can also generalize the Pell equation to

$$X^2 - DY^{2m} = 1 \ . \tag{17.15}$$

For any given $m \geq 2$, we know that there can be at most two solutions of (17.15) in positive integers. Recently Togbe, Voutier, and Walsh[9] (for the case of $m = 2$) and Bennett[10] (for the case $m \geq 3$) have produced results which provide restrictions on what these two solutions can be. However, their technique nevertheless requires us to determine whether or not U_k for some fixed value of k is a perfect mth power. The value of k is either 1 or 2, and $T_k + U_k\sqrt{D} = (t + u\sqrt{D})^k$, where t, u is the fundamental solution of (1.7). Fortunately, it seems[11] that if (17.15) has a solution, the values of X and Y are not large compared to D; thus, we can usually detect the solutions without too much trouble. On the other hand, if (17.15) does not have a solution and D is large, determining that U_k is not a perfect mth power could be difficult on account of its possible great size. However, in this case, we need only find a compact representation of $t + u\sqrt{D}$ and use this to determine the value of U_k modulo enough primes. If U_k is not a perfect mth power, it should turn out that U_k is not an mth power modulo one of these primes.[12]

Instead of changing the degree of the Pell equation, we could consider the problem of solving the simultaneous Pell equations

$$T^2 - D_1U^2 = 1 \ , \quad V^2 - D_1U^2 = 1 \tag{17.16}$$

for integers T, U, and V. For such systems it is known[13] that if D_1 and D_2 are distinct positive integers, then the system (17.16) admits at most two solutions (T, U, V) in positive integers. There is also a similar result[14] for the system

$$T^2 - D_1 U^2 = 1 , \quad U^2 - D_2 V^2 = 1 .$$

If we relax both of the conditions of requiring two variables and degree 2, we can produce another generalization of the Pell equation by observing that it can be written as

$$\begin{vmatrix} T & DU \\ U & T \end{vmatrix} = 1 .$$

This can be very easily extended to an equation of degree n in n variables $T_1, T_2, \ldots, T_n$ as

$$\begin{vmatrix} T_1 & DT_n & DT_{n-1} & \cdots & DT_2 \\ T_2 & T_1 & DT_n & \cdots & DT_3 \\ T_3 & T_2 & T_1 & \cdots & DT_4 \\ \vdots & \vdots & \vdots & \ddots & \vdots \\ T_n & T_{n-1} & T_{n-2} & \cdots & T_1 \end{vmatrix} = 1 . \tag{17.17}$$

If we let $\delta^n = D$, where $\delta > 0$, and ζ_n denote a primitive nth root of unity, we know from the theory of circulants that we can write (17.17) as[15]

$$\prod_{i=1}^{n} \left(T_1 + \zeta_n^i \delta T_2 + (\zeta_n^i \delta)^2 T_3 + \cdots + (\zeta_n^i \delta)^{n-1} T_n \right) = 1 . \tag{17.18}$$

For example, in the case of $n = 3$, (17.18) becomes[16]

$$T_1^3 + DT_2^3 + D^2 T_3^3 - 3DT_1 T_2 T_3 = 1 . \tag{17.19}$$

We can even generalize (17.18) by considering algebraic number fields. Let $\mathbb{K} = \mathbb{Q}(\theta)$ be such a field with θ a zero of an irreducible polynomial $f(x)$ of degree n. If $\alpha \in \mathbb{K}$, we generalize the concept of the norm by defining the norm of α, $N(\alpha)$, to be simply

$$N(\alpha) = \prod_{i=1}^{n} \alpha_i ,$$

where $\alpha_1 = \alpha$ and α_i $(i = 2, 3, \ldots, n)$ represent the other $n - 1$ zeros of the minimal polynomial of α. We may now write (17.18) as

$$N(\alpha) = 1 ,$$

where $\alpha = \sum_{j=1}^{n} T_j \delta^{j-i} \in \mathbb{Z}[\delta]$. Thus, a more general version of (17.18) would be

$$N(\alpha) = 1 , \tag{17.20}$$

where α is an algebraic integer of $\mathbb{K} = \mathbb{Q}(\theta)$. If $\mathcal{O}_{\mathbb{K}}$ denotes the ring of all algebraic integers of $\mathbb{K}$, then we see that α in (17.20) must be an invertible element or *unit* of $\mathcal{O}_{\mathbb{K}}$. We use $\mathcal{O}_{\mathbb{K}}^*$ to denote the group of all the units in $\mathcal{O}_{\mathbb{K}}$.

Let r_1 denote the number of real zeros of $f(x)$ and let $2r_2$ denote the number of complex zeros $(r_1 + 2r_2 = n)$. It is a fundamental result of Dirichlet,[17] generalizing the results in §4.3, that

$$\mathcal{O}_{\mathbb{K}}^* = \langle \zeta, \epsilon_1, \epsilon_2, \ldots, \epsilon_s \rangle \ ,$$

where $s = r_1 + r_2 - 1$ and ζ is some root of unity in $\mathbb{K}$. The set of units $\mathcal{U} = \{\epsilon_1, \epsilon_2, \ldots, \epsilon_s\}$ is multiplicatively independent and is called a *fundamental system of units* for $\mathcal{O}_{\mathbb{K}}$ (or $\mathbb{K}$); the quantity s is called the *unit rank* of $\mathcal{O}_{\mathbb{K}}$ (or $\mathbb{K}$). Since $N(\epsilon_i) = \sigma_i$, where $\sigma_i \in \{-1, 1\}$, we see that α is a solution of (17.20) if and only if

$$\alpha = \zeta^m \prod_{i=1}^{s} \epsilon_i^{m_i} \ ,$$

where $m, m_1, m_2, \ldots, m_s \in \mathbb{Z}$ and $\prod_{i=1}^{s} \sigma_i^{m_i} = 1$.

It follows, then, that the problem of solving (17.20) or (17.18) becomes that of determining a fundamental system of units in $\mathbb{K}$. In general, this can be a very difficult problem, particularly when the degree of $f(x)$ is large (> 20, say) or its coefficients exceed 10^6. Also, when $s > 1$, the infrastructure ideas mentioned earlier in Chapter 7 cannot be readily applied.[18] As the problem of determining $\mathcal{U}$ is of immense importance in computational algebraic number theory and the methods currently available for solving it are often quite intricate, we urge the interested reader to consult the literature on this subject.[19]

Another way to generalize the Pell equation is to change the domain of T, U, and D. For example, consider the equation

$$\xi^2 - \gamma\eta^2 = 1 \ ,$$

where $\gamma \ (\neq 0)$ is an algebraic integer in $\mathbb{K} = \mathbb{Q}(\sqrt{D})$ and ξ and η are constrained to be in $\mathcal{O}_{\mathbb{K}}$. It is known[20] that this equation has an infinitude of solutions if and only if γ is not a square in $\mathcal{O}_{\mathbb{K}}$ when $D < 0$ and γ is not totally negative (γ and $\overline{\gamma}$ not both negative) when $D > 0$. We can also consider the problem of solving

$$T^2 - DU^2 = 1 \ , \tag{17.21}$$

when $D \in \mathbb{Z}[X]$ and D is not a constant. Here, we insist that $T, U \in \mathbb{Z}[X]$ and $U \neq 0$. Clearly, in order for (17.21) to have non-trivial solutions, the degree of D must be even and its leading coefficient must be a perfect square in $\mathbb{Z}$. In Chapter 6, we discussed this problem for certain quadratic polynomials $D(X)$. Unlike the case of (1.7), it is not always true that there is a solution of (17.20) for any non-square $D(X)$ of even degree. Indeed, this is a rather rare event. For example,[21] if $D(X) = X^2 + d \ (d \in \mathbb{Z})$, there are no (non-trivial) solutions of (17.20) if $d \neq \pm 1, \pm 2$.

Now, let $D \in \mathbb{Q}[X]$, where D is not a perfect square in $\mathbb{Q}[X]$. We can extend (17.21) to the equation

$$T^2 - DU^2 = e \ , \tag{17.22}$$

where we now require that $T, U \in \mathbb{Q}[X]$ and e be a unit of $\mathbb{Q}$. As before, if (17.22) has a solution, then D must be of even degree with leading coefficient a perfect square in $\mathbb{Q}$. Certainly, any solution of (17.21) is a solution of (17.22), but it is not necessarily the case that if (17.22) has a solution, then (17.21) does. For example, consider the simple case of

$$D = a^2 X^2 + bX + c , \quad e = \frac{b^2 - 4a^2 c}{4a^2} .$$

Here, (17.22) always has the non-trivial solution $T = aX + b/2a$, $U = 1$. (The constraint that D not be a perfect square in $\mathbb{Q}[X]$ means that $e \neq 0$ and is therefore a unit in $\mathbb{Q}$.) However, as we have seen, if we take the simple case of $a = 1$, $b = 0$, and $c = d$, then (17.21) can only have a solution if $d = \pm 1$ or ± 2.

Notice that if (17.22) has a solution and we put $T' = (T^2 + DU^2)/e \in \mathbb{Q}[X]$, and $U' = 2DTU/e \in \mathbb{Q}[X]$, we get

$$T'^2 - DU'^2 = 1 .$$

Thus, if (17.22) has a solution, then so does

$$T^2 - DU^2 = 1 \tag{17.23}$$

for some $T, U \in \mathbb{Q}[X]$. An important application of (17.23) occurs in the investigation of those polynomials $D(X) \in \mathbb{Z}[X]$ for which $l(\sqrt{D(n)})$ is bounded for $n \in \mathbb{Z}$. We have already seen in §6.3 that Schinzel solved the problem in the case of $\deg D = 2$. He also showed[22] that if $\deg D \geq 2$, then $l(\sqrt{D(n)})$ will be bounded as n varies if and only if the solution T, U of (17.23), where T is of minimal positive degree, is such that $T \in \frac{1}{2}\mathbb{Z}[X]$. Subsequently, Dubois and Paysant le Roux[23] were able to generalize Louboutin's Theorem 6.9.

Unfortunately, the problem of determining whether or not (17.20) (or (17.23)) is solvable for a given D can, in general, be quite difficult[24]; however, if it is solvable, the continued fraction algorithm applied to polynomials[25] can be used to find all of its solutions.

As we have seen, the theory of quadratic number fields, especially units, is closely related to solutions of Pell's equation defined over the integers. In other domains, such as $\mathbb{Q}[X]$, the theory of quadratic function fields plays a similar role. Briefly, a quadratic function field[26] is a degree 2 extension of the rational function field $\mathbb{K}(X)$ consisting of rational functions with coefficients in a field $\mathbb{K}$. The extension is formed by adjoining the root of an irreducible quadratic polynomial[27] in $\mathbb{K}[X]$ (i.e., the function field $\mathbb{K}(X, Y) = \{a + bY \mid a, b \in \mathbb{K}(X)\}$, where $Y^2 + h(X)Y - f(X) = 0$ with $h, f \in \mathbb{K}[X]$).

Quadratic function fields have much in common with quadratic number fields. Both are quadratic extensions of fields, either the rational numbers $\mathbb{Q}$ in the case of number fields or a field of rational functions $\mathbb{K}(X)$ in the case of function fields. The definitions of orders, ideals, the class group, and units also carry over. In particular, we have the same connection between solutions of the analogue of Pell's equation and units in the field.

If $\mathbb{K} = \mathbb{Q}$, then the setting is as described above, and non-trivial units are a rare occurence. However, if $\mathbb{K}$ is a finite field of q elements, then the situation is very similar to what we have described in previous chapters. In particular, we have analogues of imaginary and real quadratic function fields that have no non-trivial units and infinitely many, respectively. The real case also has infrastructure that is very similar to that described in §7.4; this was discovered and first investigated by Stein.[28] Many of the infrastructure-based algorithms described in previous chapters generalize to the function field case without too much difficulty,[29] including NUCOMP, baby-step giant-step algorithms, and the cryptographic protocols described in §14.3.

There are also some intriguing differences between real quadratic fields and function fields. The first is that distances in the infrastructure of a real quadratic function field are defined to be the degrees of elements in the field as opposed to logarithms of real numbers. Thus, distances are integers as opposed to real numbers. This has some important consequences; for example, because distances are discrete, approximation is not an issue, and relatively expensive methods such as (f, p) representations are not required. In addition, finding the regulator, defined to be the degree of the fundamental unit of the function field, given a multiple can be done efficiently if the multiple (an integer) is factored, by using infrastructure to find the smallest divisor that is the distance of a unit. Stein and Teske showed that Pollard's kangaroo method, which has a similar complexity to the baby-step giant-step algorithm but requires only constant memory, can also be employed to compute the regulator.[30] This method, which relies on being able to follow efficiently a random walk through the infrastructure, is not amenable to the number field case because maintaining accurate approximations of the rapidly growing distances is too cumbersome.

A second interesting aspect of function fields is that whereas the index-calculus methods of Chapter 13 carry over to the function field case almost directly, they do now always work in subexponential time. Müller, Stein, and Thiel[31] showed that regulator computation and solving the principal ideal problem in a real quadratic function field over an odd characteristic finite field can be done in subexponential time as long as the genus of the function field, an invariant that is approximately equal to half the degree of the quadratic polynomial that defines the function field, is larger than $\log q$ (i.e., only for cases where the genus is large and the finite field is small). In fact, the best known algorithms for solving these problems for small genus and large finite fields have exponential complexity, suggesting that such function fields should provide similar security guarantees as for hyperelliptic curves of the same sizes. Jacobson, Scheidler, and Stein[32] have investigated the performance of such protocols and showed that they provide efficiency close to their hyperelliptic curves counterparts.

In short, there has been much interesting work in the last 15 years exploring which methods and results from quadratic fields can also be used for quadratic function fields. We fully expect this trend to continue and to have novel methods introduced in one setting which are inspired by methods which have proved successful in the other.

17.3 Some Questions

We have seen that during the last four decades a great deal of progress has been made on solving the Pell equation (1.7). It used to be that the best algorithms for solving it were of complexity $O(D^{1/2+\epsilon})$, but we now know that by making use of a number of new ideas, we can find the regulator (actually an approximation to it) by a process that requires only subexponential time. Of course, there is some uncertainty about this value, but we can verify it rigorously by a procedure which is of complexity $O(D^{1/6+\epsilon})$. We have also seen that this technique can be applied to values of D of up to 65 digits, an achievement that would have been regarded as impossible 20 years ago.

Once we know the regulator, we can very quickly find a compact representation of the fundamental solution t, u of the Pell equation, even when the values of t and u are very large, and this permits us to answer most questions that one might have concerning these numbers. We have also seen how these ideas can be extended to the problem of solving the general two-variable Diophantine equation of degree 2. In the process of developing these techniques, we have also derived efficient algorithms for conducting arithmetic in quadratic number fields. In particular, we have been able to provide a fairly efficient version of Shanks' NUCOMP algorithm for finding a reduced ideal equivalent to the product of two given reduced ideals. This is a fundamental operation which is an essential component in any collection of algorithms devoted to manipulating ideals in quadratic orders. Indeed, we have shown how it can be used in a technique for raising an ideal to a large power and, subsequently, in a cryptographic key exchange protocol.

However, in spite of the advances that have been made on the problem of solving the Pell equation, we are left with a number of questions. We will order them here in what we believe to be their level of difficulty or importance.

1. Is there a deterministic polynomial-time (in $\log \Delta$) algorithm for finding a close approximation $R'_D K$ of R_Δ?

If such an algorithm exists, then under the generalized Riemann hypothesis (GRH) there is a polynomial-time algorithm for factoring Δ, something that is not widely believed to exist.

It is interesting to note that, in the quantum model of computation, such an algorithm does exist. Hallgren[33] discovered algorithms for solving Pell's equation and the principal ideal problem that run in polynomial time on a quantum computer, a device that can achieve exponential speed-ups for certain applications by exploiting the properties of quantum mechanics. The first number-theoretic problems to benefit from quantum computing, and to highlight its potential, were integer factorization and discrete logarithm computation in finite fields, due to algorithms discovered by Shor,[34] and these ideas have been generalized to solve many problems in abelian groups, including computing discrete logarithms in class groups of quadratic fields. An explicit version and analysis of an algorithm for solving the discrete logarithm problem in the class group of an imaginary quadratic field has been published by Schmidt.[35] In addition, Hallgren's results have recently been extended, independently by Hallgren and by Schmidt and Vollmer,[36] to compute the unit group of any number field. The resulting algorithms run in polynomial time as long as the degree of the number field is fixed, but they are exponentially dependent on the degree in general.

One implication of these results is that in the event that it becomes possible to build large quantum computers, most public-key cryptosystems, including those based on arithmetic in quadratic fields presented in Chapter 14, would be broken. Although some small quantum computers have been built,[37] it seems very difficult to scale these up to the size required to solve these problems for large inputs. In fact, Schmidt[38] has proposed using the number of qubits required to implement an algorithm as a measure of the algorithm's complexity. Under that metric, he has derived estimates for key lengths of various cryptosystems, including RSA, elliptic curve cryptography, and imaginary and real quadratic field cryptography, that would provide the same level of security (i.e., require roughly the same number of qubits to implement). Surprisingly, Schmidt found that real quadratic field cryptography required the smallest key lengths and offered the best performance in this context, due largely to the fact that the quantum algorithms to break these protocols are more complex than the others.

In summary, it does not appear that quantum computers will be able to solve Pell's equation for very large values of D in the near future. However, further advances in experimental physics that would allow quantum computers to scale better may make this exciting possibility a reality.

2. Is there an unconditional, deterministic algorithm for evaluating R_Δ' that executes in subexponential time?

There may well be such an algorithm, but at present no one has any idea about how to tackle this admittedly difficult problem.

3. Is there a fast way to verify the value of R_Δ' produced by the index-calculus algorithm of Chapter 13?

We know that this can be done in $O(\Delta^{1/6+\epsilon})$ elementary operations, but this seems still to be much too large. Certainly, as mentioned in §13.6, there exists a short certificate for the value of R_Δ' which can be verified in polynomial time under the GRH. Of course, this problem could be solved if a fast method were found for determining whether or not a compact representation of ϵ_Δ is a perfect power in $\mathcal{O}$.

4. Is there an unconditional, deterministic algorithm for computing $R_\Delta{}'$ that executes in time $O(R_\Delta{}^\nu \Delta^\epsilon)$, where $\nu < 1/2$?

We have seen that this should appear to be the case, but we are still, it seems, a long way from answering this question.

5. Is there an efficient method of determing $R_\Delta{}'$ when we are given S', an approximation of an integral multiple S of R_Δ?

We have shown that the answer to this is yes if S/R_Δ is small, but if it is large, we have no idea of how to approach this problem.

6. Can the number field sieve methods be applied to the problem of computing $R_\Delta{}'$ with the success that they have enjoyed with respect to the integer factoring problem or the discrete logarithm problem in finite fields?

So far, no one knows how to do this and we have seen that it may simply not be possible.

7. Can the index-calculus methods be improved to run significantly faster than they do currently?

This is almost certainly possible. There is much room for improvement, but the problems involved, such as reducing Bach's bound and computing a value for S, are not easy. Of course, even if the Bach bound were to be reduced, we still have to find a sufficient number of relations in order to guarantee success. Thus, there is an interesting trade-off here that needs investigation. Furthermore, there is very likely much that could be done to improve the performance of the linear algebra component of this process.

8. We have seen that in the process for verifying $R_\Delta{}'$, we must compute and store a lengthy list of baby steps and that accessing this list can degrade the performance of the procedure. Is there an algorithm with expected runtime $O(R_\Delta{}^{1/2} \Delta^\epsilon)$ that requires constant storage? Alternatively, is there a better data structure that could be used to improve the storage requirements of the baby-step giant-step algorithm?

As mentioned earlier, Pollard's kangaroo does solve the first problem in the function field setting where distances are discrete, but it seems difficult to apply to number fields. Addressing the second question is a matter of paring down what is stored in order to have the least amount of data needed in order to solve the problem. This should be possible when we take into consideration that all we need to verify is that $S' = R_\Delta{}'$, something that is almost certainly true.

The next question is left over from the previous section. We list it last here, not because we think it easy but because it does not pertain directly to the problem of solving the Pell equation.

9. Given a compact representation of $x + y\sqrt{D}$, is there a fast way (polynomial in $\log D$) to determine whether or not y (or x) is a perfect mth power?

If y is not a perfect mth power, there may well be a way to tackle this, but if it is, it might not be possible to prove it, particularly when y is very large compared to D.

Given these questions, it appears that we are still a great distance from declaring the state of the art in solving the Pell equation to be completely satisfactory. Much has been done, but much more remains to be learned, guaranteeing that this will not be the last book devoted to this remarkable equation.

Notes and References

[1] An earlier version of this discussion can be found in [SSW08].

[2] The solution of this equation when reduced to a congruence modulo n is discussed in [AEM87]. See also [PS87].

[3] [Lag68].

[4] [Kor90].

[5] [Duj94] and [Dic19], Vol. II, p. 416.

[6] [Ben01].

[7] See [Mor69], p. 274.

[8] See [Mor69], p. 275.

[9] [TVW05].

[10] [Ben05].

[11] For example, in the case of $m = 2$, see the table of solutions of (17.15) in [WZ72].

[12] See [BS93].

[13] [BCMO06].

[14] [CM07].

[15] This equation is discussed in Chapters 7 and 8 of [Bar03].

[16] This equation was first investigated in 1891 by Meissel and, independently, by Mathews. See [Dic19], Vol. II, p. 594.

[17] See, for example, [Hec81], p. 109.

[18] For an investigation of this, see [Buc87b].

[19] Much information on this topic can be found in [PZ97]. Additional references are [Buc87a] and [BJP94]. Compact representations are discussed in [Thi95] and index-calculus methods are discussed in [Buc90]. The infrastructure of the principal ideal class in the case of $s = 1$ is described in [BW88b]. In the case of $n = 3$, the use of the algorithm of Voronoi (see, for example, [DF64], Ch. 4) is analogous to that of the SCF algorithm in the case of $n = 2$. A detailed implementation of this algorithm for solving (17.19) is presented in [WCS80]. A more general version, utilizing the infrastructure, is given in [WDS83].

[20] See [Niv42] and [Niv43]. Further results concerning this equation can be found in [Sko45a] and [Sko45b].

[21] [Nat76]. See also [Ste05] §§6.7 and 6.8 for further information.

[22] [Sch62].

[23] [DPLR91].

[24] See, for example, [vdP99b] and relevant sections, such as §1.5, of [Pat03].

[25] See [vdPT00] and [vdPT02]. Additional work concerning continued fraction expansions of the square root of a polynomial can be found in [WY02], [McL03a], and [McL03b].

[26] For more precise definitions and for further information on the theory of function fields, see [Sti93] or [Ros02].

[27] Quadratic function fields are an alternate representation of hyperelliptic curves $Y^2 + h(X)Y = f(X)$ defined over a field $\mathbb{K}$, so many results in this area can be described and derived using the language of algebraic geometry. See [DL06] for more information about hyperelliptic curves.

[28] Stein described the infrastructure of a real quadratic function field defined over an odd characteristic finite field in [SZ91] and [Ste96a]. Zuccherato [Zuc97] subsequently extended Stein's results to the even characteristic case.

[29] NUCOMP in function fields was first described by Jacobson and van der Poorten [JvdP02] and, subsequently, presented in a more efficient form [JSS07b]. Generalizations of the $O(\Delta^{1/4+\epsilon})$ and $O(\Delta^{1/5+\epsilon})$ baby-step giant-step algorithms for computing the regulator were described by Stein and Williams [SW99], as well as optimized versions by Stein and Teske [ST02a] [ST05]. Cryptography in the infrastructure of a real quadratic function field was first proposed by Scheidler, Stein, and Williams [SSW96]. The most recent work on the subject, including improvements that make use of the fast baby-step operation, including the basis for the idea of w-near (f, p) representations from Chapter 11, is due to Jacobson, Scheidler, and Stein [JSS07a].

[30] [ST02b].

[31] [MST99]. Although this result only holds for function fields defined over an odd characteristic function field, the ideas certainly generalize to the even characteristic case.

[32] [JSS07a].

[33] [Hal02].

[34] [Sho97a].

[35] [Sch06].

[36] [Hal05] and [SV05].

[37] At the time of writing this book, the largest quantum computer of which we are aware was able to compute with 8 qubits [HHR+05]. A qubit is the quantum analogue of a bit, the basic unit of computation in a quantum computer.

[38] [Sch06] and [Sch07].

Appendix

A.1 NUCOMP

As mentioned in §5.4, we use this appendix as the place to record the pseudocode for some of our most needed algorithms, together with the proofs of correctness of these routines. We will begin by deriving some useful material needed in establishing our version of NUCOMP for $D > 0$.

Theorem A.1. *Let* $\Delta > 0$. *If* $\mathfrak{a} = [a, b + \omega]$ *is a primitive* $\mathcal{O}$-*ideal and* $\sqrt{\Delta} < a < 2\sqrt{\Delta}$, *then either* $\rho(\mathfrak{a})$ *or* $\rho^2(\mathfrak{a})$ *is a reduced* $\mathcal{O}$-*ideal. Furthermore, if* $\rho(\mathfrak{a})$ *is not a reduced ideal, then* $\rho^2(\mathfrak{a}) = \psi\mathfrak{a}$, *where* $|\psi|, |\overline{\psi}| < 1$.

Proof. By Theorem 5.13, we know that if $a < 3\sqrt{\Delta}/2$, then $\rho(\mathfrak{a})$ is reduced and we are done. We will now assume that $\rho(\mathfrak{a})$ is not a reduced $\mathcal{O}$-ideal. Since $\mathfrak{a}' = \rho(\mathfrak{a}) = [a', b' + \omega]$ and $b' + \omega = \sqrt{\Delta} - \eta a$ $(0 < \eta < 1)$, we have $-\sqrt{\Delta} < b' + \omega < \sqrt{\Delta}$. If $\eta \le 1/2$, then $|a'| = \eta |b' + \omega| < \sqrt{\Delta}/2$, which is not possible because $\mathfrak{a}'$ is not a reduced $\mathcal{O}$-ideal. If $\eta > 1/2$ and $\eta a < \sqrt{\Delta}$, then $0 < b' + \omega = \sqrt{\Delta} - \eta a < \sqrt{\Delta}/4$ and $0 < a' < \sqrt{\Delta}/4$, which we have just argued is impossible. Thus, we can only have $\eta > 1/2$, $\eta a > \sqrt{\Delta}$ and $-\sqrt{\Delta} < a' < -\sqrt{\Delta}/2$ by (5.15) and Theorem 5.9.

Now, in this case, $\mathfrak{a}'' = \rho^2(\mathfrak{a}) = [a'', b'' + \omega]$, where $b'' + \omega = \sqrt{\Delta} - \eta' a'$ and $a'' = \eta'(\sqrt{\Delta} - \eta' a') > 0$. Also, $b'' = q'a' - b' - T(\omega)$, $q' = \lfloor (b' + \omega)/a' \rfloor = \lfloor 1/\eta \rfloor = 1$ by (5.5), $0 < \eta' < 1$, and $2b'' + T(\omega) = \sqrt{\Delta} - 2\eta' a' > \sqrt{\Delta}$. Hence,

$$2b' + T(\omega) = 2a' - 2b'' - T(\omega) < 2a' - \sqrt{\Delta}\,.$$

Since
$$a'' = -N(a' - b' - T(\omega) + \omega)/a' = -a' + a + 2b' + T(\omega)$$
and $a' < -\sqrt{\Delta}/2$, we get

$$0 < a'' < a' + a - \sqrt{\Delta} < \sqrt{\Delta}/2\,.$$

It follows that $\rho^2(\mathfrak{a})$ is reduced. In this case, we get $\mathfrak{a}'' = \psi\mathfrak{a}$, where $\psi = (b'+\omega)(b''+\omega)/aa'$. We must have $|\overline{\psi}| = |\eta\eta'| < 1$ by (5.14). Also,

$$\psi = \frac{b'+\omega}{a}\left[1 - \frac{b'+\overline{\omega}}{a'}\right] = \frac{b'+\omega}{a} + 1 \ .$$

Since $(b'+\omega)/a < 0$ and $-(b'+\omega) = \eta a - \sqrt{\Delta} < \sqrt{\Delta}$, we get $-(b'+\omega)/a < \sqrt{\Delta}/a < 2$; thus, $(b'+\omega)/a > -2$ and $-1 < \psi < 1$.   □

Suppose that for some $i \in \mathbb{Z}^{\geq 0}$, we have

$$\frac{P+\sqrt{D}}{Q} = \langle q_0, q_1, q_2, \ldots, q_i, \phi_{i+1}\rangle \ ,$$

where $\phi_{i+1} = (P_{i+1} + \sqrt{D})/Q_{i+1}$. Put $\sigma = \text{sign}\, Q_{i+1}$, $P_0{}^* = P_{i+1}$, and $Q_0{}^* = \sigma Q_{i+1} = |Q_{i+1}|$, and let

$$\frac{P^* + \sqrt{D}}{Q^*} = \langle q_0{}^*, q_1{}^*, \ldots, q_j{}^*, \phi_{j+1}^*\rangle \ ,$$

where we define $\phi_k{}^*$, $Q_k{}^*$, $P_k{}^*$, and $\theta_k{}^*$ analogously to the definitions of ϕ_k, Q_k, P_k, and θ_k in the continued fraction expansion of $(P+\sqrt{D})/Q$.

Theorem A.2. *If we put $\tilde{B}_{-2} = \sigma B_{i-1}$, and $\tilde{B}_{-1} = B_i$ and define*

$$\tilde{B}_{k+1} = q_{k+1}^* \tilde{B}_k + \tilde{B}_{k-1} \quad (k = -1, 0, 1, \ldots) \ ,$$

then

$$|\theta_{j+1}^* \theta_{i+2}|^{-1} = |\tilde{B}_{j-2} + \overline{\phi}_j^* \tilde{B}_{j-1}| \quad (j = 0, 1, 2, \ldots) \ .$$

Proof. For $j = 0$, $|\tilde{B}_{-2} + \overline{\phi}_j^* \tilde{B}_{-1}| = |\sigma B_{i-1} + \overline{\phi}_0^* B_i| = |B_{i-1} + \overline{\phi}_{i+1} B_i| = 1/|\theta_{i+2}|$ by (3.19). Since $\theta_1{}^* = 1$, the theorem is true for $j = 0$. For $k \geq 1$, we have

$$Q_{k-1}^* \tilde{B}_{k-3} + P_{k-1}^* \tilde{B}_{k-2} = Q_{k-1}^*(\tilde{B}_{k-1} - q_{k-1}^* \tilde{B}_{k-2}) + P_{k-1}^* \tilde{B}_{k-2}$$
$$= Q_{k-1}^* \tilde{B}_{k-1} - P_k^* \tilde{B}_{k-2} \ .$$

Hence,

$$-\tilde{B}_{k-2}\psi_k{}^* + \tilde{B}_{k-1} = \tilde{B}_{k-3} + \tilde{B}_{k-2}\overline{\phi}_{k-1}^* \ .$$

Since $\psi_k{}^* \overline{\phi}_k^* = -1$, we have

$$\tilde{B}_{k-2} + \overline{\phi}_k^* \tilde{B}_{k-1} = -(\tilde{B}_{k-3} + \tilde{B}_{k-2}\overline{\phi}_{k-1}^*)/\psi_k{}^* \ .$$

Putting $k = j$ and iterating the above, we get

$$\tilde{B}_{j-2} + \overline{\phi}_j^* \tilde{B}_{j-1} = \frac{(-1)^j(\tilde{B}_{-2} + \tilde{B}_{-1}\overline{\phi}_0^*)}{\theta_{j+1}^*}$$
$$= \frac{(-1)^i \sigma(B_{i-1} + \overline{\phi}_{i+1}B_i)}{\theta_{j+1}^*} \ .$$

Thus,

$$|\tilde{B}_{j-2} + \overline{\phi}_j^* \tilde{B}_{j-1}| = |\theta_{j+1}^* \theta_{i+2}|^{-1} .$$

$\square$

Now suppose we are given reduced $\mathcal{O}$-ideals $\mathfrak{a}'$ and $\mathfrak{a}''$. Write $\mathfrak{a} = S[Q/r, (P + \sqrt{D})/r] = \mathfrak{a}'\mathfrak{a}''$ as in §5.4. We will now assume that $Q > 2\sqrt{D}$; in this case, because $Q'' < Q'$, we have $R_{-1} = Q'/S > \sqrt{2r}D^{1/4}$. If $Q < 2\sqrt{D}$, then either $\mathfrak{a}$ or $\rho(\mathfrak{a})$ is reduced by Theorem 5.13, so this is not a severe restriction. We proceed as in the version of NUCOMP given in §5.4 except that we compute R_i, C_i, and C_{i-1} such that

$$R_i < \sqrt{2r}D^{1/4} < R_{i-1} .$$

In this case, we have $0 < x < 2\sqrt{D}$, $|y| < 2\sqrt{D}$, and

$$B_i < \frac{Q'/S}{\sqrt{2r}D^{1/4}} .$$

We may assume with no loss of generality that because $\mathfrak{a}''$ is reduced, we have $0 < (\sqrt{D} - P'')/Q'' < 1$ and $(\sqrt{D} + P'')/Q'' > 1$. Hence,

$$0 < R'' = [(\sqrt{D} - P'')/Q''](\sqrt{D} + P'') < \sqrt{D} + P'' < 2\sqrt{D} .$$

Thus, since $Q' < 2\sqrt{D}$, we get

$$0 < z = rSR''C_i^2/(Q'/S) < R'' < 2\sqrt{D} .$$

It follows that

$$|Q_{i+1}| < 4\sqrt{D} .$$

In fact, empirical studies suggest that $\mathfrak{a}_{i+2}$ is reduced about 88% of the time. However, if not, $\rho(\mathfrak{a}_{i+2})$ turns out to be reduced for most of the remaining 12% of the cases. Only very infrequently do we have to go as far as $\rho^2(\mathfrak{a}_{i+2})$ to find the reduced ideal that we need. That at least one of $\mathfrak{a}_{i+2}$, $\rho(\mathfrak{a}_{i+2})$ or $\rho^2(\mathfrak{a}_{i+2})$ must be reduced follows from Theorem A.1.

We are now ready to present the NUCOMP algorithm for $D > 0$. We implicitly assume in this and all subsequent algorithms involving arithmetic in $\mathbb{Q}(\sqrt{D})$ that D is included in the input.

Algorithm 5.1: NUCOMP

Input: $\mathfrak{a}' = [Q'/r, (P'+\sqrt{D})/r]$, $\mathfrak{a}'' = [Q''/r, (P''+\sqrt{D})/r]$ reduced invertible $\mathcal{O}$-ideals with $Q' > Q'' > 0$.

Output: A reduced $\mathcal{O}$-ideal $\mathfrak{b} = [Q/r, (P + \sqrt{D})/r]$ such that $\mathfrak{b} \sim \mathfrak{a}'\mathfrak{a}''$. (Optional output: A, B, C, where $\mu = |(A + B\sqrt{D})/C|$ and $\mathfrak{a}'\mathfrak{a}'' = \mu\mathfrak{b}$.)

1: Compute $G = (Q'/r, Q''/r)$ and solve $(Q''/r)X \equiv G \pmod{Q'/r}$ for $X \in \mathbb{Z}$, $0 \leq X < Q'/r$.

2: Compute $S = ((P' + P'')/r, G)$ and solve $Y(P' + P'')/r + ZG = S$ for $Y, Z \in \mathbb{Z}$.

3: Put $R'' = (D - P''^2)/Q''$, $U \equiv XZ(P' - P'') + YR''$ (mod Q'/S), where $0 \le U < Q'/S$.

4: Put $R_{-1} = Q'/S$, $R_0 = U$, $C_{-1} = 0$, $C_0 = -1$, $i = -1$.

5: **if** $R_{-1} < \lfloor \sqrt{2r}D^{1/4} \rfloor$ **then**

6:     Put

$$Q_{i+1} = Q'Q''/rS^2 \, ,$$
$$P_{i+1} \equiv P'' + UQ''/rS \pmod{Q_{i+1}} \, .$$

    $(B_{-2} = 1, \, B_{-1} = 0.)$

7:     Go to 16.

8: **end if**

9: **while** $R_i > \lfloor \sqrt{2r}D^{1/4} \rfloor$ **do**

10:     $i \leftarrow i + 1$

11:     $q_i = \lfloor R_{i-2}/R_{i-1} \rfloor$

12:     $C_i = C_{i-2} - q_i C_{i-1}$

13:     $R_i = R_{i-2} - q_i R_{i-1}$

14: **end while**

15: Put

$$M_1 = \frac{(Q''/rs)R_i + (P' - P'')C_i}{Q'/S} \, ,$$
$$M_2 = \frac{(P' + P'')R_i + rSR''C_i}{Q'/S} \, ,$$
$$Q_{i+1} = (-1)^{i+1}(R_i M_1 - C_i M_2) \, ,$$
$$P_{i+1} = \frac{(Q''/rS)R_i + Q_{i+1}C_{i-1}}{C_i} - P'' \, .$$

16: Put $j = 1$,

$$Q'_{i+1} = |Q_{i+1}| \, ,$$
$$k_{i+1} = \left\lfloor \left| \frac{\lfloor \sqrt{D} \rfloor - P_{i+1}}{Q'_{i+1}} \right| \right\rfloor \, ,$$
$$P'_{i+1} = k_{i+1}Q'_{i+1} + P_{i+1} \, .$$

    $(\sigma = \text{sign}(Q_{i+1}), \, B_{i-1} = \sigma|C_{i-1}|, \, B_{i-2} = |C_{i-2}|.)$

17: **if** $P'_{i+1} + \lfloor \sqrt{D} \rfloor \ge Q'_{i+1}$ **then**

18:     Go to 27.

19: **else**

20:     Put $j = 2$ and

$$q_{i+1} = \left\lfloor \frac{P_{i+1} + \lfloor \sqrt{D} \rfloor}{Q'_{i+1}} \right\rfloor ,$$

$$P_{i+2} = q_{i+1} Q'_{i+1} - P_{i+1} ,$$

$$Q_{i+2} = \frac{D - P_{i+2}^2}{Q'_{i+1}} ,$$

$$Q'_{i+2} = |Q_{i+2}| ,$$

$$k_{i+2} = \left\lfloor \frac{\lfloor \sqrt{D} \rfloor - P_{i+2}}{Q'_{i+2}} \right\rfloor ,$$

$$P'_{i+2} = k_{i+2} Q'_{i+1} + P_{i+2} .$$

$$(B_{i+1} = q_{i+1} B_i + B_{i-1} .)$$

21:     **if** $P'_{i+2} + \lfloor \sqrt{D} \rfloor \geq Q'_{i+2}$ **then**
22:         Go to 27.
23:     **else**
24:         Put $j = 3$ and

$$Q_{i+3} = Q'_{i+3} = Q_{i+1} - Q_{i+2} + 2P_{i+2} ,$$

$$P_{i+3} = Q_{i+2} - P_{i+2} ,$$

$$P'_{i+2} = P_{i+3} - Q_{i+3} .$$

$$(B_{i+2} = B_{i+1} + B_i .)$$

25:     **end if**
26: **end if**
27: Put $\mathfrak{b} = [Q'_{i+j}/r, (P'_{i+j} + \sqrt{D})/r]$. $(A = S(Q_{i+j}B_{i+j-2} + P_{i+j}B_{i+j-1}),$
    $B = -SB_{i+j-1}, C = Q_{i+j}.)$

Proof (of correctness of NUCOMP). We have seen by (5.36) and (5.37) that

$$\mathfrak{a}'\mathfrak{a}'' = S \left[\frac{Q}{r}, \frac{P + \sqrt{D}}{r} \right] .$$

By the variation of NUCOMP presented in §5.4, (5.44), and Theorem A.1, we know that

$$\mathfrak{a}'\mathfrak{a}'' = \mu \mathfrak{b} ,$$

where $\mathfrak{b} = [Q'_{i+j}/r, (P'_{i+j} + \sqrt{D})/r]$ is a reduced ideal with $(P'_{i+j} + \sqrt{D})/Q'_{i+j} > 1$ and $-1 < (P'_{i+j} - \sqrt{D})/Q'_{i+j} < 0$ (see Theorem 5.10). In the case where $j = 3$, we know by Theorem A.1 that $q_{i+2} = 1$. Also, it is not difficult to show that $0 < (P_{i+2} - \sqrt{D})/Q_{i+2} < 1$; hence, $k_{i+2} = -1$ and the formulas for $P_{i+3}, P'_{i+3}, Q_{i+3}$, and Q'_{i+3} in NUCOMP follow easily.

Since we can put $\mu = S/|\psi\theta_{i+2}|$, where $\psi = \theta^*_{j+1}$ for $j = 0, 1$, or 2, we can easily compute μ once we have B_i, B_{i-1}, Q_{i+1} and P_{i+1}. Indeed, by Theorem A.2,

$$|Q_{i+j}|\mu = S|Q_{i+j}B_{i+j-2} + (P_{i+j} - \sqrt{D})B_{i+j-1}| \,. \qquad (A.1)$$

Since $N(\mathfrak{a}'), N(\mathfrak{a}'') < \sqrt{\Delta}$, we see that NUCOMP will execute in $O(\log \Delta)$ elementary operations.

We now derive some results concerning the size of μ. As above, we have

$$\mu = S/|\psi\theta_{i+2}| \,. \qquad (A.2)$$

By Theorems 5.13 and A.1, we know that $|\psi|, |\overline{\psi}| < 1$. Since

$$\theta_{i+2} = (G_i + \sqrt{D}B_i)/Q$$

and

$$G_i = (-1)^{i+1} \left(Q''R_i/rS - P''C_i\right) \,,$$

we get

$$\theta_{i+2} = \frac{(-1)^{i+1} \left(R_i + rSC_i(\sqrt{D} - P'')/Q''\right)}{Q'/S} \,. \qquad (A.3)$$

Hence,

$$|\theta_{i+2}| < \frac{R_i + rS|C_i|}{Q'/S} \,. \qquad (A.4)$$

If $i = 0$, then $C_0 = -B_0 = -1$ and

$$\theta_2 = \frac{rS(\sqrt{D} - P'')/Q'' - R_0}{Q'/S} < \frac{rS}{Q'/S} \,.$$

Since $Q'/rS \geq 1$, we have $\theta_2 < S$. Also, since

$$\theta_2 > -R_0/(Q'/S) = -R_0/R_{-1} > -1 \,,$$

we get $|\theta_2| < S$. If $i = 1$, then

$$\theta_3 = \frac{R_1 + rSB_1(\sqrt{D} - P'')/Q''}{Q'/S} > 0 \,.$$

Since $\theta_3 = q_1\theta_2 + 1$ and $q_1 = \lfloor R_{-1}/R_0 \rfloor$, we get

$$\theta_3 < \frac{Q'/S}{R_0} \left(\frac{rS(\sqrt{D} - P'')/Q'' - R_0}{Q'/S}\right) + 1$$

$$= \frac{rS(\sqrt{D} - P'')}{Q''R_0} < \frac{rS}{R_0} \,.$$

Since $r \mid R_{-1}$ and $r \mid R_{-2}$, we must have $r \mid R_0$ and, therefore, $0 < \theta_3 < S$.
If $i > 1$, then $B_i \geq 2$ and

$$|\theta_{i+2}| < \frac{R_i + rS|C_i|}{R_i|C_{i-1}| + |C_i|R_{i-1}}$$

by (A.4) and (3.23). Thus,

$$|\theta_{i+2}| < \frac{R_i + rS|C_i|}{|C_i|R_{i-1}}$$

$$= \frac{R_i}{|C_i|R_{i-1}} + \frac{rS}{R_{i-1}}$$

$$< 1/2 + rS/R_{i-1} \, .$$

Now, $R_{i-1} > \sqrt{2r}D^{1/4}$ means that $R_{i-1}/r > D^{1/4} > 2$ (we will assume that $D > 16$). Thus,

$$|\theta_{i+2}| < 1/2 + S/2 \leq S \, .$$

By (A.2), we see that $\mu > 1$. In the case that $Q < 2\sqrt{D}$, we get $\mathfrak{b} = \mathfrak{a}$ or $\rho(\mathfrak{a})$ and $\mu = 1$ or ψ. Hence, we can say that $\mu \geq 1$.

We will now attempt to get some idea of what the value of μ will usually be. We may assume that since $\mathfrak{a}'$ is reduced, we have Q' and P' such that $(P' + \sqrt{D})/Q > 1$ and $-1 < (P' - \sqrt{D})/Q' < 0$. Hence, $P' > \sqrt{D} - Q'$ and $P' < \sqrt{D}$. Since $(P' + \sqrt{D})/Q' = q' + \eta$ $(0 < \eta < 1)$, it is easy to deduce that

$$q' < 2\sqrt{D}/Q' < q' + 2 \, . \tag{A.5}$$

By (A.4), we have

$$|\theta_{i+2}|/S < R_i/Q' + rS|C_i|/Q' \, ;$$

also,

$$rS|C_i|/Q' < r/R_{i-1} < \sqrt{r/2}D^{-1/4} \, .$$

Since, by (A.5),

$$R_i/Q' < \sqrt{2r}D^{1/4}/Q' < \sqrt{r/2}D^{-1/4}(q' + 2) \, ,$$

we get

$$\frac{S}{|\theta_{i+2}|} > \frac{\sqrt{2/r}D^{1/4}}{q' + 3}$$

and

$$\mu > \sqrt{2/r}D^{1/4}/(q' + 3) \, . \tag{A.6}$$

By (A.3), we have

$$|\bar{\theta}_{i+2}| = \frac{|R_i - rSC_i(P'' + \sqrt{D})/Q''|}{Q'/S} \, ;$$

thus, by (3.19), we get

$$\frac{S}{|\theta_{i+2}|} = \frac{|(Q''/r)R_i - S(\sqrt{D} + P'')C_i|}{|Q_{i+1}|}$$

$$< \frac{(Q''/r)R_i + 2\sqrt{D}S|C_i|}{|Q_{i+1}|} \, .$$

We know that

$$S|C_i| < Q'/R_{i-1} < (2\sqrt{D}/q')(\sqrt{2r}D^{1/4})^{-1} = \sqrt{2/r}D^{1/4}/q'$$

and

$$(Q''/r)R_i \le (Q'/r)R_i < (2\sqrt{D}/rq')\sqrt{2r}D^{1/4} = 2\sqrt{2/r}D^{3/4}/q' \ .$$

It follows that

$$\frac{S}{|\theta_{i+2}|} < \frac{4\sqrt{2/r}D^{3/4}}{q'|Q_{i+1}|} \ . \tag{A.7}$$

Since $|\psi\overline{\psi}| = N(\mathfrak{b})/N(\mathfrak{a}_{i+2})$, we get

$$|\psi|^{-1} = \frac{N(\mathfrak{a}_{i+2})|\overline{\psi}|}{N(\mathfrak{b})} \le \frac{N(\mathfrak{a}_{i+2})}{N(\mathfrak{b})} = \frac{|Q_{i+1}|}{rN(\mathfrak{b})} \ . \tag{A.8}$$

Hence,

$$\mu < \frac{2\left(\sqrt{2/r}D^{1/4}\right)^3}{q'N(\mathfrak{b})} \le 2\Delta^{3/4}. \tag{A.9}$$

Now, $\mathfrak{b} = [\tilde{Q}/r, (\tilde{P} + \sqrt{D})/r]$ is a reduced ideal equivalent to $\mathfrak{a}$, and we may assume that $2\sqrt{D}/\tilde{Q} < \tilde{q} + 2$, where $N(\mathfrak{b}) = \tilde{Q}/r$ and $\tilde{q}$ is a partial quotient in the simple continued fraction expansion of $(P + \sqrt{D})/Q$. Thus, since $\sqrt{2/r}D^{1/4} = \Delta^{1/4}$, we can easily derive (5.45) from (A.6), (A.7), and (A.8).

A.2 NUMULT

In order to develop the pseudocode for NUMULT, we need an algorithm which permits us to remove factors in $\mathbb{K}$ from (f, p) representations. In order to prove the correctness of this algorithm, we will require two preliminary lemmas.

Lemma A.3. *Let $\gamma = |(a + b\sqrt{D})/c| > 1$, where $a, b, c \in \mathbb{Z}$, and $\sqrt{D} \notin \mathbb{Q}$. Put $t = 2^s a + b\lfloor 2^s \sqrt{D} \rfloor$ with $s \in \mathbb{Z}^{\ge 0}$ and $2^s|c| > 2^{p+4}|b|$. If we put $e = \lfloor 2^{p+3-s}\lceil t/c \rceil \rceil$, then $e \ge 2^{p+3}$ and*

$$\left| \frac{2^{p+3}\gamma}{e} - 1 \right| < \frac{1}{2^{p+3}} \ .$$

Proof. Since $2^s|c| > 2^{p+4}|b|$, we have

$$\left| \frac{a + b\sqrt{D}}{c} - \frac{t}{2^s c} \right| < \frac{|b|}{2^s|c|} < \frac{1}{2^{p+4}} \ .$$

Also, since $|a + b\sqrt{D}/c| > 1$, we must have $\text{sign}(t) = \text{sign}(a + b\sqrt{D})$ and

$$\left| 2^{p+3}\gamma - 2^{p-s+3}\frac{|t|}{|c|} \right| < \frac{1}{2} \ .$$

It follows that $|2^{p+3}\gamma - e| < 1$ and so $e > 2^{p+3}\gamma - 1 > 2^{p+3} - 1$. Hence, $e \ge 2^{p+3}$ and $|2^{p+3}\gamma/e - 1| < 1/2^{p+3}$. $\qquad\square$

Lemma A.4. *Suppose* $p, k, d, f, e, \in \mathbb{Z}$, $2^p < d \le 2^{p+1}$, $1 \le f < 2^p$, $e \ge 2^{p+3}$, $p \ge 2$, *and* $\theta, \gamma \in \mathbb{K}$ *such that*

$$\left| \frac{2^p \theta}{2^k d} - 1 \right| < \frac{f}{2^p}, \quad \left| \frac{2^{p+3} \gamma}{e} - 1 \right| < \frac{1}{2^{p+3}} .$$

If $t \in \mathbb{Z}^{\ge 0}$ *is defined by*

$$2^{t-1} \le \frac{e}{8d} < 2^t ,$$

then

$$\left| \frac{2^p \theta}{2^h g \gamma} - 1 \right| < \frac{f + 9/8}{2^p} ,$$

where $g = \lceil 2^{p+3+t} d/e \rceil$ *and* $h = k - t$. *Furthermore,* $2^p < g \le 2^{p+1}$.

Proof. We have

$$\left| \frac{2^{p-k} \theta}{d} - 1 \right| < \frac{f}{2^p}, \quad \left| \frac{2^{p+3} \gamma}{e} - 1 \right| < \frac{1}{2^{p+3}} ,$$

so

$$\frac{1 - 2^{-p} f}{1 + 2^{-(p+3)}} < 2^{-(k+3)} \frac{\theta e}{\gamma d} < \frac{1 + 2^{-p} f}{1 - 2^{-(p+3)}} . \tag{A.10}$$

Also,

$$1 + \frac{f}{2^p} < 1 + \frac{f}{2^p} + \frac{1}{2^p} \left(1 - \frac{f + 9/8}{2^{p+3}} \right) = \left(1 + \frac{f + 9/8}{2^p} \right) \left(1 - \frac{1}{2^{p+3}} \right) .$$

Since, by definition of g, $2^{p-h} \theta / \gamma g \le 2^{-(k+3)} \theta e / \gamma d$, we get

$$\frac{2^{p-h} \theta}{\gamma g} < 1 + \frac{f + 9/8}{2^p} .$$

By definition of g and t, we see that $2^p < g \le 2^{p+1}$. Now, $2^p < g < 2^{p+t+3} d/e + 1$ implies

$$1 - \frac{1}{2^p} < 1 - \frac{1}{g} < \frac{2^{p+t+3} d}{eg} ,$$

so

$$\frac{2^{p-h} \theta}{\gamma g} = 2^{-(k+3)} \frac{\theta e}{\gamma d} \frac{2^{p+t+3} d}{eg} > 2^{-(k+3)} \frac{\theta e}{\gamma d} \left(1 - \frac{1}{2^p} \right) > \frac{(1 - 2^{-p})(1 - 2^{-p} f)}{1 + 2^{-(p+3)}} ,$$

where the last inequality follows from (A.10). Furthermore,

$$\left(1 + \frac{1}{2^{p+3}} \right) \left(1 - \frac{f + 9/8}{2^p} \right) = 1 - \frac{f+1}{2^p} - \frac{f + 9/8}{2^{2p+3}} < \left(1 - \frac{1}{2^p} \right) \left(1 - \frac{f}{2^p} \right) ,$$

from which we see that

$$\frac{2^{p-h} \theta}{\gamma g} > 1 - \frac{f + 9/8}{2^p} .$$

$\square$

We are now able to present our first routine.

Algorithm A.1: REMOVE

Input: $(\mathfrak{b}, d, k), T, C, s, p$, where $(\mu\mathfrak{b}, d, k)$ is an (f, p) representation of some
ideal $\mathfrak{a}$ with $\mu = |(A + B\sqrt{D})/C| \geq 1$ $(A, B, C \in \mathbb{Z}$ with $C \neq 0)$, $T = 2^s A + B\lfloor 2^s \sqrt{D}\rfloor$, and $s \in \mathbb{Z}^{\geq 0}$ with $2^s |C| > 2^{p+4}|B|$.

Output: An $(f + 9/8, p)$ representation $(\mathfrak{b}, d', k')$ of $\mathfrak{a}$.

1: Set $e = \lfloor 2^{p+3-s}|T/C|\rfloor$.
2: Find $t \in \mathbb{Z}^{\geq 0}$ with $2^{t-1} \leq e/8d < 2^t$.
3: Set $d' = \lceil 2^{p+3+t}d/e\rceil$ and $k' = k - t$.

The correctness of this algorithm follows immediately from Lemmas A.3
and A.4. Notice that since

$$\frac{e}{8d} < \mu + \frac{1}{2^{p+4}},$$

we get

$$t - 1 < \log_2\left(\mu + \frac{1}{2^{p+4}}\right)$$

and

$$t \leq \lfloor \log_2 \mu\rfloor + 2. \tag{A.11}$$

Algorithm NUMULT is essentially NUCOMP performed on (f, p) representations.

Algorithm 11.1: NUMULT

Input: $(\mathfrak{b}', d', k')$, $(\mathfrak{b}'', d'', k'')$, p, where $(\mathfrak{b}', d', k')$ is a reduced (f', p) representation of an invertible $\mathcal{O}$-ideal $\mathfrak{a}'$ and $(\mathfrak{b}'', d'', k'')$ is reduced (f'', p) representation of an invertible $\mathcal{O}$-ideal $\mathfrak{a}''$. Here,

$$\mathfrak{b}' = \left[\frac{Q'}{r}, \frac{P' + \sqrt{D}}{r}\right], \quad \mathfrak{b}'' = \left[\frac{Q''}{r}, \frac{P'' + \sqrt{D}}{r}\right], \quad Q' \geq Q'' > 0.$$

Output: A reduced (f, p) representation $(\mathfrak{b}, d, k)$ of $\mathfrak{a}'\mathfrak{a}''$, where

$$\mathfrak{b} = \left[\frac{Q}{r}, \frac{P + \sqrt{D}}{r}\right],$$

$(P + \sqrt{D})/Q > 1$, $-1 < (P - \sqrt{D})/Q < 0$, $k \leq k' + k'' + 1$, $f = f^* + 17/8$
with $f^* = f' + f'' + 2^{-p}f'f''$. (Optional output: $a, b \in \mathbb{Z}$, where $\nu = (a + b\sqrt{D})/r \in \mathcal{O}$, and $\mathfrak{b} = \nu/(N(\mathfrak{b}')N(\mathfrak{b}''))\mathfrak{b}'\mathfrak{b}''$.)

1: Compute $(\mathfrak{b}, A, B, C) = \text{NUCOMP}(\mathfrak{b}', \mathfrak{b}'')$, where $\mathfrak{b} = [Q/r, (P + \sqrt{D})/r]$.
2: **if** $d'd'' \leq 2^{2p+1}$ **then**
3:     Put $e = \lfloor d'd''/2^p\rfloor$, $h = k' + k''$
4: **else**

5:      Put $e = \lfloor d'd''/2^{p+1} \rfloor$, $h = k' + k'' + 1$

6: **end if**

7: Find $s \geq 0$ such that $2^s Q > 2^{p+4} B$.

8: Put $T = 2^s A + B \lfloor 2^s \sqrt{D} \rfloor$. $(a = A, b = -B.)$

9: $(\mathfrak{b}, d, k) = \text{REMOVE}((\mathfrak{b}, e, h), T, C, s, p)$.

Proof (of correctness of NUMULT). By Theorem 11.2, $(\mathfrak{b}'\mathfrak{b}'', e, h)$ is an (f, p) representation of $\mathfrak{a}'\mathfrak{a}''$ with $f = 1 + f^*$. Also, since $\mu \geq 1$ and $\mu \mathfrak{b} = \mathfrak{b}'\mathfrak{b}''$, REMOVE will compute an (f, p) representation $(\mathfrak{b}, d, k)$ of $\mathfrak{a}'\mathfrak{a}''$ with $f = 1 + f^* + 9/8 = 17/8 + f^*$. Furthermore, $|N(\mu)| = N(\mathfrak{b}')N(\mathfrak{b}'')/N(\mathfrak{b})$, $N(\mathfrak{b}) = |Q_{i+j}|/r$,

$$\nu = \frac{N(\mathfrak{b}')N(\mathfrak{b}'')}{\mu} = N(\mathfrak{b})|\overline{\mu}| = \left| \frac{a + b\sqrt{D}}{r} \right| \in \mathcal{O},$$

by (A.1). Also, since $N(\mathfrak{b})|\overline{\mu}| < N(\mathfrak{b}')N(\mathfrak{b}'') < \Delta$ and, by (A.9),

$$\frac{2\sqrt{D}}{r}|B| \leq N(\mathfrak{b})|\overline{\mu}| + N(\mathfrak{b})\mu < \Delta + 2\Delta^{3/4},$$

we see that $|B| < \Delta^{1/2} + 2\Delta^{1/4}$; hence, $2^s = O(2^p \Delta^{1/2})$.

Notice that NUMULT also executes in $O(\log \Delta)$ elementary operations.

A.3 Theoretical Background for WNEAR

We will devote this section to developing the theoretical background needed for devising an algorithm which will produce a w-near representation for an $\mathcal{O}$-ideal $\mathfrak{a}$, given a reduced (f, p) representation $(\mathfrak{b}, d, k)$ of $\mathfrak{a}$. We will assume that $\mathfrak{b} = [Q/r, (P + \sqrt{D})/r]$, where $(P + \sqrt{D})/Q > 1$ and $-1 < (P - \sqrt{D})/Q < 0$.

We require two lemmas. The first of these is a refinement of Lemma A.3 and the second is a version of Lemma A.4 in which multiplication is featured instead of division.

Lemma A.5. *Let* $\gamma = (a + b\sqrt{D})/c$, *where* $D, a, b, c \in \mathbb{Z}$, $\sqrt{D} \notin \mathbb{Q}$, $1 \leq c < 2\sqrt{D}$, $b \geq 0$, $\gamma \geq 1$, *and* $|\overline{\gamma}| \leq 1$. *If we compute* $s \in \mathbb{Z}^{>0}$ *such that* $2^s c \geq 2^{p+4}$ *and put* $t = 2^s a + b \lfloor 2^s \sqrt{D} \rfloor$, *and* $e = \lceil 2^{p+3-s} t/c \rceil$, *then* $e \geq 2^{p+3}$ *and*

$$\left| \frac{2^{p+3}\gamma}{e} - 1 \right| < \frac{1}{2^{p+3}}.$$

Proof. We have

$$0 < \gamma - 2^{-s}t/c < 2^{-s}b/c \leq 2^{-p-4}b.$$

Since

$$e - 1 < 2^{p+3-s}t/c \leq e,$$

we get

$$-1 < 2^{p+3}\gamma - e < b/2 \, .$$

If $b \leq 2$, then $e > 2^{p+3}\gamma - b/2 > 2^{p+3} - 1$ and $e \geq 2^{p+3}$, $|2^{p+3}\gamma/e - 1| < 2^{-p-3}$. Suppose $b \geq 3$. Since $\gamma + \overline{\gamma} = 2b\sqrt{D}/c$, we get $\gamma > 2b\sqrt{D}/c - |\overline{\gamma}| > b - 1$. Hence,

$$e > 2^{p+3}(b-1) - b/2 = (2^{p+3} - 1/2)b - 2^{p+3} > 2^{p+2}b \, .$$

It follows that $e > 2^{p+3}$ and

$$|2^{p+3}\gamma/e - 1| < b/2e < 2^{-(p+3)} \, .$$

$\square$

Lemma A.6. *Suppose that $(\mathfrak{b}, d, k)$ is an (f, p) representation of $\mathfrak{a}$ with $\mathfrak{b} = \theta\mathfrak{a}$ and suppose further that e and γ obey the same conditions as those in Lemma A.5. If $t \in \mathbb{Z}$ is defined by*

$$2^t < \frac{ed}{2^{2p+3}} \leq 2^{t+1} \, ,$$

then

$$\left| \frac{2^p \theta \gamma}{2^h g} - 1 \right| < \frac{f + 1/4}{2^p} \, ,$$

where $g = \lceil ed/2^{p+t+3} \rceil$ and $h = k + t$. Furthermore, $2^p < g \leq 2^{p+1}$.

Proof. We note that

$$ed = 2^{p+t+3}(g - \eta) \, ,$$

where $0 \leq \eta < 1$. The lemma now follows on using the same kind of reasoning as that employed in the proof of Theorem 11.2. $\square$

We now let $\mathfrak{b} = \theta\mathfrak{a}$ and put $\mathfrak{b}_1 = \mathfrak{b}$, $Q_0 = Q$, and $P_0 = P$. Suppose that $w > k$. In this case, we will now show how to use the simple continued fraction (SCF) (cf. (5.10)) algorithm to find some $n \geq 2$, $\mathfrak{b}_{n-1} \, (= \rho^{n-2}(\mathfrak{b}_1) = \theta_{n-1}\mathfrak{b}_1)$, h, g, and $\mathfrak{b}_n \, (= \rho^{n-1}(\mathfrak{b}_1) = \theta_n\mathfrak{b}_1)$, h', g' such that

$$\left| \frac{2^p \theta_{n-1}\theta}{2^h g} \right| < \frac{f + 1/4}{2^p} \, , \quad \left| \frac{2^p \theta_n \theta}{2^{h'} g'} - 1 \right| < \frac{f + 1/4}{2^p} \, ,$$

$2^p < g, g' \leq 2^{p+1}$, $h < w$, and $h' \geq w$.

We begin by finding $s \in \mathbb{Z}^{>0}$ such that $2^s Q_0 > 2^{p+4}$. If we define

$$T_j = 2^s G_j + B_j \lfloor 2^s \sqrt{D} \rfloor \, ,$$

we find that $T_{-2} = \lfloor 2^s \sqrt{D} \rfloor - 2^s P_0 \geq 0$ and $T_{-1} = 2^s Q_0$; since

$$T_{j+1} = q_{j+1} T_j + T_{j-1} \, , \tag{A.12}$$

we have
$$T_{j+1} > T_j \quad (j \geq 1).$$

Put $M = \lceil 2^{p+s-k+w}Q_0/d \rceil$; then
$$M \geq 2^{p+s-k+w}Q_0/2^{p+1} = 2^s Q_0 2^{w-k-1} \geq 2^s Q_0 = T_{-1}.$$

Thus, for some $i \ (\geq 2)$, we must find $T_{i-2} > M$ and $T_{i-3} \leq M$.

We next define
$$e_j = \left\lceil \frac{2^{p-s+3}T_{j-2}}{Q_0} \right\rceil.$$

For $j \geq 3$, we have $T_{j-2} \geq T_{j-3} + T_{-1}$; hence,
$$e_j \geq 2^{p-s+3}(T_{j-3} + 2^s Q_0)/Q_0 = 2^{p-s+3}T_{j-3}/Q_0 + 2^{p+3}$$
$$> e_{j-1} + 2^{p+3} - 1.$$

Also, by definition of θ_j and Lemma A.6, we have
$$\left| \frac{2^{p+3}\theta_j}{e_j} - 1 \right| < \frac{1}{2^{p+3}}.$$

We next observe that
$$e_i > 2^{p-s+3}T_{i-2}/Q_0 > 2^{p-s+3}M/Q_0 \geq 2^{2p-k+w+3}/d.$$

Also, $e_1 = 2^{p+3}$ and $de_1 \leq 2^{2p+4} \leq 2^{2p-k+w+3}$. Thus, if $de_{i-1} > 2^{2p-k+w+3}$, then $i \geq 3$. Furthermore,
$$e_{i-2} \leq e_{i-1} - 2^{p+3} + 1$$
$$\leq 2^{p-s+3}T_{i-3}/Q_0 - 2^{p+3} + 1$$
$$< \frac{2^{p-s+3}}{Q_0}\left(\frac{2^{p+s-k+w}Q_0}{d} + 1 \right) - 2^{p+3} + 1$$
$$\leq 2^{2p-k+w+3}/d + \frac{3}{2} - 2^{p+3}$$
$$< 2^{2p-k+w+3}/d.$$

Thus, for $n = i$ or $n = i - 1$, we must have
$$\frac{de_{n-1}}{2^{2p+3}} \leq 2^{w-k},$$
$$\frac{de_n}{2^{2p+3}} > 2^{w-k}.$$

If we define t and t' by
$$2^t < \frac{de_{n-1}}{2^{2p+3}} \leq 2^{t+1},$$
$$2^{t'} < \frac{de_n}{2^{2p+3}} \leq 2^{t'+1}$$

and put $h = k + t$, $h' = k + t'$, $g = \lceil de_{n-1}/2^{p+t+3} \rceil$, and $g' = \lceil de_n/2^{p+t+3} \rceil$, then by Lemma A.6 we get

$$\left| \frac{2^p \theta_{n-1} \theta}{2^h g} - 1 \right|, \left| \frac{2^p \theta_n \theta}{2^{h'} g'} - 1 \right| < \frac{f + 1/4}{2^p}$$

and $2^p < g, g' \le 2^{p+1}$. Since $de_{n-1}/2^{2p+3} \le 2^{w-k}$, we must have $2^t < 2^{w-k}$ and $w > k + t = h$; in addition, $de_n/2^{2p+3} > 2^{w-h}$, so we get $2^{t'+1} > 2^{w-k}$ and $w < k + t' + 1$ or $h' \ge w$.

The case of $w \le k$ is similar except that we need certain Q_n^* and Q_{n-1}^*, which we do not know *a priori*. We overcome this by selecting $s = p + 4$. In this case, we apply the SCF algorithm backward (see §5.3) to find g, g', h, h',

$$\mathfrak{c}_n = \rho^{-n}(\mathfrak{b}_1) = \chi_n \mathfrak{b}_1 \quad \text{(cf. (5.27))} ,$$

and

$$\mathfrak{c}_{n-1} = \rho^{-(n-1)}(\mathfrak{b}_1) = \chi_{n-1} \mathfrak{b}_1 ,$$

where

$$\left| \frac{2^p \theta \chi_n}{2^h g} - 1 \right|, \left| \frac{2^p \theta \chi_{n-1}}{2^{h'} g'} - 1 \right| < \frac{f + 9/8}{2^p} ,$$

$2^p < g, g' \le 2^{p+1}$, $h < w$, and $h' \ge w$.

Note that since $0 < \xi_j < 1$ and $-\bar{\xi}_j > 1$ ($j \ge 1$), we get $0 < \chi_k \le 1$ and $|\bar{\chi}_k| \ge 1$ for $k \ge 0$. Put $Q_0^* = Q$ and $P_0^* = P$ and define

$$T_j^* = 2^s G_j^* + A_j^* \lfloor 2^s \sqrt{D} \rfloor .$$

We have $T_{-2}^* = 2^s Q_0^*$ and $T_{-1}^* = 2^s P_0^* + \lfloor 2^s \sqrt{D} \rfloor \ge 2^s Q_0^*$. Since

$$T_{j+1}^* = q_{j+2}^* T_j^* + T_{j-1}^* , \tag{A.13}$$

we see that $T_{j+1}^* > T_j^*$ ($j \ge -1$). Put $M^* = d2^{k-w+4} = d2^{k-w-p+s} > 2^s$. It follows that $T_{-2} = 2^s Q_0^* < Q_0^* M^*$. Since the value of Q_j^* is bounded above by $2\sqrt{D}$ and the integers $T_{-1}^*, T_0^*, T_1^*, \dots$ strictly increase, there will be some minimal i such that

$$T_{i-2}^* \ge Q_i^* M^* .$$

We define

$$e_j = \left\lceil \frac{T_{j-1}^*}{2Q_{j+1}^*} \right\rceil .$$

Now, $1 \le Q_{j+1}^* < 2\sqrt{D}$, $1/\chi_{j+1} \ge 1$, and $1/|\bar{\chi}_{j+1}| \le 1$ ($j \ge -1$); hence, by Lemma A.5 and (5.31), we get

$$\left| \frac{2^{p+3}}{\chi_{j+1} e_j} - 1 \right| < \frac{1}{2^{p+3}} .$$

Also, $e_{-1} = 2^{p+3}$, $e_{-1}/8d = 2^p/d < 1 \le 2^{k-w}$,

$$e_{i-1} \geq \frac{T_{i-2}^*}{2Q_i^*} \geq \frac{M^*}{2} = 8d2^{k-w} \, ,$$

and

$$e_{i-2} < \frac{T_{i-3}^*}{2Q_{i-1}^*} + 1 < \frac{M^*}{2} + 1 = 8d2^{k-w} + 1 \, .$$

Put $\eta = 2^s \sqrt{D} - \lfloor 2^s \sqrt{D} \rfloor$. By (5.31), we have

$$\frac{1}{\chi_j} - \frac{T_{j-2}^*}{2^s Q_j^*} = \frac{\eta A_{j-2}^*}{2^s Q_j^*}$$

and

$$\frac{1}{\chi_j} - \frac{1}{\overline{\chi}_j} = \frac{2\sqrt{D} A_{j-2}^*}{Q_j^*} \, .$$

From these results we can easily deduce that

$$\frac{1}{\chi_j}\left(1 - \frac{\eta}{2^{s+1}\sqrt{D}}\right) = \frac{T_{j-2}^*}{2^s Q_j^*} - \frac{\eta}{2^{s+1}\sqrt{D}\,\overline{\chi}_j} \, . \tag{A.14}$$

Since $\chi_{j+1} = \xi_{j+1}\chi_j$, we can use (A.14) and (A.14) with j replaced by $j+1$ to discover that

$$\frac{T_{j-1}^*}{2^s Q_{j+1}^*} = \frac{1}{\xi_{j+1}}\left(\frac{T_{j-2}^*}{2^s Q_j^*} - \frac{\eta}{2^{s+1}\sqrt{D}\,\overline{\chi}_j}\right) + \frac{\eta}{2^{s+1}\sqrt{D}\,\overline{\chi}_j \overline{\xi}_{j+1}} \, .$$

Also, $\xi_{j+1}^{-1} = (\sqrt{D} + P_j^*)/Q_{j+1}^*$; hence,

$$\frac{T_{j-1}^*}{2Q_{j+1}^*} = \frac{1}{\xi_{j+1}}\frac{T_{j-2}^*}{2Q_j^*} - \frac{\eta}{2^s Q_{j+1}^* \overline{\chi}_j} \, .$$

It follows that since $0 < \eta < 1$, $|\overline{\chi}_j| \geq 1$, and $Q_{j+1}^* \geq 1$, we get

$$e_j > \frac{1}{\xi_{j+1}}(e_{j-1} - 1) - \frac{1}{2} \tag{A.15}$$

and

$$\frac{e_j + 1/2}{e_{j-1} - 1} > \frac{1}{\xi_{j+1}} \, .$$

If $3 \leq e_j \leq e_{j-1}$, then

$$\frac{1}{\xi_{j+1}} < 2 \, .$$

If we replace j by $j-1$ in (A.15), we get

$$e_{j-1} > \frac{1}{\xi_j}(e_{j-2} - 1) - \frac{1}{2} \, ,$$

and on substituting this into (A.15), we find that

$$e_j > \frac{1}{\xi_j \xi_{j+1}}(e_{j-2} - 1) - \frac{7}{2}$$

when $1/\xi_{j+1} < 2$. Hence,

$$\frac{1}{\xi_j \xi_{j+1}} < \frac{e_j + 7/2}{e_{j-2} - 1} .$$

If $e_{j-2} \geq e_j \geq 6$, then $1/\xi_j \xi_{j+1} < 2$, but since, by (5.29),

$$\frac{1}{\xi_j} \frac{1}{\xi_{j+1}} = q_j^* \frac{1}{\xi_{j+1}} + 1 > 2 ,$$

this is a contradiction. Thus, if we have $e_j = 8d2^{k-w}$ for some j, then either e_{j-1} or e_{j-2} must be less than $8d2^{k-w}$.

It follows that we must be able to find some $n \geq 1$ and $n \in \{i-2, i-1, i\}$ such that

$$\frac{e_{n-1}}{8d} \geq 2^{k-w} \text{ and } \frac{e_{n-2}}{8d} < 2^{k-w} .$$

If we define t and t' by

$$2^{t-1} \leq \frac{e_{n-1}}{8d} < 2^t , \quad 2^{t'-1} \leq \frac{e_{n-2}}{8d} < 2^{t'} ,$$

respectively, then by Lemma A.4 with $\gamma = 1/\chi_n$ or $1/\chi_{n-1}$, we have

$$\left| \frac{2^p \theta \chi_n}{2^h g} - 1 \right| , \left| \frac{2^p \theta \chi_{n-1}}{2^{h'} g'} - 1 \right| < \frac{f + 9/8}{2^p} ,$$

where $g = \lceil 2^{p+3+t} d/e_{n-1} \rceil$, $g' = \lceil 2^{p+3+t'} d/e_{n-2} \rceil$, $h = k - t$, $h' = k - t'$, and $2^p < g, g' \leq 2^{p+1}$. Since $e_{n-1}/8d \geq 2^{k-w}$, we have $2^t > 2^{k-w}$; hence, $t > k - w$ and $h < w$. Since $e_{n-2}/8d < 2^{k-w}$, we have $2^{t'-1} < 2^{k-w}$. In this case, $t' - 1 < k - w$, $w < k - t' + 1$, and $h' \geq w$.

A.4 WNEAR

We can now consolidate these various observations to produce Algorithm WNEAR.

Algorithm 11.2: WNEAR

Input: $(\mathfrak{b}, d, k), w, p$, where $(\mathfrak{b}, d, k)$ is a reduced (f, p) representation of some $\mathcal{O}$-ideal $\mathfrak{a}$. Here $\mathfrak{b} = [Q/r, (P + \sqrt{D})/r]$, where $P + \lfloor \sqrt{D} \rfloor \geq Q$, $0 \leq \lfloor \sqrt{D} \rfloor - P \leq Q$.

Output: $(\mathfrak{c}, g, h)$ a w-near $(f + 9/8, p)$ representation of $\mathfrak{a}$. (Optional output: a reduced $(f + 9/8, p)$ representation $(\rho(\mathfrak{c}), g', h')$ of $\mathfrak{a}$.)

/*There are two possible cases.*/

1: **case 1:** $k < w$

2:      Find $s \in \mathbb{Z}^{\geq 0}$ such that $2^s Q \geq 2^{p+4}$. Put $Q_0 = Q$, $P_0 = P$, $M = \lceil 2^{p+s-k+w} Q_0/d \rceil$, $Q_{-1} = (D - P^2)/Q$, $T_{-2} = -2^s P_0 + \lfloor 2^s \sqrt{D} \rfloor$, $T_{-1} = 2^s Q_0$, $i = 1$.

3:      **while** $T_{i-2} \leq M$ **do**

4:          $q_{i-1} = \lfloor (P_{i-1} + \lfloor \sqrt{D} \rfloor)/Q_{i-1} \rfloor$

5:          $P_i = q_{i-1} Q_{i-1} - P_{i-1}$

6:          $Q_i = Q_{i-2} - q_{i-1}(P_i - P_{i-1})$

7:          $T_{i-1} = q_{i-1} T_{i-2} + T_{i-3}$

8:          $i \leftarrow i + 1$

9:      **end while**

10:     Put $e_{i-1} = \lceil 2^{p-s+3} T_{i-3}/Q_0 \rceil$

11:     **if** $de_{i-1} \leq 2^{2p-k+w+3}$ **then**

12:         Put $\mathfrak{c} = [Q_{i-2}/r, (P_{i-2} + \sqrt{D})/r]$, $e = e_{i-1}$.
            $(\rho(\mathfrak{c}) = [Q_{i-1}/r, (P_{i-1} + \sqrt{D})/r]$, $e' = \lceil 2^{p-s+3} T_{i-2}/Q_0 \rceil.)$

13:     **else**

14:         Put $\mathfrak{c} = [Q_{i-3}/r, (P_{i-3} + \sqrt{D})/r]$, $e = \lceil 2^{p-s+3} T_{i-4}/Q_0 \rceil$.
            $(\rho(\mathfrak{c}) = [Q_{i-2}/r, (P_{i-2} + \sqrt{D})/r]$, $e' = e_{i-1}.)$

15:     **end if**

16:     Find t (t') such that

$$2^t < \frac{ed}{2^{2p+3}} \leq 2^{t+1} \; . \quad \left(2^{t'} < \frac{e'd}{2^{2p+3}} \leq 2^{t'+1} \; . \right)$$

17:     Put

$$g = \left\lceil \frac{ed}{2^{p+t+3}} \right\rceil \; , \quad h = k + t \; . \quad \left(g' = \left\lceil \frac{e'd}{2^{p+t+3}} \right\rceil \; , \quad h' = k + t' \; . \right)$$

18: **end case**

19: **case 2:** $k \geq w$

20:     Put $s = p + 4$, $Q_0^* = Q$, $P_0^* = P$, $M^* = d2^{k-w+4}$, $Q_1^* = (D - P^2)/Q$, $T_{-2}^* = 2^s Q_0^*$, $T_{-1}^* = 2^s P_0^* + \lfloor 2^s \sqrt{D} \rfloor$, $i = 1$.

21:     **while** $T_{i-2}^* < Q_i^* M^*$ **do**

22:         $q_i^* = \lfloor (P_{i-1}^* + \lfloor \sqrt{D} \rfloor)/Q_i^* \rfloor$

23:         $P_i^* = q_i^* Q_i^* - P_{i-1}^*$

24:         $Q_{i+1}^* = Q_{i-1}^* - q_i^*(P_i^* - P_{i-1}^*)$

25:         $T_{i-1}^* = q_i^* T_{i-2}^* + T_{i-3}^*$

26:         $i \leftarrow i + 1$

27:     **end while**

28:     Put $q_i^* = \lfloor (P_{i-1}^* + \lfloor \sqrt{D} \rfloor)/Q_i^* \rfloor$, $P_i^* = q_i^* Q_i^* - P_{i-1}^*$, $e = \lceil T_{i-2}^*/2Q_i^* \rceil$, $e' = \lceil T_{i-3}^*/2Q_{i-1}^* \rceil$, $j = 3$.

29:     **while** $e' \geq d2^{k-w+3}$ **do**

30:         $e \leftarrow e'$

31:         $e' \leftarrow \lceil T_{i-2-j}^*/2Q_{i-j}^* \rceil$

32:                $j \leftarrow j + 1$
33:        **end while**
34:        Find t (t') such that

$$2^{t-1} \le \frac{e}{8d} < 2^t . \quad \left(2^{t'-1} \le \frac{e'}{8d} < 2^{t'} . \right)$$

35:        Put $\mathfrak{c} = [Q^*_{i-j+3}/r, (P^*_{i-j+3} + \sqrt{D})/r]$, $g = \lceil 2^{p+3+t}d/e \rceil$, $h = k - t$.
          $(\rho(\mathfrak{c}) = [Q^*_{i-j+2}/r, (P^*_{i-j+2} + \sqrt{D})/r]$, $g' = \lceil 2^{p+3+t}d/e' \rceil$, $h' = k - t'.)$
36: **end case**

In order to determine how quickly WNEAR will execute, we first note that in Case 1, we perform $O(i)$ steps where

$$T_{i-3} \le M < T_i .$$

In Case 2, we perform $O(i)$ steps where

$$T^*_{i-2} \ge Q^*_i M^* \text{ and } T^*_{i-3} < Q^*_{i-1} M^* .$$

From (A.12) and (A.13), we can easily establish by induction that

$$T_n > \tau^n T_{-1} , \quad T^*_n > \tau^n T^*_{-1} ,$$

where $\tau = (1 + \sqrt{5})/2$. Now, $T_{-1} = 2^s Q_0$ and $T^*_{-1} \ge \lfloor 2^s \sqrt{D} \rfloor + 2^s > 2^s \sqrt{D}$. Also, $M < 2^{p+3-k+w} Q_0/d + 1 < 2^s Q_0 2^{w-k} + 1$; hence, for Case 1 of WNEAR,

$$\tau^{i-3} < T_{i-3}/T_{-1} < 2^{w-k} + 1 .$$

Since $M^* < d2^{k-w+4} + 1$ and $Q^*_{i-1} < 2\sqrt{D}$, we get

$$\tau^{i-3} < T^*_{i-3}/T^*_{-1} < 2^{k-w+2} + 1$$

for Case 2. It follows from these observations that, in either case, $i = O(|k-w|)$ and that the number of steps needed to execute WNEAR is $O(|k - w|)$.

We compare T_{i-2} (T^*_{i-2}) to M $(Q^*_{i-1} M^*)$ instead of directly examining de $(e/8d)$ in order to avoid performing the divides needed to compute these numbers until we are close to the correct values. This is because on a computer, the division process is much more expensive than that for multiplication.

Because $\mathfrak{b} \sim \mathfrak{a}$, in WNEAR there must exist some $\kappa \in \mathbb{K}$ such that $N(\mathfrak{b})\kappa \in \mathcal{O}$ and $\mathfrak{c} = \kappa\mathfrak{b}$. In Chapter 12, it is necessary to have a technique for computing this κ. We can do this by making some very simple modifications to WNEAR, and we will call this extended algorithm EWNEAR. For our purposes it will only be necessary to consider Case 1. step 2 of this case of WNEAR should also initialize $B_{-2} = 1$, and $B_{-1} = 0$ and step 7 should also compute $B_{i-1} = q_{i-1} B_{i-2} + B_{i-3}$. In step 16, when $de_{i-1} \le 2^{2p-k+w+3}$, put $a = (T_{i-3} - \lfloor 2^s \sqrt{D} \rfloor B_{i-2})/2^s = G_{i-3}$, and $b = B_{i-3}$; otherwise put

$a = (T_{i-4} - \lfloor 2^s \sqrt{D} \rfloor B_{i-4})/2^s = G_{i-4}$, and $b = B_{i-4}$. Then $\kappa = (a + b\sqrt{D})/Q_0$. This follows very easily by recalling that

$$\left[\frac{Q_j}{r}, \frac{P_j + \sqrt{D}}{r} \right] = \frac{G_{j-1} + B_{j-1}\sqrt{D}}{Q_0} \left[\frac{Q_0}{r}, \frac{P_0 + \sqrt{D}}{r} \right].$$

Also, note that $\kappa > 1$ in this case.

Algorithm 12.1: EWNEAR

Input: $(\mathfrak{b}, d, k), w, p$, where $(\mathfrak{b}, d, k)$ is a reduced (f, p) representation of some $\mathcal{O}$-ideal $\mathfrak{a}$ and $k < w$. Here $\mathfrak{b} = [Q/r, (P + \sqrt{D})/r]$, where $P + \lfloor \sqrt{D} \rfloor \geq Q$, $0 \leq \lfloor \sqrt{D} \rfloor - P \leq Q$.

Output: $(\mathfrak{c}, g, h)$ a w-near $(f + 9/8, p)$ representation of $\mathfrak{a}$ and a, b, where $\kappa = (a + b\sqrt{D})/Q$ and $\mathfrak{c} = \kappa \mathfrak{b}$.

1: Put $B_{-2} = 1$, $B_{-1} = 0$.
2: Find $s \in \mathbb{Z}^{\geq 0}$ such that $2^s Q \geq 2^{p+4}$. Put $Q_0 = Q$, $P_0 = P$, $M = \lceil 2^{p+s-k+w} Q_0/d \rceil$, $Q_{-1} = (D - P^2)/Q$, $T_{-2} = -2^s P_0 + \lfloor 2^s \sqrt{D} \rfloor$, $T_{-1} = 2^s Q_0$, $i = 1$.
3: **while** $T_{i-2} \leq M$ **do**
4: $\quad q_{i-1} = \lfloor (P_{i-1} + \lfloor \sqrt{D} \rfloor)/Q_{i-1} \rfloor$
5: $\quad P_i = q_{i-1} Q_{i-1} - P_{i-1}$
6: $\quad Q_i = Q_{i-2} - q_{i-1}(P_i - P_{i-1})$
7: $\quad T_{i-1} = q_{i-1} T_{i-2} + T_{i-3}$
8: $\quad B_{i-1} = q_{i-1} B_{i-2} + B_{i-3}$
9: $\quad i \leftarrow i + 1$
10: **end while**
11: Put $e_{i-1} = \lceil 2^{p-s+3} T_{i-3}/Q_0 \rceil$
12: **if** $de_{i-1} \leq 2^{2p-k+w+3}$ **then**
13: $\quad$ Put $\mathfrak{c} = [Q_{i-2}/r, (P_{i-2} + \sqrt{D})/r]$, $e = e_{i-1}$, $a = (T_{i-3} - \lfloor 2^s \sqrt{D} \rfloor)/2^s$, $b = B_{i-3}$.
14: **else**
15: $\quad$ Put $\mathfrak{c} = [Q_{i-3}/r, (P_{i-3} + \sqrt{D})/r]$, $e = \lceil 2^{p-s+3} T_{i-4}/Q_0 \rceil$, $a = (T_{i-4} - \lfloor 2^s \sqrt{D} \rfloor)/2^s$, $b = B_{i-4}$.
16: **end if**
17: Find t such that

$$2^t < \frac{ed}{2^{2p+3}} \leq 2^{t+1}.$$

18: Put

$$g = \left\lceil \frac{ed}{2^{p+t+3}} \right\rceil, \quad h = k + t.$$

By employing somewhat similar ideas to those used in developing WNEAR, we can produce the FIND algorithm.

Algorithm 12.5: FIND

Input: $(\mathfrak{b}, d, k)$ a reduced (f, p) representation of a primitive $\mathcal{O}$-ideal $\mathfrak{a}$ with $\mathfrak{b} = \mathfrak{a}_i = [Q_{i-1}/r, (P_{i-1} + \sqrt{D})/r]$, where $(P_{i-1} + \sqrt{D})/Q_{i-1} > 1$, $-1 < (P_{i-1} - \sqrt{D})/Q_{i-1} < 0$; $\mathfrak{c} = \mathfrak{a}_j = [Q_{j-1}/r, (P_{j-1} + \sqrt{D})/r]$ with

$$i \le j \le i + c ,$$

where $c \in \mathbb{Z}^{\ge 0}$.

Output: $(\mathfrak{c}, g, h)$ a reduced $(f + 1/4, p)$ representation of $\mathfrak{a}$ (optional output $m, n \in \mathbb{Z}$, where $\lambda = (m + n\sqrt{D})/r \in \mathcal{O}$ and $N(\mathfrak{b})\mathfrak{c} = \lambda\mathfrak{b}$.)

1: Find $s \in \mathbb{Z}^{\ge 0}$ such that $2^s Q > 2^{p+4}$. Put $Q_0' = Q_{i-1}$, $P_0' = P_{i-1}$, $Q_{-1}' = (D - P_0'^2)/Q_0'$, $G_{-2}' = -P_0'$, $G_{-1}' = Q_0'$, $B_{-2}' = 1$, $B_{-1}' = 0$, $l = 0$.
2: **while** $Q_l' \ne Q_{j-1}$ or $P_l' \ne P_{j-1}$ **do**
3:      Put

$$q_l' = \left\lfloor \frac{P_l' + \lfloor \sqrt{D} \rfloor}{Q_l'} \right\rfloor ,$$

$$P_{l+1}' = q_l' Q_l' - P_l' ,$$
$$Q_{l+1}' = Q_{l-1}' - q_l'(P_{l+1}' - P_l') ,$$
$$B_l' = q_l' B_{l-1}' + B_{l-2}' ,$$
$$G_l' = q_l' G_{l-1}' + G_{l-2}' ,$$
$$l \leftarrow l + 1 .$$

4: **end while**
5: Put $\mathfrak{c} = [Q_l'/r, (P_l' + \sqrt{D})/r]$.
6: Put $T = 2^s G_{l-1}' + B_{l-1}' \lfloor 2^s \sqrt{D} \rfloor$.
7: Put $e = \lceil 2^{p+3-s} T/Q_0' \rceil$ and find t such that

$$2^t < \frac{ed}{2^{2p+3}} \le 2^{t+1} .$$

8: Put $g = \lceil ed/2^{p+t+3} \rceil$, $h = k + t$. ($m = G_{l-1}'$, $n = B_{l-1}'$.)

Proof (of correctness of FIND). If we define

$$\theta_q' = \frac{G_{q-2}' + \sqrt{D}B_{q-2}'}{Q_0'} \quad (q = 0, 1, 2, \dots) ,$$

then after step 3 has executed we have

$$\mathfrak{c} = \mathfrak{a}_j = \left[\frac{Q_l'}{r}, \frac{P_l' + \sqrt{D}}{r} \right] = \theta_{l+1}' \mathfrak{a}_i = \left(\frac{\lambda}{N(\mathfrak{b})} \right) \mathfrak{b} . \tag{A.16}$$

By Lemmas A.5 and A.6, we know that $(\mathfrak{c}, g, h)$ is a reduced $(f + 1/4, p)$ representation of $\mathfrak{a}$. Clearly, the complexity of this algorithm is $O(c)$ elementary operations.

We remark here that since FIND searches for the least value of l for which (A.16) holds, we must have

$$1 \leq \frac{\lambda}{N(\mathfrak{b})} < \epsilon_\Delta ,$$

where ϵ_Δ is the fundamental unit of $\mathcal{O}$.

Algorithm FIND will work resaonably well as long as the values of A'_l and B'_l do not become very large (i.e., c is not very large). In the case of larger values of c, we need to modify the algorithm. As before, we let $\mathfrak{a} = [Q/r, (P + \sqrt{D})/r]$ be any reduced $\mathcal{O}$-ideal and let

$$\mathfrak{a}_1 (= \mathfrak{a}), \mathfrak{a}_2, \mathfrak{a}_3, \ldots, \mathfrak{a}_j, \ldots$$

denote the sequence of reduced ideals equivalent to $\mathfrak{a}$ produced from the SCF expansion of $(P + \sqrt{D})/Q$. Here, as usual, $\mathfrak{a}_j = [Q_{j-1}/r, (P_{j-1} + \sqrt{D})/r]$. Now, let $(\mathfrak{a}_j, d_j, k_j)$ denote a reduced (f_j, p) representation of $\mathfrak{a}$ and put $E = \lceil 2^p \sqrt{D} \rceil$. We now need the following simple theorem.

Theorem A.7. *If we put*

$$e_i = \left\lceil \frac{2^p P_i + E}{Q_{i-1}} \right\rceil , \quad d'_i = \left\lceil \frac{e_i d_i}{2^p} \right\rceil$$

and find t such that

$$2^{p+t} < d'_i \leq 2^{p+t+1} ,$$

then $(\mathfrak{a}_{i+1}, d_{i+1}, k_{i+1})$ is a reduced (f_{i+1}, p) representation of $\mathfrak{a}$, where

$$d_{i+1} = \left\lceil \frac{d'_i}{2^t} \right\rceil = \left\lceil \frac{e_i d_i}{2^{p+t}} \right\rceil , \quad k_{i+1} = k_i + t , \quad f_{i+1} < f_i + 2 .$$

Proof. We recall that $\mathfrak{a}_i = \theta_i \mathfrak{a}_1$, $\mathfrak{a}_{i+1} = \theta_{i+1} \mathfrak{a}_1$, and $\theta_{i+1} = \psi_i \theta_i$. Also, $e_i > 2^p \psi_i > 2^p$ and, therefore, $d'_i > 2^p$, $t \geq 0$. That

$$2^p < d_{i+1} \leq 2^{p+1}$$

follows easily from the definition of d_{i+1} and t. Since

$$\left| \frac{2^p \theta_i}{2^{k_i} d_i} - 1 \right| < \frac{f_i}{2^p}, \quad \left| \frac{2^p \psi_i}{e_i} - 1 \right| < \frac{1}{2^p} ,$$

and

$$\frac{e_i d_i}{2^{p+t}} = d_{i+1} - \eta \quad (0 \leq \eta < 1) ,$$

we get

$$\left(1 - \frac{\eta}{d_{i+1}} \right) \left(1 - \frac{f_i}{2^p} \right) \left(1 - \frac{1}{2^p} \right) < \frac{2^p \theta_{i+1}}{2^{k_{i+1}} d_{i+1}} < \left(1 + \frac{f_i}{2^p} \right) \left(1 + \frac{1}{2^p} \right) .$$

Hence,

$$1 - \frac{f_i + 2}{2^p} < \frac{2^p \theta_{i+1}}{2^{k_{i+1}} d_{i+1}} < 1 + \frac{f_i + 2}{2^p} .$$

$\square$

We are now able to present algorithm FPCF. This algorithm is a version of FIND for larger values of c than those used in the FIND algorithm, but it obtains its output at the expense of less precision. Its proof of correctness follows by induction from Theorem A.7.

Algorithm A.2: FPCF

Input: p and $(\mathfrak{b}, d_i, k_i)$ a reduced (f, p) representation of a primitive $\mathcal{O}$-ideal $\mathfrak{a}$ with $\mathfrak{b} = \mathfrak{a}_i = [Q_{i-1}/r, (P_{i-1} + \sqrt{D})/r]$, where $(P_{i-1} + \sqrt{D})/Q_{i-1} > 1$, $-1 < (P_{i-1} - \sqrt{D})/Q_{i-1} < 0$; $\mathfrak{c} = \mathfrak{a}_j = [Q_{j-1}/r, (P_{j-1} + \sqrt{D})/r]$ with $j = i + c$, where $c \in \mathbb{Z}^{\geq 0}$.

Output: $(\mathfrak{c}, d_j, k_j)$ a reduced $(f + 2c, p)$ representation of $\mathfrak{a}$.

1: Put $Q_0' = Q_{i-1}$, $P_0' = P_{i-1}$, $Q_{-1}' = (D - P_0'^2)/Q_0'$, $d_0' = d_i$, $k_0' = k_i$, $l = 0$, $E = \lceil 2^p \sqrt{D} \rceil$.

2: **while** $Q_l' \neq Q_{j-1}$ or $P_l' \neq P_{j-1}$ **do**

3:     Put

$$
q_l' = \left\lfloor \frac{P_l' + \lfloor \sqrt{D} \rfloor}{Q_l'} \right\rfloor ,
$$

$$
P_{l+1}' = q_l' Q_l' - P_l' ,
$$

$$
Q_{l+1}' = Q_{l-1}' - q_l'(P_{l+1}' - P_l') ,
$$

$$
e_l' = \left\lceil \frac{2^p P_{l+1}' + E}{Q_l'} \right\rceil ,
$$

$$
d = \left\lceil \frac{e_l' d_l'}{2^p} \right\rceil .
$$

4:     Compute t such that

$$
2^{p+t} < d < 2^{p+t+1} .
$$

5:     Put

$$
d_{l+1}' = \lceil d/2^t \rceil , \quad k_{l+1}' = k_l' + t .
$$

6:     $l \leftarrow l + 1$

7: **end while**

8: Put

$$
\mathfrak{c} = \left[\frac{Q_l'}{r}, \frac{P_l' + \sqrt{D}}{r} \right] , \quad d_j = d_l' , \quad k_j = k_l' .
$$

References

[AAC52]   N. C. Ankeny, E. Artin, and S. Chowla, *The class-number of real quad-ratic number fields*, Annals of Math. **56** (1952), no. 3, 479–493.

[Abe94]   C. S. Abel, *Ein Algorithmus zur Berechnung der Klassenzahl und des Regulators reellquadratischer Ordnungen*, Ph.D. thesis, Universität des Saarlandes, Saarbrücken, Germany, 1994.

[ABR01]   M. Abdalla, M. Bellare, and P. Rogaway, *DHIES: An encryption scheme based on the Diffie-Hellman problem*, Topics in Cryptology CT-RSA 2001, Lecture Notes in Computer Science, vol. 2020, Springer, Berlin, 2001, pp. 143–158.

[AC60]   N. C. Ankeny and S. Chowla, *A note on the class number of real quadratic fields*, Acta Arith. **6** (1960), 145–147.

[AC62]   ———, *A further note of the class number of real quadratic fields*, Acta Arith. **7** (1962), 271–272.

[ACL91]   American Council of Learned Societies, *Bibliographical Dictionary of Mathematics*, Vol. 4, Scribner's Sons, New York, 1991.

[Ada79]   W. W. Adams, *On the relationship between the convergents of the nearest integer and regular continued fractions*, Math. Comp. **33** (1979), 1321–1331.

[AEM87]   L. Adelman, D. Estes, and K. McCurley, *Solving bivariate quadratic congruences in random polynomial time*, Math. Comp. **48** (1987), 17–28.

[AHU74]   A. V. Aho, J. E. Hopcroft, and J. D. Ullman, *The Design and Analysis of Computer Algorithms*, Addison-Wesley, Reading, MA, 1974.

[AKS04]   M. Agrawal, N. Kayal, and N. Saxena, *PRIMES is in P*, Annals of Math **160** (2004), 781–793.

[Amt80]   A. Amthor, *Das Problema bovinum des Archimedes*, Zeitschrift für Math. u. Physik (Hist. Litt. Abtheilung) **25** (1880), 153–171.

[AP95]   W. R. Alford and C. Pomerance, *Implementing the self-initializing quadratic sieve on a distributed network*, Proceedings of International Conference "Number Theoretic and Algebraic Methods in Computer Science" (Moscow, 1993) (A. J. van der Poorten, I. Shparlinski, and H. G. Zimmer, eds.), World Scientific, Singapore, 1995, pp. 163–174.

[Apo98]   T. Apostol, *Introduction to Analytic Number Theory*, Springer, New York, 1998.

[Arc71]   Archimède, *Oeuvres, 3 vol.*, texte établi et traduit par C. Mugler, Les Belles Lettres, Paris, 1970–71.

[Arc99]   Archimedes, *The Cattle Problem*, in English verse by S. J. P. Hillion & H. W. Lenstra Jr., Mercator, Santpoort, 1999.

[Arn92]   S. Arno, *The imaginary quadratic fields of class number 4*, Acta Arithmetica **60** (1992), 321–334.

[ARW98]   S. Arno, M. Robinson, and F. Wheeler, *Imaginary quadratic fields with small odd class number*, Acta Arithmetica **83** (1998), 295–330.

[AW04]   Ş. Alaca and K. S. Williams, *Introductory Algebraic Number Theory*, Cambridge University Press, Cambridge, 2004.

[Ayo63]   R. Ayoub, *An Introduction to the Analytic Theory of Numbers*, American Mathematical Society, Providence, RI, 1963.

[Ayy40]   A. A. Krishnaswami Ayyangar, *Theory of the nearest square continued fraction*, J. Mysore Univ. Sect A. **1** (1940), 21–32, (1941), 97–117.

[Azu84]   T. Azuhata, *On the fundamental units and the class numbers of real quadratic fields*, Nagoya Math. J. **95** (1984), 125–135.

[Azu87]   ———, *On the fundamental units and the class numbers of real quadratic fields II*, Tokyo J. Math **10** (1987), no. 2, 259–270.

[Bac90]   E. Bach, *Explicit bounds for primality testing and related problems*, Math. Comp. **55** (1990), no. 191, 355–380.

[Bac95]   ———, *Improved Approximations for Euler Products*, Number Theory: CMS Proc. Vol. 15, American Mathematical Society, Providence, RI, 1995, pp. 13–28.

[Bak66]   A. Baker, *Linear forms in the logarithms of algebraic numbers*, Mathematika **13** (1966), 204–216.

[Bak71]   ———, *Imaginary quadratic fields with class number two*, Annals of Math **94** (1971), 139–152.

[Bar03]   E. J. Barbeau, *Pell's Equation*, Springer, New York, 2003.

[BB94]   I. Biehl and J. Buchmann, *Algorithms for quadratic orders*, Proceedings of Symposium in Applied Mathematics Vol. 48, American Mathematical Society, Providence, RI, 1994, pp. 425–451.

[BB97]   ———, *An analysis of the reduction algorithms for binary quadratic forms*, Tech. Report No. TI-26/97, Technische Universität Darmstadt, 1997.

[BB98]   ———, *An analysis of the reduction algorithm for binary quadratic forms*, Voronoi's Impact on Modern Science (P. Engel and H. Syta, eds.) Vol. 1, Institute of Mathematics of National Academy of Sciences, Kyiv, Ukraine, 1998, pp. 71–98.

[BBFK05]   F. Bahr, M. Boehm, J. Franke, and T. Kleinjung, *RSA-200*, Email announcement, 2005, http://www.crypto-world/com/announcements/rsa200.txt.

[BBHM02]   I. Biehl, J. Buchmann, S. Hamdy, and A. Meyer, *A signature scheme based on the intractability of extracting roots*, Designs, Codes and Cryptography **25** (2002), 223–236.

[BBT95]   I. Biehl, J. Buchmann, and C. Thiel, *Cryptographic protocols based on discrete logarithms in real-quadratic orders*, Advances in Cryptology — CRYPTO '94, Lecture Notes in Computer Science Vol. 839, Springer, Berlin, 1995, pp. 56–60.

[BC70]   P. Barrucand and H. Cohn, *A rational genus, class number divisibility, and unit theory for pure cubic fields*, J. Number Theory **2** (1970), 7–21.

[BCMO06]   M. A. Bennett, M. Cipu, M. Mignotte, and R. Okazaki, *On the number of solutions of simultaneous Pell equations II*, Acta Arith. **122** (2006), 407–417.

[BD91a]   J. Buchmann and S. Düllmann, *On the computation of discrete logarithms in class groups*, Advances in Cryptology — CRYPTO '90, Lecture Notes in Computer Science Vol. 537, Springer, Berlin, 1991, pp. 134–139.

[BD91b]   ———, *A probabilistic class group and regulator algorithm and its implementation*, Computational Number Theory, Walter de Gruyter & Co., New York, 1991, pp. 53–72.

[BD92]   ———, *Distributed class group computation*, Festschrift aus Anlaß des sechzigsten Geburtstages von Herrn Prof. Dr. G. Hotz, Universität des Saarlandes, 1991 / and Teubner, Stuttgart, 1992, pp. 69–79.

[BDW90]   J. Buchmann, S. Düllmann, and H. C. Williams, *On the complexity and efficiency of a new key exchange system*, Advances in Cryptology - EUROCRYPT '89, Lecture Notes in Computer Science Vol. 434, Springer, Berlin, 1990, pp. 597–616.

[Bei64]   A. H. Beiler, *Recreations in the Theory of Numbers*, Dover, New York, 1964.

[Ben01]   M. A. Bennett, *Rational approximation to algebraic numbers of small height: The Diophantine equation $|ax^n - by^n| = 1$*, J. Reine Angew Math. **535** (2001), 1–49.

[Ben05]   ———, *Powers in recurrence sequences: Pell equations*, Trans. Amer. Math. Soc. **357** (2005), 1675–1691.

[Ber]   D. Bernstein, *How to find small factors of integers*, Math. Comp., to appear.

[Ber28]   W. E. H. Berwick, *The arithmetic of quadratic number-fields*, Math. Gazette **14** (1928), 1–11.

[Ber76a]   L. Bernstein, *Fundamental units and cycles I*, J. of Number Theory **8** (1976), no. 4, 446–491.

[Ber76b]   ———, *Fundamental units and cycles in the period of real quadratic numbers fields I, II*, Pac. J. Math. **63** (1976), 37–61, 63–78.

[BF03]   D. Boneh and M. Franklin, *Identity based encryption from the Weil pairing*, SIAM Journal of Computing **32** (2003), no. 3, 586–615.

[BG07]   A. Biró and A. Granville, *Zeta functions for ideal classes in real quadratic fields, at $s = 0$*, submitted for publication, 2007.

[BH01]   J. Buchmann and S. Hamdy, *A survey on IQ-cryptography*, Public-Key Cryptography and Computational Number Theory, de Gruyter, Berlin, 2001, pp. 1–15.

[BH03]   M. Bauer and S. Hamdy, *On class group computations using the number field sieve (extended abstract)*, Advances in Cryptology - ASIACRYPT 2003, Lecture Notes in Computer Science Vol. 2894, Springer, Berlin, 2003, pp. 311–325.

[BH96]   J. Buchmann and C. S. Hollinger, *On smooth ideals in number fields*, J. Number Theory **59** (1996), no. 1, 82–87.

[Bil04]   Y. F. Bilu, *Catalan's conjecture (after Mihailescu)*, Astérisque (2004), no. 294, 1–26.

[Bil05]   ———, *Catalan without logarithmic forms (after Bugeaud, Hanrot and Mihailescu)*, J. Théor. Nombres Bordeaux **17** (2005), 69–85.

[Bir03a]   A. Biró, *Chowla's conjecture*, Acta Arith. **107** (2003), 178–194.

[Bir03b]   _____, *Yokoi's conjecture*, Acta Arith. **106** (2003), 85–104.

[BJN00]   D. Boneh, A. Joux, and P. Nguyen, *Why textbook El Gamal and RSA encryption are insecure*, Advances in Cryptology - ASIACRYPT 2000, Lecture Notes in Computer Science Vol. 1976, Springer, Berlin, 2000, pp. 30–44.

[BJP94]   J. Buchmann, M. Jüntgen, and M. Pohst, *A practical version of the generalized Lagrange algorithm*, Exper. Math. **3** (1994), 200–207.

[BJT97]   J. Buchmann, M. J. Jacobson, Jr., and E. Teske, *On some computational problems in finite abelian groups*, Math. Comp. **66** (1997), no. 220, 1663–1687.

[BM98]   J. Buchmann and M. Maurer, *Approximate evaluation of $L(1, \chi_d)$*, Tech. Report TI-6/98, Department of Computer Science, Technical University of Darmstadt, Darmstadt, Germany, 1998.

[BMM00]   J. Buchmann, M. Maurer, and B. Möller, *Cryptography based on number fields with large regulator*, Journal de Théorie des Nombres de Bordeaux **12** (2000), 293–307.

[Bon99]   D. Boneh, *Twenty years of attacks on the RSA cryptosystem*, Notices of the AMS **46** (1999), no. 2, 203–213.

[Boo06]   A. Booker, *Quadratic class numbers and character sums*, Math. Comp. **75** (2006), no. 255, 1481–1492.

[BP97]   J. Buchmann and S. Paulus, *A one way function based on ideal arithmetic in number fields*, CRYPTO '97, Lecture Notes in Computer Science Vol. 1294, Springer, Berlin, 1997, pp. 385–394.

[BPT04]   I. Biehl, S. Paulus, and T. Takagi, *Efficient undeniable signatures based on ideal arithmetic in quadratic orders*, Designs, Codes and Cryptography **31** (2004), 99–123.

[Bre00]   R. P. Brent, *Recent progress and prospects for integer factorization algorithms*, Computing and Combinatorics, Lecture Notes in Computer Science Vol. 1858, Springer-Verlag, Berlin, 2000, pp. 3–22.

[Bre76]   _____, *Fast multiple-precision evaluation of elementary functions*, Journal of the ACM **23** (1976), 242–251.

[Bre80]   C. Brezinski, *History of Continued Fractions and Padé Approximants*, Springer-Verlag, New York, 1980.

[BS05]   J. Buchmann and A. Schmidt, *Computing the structure of a finite abelian group*, Math. Comp. **74** (2005), 2017–2026.

[BS07]   D. Bernstein and J. Sorenson, *Modular exponentiation via the explicit Chinese Remainder Theorem*, Math. Comp. **76** (2007), 443–454.

[BS93]   E. Bach and J. Sorenson, *Sieve algorithms for perfect power testing*, Algorithmica **9** (1993), 313–328.

[BS96a]   E. Bach and J. O. Shallit, *Algorithmic Number Theory*, MIT Press, Cambridge, MA, 1996.

[BS96b]   W. Bosma and P. Stevenhagen, *Density computations for real quadratic units*, Math. Comp. **65** (1996), no. 215, 1327–1337.

[BS96c]   _____, *On the computation of quadratic 2-class groups*, Journal de Théorie des Nombres de Bordeaux **8** (1996), no. 2, 283–313.

[BS97]   _____, *Erratum: On the computation of quadratic 2-class groups*, Journal de Théorie des Nombres de Bordeaux **9** (1997), no. 1, 249.

[Bsh99]   N. H. Bshouty, *Lower bounds for the complexity of functions in a realistic RAM model*, J. of Algorithms **32** (1999), 1–20.

[BST02]   J. Buchmann, K. Sakurai, and T. Takagi, *An IND-CCA2 public-key cryptosystem with fast decryption*, Information Security and Cryptology — ICISC 2001, Lecture Notes in Computer Science Vol. 2288, Springer, Berlin, 2002, pp. 51–71.

[BT02]   I. Biehl and T. Takagi, *A new distributed primality test for shared RSA keys using quadratic fields*, ACISP 2002, Lecture Notes in Computer Science Vol. 2384, Springer, Berlin, 2002, pp. 1–16.

[BTV04]   J. Buchmann, T. Takagi, and U. Vollmer, *Number field cryptography*, High Primes and Misdemeanors: Lectures in Honour of the 60th Birthday of Hugh Cowie Williams, Fields Institute Communications Vol. 41, American Mathematical Society, Providence, RI, 2004, pp. 111–125.

[BTW95]   J. Buchmann, C. Thiel, and H. C. Williams, *Short representation of quadratic integers.*, Computational Algebra and Number Theory, Mathematics and its Applications 325, Kluwer Academic Publishers, Amsterdam, 1995, pp. 159–185.

[Buc87a]   J. Buchmann, *On the computation of units and class numbers by a generalization of Lagrange's algorithm*, Journal of Number Theory **26** (1987), 8–30.

[Buc87b]   ———, *Zur Komplexität der Berechnung von Einheiten und Klassenzahlen algebraischer Zahlkörper*, Habilitationsschrift, Universität Düsseldorf, 1987.

[Buc90]   ———, *A subexponential algorithm for the determination of class groups and regulators of algebraic number fields*, Séminaire de Théorie des Nombres, Paris 1988–1989, Progress in Mathematics Vol. 91, Birkhäuser, Boston, 1990, pp. 27–41.

[Bue76]   D. A. Buell, *Class groups of quadratic fields*, Math. Comp. **30** (1976), no. 135, 610–623.

[Bue87]   ———, *Class groups of quadratic fields. II*, Math. Comp. **48** (1987), no. 177, 85–93.

[Bue89]   ———, *Binary quadratic forms, Classical Theory and Modern Computations*, Springer, New York, 1989.

[Bue99]   ———, *The last exhaustive computation of class groups of complex quadratic number fields*, Number theory, CRM Proceedings and Lecture Notes Vol. 19, American Mathematical Society, Providence, RI, 1999, pp. 35–53.

[Bur57]   D. A. Burgess, *The distribution of quadratic residues and non-residues*, Mathematika **4** (1957), 106–112.

[BV06]   J. Buchmann and U. Vollmer, *A Terr algorithm for computations in the infrastructure of real-quadratic number fields*, Journal de Théorie des Nombres de Bordeaux **18** (2006), no. 3, 559–572.

[BV07]   ———, *Binary Quadratic Forms: An Algorithmic Approach*, Algorithms and Computation in Mathematics Vol. 20, Springer-Verlag, Berlin, 2007.

[BV99]   D. Boneh and R. Venkatesan, *Breaking RSA may not be equivalent to factoring*, Advances in Cryptology - EUROCRYPT '98, Lecture Notes in Computer Science Vol. 1233, Springer-Verlag, Berlin, 1999, pp. 59–71.

[BW72]   B. D. Beach and H. C. Williams, *A numerical investigation of the Diophantine equation* $x^2 - dy^2 = -1$, Proc. 3rd S.-E. Conf. Combinatories, Graph Theory and Computing, 1972, pp. 37–52.

[BW88a]   J. Buchmann and H. C. Williams, *A key exchange algorithm based on imaginary quadratic fields*, J. Cryptology **1** (1988), 107–118.

[BW88b]   _____, *On the infrastructure of the principal ideal class of an algebraic number field of unit rank one*, Math. Comp. **50** (1988), no. 182, 569–579.

[BW89a]   _____, *A key-exchange system based on real quadratic fields*, CRYPTO '89, Lecture Notes in Computer Science, Vol. 435, Springer, Berlin, 1989, pp. 335–343.

[BW89b]   _____, *On the existence of a short proof for the value of the class number and regulator of a real quadratic field*, Proc. NATO ASI on Number Theory and Applications (R. A. Mollin, ed.), Kluwer Academic Press, Amsterdam, 1989, pp. 327–345.

[BW90]   _____, *Quadratic fields and cryptography*, Number Theory and Cryptography, London Math. Soc. Lecture Note Series **154** (1990), 9–26.

[BW91]   _____, *Some remarks concerning the complexity of computing class groups of quadratic fields*, Journal of Complexity **7** (1991), 311–315.

[BW94]   N. Buck and K. Williams, *Comparison of the lengths of the continued fractions of $\sqrt{D}$ and $\frac{1+\sqrt{D}}{2}$*, Proc. Amer. Math. Soc. **120** (1994), no. 4, 992–1002.

[Car53]   L. Carlitz, *Note on the class number of real quadratic fields*, Proc. Amer. Math. Soc. **4** (1953), 535–537.

[CDyDO93]   H. Cohen, F. Diaz y Diaz, and M. Olivier, *Calculs de nombres de classes et de régulateurs de corps quadratiques en temps sous-exponentiel*, Séminaire de Théorie des Nombres, Paris 1990–1991, Progress in Mathematics Vol. 108, Birkhäuser, 1993, pp. 35–46.

[CDyDO97]   H. Cohen, F. Diaz y Diaz, and M. Olivier, *Subexponential algorithms for class and unit group computations*, J. Symb. Comp. **24** (1997), 433–441.

[CDyDO98]   _____, *Computing ray class groups, conductors, and discriminants*, Math. Comp. **67** (1998), no. 222, 773–795.

[CF76]   S. Chowla and J. Friedlander, *Class numbers and quadratic residues*, Glasgow Math. J. **17** (1976), 47–52.

[CG90]   B. Chor and O. Goldreich, *An improved parallel algorithm for integer gcd*, Algorithmica **5** (1990), 1–10.

[Che03]   K. H. F. Cheng, *Some results concerning periodic continued fractions*, Ph.D. thesis, University of Calgary, Calgary, 2003.

[Che94]   J. H. Chen, *A new solution of the Diophantine equation $x^2 + 1 = 2y^4$*, J. Number Theory **48** (1994), no. 1, 62–74.

[Cho49]   S. Chowla, *Improvement of a theorem of Linnik and Walfisz*, Proc. London Math. Soc. **50** (1949), 423–429.

[Chr61]   G. Chrystal, *Algebra, Part II*, Dover Publications, New York, 1961.

[CL83]   H. Cohen and H. W. Lenstra, Jr., *Heuristics on class groups of number fields*, Number Theory, Lecture Notes in Mathematics Vol. 1068, Springer-Verlag, New York, 1983, pp. 33–62.

[CL84]   _____, *Heuristics on class groups*, Number Theory, Noordwijkerhout, 1983, Lecture Notes in Mathematics Vol. 1052, Springer-Verlag, New York, 1984, pp. 26–36.

[CM05]   A. C. Cojocaru and M. R. Murty, *An Introduction to Sieve Methods and their Application*, Cambridge University Press, Cambridge, 2005.

[CM07]   M. Cipu and M. Mignotte, *On the number of solutions to systems of Pell equations*, Journal of Number Theory **125** (2007), 356–392.

[CM87]   H. Cohen and J. Martinet, *Class groups of number fields: Numerical heuristics*, Math. Comp. **48** (1987), no. 177, 123–137.

[Coh62]   H. Cohn, *A Second Course in Number Theory*, John Wiley and Sons, New York, 1962.

[Coh93]   H. Cohen, *A course in Computational Algebraic Number Theory*, Springer-Verlag, Berlin, 1993.

[Col17]   H. T. Colbrooke, *Algebra with Arithmetic and Mensuration from the Sanscrit of Brahmegupta and Bhascara*, John Murray, London, 1817.

[Con03]   J. B. Conrey, *The Riemann hypothesis*, Notices of the AMS **50** (2003), 341–35.

[Con97]   S. Contini, *Factoring integers with the self-initializing quadratic sieve*, Master's thesis, University of Georgia, Athens, Georgia, 1997.

[Cor08]   G. Cornacchia, *Sur di un metodo per la risoluzione in numeri interi dell' equazione $\sum_{h=0}^{n} C_h x^{n-h} y^h = P$*, Giornale di Matematiche di Battaglini **46** (1908), 33–90.

[Cox89]   D. A. Cox, *Primes of the form $x^2 + ny^2$*, John Wiley and Sons, New York, 1989.

[CP05]   R. Crandall and C. Pomerance, *Prime numbers: A computational perspective*, 2nd ed., Springer-Verlag, New York, 2005.

[CW05]   K. Cheng and H. Williams, *Some results concerning certain periodic continued fractions*, Acta. Arith. **117** (2005), no. 3, 247–264.

[Dav00]   H. Davenport, *Multiplicative Number Theory*, rev. ed., Springer-Verlag, New York, 2000.

[Dav60]   H. Davenport, *The Higher Arithmetic*, Harper and Brothers, New York, 1960.

[Ded96]   R. Dedekind, *Theory of Algebraic Integers*, J. Stillwell (trans.), Cambridge University Press, Cambridge, 1996.

[Deg58]   G. Degert, *Über die Bestimmung der Grundeinheit gewisser reellquadratischer Zahlkörper*, Math. Sem. Univ. Hamburg **22** (1958), 92–97.

[deM88]   I. G. deMille, *The continued fraction of $(1 + \sqrt{D})/2$ for certain infinite classes of D with applications to units and class numbers*, Master's thesis, Carleton University, Ottawa, 1988.

[DF04]   D. S. Dummit and R. M. Foote, *Abstract algebra*, 3rd ed., John Wiley and Sons, New York, 2004.

[DF64]   B. N. Delone and D. K. Faddeev, *The Theory of Irrationalities of the Third Degree*, Translations of Mathematical Monographs Vol. 10, American Mathematical Society, Providence, RI, 1964.

[dH04]   R. de Haan, *A fast, rigorous technique for verifying the regulator of a real quadratic field*, Master's thesis, University of Amsterdam, 2004.

[DH76]   W. Diffie and M. Hellman, *New directions in cryptography*, IEEE Transactions on Information Theory **22** (1976), 472–492.

[dHJW07]   R. de Haan, M. J. Jacobson, Jr., and H. C. Williams, *A fast, rigorous technique for computing the regulator of a real quadratic field*, Math. Comp. **76** (2007), no. 260, 2139–2160.

[Dic19]   L. E. Dickson, *History of the Theory of Numbers*, Carnegie Institution of Washington, Publication No. 256, 1919; Dover Publications, New York, 2005.

[Dij87]   E. J. Dijksterhuis, *Archimedes*, Princeton University Press, Princeton, NJ, 1987.

[Dir93]   P. G. L. Dirichlet, *Vorlesungen über zahlentheorie*, 4th ed., Vieweg, Braunschweig, 1893.

[Dir99]   _____, *Lectures on Number Theory*, American Mathematical Society, Providence, RI, 1999.

[DK02]   I. Damgård and M. Koprowski, *Generic lower bounds for root extraction and signature schemes in general groups*, Advances in Cryptology - EUROCRYPT 2002, Lecture Notes in Computer Science Vol. 2332, Springer, Berlin, 2002, pp. 256–271.

[DL06]   C. Doche and T. Lange, *Arithmetic of elliptic curves*, Handbook of Elliptic and Hyperelliptic Curve Cryptography (H. Cohen and G. Frey, eds.), Chapman and Hall/CRC, New York, 2006, pp. 267–302.

[DPLR91]   E. Dubois and R. Paysant-Le Roux, *Sur la longueur du développement en fraction continue de $\sqrt{f(n)}$*, Journées Arithmétiques, 1989 (Luminy, 1989), Astérisque Vol. 198–200, 1991, pp. 107–119 (1992).

[DS62]   B. Datta and A. N. Singh, *History of Hindu Mathematics*, Asia Publishing House, Bombay, 1962, Part II.

[Duj94]   M. Dujardin, *Sur une erreur relevée dans la "Théorie des nombres" de Legendre*, Computes Rendus, Acad. des Sciences, Paris **119** (1894), 843–844.

[Dül91]   S. Düllmann, *Ein Algorithmus zur Bestimmung der Klassengruppe positiv definiter binärer quadratischer Formen*, Ph.D. thesis, Universität des Saarlandes, Saarbrücken, Germany, 1991.

[Edw05]   H. M. Edwards, *Essays in Constructive Mathematics*, Springer, New York, 2005.

[Edw07]   _____, Composition of binary quadratic forms and the foundations of mathematics, Springer, New York, 2007.

[Edw74]   _____, *Riemann's zeta function*, Academic Press, New York, 1974.

[EGG⁺06]   W. Eberly, M. Giesbrecht, P. Giorgi, A. Storjohann, and G. Villard, *Solving sparse rational linear systems*, International Symposium on Symbolic and Algebraic Computation (ISSAC'06) (Genova, Italy), ACM Press, New York, 2006, pp. 63–70.

[EGG⁺07]   _____, *Faster inversion and other black box matrix computations using efficient block projections*, International Symposium on Symbolic and Algebraic Computation (ISSAC'07) (Waterloo, Ontario, Canada), ACM Press, New York, 2007, pp. 143–150.

[Eis44a]   G. Eisenstein, *Allgemeine Untersuchengen über dir Formen dritten Grades mit drei variabeln, welche der Kreistheilung ihre Enstehung verdanken*, J. Reine Angew. Math. **28** (1844), 289–374.

[Eis44b]   _____, *Aufgaben*, J. Reine Angew. Math. **27** (1844), 86–87.

[Ell69]   P. D. T. A. Elliott, *On the size of $L(1,\chi)$*, J. Reine Angew. Math. **236** (1969), 26–36.

[Ell70]   _____, *The distribution of the quadratic class number*, Litovsk. Math. Sb. **10** (1970), 189–197.

[Ell73]   _____, *On the distribution of the values of quadratic L-series in the half-plane $\sigma > 1/2$*, Invent. Math. **21** (1973), 319–338.

[Ell80]   _____, *Probabilistic Number Theory II*, Springer-Verlag, Berlin, 1980.

[Ell87]   J. H. Ellis, *The History of Non-Secret Encryption*, CESG, 1987, Declassified and available at http://www.cseg.gov.uk/site/publications/media/ellis.pdf.

[Ene02]   G. Eneström, *Über der Ursprung der Benennung "Pellsche Gleichung,"* Bibliotheca Math. **3** (1902), pp. 204–207.

[ET02]   P. Ebinger and E. Teske, *Factoring $N = pq^2$ with the Elliptic Curve Method*, Algorithmic Number Theory - ANTS-V (Sydney, Australia), Lecture Notes in Computer Science Vol. 2369, Springer-Verlag, Berlin, 2002, pp. 475–490.

[Eul43]   L. Euler, *Correspondence, Mathématique et Physique* P. H. Fuss (ed.), Vol. T1, Imperial Academy of Science, St. Petersbourg, 1843, pp. 35–39.

[Far94]   A. Farhane, *Minoration de le période du développement de $\sqrt{a^2n^2 + bn + c}$ en fraction continue*, Acta. Arith. **67** (1994), no. 1, 63–67.

[Fer12]   P. Fermat, *Oeuvres de fermat, Vol. II*, Gauthier-Villars, Paris, 1891–1912.

[FK06]   É. Fouvry and J. Klüners, *Cohen-Lenstra heuristics of quadratic number fields*, Algorithmic Number Theory - ANTS-VII (Berlin, Germany), Lecture Notes in Computer Science Vol. 4076, Springer-Verlag, Berlin, 2006, pp. 40–55.

[FOPS04]   E. Fujisaki, T. Okamoto, D. Pointcheval, and J. Stern, *RSA-OAEP is secure under the RSA assumption*, Journal of Cryptology **17** (2004), 81–104.

[Fow87]   D. H. Fowler, *The Mathematics of Plato's Academy: A new reconstruction*, 2nd ed., Clarendon Press, Oxford, 1987, 1999.

[FP03]   P.-A. Fouque and G. Poupard, *On the security of RDSA*, Advances in Cryptology - EUROCRYPT 2003, Lecture Notes in Computer Science Vol. 2656, Springer, Berlin, 2003, pp. 462–476.

[Fra72]   P. M. Fraser, *Ptolemaic Alexandria*, The Clarendon Press, Oxford, 1972.

[Fri88]   C. Friesen, *On continued fractions of given period*, Proc. Amer. Math. Soc. **103** (1988), no. 1, 9–14.

[Fun90]   G. W. Fung, *Computational problems in complex cubic fields*, Ph.D. thesis, University of Manitoba, Winnipeg, Manitoba, 1990.

[Gag03]   M. Gagné, *Identity-based encryption: A survey*, RSA Laboratories Cryptobytes **6** (2003), no. 1, 10–19.

[Gam85]   T. El Gamal, *A public key cryptosystem and a signature scheme based on discrete logarithms*, IEEE Transactions on Information Theory **31** (1985), 469–472.

[Gau86]   C. F. Gauss, *Disquisitiones Arithmeticae*, A. A. Clarke (trans.), Springer-Verlag, New York, 1986.

[GJS01]   M. Giesbrecht, M. J. Jacobson, Jr., and A. Storjohann, *Algorithms for large integer matrix problems*, Applied Algebra, Algebraic Algorithms and Error-Correcting Codes - AAECC-14 (Melbourne, Australia), Lecture Notes in Computer Science Vol. 2227, Springer, Berlin, 2001, pp. 297–307.

[Gol04]   O. Goldreich, *The Foundations of Cryptography - Basic Applications*, Cambridge University Press, Cambridge, 2004.

[Gol76]   D. Goldfeld, *The class number of quadratic fields and the conjectures of Birch and Swinnerton-Dyer*, Ann. Scuola Norm. Sup. Pisa Cl. Sci. **3** (1976), no. 4, 623–663.

[Gol85]   ———, *Gauss's class number problem for imaginary quadratic fields*, Bulletin of the AMS (N.S.) **13** (1985), no. 1, 23–37.

[Gor93]   D. Gordon, *Discrete logarithms using the number field sieve*, SIAM J. Discrete Math. **6** (1993), 124–138.

[Gor98]   ———, *A survey of fast exponentiation methods*, J. Algorithms **27** (1998), 129–146.

[Gou04]   X. Gourdon, *The 10^{13} first zeros of the Riemann zeta function, and zeros computation at very great height*, Available at http://numbers.computation.free.fr/Constants/Miscellaneous/zetazeros1e13-1e24.pdf, 2004.

[GQ88]   L. C. Guillou and J.-J. Quisquater, *A practical zero-knowledge protocol fitted to security microprocessors minimizing both transmission and memory*, Advances in Cryptology - EUROCRYPT'88, Lecture Notes in Computer Science Vol. 330, Springer, Berlin, 1988, pp. 123–128.

[Gra07]   T. Granlund, *GNU MP: The GNU Multiprecision Arithmetic Library*, Edition 4.2.2, http://gnulib.org/manual/, 2007.

[Gra91]   A. Granville, *The lattice points of an n-dimensional tetrahedron*, Aequationes Math. **41** (1991), 234–241.

[GS02]   A. Granville and K. Soundararajan, *Upper bounds for $|L(1,\chi)|$*, Quarterly J. Math. **53** (2002), 265–284.

[GS03]   ———, *The distribution of values of $L(1,\chi_d)$*, GAFA, Geometric and Functional Analysis **13** (2003), 992–1028.

[GTTD07]   P. Gaudry, E. Thomé, N. Thériault, and C. Diem, *A double large prime variation for small genus hyperelliptic index calculus*, Math. Comp. **76** (2007), no. 257, 475–492.

[GW08]   J. E. Gower and S. S. Wagstaff, Jr., *Square form factorization*, Math. Comp. **77** (2008), no. 261, 551–588.

[GZ86]   B. Gross and D. Zagier, *Heegner points and derivatives of L-series*, Invent. Math. **84** (1986), 225–320.

[Hal02]   S. Hallgren, *Polynomial-time quantum algorithms for Pell's equation and the principal ideal problem*, STOC '02: Proceedings of the Thirty-Fourth Annual ACM Symposium on Theory of Computing, 2002, pp. 653–658.

[Hal05]   ———, *Fast quantum algorithms for computing the unit group and class group of a number field*, STOC '05: Proceedings of the Thirty-Seventh Annual ACM Symposium on Theory of Computing, 2005, pp. 468–474.

[Ham02]   S. Hamdy, *Über die Sicherheit und Effizienz kryptografischer Verfahren mit Klassengruppen imaginär-quadratischer Zahlkörper*, Ph.D. thesis, Technische Universität Darmstadt, Darmstadt, Germany, 2002.

[Ham08]   ———, *libiq: A library for arithmetic in class groups of imaginary quadratic orders*, Software, 2008, http://faculty.uaeu.ac.ae/s_hamdy/libiq.html.

[Han65]   H. Hankel, *Zur Geschichte der Mathematik in Altertum und Mittelalter*, 2nd ed., Georg Olms Verlag, Hildesheim, 1965.

[Has65]   H. Hasse, *Über mehrklassige, aber eingeschlechtige reell-quadratische Zahlkörper*, Elem. Math. **20** (1965), 49–59.

[Hea12]   T. L. Heath, *The Works of Archimedes with the Method of Archimedes*, Cambridge University Press, Cambridge, 1912; reprinted by Dover publication, New York, undated.

[Hea56]   _____, *The Thirteen Books of Euclid's Elements*, Dover, New York, 1956.

[Hea64]   _____, *Diophantus of Alexandria: A Study in the History of Greek Algebra*, Dover, New York, 1964.

[Hea81]   _____, *A History of Greek Mathematics, Vols. I and II*, Dover, New York, 1981.

[Hec81]   E. Hecke, *Lectures on the Theory of Algebraic Numbers*, Springer-Verlag, New York, 1981.

[Hee52]   K. Heegner, *Diophantische Analysis und Modulfunctionen*, Math. Z. **56** (1952), 227–253.

[Hen74]   M. D. Hendy, *Applications of a continued fraction algorithm to some class number problems*, Math. Comp. **28** (1974), 267–277.

[HHR⁺05]   H. Häffner, W. Hänsell, C. F. Roos, J. Benhelm, D. Chek al kar, M. Chwalla, U. D. Rapol T. Körber and, M. Riebe, P. O. Schmidt, C. Becher, O. Gühne, W. Dür, and R. Blatt, *Scalable multiparticle entanglement of trapped ions*, Nature **438** (2005), 643–646.

[HJPT98]   D. Hühnlein, M. J. Jacobson, Jr., S. Paulus, and T. Takagi, *A cryptosystem based on non-maximal imaginary quadratic orders with fast decryption*, Advances in Cryptology - EUROCRYPT '98, Lecture Notes in Computer Science Vol. 1403, Springer, Berlin, 1998, pp. 294–307.

[HJW01]   D. Hühnlein, M. J. Jacobson, Jr., and D. Weber, *Towards practical non-interactive public-key cryptosystems using non-maximal imaginary quadratic orders (extended abstract)*, Selected Areas in Cryptography — SAC2000, Lecture Notes in Computer Science Vol. 2012, Springer, Berlin, 2001, pp. 275–287.

[HJW03]   _____, *Towards practical non-interactive public-key cryptosystems using non-maximal imaginary quadratic orders*, Designs, Codes and Cryptography **30** (2003), no. 3, 281–299.

[HK89a]   F. Halter-Koch, *Einige periodische Kettenbruchentwicklungen und Grundeinheiten quadratischer Ordnungen*, Abh. Math. Sem. Univ. Hamburg **59** (1989), 157–169.

[HK89b]   _____, *Reell-quadratische Zahlkörper mit großer Grundeinheit*, Abh. Math. Sem. Univ. Hamburg **59** (1989), 171–181.

[HK90]   _____, *Quadratische Ordnungen mit grosser Klassenzahl*, J. Number Theory **34** (1990), 82–94.

[HK91]   _____, *Continued fractions of given symmetric period*, Fibonacci Quart. **29** (1991), no. 4, 298–303.

[HL23]   G. H. Hardy and J. E. Littlewood, *Partitio numerorum III: On the expression of a number as a sum of primes*, Acta Mathematica **44** (1923), 1–70.

[HM00a]   S. Hamdy and B. Möller, *Security of cryptosystems based on class groups of imaginary quadratic orders*, Advances in Cryptology - ASIACRYPT 2000, Lecture Notes in Computer Science Vol. 1976, Springer, Berlin, 2000, pp. 234–247.

[HM00b]   D. Hühnlein and J. Merkle, *An efficient NICE-Schnorr-type signature scheme*, Proceedings of PKC 2000, Lecture Notes in Computer Science Vol. 1751, Springer, Berlin, 2000.

[HM89]   J. L. Hafner and K. S. McCurley, *A rigorous subexponential algorithm for computation of class groups*, J. Amer. Math. Soc. **2** (1989), 837–850.

[HMU07]   J. E. Hopcroft, R. Motwani, and J. D. Ullman, *Introduction to Automata Theory, Languages, and Computation*, 3rd ed., Addison-Wesley, Reading, MA, 2007.

[HMV04]   D. Hankerson, A. Menezes, and S. Vanstone, *Guide to Elliptic Curve Cryptography*, Springer-Verlag, New York, 2004.

[Hof80]   J. Hoffstein, *On the Siegel-Tatuzawa theorem*, Acta Arith. **38** (1980), 167–174.

[Hof94]   J. E. Hofmann, *Studien zur Zahlentheorie Fermats*, Abh. Preuss. Akad. Wiss., 1994, no. 7.

[Hom46]   Homer, *The Odyssey*, E. V. Rieu (trans.) Penguin Books, Baltimore, 1946.

[Hoo84]   C. Hooley, *On the Pellian equation and the class number of indefinite binary quadratic forms*, J. Reine Angew. Math. **353** (1984), 98–131.

[HP00]   D. Hühnlein and S. Paulus, *On the implementation of cryptosystems based on real quadratic number fields*, Seventh Annual Workshop on Selected Areas in Cryptography SAC(2000), Lecture Notes in Computer Science Vol. 2012, Springer, New York, 2000, pp. 288–302.

[Hua82]   L. K. Hua, *Introduction to Number Theory*, Springer-Verlag, New York, 1982.

[Hüh00]   D. Hühnlein, *Quadratic orders for NESSIE – overview and parameter sizes of three public key families*, Technical Report TI-3/00, Department of Computer Science, TU Darmstadt, Germany, 2000, Available at http://www.informatik.tu-darmstadt.de/TI/Welcome.html.

[Hüh01]   ———, *Faster generation of NICE-Schnorr signatures*, Topics in Cryptology - CT-RSA 2001, The Cryptographer's Track at RSA Conference 2001, Lecture Notes in Computer Science Vol. 2020, Springer, Berlin, 2001, pp. 1–12.

[Hun74]   T. W. Hungerford, *Algebra*, Springer-Verlag, New York, 1974.

[Hur89]   A. Hurwitz, *Über eine besondere Art der Kettenbruchentwicklung reeller Grössen*, Acta. Math. **12** (1889), 367–405.

[IK04]   H Iwaniec and E. Kowalski, *Analytic number theory*, American Mathematical Society, Providence, RI, 2004.

[IKW90]   N. Ishii, P. Kaplan, and K. S. Williams, *On Eisenstein's problem*, Acta Arith. **54** (1990), 323–345.

[Inc34]   E. L. Ince, *Cycles of Reduced Ideals in Quadratic Fields*, Mathematical Tables Vol. IV, British Association for the Advancement of Science, London, 1934.

[IR82]   K. Ireland and M. Rosen, *A Classical Introduction to Modern Number Theory*, Springer, New York, 1982.

[Jac00]   M. J. Jacobson, Jr., *Computing discrete logarithms in quadratic orders*, Journal of Cryptology **13** (2000), 473–492.

[Jac98]   ———, *Experimental results on class groups of real quadratic fields (extended abstract)*, Algorithmic Number Theory - ANTS-III

(Portland, Oregon), Lecture Notes in Computer Science Vol. 1423, Springer-Verlag, Berlin, 1998, pp. 463–474.

[Jac99a] _____, *Applying sieving to the computation of quadratic class groups*, Math. Comp. **68** (1999), no. 226, 859–867.

[Jac99b] _____, *Subexponential class group computation in quadratic orders*, Ph.D. thesis, Technische Universität Darmstadt, Darmstadt, Germany, 1999.

[Jeb93] T. Jebelean, *A Generalization of the Binary GCD Algorithm*, ACM International Symposium on Symbolic and Algebraic Computation (M. Bronstein, ed.), ACM Press, New York, 1993, pp. 111–116.

[Jeb95] _____, *A double-digit Lehmer-Euclid algorithm for finding the GCD of long integers*, J. Symbolic Comput. **19** (1995), no. 1–3, 145–157.

[JJ00] É. Jaulmes and A. Joux, *A NICE cryptanalysis*, Advances in Cryptology - EUROCRYPT 2000, Lecture Notes in Computer Science Vol. 1807, Springer, Berlin, 2000, pp. 382–391.

[JLW95] M. J. Jacobson, Jr., R. F. Lukes, and H. C. Williams, *An investigation of bounds for the regulator of quadratic fields*, Experimental Mathematics **4** (1995), no. 2, 211–225.

[JPW03] M. J. Jacobson, Jr., Á. Pintér, and P. G. Walsh, *A computational approach for solving $y^2 = 1^k + 2^k + \cdots + x^k$*, Math. Comp. **72** (2003), no. 244, 2099–2110.

[JRW06] M. J. Jacobson, Jr., S. Ramachandran, and H. C. Williams, *Numerical results on class groups of imaginary quadratic fields*, Algorithmic Number Theory - ANTS-VII (Berlin, Germany), Lecture Notes in Computer Science Vol. 4076, Springer-Verlag, Berlin, 2006, pp. 87–101.

[JSS07a] M. J. Jacobson, Jr., R. Scheidler, and A. Stein, *Cryptographic protocols on real hyperelliptic curves*, Advances in Mathematics of Communications **1** (2007), no. 2, 197–221.

[JSS07b] _____, *Fast arithmetic on hyperelliptic curves via continued fraction expansions*, Advances in Coding Theory and Cryptology (T. Shaska, W. C. Huffman, D. Joyner, and V. Ustimenko, eds.), Series on Coding Theory and Cryptology Vol. 3, World Scientific Publishing, Singapore, 2007, pp. 201–244.

[JSW01] M. J. Jacobson, Jr., R. Scheidler, and H. C. Williams, *The efficiency and security of a real quadratic field based key exchange protocol*, Public Key cryptography and Computational Number Theory, Walter de Gruyter, Berlin, 2001, pp. 89–112.

[JSW06a] M. J. Jacobson, Jr., R. E. Sawilla, and H. C. Williams, *Efficient ideal reduction in quadratic fields*, International Journal of Mathematics and Computer Science **1** (2006), 83–116.

[JSW06b] M. J. Jacobson, Jr., R. Scheidler, and H. C. Williams, *An improved real quadratic field based key-exchange procedure*, J. Cryptology **19** (2006), 211–239.

[JSW08] M. J. Jacobson, Jr., R. Scheidler, and D. Weimer, *An adaptation of the NICE cryptosystem to real quadratic orders*, Progress in Cryptology - AFRICACRYPT 2008, Lecture Notes in Computer Science Vol. 5023, Springer-Verlag, Berlin, 2008, pp. 191–208.

[JvdP02] M. J. Jacobson, Jr. and A. J. van der Poorten, *Computational aspects of NUCOMP*, Proc. ANTS-V, Lecture Notes in Computer Science Vol. 2369, Springer, Berlin, 2002, pp. 120–133.

[JW00]   M. J. Jacobson, Jr. and H. C. Williams, *The size of the fundamental solutions of consecutive Pell equations*, Experimental Mathematics **9** (2000), no. 4, 631–640.

[JW02]   _____, *Modular arithmetic on elements of small norm in quadratic fields*, Designs, Codes and Cryptography **27** (2002), 93–110.

[JW03]   _____, *New quadratic polynomials with high densities of prime values*, Math. Comp. **72** (2003), no. 241, 499–519.

[Kah96]   D. Kahn, *The codebreakers*, Scribner, New York, 1996.

[Kap98]   I. Kaplansky, Letter to Richard Mollin, Kenneth Williams and Hugh Williams, November 23, 1998.

[Khi36]   A. Y. Khintchine, *Zur metrischen Theorie der Diophantischen Approximatisnen*, Math. Z. **24** (1936), 706–714.

[Khi97]   _____, *Continued fractions*, Dover, New York, 1997.

[Kis48]   A. A. Kiselev, *An expression for the number of classes of ideals of real quadratic fields by means of Bernoulli numbers*, Doklady Akad. Nauk SSSR (N.S.) **61** (1948), 777–779, (in Russian).

[Kle06]   T. Kleinjung, *On polynomial selection for the general number field sieve*, Math. Comp. **75** (2006), no. 256, 2037–2047.

[KM07]   N. Koblitz and A. Menezes, *Another look at "provable security,"* Journal of Cryptology **20** (2007) 3–37.

[KM61]   R. Kortum and G. McNeil, *A Table of Periodic Continued Fractions*, Lockheed Aircraft Corp., Sunnyvale, CA, 1961.

[Kno75a]   W. Knorr, *Archimedes and the measurement of the circle: A new interpretation*, Arch. Hist. Exact Sci. **15** (1975), 115–140.

[Kno75b]   _____, *The Evolution of the Euclidean Elemets*, D. Reidel, Dordrecht, Boston, 1975.

[Kno86]   _____, *The Ancient Tradition of Geometric Problems*, Birkhäuser, Boston, 1986.

[Kno93]   _____, *Arithmetike stoicheiosis: on Diophantus and Hero of Alexandria*, Historia Math. **20** (1993), no. 2, 180–192.

[Knu97]   D. E. Knuth, *The Art of Computer Programming, Vol. 1: Fundamental algorithms*, 3rd ed., Addison-Wesley, Reading, MA, 1997.

[Knu98]   _____, *The Art of Computer Programming, Vol. 2: Seminumerical Algorithms*, 3rd ed., Addison-Wesley, Reading, MA, 1998, pp. 333–379.

[Kob87]   N. Koblitz, *Elliptic curve cryptosystems*, Math. Comp. **48** (1987), 203–209.

[Kon01]   H. Konen, *Geschichte der Gleichung $t^2 - du^2 = 1$*, S. Hirzel, Leipzig, 1901.

[Kor90]   D. M. Kornhauser, *On the smallest solution to the general binary quadratic Diophantine euqation*, Acta Arith. **55** (1990), 83–94.

[Kra22]   M. Kraitchik, *Théorie des Nombres*, Vol. T. I, Gauthier-Villars, Paris, 1922.

[Kra26]   _____, *Théorie des Nombres*, Vol. T. II, Gauthier-Villars, Paris, 1926.

[Kru80]   B. Krumbiegel, *Das problema bovinum des Archimedes*, Zeitschrift für Math. u. Physik (Hist. Litt. Abtheilung) **25** (1880), 121–136.

[Lag68]   J. L. Lagrange, *Sur la solution des problèmes indétermine's du second degré*, Oeuvres, Vol. II Gauthier-Villars, Paris, 1868, pp. 377–535.

[Lag79]   J. C. Lagarias, *Succinct certificates for the solvability of binary quadratic Diophantine equations (extended abstract)*, Proc. 20th IEEE Symp. on Foundations of Computer Science, 1979, pp. 47–54.

[Lag80a]   _____, *On the computational complexity of determining the solvability of the equation* $X^2 - DY^2 = -1$, Trans. Amer. Math. Soc. **260** (1980), 485–508.

[Lag80b]   _____, *Worst-case complexity bounds for algorithms in the theory of integral quadratic forms*, J. Algorithms **1** (1980), 142–186.

[Lag81]   _____, *Succinct certificates for the solvability of binary quadratic Diophantine equations*, Technical Memorandum 81-11216-54, Bell Labs, 1981.

[Lan18]   E. Landau, *Über die Klassenzahl imaginär-quadratischer Zahlkörper*, Gött. Nachr. (1918), 285–295.

[Lan36]   _____, *On a Titchmarsh-Estermann sum*, J. London Math. Soc. **II** (1936), 242–245.

[Lan50]   _____, *Elementare Zahlentheorie*, Chelsea, New York, 1950.

[Lan91]   S. Lang, *Algebraic Number Theory*, 2nd ed., Springer, Berlin, 1991.

[Le94]   M. Le, *Upper bounds for class numbers of real quadratic fields*, Acta. Arith. **68** (1994), 141–144.

[Le95]   _____, *Some exponential Diophantine equations I: The equation* $D_1 x^2 - D_2 y^2 = \lambda k^z$, Journal of Number Theory **55** (1995), 209–221.

[Leh26a]   D. H. Lehmer, *A list of errors in tables of the Pell equation*, Bull. Amer. Math. Soc. **32** (1926), 545–550.

[Leh26b]   _____, *On the indeterminate equation* $y^2 - p^2 du^2 = 1$, Annals of Math. **27** (1926), 471–476.

[Leh28]   _____, *On the multiple solutions of the Pell equation*, Annals of Math. **30** (1928), 66–72.

[Leh38]   _____, *Euclid's algorithm for large numbers*, The American Mathematical Monthly **45** (1938), no. 4, 227–233.

[Leh40]   _____, *The lattice points of an n-dimensional tetrahedron*, Duke Math. J. **7** (1940), 341–353.

[Leh41]   _____, *Guide to Tables in the Theory of Numbers*, National Research Council, Washington, DC, 1941.

[Leh56]   _____, *On the Diophantine equation* $x^3 + y^3 + z^3 = 1$, J. London Math. Soc. **31** (1956), 275–280.

[Leh64]   _____, *On a problem of Störmer*, Illinois Journal of Mathematics **8** (1964), 57–79.

[Len00]   A. K. Lenstra, *Integer factoring*, Designs, Codes and Cryptography **19** (2000), 101–128.

[Len02]   H. W. Lenstra, Jr., *Solving the Pell equation*, Notices of the the AMS **49** (2002), no. 2, 182–192.

[Len82]   _____, *On the calculation of regulators and class numbers of quadratic fields*, London Math. Soc. Lecture Note Series **56** (1982), 123–150.

[Len87]   _____, *Factoring integers with elliptic curves*, Annals of Math. (2) **126** (1987), 649–673.

[Les73]   G. E. Lessing, *Zur Geschichte der Literatur. aus den schatzen der herz. Bibliothek zu Wolfenbüttel.*, Zweiter Beitrag. Braunschweig, 1773.

[Les97]   _____, *Sämmtliche Schriften*, Lerausgegeben von K. Lachmannm, besorgt durch F. Munker, Göschen, Leipzig, Band 12, 1897, pp. 100–107, 110–115.

[Lev02]   S. Levy, *Crypto: How the code rebels beat the government*, Penguin, London, 2002.

[Lév36]   P. Lévy, *Sur le développement en fraction continue d'un nombre choisi au hasard*, Compositio Math. **3** (1936), 286–303 reprinted in *Oemres de Paul Lévy* Vol. 6, Gauthier-Villars, Paris, 1980, pp. 285–302.

[Lev88]   C. Levesque, *Continued fraction expansions and fundamental units*, J. Math. Phys. Sci. **22** (1988), no. 1, 11–44.

[Lit28]   J. E. Littlewood, *On the class number of the corpus $P(\sqrt{-k})$*, Proc. London Math. Soc. **27** (1928), 358–372.

[Lju42]   W. Ljunggren, *Zur Theorie der Gleichung $x^2 + 1 = dy^4$*, Avh. Norske Vid. Akad. Olso (1942), no. 5, 1–27.

[LL90]   A. K. Lenstra and H. W. Lenstra, Jr., *Algorithms in number theory*, Handbook of theoretical computer science (J. van Leeuwen, ed.), Elsevier Science Publishers, Amsterdam, 1990, pp. 673–715.

[LL93]   A. K. Lenstra and H. W. Lenstra, Jr. (eds.), *The Development of the Number Field Sieve*, A. K. Lenstra and H. W. Lenstra, Jr. (eds.), Lecture Notes in Mathematics Vol. 1554, Springer-Verlag, New York, 1993.

[LO77]   J. C. Lagarias and A. M. Odlyzko, *Effective versions of the Chebotarev density theorem*, Algebraic Number Fields (A. Frohlich, ed.), Academic Press, New York, 1977, pp. 409–464.

[Lou02a]   S. Louboutin, *Computation of class numbers of quadratic number fields*, Math. Comp. **71** (2002), no. 240, 1735–1743.

[Lou02b]   _____, *Explicit upper bounds for $|L(1, \chi)|$ for primitive even Dirichlet characters*, Acta Arith. **111** (2002), 1–18.

[Lou89]   _____, *Une version effective d'un théorèm de A. Schinzel sur les longueurs des périodes de certains développements en fractions continues*, C. R. Acad. Sci. Paris. Sér. I Math. **308** (1989), no. 17, 511–513.

[LP92]   H. W. Lenstra, Jr. and C. Pomerance, *A rigorous time bound for factoring integers*, J. Amer. Math. Soc. **5** (1992), 483–516.

[LR86]   C. Levesque and G. Rhin, *A few classes of periodic continued fractions*, Util. Math. **30** (1986), 79–107.

[LW92]   L. Lorentzen and H. Waadeland, *Continued Fractions with Applications*, North-Holland, Amsterdam, Netherlands, 1992.

[Mad01]   D. Madden, *Constructing families of long continued fractions*, Pac. J. Math. **198** (2001), no. 1, 123–147.

[Mah94]   M. S. Mahoney, *The Mathematical Career of Pierre de Fermat*, 2nd ed., Princeton University Press, NJ, 1994.

[Mar07]   A. A. Martinez, *Euler's "mistake?" The radical product rule in historical perspective*, Amer. Math. Monthly **114** (2007), 273–285.

[Mar77a]   D. A. Marcus, *Number fields*, Springer-Verlag, New York, 1977.

[Mar77b]   A. Martin, *Solution*, The Analyst **4** (1877), 154–155.

[Mas79]   J. Masley, *Where are number fields with small class numbers?*, Number Theory, Proceedings of the Southern Illinois Conference (Southern University, Carbondale, Illinois), Lecture Notes in Mathematics, vol. 751, Springer, Berlin, 1979, pp. 221–242.

[Mat00]   K. R. Matthews, *The diophantine equation $x^2 - Dy^2 = N$, $D > 1$, in integers*, Expositiones Math. **18** (2000), 323–331.

[Mat93]   Y. Y. Matiyasevich, *Hilbert's tenth problem*, MIT Press, Cambridge, MA, 1993.

[Mau00]   M. Maurer, *Regulator approximation and fundamental unit compu-tation for real-quadratic orders*, Ph.D. thesis, Technische Universität Darmstadt, Darmstadt, Germany, 2000.

[MC02a]   R. A. Mollin and K. Cheng, *Beepers, creepers, and sleepers*, Int. Math. J. **2** (2002), no. 9, 951–956.

[MC02b]   ———, *Continued fractions beepers and Fibonacci numbers*, C. R. Math. Acad. Sci. Soc. R. Can. **24** (2002), no. 3, 102–108.

[MC04]   ———, *Period lengths of continued fractions involving Fibonacci num-bers*, Fibonacci Quart. **42** (2004), 161–169.

[McC89]   K. S. McCurley, *Cryptographic key distribution and computation in class groups*, Proc. NATO ASI on Number Theory and Applica-tions (R. A. Mollin, ed.), Kluwer Academic Press, Amsterdam, 1989, pp. 459–479.

[MCG02]   R. A. Mollin, K. Cheng, and B. Goddard, *Pellian polynomials and period lengths of continued fractions*, JP J. Algebra Number Theory Appl. (2002), no. 1, 47–60.

[McL03a]   J. McLaughlin, *Multi-variable polynomial solutions to Pell's equation and fundamental units in real quadratic fields*, Pacific J. Math. **210** (2003), no. 2, 335–349.

[McL03b]   ———, *Polynomial solutions to Pell's equation and fundamental units in real quadratic fields*, J. London Math. Soc. (2) **67** (2003), no. 1, 16–28.

[Mih03]   P. Mihailescu, *A class number free criterion for Catalan's conjecture*, J. Number Theory **99** (2003), 225–231.

[Mih04]   ———, *Primary cyclotomic units and a proof of Catalan's conjecture*, J. Reine Angew. Math. **572** (2004), 167–195.

[Mil86]   V. Miller, *Use of elliptic curves in cryptography*, Advances in Cryp-tology — CRYPTO '85, Lecture Notes in Computer Science Vol. 218, Springer, Berlin, 1986, pp. 417–426.

[MM05]   A. Magidin and D. McKinnon, *Gauss's Lemma for number fields*, Amer. Math. Monthly **112** (2005), 385–416.

[Mol05]   R. A. Mollin, *Codes*, Chapman and Hall/CRC Press, Boca Raton, FL, 2005.

[Mol07]   ———, *An Introduction to Cryptography*, 2nd ed., Chapman and Hall/CRC Press, Boca Raton, FL, 2007.

[Mol95]   ———, *Quadratics*, CRC Press, Boca Raton, FL, 1995.

[Mor60]   L. J. Mordell, *On a Pellian equation conjecture*, Acta Arith. **6** (1960), 137–144.

[Mor61]   ———, *On a Pellian equation conjecture (II)*, J. London Math. Soc. **36** (1961), 282–288.

[Mor69]   ———, *Diophantine Equations*, Academic Press, London, 1969.

[MS06]   J. A. Muir and D. R. Stinson, *Minimality and other properties of the width-w nonadjacent form*, Math. Comp. **75** (2006), 369–384.

[MS99]   T. Mulders and A. Storjohann, *Diophantine linear system solving*, Proc. International Symposium on Symbolic and Algebraic Compu-tation: ISSAC'99, 1999, pp. 281–288.

[MST99]   V. Müller, A. Stein, and C. Thiel, *Computing discrete logarithms in real quadratic congruence function fields of large genus*, Math. Comp. **68** (1999), 807–822.

[Mül06]   S. Müller, *Some remarks on Williams' public-key crypto functions*, Fibonacci Quart. **44** (2006), 224–234.

[MV99]   H. L. Montgomery and R. C. Vaughan, *Extreme values of Dirichlet L-functions at 1*, Number Theory in progress, de Gruyter, Berlin, 1999, pp. 1039–1052.

[MvdPW94]   R. A. Mollin, A. J. van der Poorten, and H. C. Williams, *Halfway to a solution of $X^2 - DY^2 = -3$*, J. Théorie des Nombres Bordeaux **6** (1994), 421–459.

[MvOV96]   A. Menezes, P. van Oorschot, and S. Vanstone, *Handbook of Applied Cryptograph*, Series on Discrete Mathematics and Its Applications, CRC Press, Boca Raton, FL, 1996.

[MW74]   H. L. Montgomery and P. J. Weinberger, *Notes on small class numbers*, Acta Arith. **24** (1974), 529–542.

[MW90]   R. A. Mollin and H. C. Williams, *Class number problems for real quadratic fields*, Number Theory and Cryptograph, LMS Lecture Note Series 154, 177–195, LMS Lecture Note Series 154, Cambridge University Press, Cambridge, 1990, pp. 177–195.

[MW91a]   ———, *Affirmative solution of a conjecture related to a sequence of Shanks*, Proc. Japan Academy **67** (1991), 70–71.

[MW91b]   ———, *On a determination of real quadratic fields of class number one and related continued fraction period length less than 25*, Proceed. Japan Acad. **67** (1991), 20–25.

[MW91c]   ———, *On real quadratic fields of class number two*, Math. Comp. **67** (1991), 20–25.

[MW92a]   ———, *Computation of the class number of a real quadratic field*, Utilitas Mathematica **41** (1992), 259–308.

[MW92b]   ———, *Consecutive powers in continued fractions*, Acta. Arith. **61** (1992), no. 3, 233–264.

[MW92c]   ———, *On the period length of some special continued fractions*, J. Théor. Nombres Bordeaux **4** (1992), no. 1, 19–42.

[MW94]   ———, *Quadratic residue covers in certain real quadratic fields*, Math. Comp. **62** (1994), 885–897.

[MZ]   J. McLaughlin and P. Zimmer, *Some more long continued fractions, I*, Acta. Arith., to appear.

[Nag48]   T. Nagell, *Laste oppgaver*, Nordisk Mat. Tidskr. **30** (1948), 62–64.

[Nag61]   ———, *The Diophantine equation $x^2 + 7 = 2^n$*, Arkiv. for Mat. **4** (1961), 185–187.

[Nag64]   ———, *Introduction to Number Theory*, Chelsea, New York, 1964.

[Nar04]   W. Narkiewicz, *Elementary and Analytic Theory of Algebraic Numbers*, 3rd ed., Springer-Verlag, Berlin, 2004.

[Nat00]   National Institute of Standards and Technology (NIST), *Digital signature standard (DSS)*, Federal Information Processing Standard, FIPS PUB 186-2, Jan. 2000.

[Nat07]   ———, *Recommendation for key management - part 1: General (revised)*, NIST Special Publication 800-57, March, 2007, Available at http://csrc.nist.gov/groups/ST/toolkit/documents/SP800-57Part1_3-8-07.pdf.

[Nat76]   M. B. Nathanson, *Polynomial Pell equations*, Proc. Amer. Math. Soc. **56** (1976), 89–92.

[Nel81]   H. L. Nelson, *A solution to Archimedes' Cattle Problem*, J. Recreational Math. **13** (1981), 162–176.

[Nit95]   A. Nitaj, *L'algorithme de Cornacchia*, Expositiones Math. **13** (1995), 358–365.

[Niv42]   I. Niven, *Quadratic Diophantine equations in the rational and quadratic fields*, Trans. Amer. Math. Soc. **52** (1942), 1–11.

[Niv43]   _____, *The Pell equation in quadratic fields*, Bull. Amer. Math. Soc. **49** (1943), 413–416.

[Niv56]   _____, *Irrational Numbers*, The Carus Mathematical Monographs, No. 11, MAA, New York, 1956.

[Nyb49]   M. Nyberg, *Culminating and almost culminating continued fractions*, Norsk. Math. Tidsskr. **31** (1949), 95–99, (in Norwegian).

[NZM91]   I. Niven, H. S. Zuckerman, and H. Montgomery, *An Introduction to the Theory of Numbers*, 5th ed., John Wiley and Sons, New York, 1991.

[Odl01]   A. M. Odlyzko, *The 10^{22} nd zero of the Riemann Zeta function*, Dynamical Spectral and Arithmetic Zeta Function, American Mathematical Society Contemporary Math Series, no. 290, 139–144, American Mathematical Society Contemporary Math Series, no. 290, American Mathematical Society, Providence, RI, 2001, pp. 139–144.

[Oes79]   J. Oesterlé, *Versions effectives du théorèm de Chebotarev sous l'hypothèse de Riemann généralisee*, Astérisque **61** (1979), 165–167.

[Oes85]   _____, *Nombres de classes des corps quadratiques imaginaires*, Séminaire N. Bourbaki (1983–84), Astérisque **121–122** (1985), 309–323.

[Old63]   C. D. Olds, *Continued fractions*, Random House, New York, 1963.

[OSS84]   H. Ong, C. P. Schnorr, and A. Shamir, *An efficient signature scheme based on quadratic equations*, STOC '84: Proceedings of the Sixteenth Annual ACM symposium on Theory of computing, ACM, Washington, DC, 1984, pp. 208–216.

[Pat03]   R. D. Patterson, *Creepers: Real quadratic number fields with large class numbers*, Ph.D. thesis, Macquarie University, Sydney, 2003.

[Pat41]   W. Patz, *Tafel der Regelmässigen Kettenbrüche für die Quadratwirzeln aus den Natürlichen Zahlen von 1–10000*, Becker and Erler, Leipzig, 1941.

[Pat55]   _____, *Tafel der Regelmässigen, Kettenbrüche und ihres vollständigen Quotienten für Quadratwuzeln aus den Natürlichen Zahlen von 1–10000*, Akademie-Verlag, Berlin, 1955.

[Pau96a]   S. Paulus, *An algorithm of subexponential type computing the class group of quadratic orders over principal ideal domains*, Algorithmic Number Theory - ANTS-II (Université Bordeaux I, Talence, France), Lecture Notes in Computer Science, vol. 1122, Springer-Verlag, Berlin, 1996, pp. 243–257.

[Pau96b]   A. F. Pauly, *Paulys Real-Encyclopädie der classischen Altertumswissenschaft*, Neue Bearbeitung begonnen von G. Wissowa fortgefürht von W. Kroll und K. Mittelhaus, J. B. Metzler, Stuttgart, 1894–1896.

[PD98]   H. Pollard and H. G. Diamond, *The Theory of Algebraic Numbers*, 3rd ed., Dover, Mineola, NY, 1998.

[Per57]   O. Perron, *Die Lehre von den Kettenbrüchen, 3. verb. und erweiterte Aufl.*, Teubner, Stuttgart, 1954–57.

480   References

[PH78]   S. C. Pohlig and M. E. Hellman, *An improved algorithm for comput-
            ing logarithms over GF(p) and it's cryptographic significance*, IEEE
            Transactions on Information Theory **24** (1978), 106–110.

[PO96]   R. Peralta and E. Okamoto, *Faster factoring of integers of a special
            form*, IEICE Transactions Fundamentals **E79-A** (1996), no. 4, 489–
            493.

[Poe41]   E. A. Poe, *A few words on secret writing*, Graham's Magazine **19**
            (1841), 33–38.

[Pom83]   C. Pomerance, *Analysis and comparison of some integer factoring al-
            gorithms*, Computational Methods in Number Theory, (H. W. Lenstra,
            Jr. and R. Tijdeman, eds.), Math. Centre Tracts, Number 154, Part I,
            Mathematisch Centum, Amsterdam, 1983, pp. 89–139.

[Pom94]   _____ , *The number field sieve*, Proceedings of Symposia in Applied
            Mathematics **48** (1994), 465–480.

[PS87]   J. M. Pollard and C-P. Schnorr, *An efficient solution of the congruence
            $x^2 + ky^2 \equiv m \pmod{n}$*, IEEE Trans. Inform. Theory **33** (1987), no. 5,
            702–709.

[PS98]   G. Poupard and J. Stern, *Security analysis of a practical on the
            fly authentication and signature generation*, Advances in Cryptology
            - EUROCRYPT'98, Lecture Notes in Computer Science, vol. 1403,
            Springer, Berlin, 1998, pp. 422–436.

[PT00]   S. Paulus and T. Takagi, *A new public-key cryptosystem over a quad-
            ratic order with quadratic decryption time*, Journal of Cryptology **13**
            (2000), 263–272.

[PvdPW07]   R. D. Patterson, A. J. van der Poorten, and H. C. Williams, *Charac-
            terization of a generalized Shanks sequence*, Pac. J. Math. **230** (2007),
            185–216.

[PW85]   C. D. Patterson and H. C. Williams, *Some periodic continued fractions
            with long periods*, Math. Comp. **44** (1985), 523–532.

[PZ79]   M. Pohst and H. Zassenhaus, *On unit computation in real quadratic
            fields*, Symbolic and Algebraic Computation, Lecture Notes in Com-
            puter Science, vol. 72, Springer, Berlin, 1979.

[PZ85]   _____ , *Über die Berechnung von Klassenzahlen und Klassengruppen
            algebraischer Zahlkörper*, J. Reine Angew. Math. **365** (1985), 50–72.

[PZ97]   _____ , *Algorithmic Algebraic Number Theory*, Cambridge University
            Press, Cambridge, 1997.

[Rab79]   M. O. Rabin, *Digitalized signatures and public-key functions as in-
            tractable as factorization*, MIT Laboratory for Computer Science
            MIT/LCS/TR-212, 1979.

[Rah68]   J. H. Rahn, *An Introduction to Algebra*, T. Brancker (trans.), Moses
            Pitt, London, 1668.

[Ram00]   S. Ramanujan, *Collected papers of Srinivasa Ramanujan*, G. H. Hardy,
            P. V. S. Alyar, and B. M. Wilson (eds.), American Mathematical So-
            ciety, Providence, RI, 2000.

[Ram01]   O. Ramaré, *Approximate formulae for $L(1,\chi)$*, Acta. Arith. **100**
            (2001), 245–266.

[Ram06]   S. Ramachandran, *Class Groups of Quadratic Fields*, Master's thesis,
            University of Calgary, Calgary, Alberta, 2006.

[Rib96]   P. Ribenboim, *The New Book of Prime Number Records*, 3rd ed.,
            Springer, New York, 1996.

[Ric66]   C. Richaud, *Sue la résolution des équations* $x^2 - Ay^2 = \pm 1$, Atti Accad. pontif. Nuovi Lincei (1866), 177–182.

[Rie94]   H. Riesel, *Prime Numbers and Computer Methods for Factorization*, 2nd ed., Birkhäuser, Berlin, 1994.

[Rio68]   J. Riordan, *Combinatorial Identities*, John Wiley and Sons, New York, 1968.

[Ros02]   M. Rosen, *Number Theory in Function Fields*, Graduate Texts in Mathematics, vol. 210, Springer-Verlag, Berlin, 2002.

[RS62]   B. Rosser and L. Schoenfeld, *Approximate formulas for some functions of prime numbers*, Illinois J. Math. **6** (1962), 64–94.

[RS92]   A. M. Rockett and P. Szüsz, *Continued Fractions*, World Scientific, New York, 1992.

[RSA78]   R. Rivest, A. Shamir, and L. Adelman, *A method for obtaining digital signatures and public-key cryptosystems*, Communications of the ACM **21** (1978), 120–126.

[Rum93]   R. Rumely, *Numerical computations concerning the ERH*, Math. Comp. **61** (1993), 415–440.

[Sal90]   A. Salomaa, *Public key cryptography*, Springer, Berlin, 1990.

[SBW94]   R. Scheidler, J. Buchmann, and H. C. Williams, *A key-exchange protocol using real quadratic fields*, Journal of Cryptology **7** (1994), 171–199.

[Sch00]   O. Schirokauer, *Using number fields to compute logarithms in finite fields*, Math. Comp. **69** (2000), 1267–1283.

[Sch06]   A. Schmidt, *Quantum algorithm for solving the discrete logarithm problem in the class group of an imaginary quadratic field and security comparison of current cryptosystems at the beginning of the quantum computer age*, ETRICS 2006, Lecture Notes in Computer Science, vol. 3995, Springer, Berlin, 2006, pp. 481–493.

[Sch07]   ———, *Zur Lösung von zahlentheoretischen Problemen mit klassischen und Quantencomputern*, Ph.D. thesis, Technische Universität Darmstadt, Darmstadt, Germany, 2007.

[Sch08]   R. J. Schoof, *Computing Arakelov class groups*, Surveys in algorithmic number theory, MSRI Publications, vol. 44, Cambridge University Press, Cambridge, 2008, pp. 447–495.

[Sch61]   A. Schinzel, *On some problems of the arithmetical theory of continued fractions*, Acta. Arith. **6** (1961), 393–413.

[Sch62]   A. Schinzel, *On some problems of the arithmetical theory of continued fractions II*, Acta. Arith. **7** (1962), 187–298, Corrigendum, ibid, **47** (1986), 295.

[Sch71]   A. Schönhage, *Schnelle Berechnung von Kettenbruchentwicklungen*, Acta. Arith. **1** (1971), 139–144.

[Sch83]   R. J. Schoof, *Quadratic fields and factorization*, Computational Methods in Number Theory (H. W. Lenstra, Jr. and R. Tijdeman, eds.), Math. Centre Tracts, Number 155, Part II, Mathematisch Centrum, Amsterdam, 1983, pp. 235–286.

[Sch91]   A. Schönhage, *Fast reduction and composition of binary quadratic forms*, ISSAC: International Symposium on Symbolic and Algebraic Computation, ACM, New York, 1991, pp. 128–133.

[Sch93]   P. Schreiber, *A note on the Cattle Problem of Archimedes*, Historia Math. **20** (1993), 304–306.

[Sch96]   B. Schneier, *Applied Cryptography*, Wiley, New York, 1996.

[Sco38]   J. F. Scott, *The Mathematical Work of John Wallis*, Taylor and Francis, London, 1938.

[Scr74]   C. J. Scriba, *John Pell's English edition of J. H. Rahn's Teutsche Algebra*, R. S. Cohen et al. (eds.), For Dirk Struick, Reidel, Dordrecht, 1974.

[Sel63]   C.-O. Selenius, *Kettenbruchtheoretische Erklärung der zyklischen Methode zur Lösung der Bhaskara-Pell-Gleichung*, Acta acad. Aboensis, math. phys. **23** (1963), no. 10.

[Sel75]   _____, *Rationale of the chakravala process of Jayadeva and Bhaskara II*, Historia Math. **2** (1975), 167–184.

[Ses82]   J. Sesiano, *Books IV to VII of Diophantus' Arithmetica*, Springer-Verleg, New York, 1982.

[Sey87]   M. Seysen, *A probabilistic factorization algorithm with quadratic forms of negative discriminant*, Math. Comp. **48** (1987), 757–780.

[Sha62]   D. Shanks, *Review RMT30*, Math. Comp. **16** (1962), 377–379; also **23** (1969), 217–219.

[Sha69]   _____, *On Gauss's class number problems*, Math. Comp. **23** (1969), 151–163.

[Sha71]   _____, *Class number, a theory of factorization and genera*, Proc. Sympos. Pure Mathematics, vol. 29, American Mathematical Society, Providence, RI, 1971, pp. 415–440.

[Sha72]   _____, *The infrastructure of a real quadratic field and its applications*, Proc. 1972 Number Theory Conference, Boulder, CO, University of Colorado, Boulder, 1972, pp. 217–224.

[Sha73]   _____, *Systematic examination of Littlewood's bounds on $L(1, \chi)$*, Analytic Number Theory, Proc. Symp. Pure Math., vol. 24, American Mathematical Society, Providence, RI, 1973, pp. 267–283.

[Sha74]   _____, *Review of RMT 11*, Math. Comp. **28** (1974), 333–334.

[Sha89]   _____, *On Gauss and composition I, II*, Number Theory and Applications, NATO ASI Series C, vol. 265, Kluwer, Dordrecht, 1989.

[Sha94]   J. O. Shallit, *Origins of the analysis of the Euclidean algorithm*, Historia Mathematica **21** (1994), 401–419.

[Sho01]   _____, *NTL: A library for doing number theory*, Software, 2001, Available at http://www.shoup.net/ntl.

[Sho97a]   P. Shor, *Polynomial-time algorithms for prime factorization and discrete logarithms on a quantum computer*, SIAM Journal of Computing **26** (1997), no. 5, 1484–1509.

[Sho97b]   V. Shoup, *Lower bounds for discrete logarithms and related problems*, Advances in Cryptology - EUROCRYPT'97, Lecture Notes in Computer Science, vol. 1233, Springer, Berlin, 1997, pp. 256–266.

[Shu54]   K. Shankar Shukla, *Acarya Jayadeva, the mathematician*, Ganita **5** (1954), 1–20.

[Sie35]   C. L. Siegel, *Über die Klassenzahl quadratischer Zahlkörper*, Acta Arithmetica **1** (1935), 83–86.

[Sil06]   A. K. Silvester, *Fast and unconditional principal ideal testing*, Master's thesis, University of Calgary, 2006, Available at http://math.ucalgary.ca/~aksilves/papers/msc-thesis.pdf.

[Sil87]   R. D. Silverman, *The multiple polynomial quadratic sieve*, Math. Comp. **48** (1987), 329–339.

[Sin99]   S. Singh, *The code book*, Doubleday, New York, 1999.

[Sko45a]   Th. Skolem, *A remark on the equation* $\zeta^2 - \delta\eta^2 = 1$, $\delta > 0$, $\delta', \delta'', \cdots <$ 0, *where* δ, ζ, η *belong to a total real number field*, Avh. Norske Vid. Akad. Olso I (1945), no. 12, 1–15.

[Sko45b]   ———, *A theorem on the equation* $\zeta^2 - \delta\eta^2 = 1$ *where* δ, ζ, η *are integers in an imaginary field*, Avh. Norske. Vid. Akad. Oslo I (1945), no. 1, 1–13.

[SL84]   C. P. Schnorr and H. W. Lenstra, Jr., *A Monte Carlo factoring algorithm with linear storage*, Math. Comp. **43** (1984), no. 167, 289–311.

[SL96a]   P. Stevenhagen and H. W. Lenstra, Jr., *Chebotarëv and his density theorem*, Math Intelligencer **18** (1996), no. 2, 26–37.

[SL96b]   A. Storjohann and G. Labahn, *Asymptotically fast computation of Hermite normal forms of integer matrices*, Proceedings of the 1996 International Symposium on Symbolic and Algebraic Computation— ISSAC'96 (Zürich, Switzerland), ACM Press, New York, 1996, pp. 259– 266.

[Sla65]   I. S. Slavutskii, *On Mordell's theorem*, Acta Arith. **11** (1965), 57–66.

[Sla69]   ———, *Upper bounds and numerical calculation of the number of ideal classes of real quadratic fields*, Amer. Math. Soc. Transl. (2) **82** (1969), 67–71.

[Sma98]   N. P. Smart, *The Algorithmic Resolution of Diophantine Equations*, LMS Student Text, vol. 41, Cambridge University Press, Cambridge, 1998.

[Smi65]   H. J. S. Smith, *Report on the Theory of Numbers*, Chelsea, New York, 1965.

[Sor04a]   J. P. Sorenson, *An analysis of the generalized binary gcd algorithm*, High Primes and Misdemeanors: Lectures in Honour of the 60th Birthday of Hugh Cowie Williams (Banff, AB, Canada) (A. van der Poorten and A. Stein, eds.), American Mathematical Society, Providence, RI, 2004, pp. 327–340.

[Sor04b]   ———, *Lehmer's algorithm for very large numbers*, abstract appeared in SIGSAM Bulletin **38** (2004), no. 3, 102–104.

[Sor94]   ———, *Two fast GCD algorithms*, Journal of Algorithms **16** (1994), 110–144.

[Sor95]   J. P. Sorenson, *An analysis of Lehmer's Euclidean GCD algorithm*, 1995 ACM International Symposium of Symbolic and Algebraic Computation (A. H. M. Levelt, ed.), Montreal, Canada, 1995, pp. 254–258.

[SP05]   D. Schielzeth and M. E. Pohst, *On real quadratic number fields suitable for cryptography*, Experimental Mathematics **14** (2005), no. 2, 189– 197.

[Spr93]   V. G. Sprindzuk, *Classical Diophantine Equations*, Lecture Notes in Mathematics, vol. 1559, Springer-Verlag, Berlin, 1993.

[Sri67]   C. N. Srinivasiengar, *The History of Anicent Indian Mathematics*, The World Press Private Ltd., Calcutta, 1967.

[Sri98]   A. Srinivasan, *Computations of class numbers of real quadratic fields*, Math. Comp. **67** (1998), no. 223, 1285–1308.

[SS71]   A. Schönhage and V. Strassen, *Schnelle Multiplikation großer Zahlen*, Computing (Arch. Elektron. Rechnen) **7** (1971), 281–292.

[SSW08]   R. Sawilla, A. Silvester, and H. C. Williams, *A new look at an old equation*, Algorithmic Number Theory, proceedings of ANTS-VIII, Lecture Notes in Computer Science, vol. 5011, 39–59, Lecture Notes in Computer Science, Springer, Berlin, 2008, pp. 39–59.

[SSW76]   R. G. Stanton, C. Sudler, and H. C. Williams, *An upper bound for the period of the simple continued fraction for $\sqrt{D}$*, Pacific Jounal of Math. **67** (1976), 525–536.

[SSW96]   R. Scheidler, A. Stein, and H. C. Williams, *Key-exchange in real quadratic congruence function fields*, Designs, Codes and Cryptography **7** (1996), 153–174.

[ST02a]   A. Stein and E. Teske, *Explicit bounds and heuristics on class numbers in hyperelliptic function fields*, Math. Comp. **71** (2002), no. 238, 837–861.

[ST02b]   ———, *The parallelized Pollard kangaroo method in real quadratic function fields*, Math. Comp. **71** (2002), no. 238, 793–814.

[ST05]   ———, *Optimized baby step-giant step methods in hyperelliptic function fields*, J. Ramanujan Math. Soc. **20** (2005), 1–32.

[ST86]   T. N. Shorey and R. Tijdeman, *Exponential Diophantine Equations*, Cambridge Tracts in Mathematics, vol. 87, Cambridge University Press, Cambridge, 1986.

[ST87]   N. Stewart and D. O. Tall, *Algebraic Number Theory*, 2nd ed., Chapman and Hall/CRC Press, New York, 1987.

[ST91]   R. Steiner and N. Tzanakis, *Simplifying the solution of Ljunggren's equation $x^2 + 1 = 2y^4$*, J. Number Theory **37** (1991), no. 2, 123–132.

[Sta67]   H. M. Stark, *A complete determination of the complex quadratic fields of class number one*, Michigan Math. J. **14** (1967), 1–27.

[Sta69a]   ———, *A historical note on complex quadratic fields with class-number one*, Proc. Amer. Math. Soc. **21** (1969), 254–255.

[Sta69b]   ———, *On the "gap" in a theorem of Heegner*, J. Number Theory **1** (1969), 16–27.

[Sta70]   ———, *An Introduction to Number Theory*, Markham Publishing, Chicago, 1970.

[Sta71]   ———, *A transcendence theorem for class number problems*, Annals of Math **94** (1971), 153–173.

[Sta75]   ———, *On complex quadratic fields with class number two*, Math. Comp. **29** (1975), 289–302.

[Ste05]   J. Steuding, *Diophantine Analysis*, Chapman and Hall/CRC, Boca Raton, FL, 2005.

[Ste34]   M. A. Stern, *Theorie der Kettenbrüche und ihre Anwendung*, J. Reine Angew. Math. **11** (1834), 327–341.

[Ste79]   H.-J. Stender, *Über die Grundeinheit der reell-quadratischen Zahlkörper $Q(\sqrt{A^2 N^2 + BN + C})$*, J. Reine Angew. Math. **311–312** (1979), 302–306.

[Ste95]   P. Stevenhagen, *A density conjecture for the negative Pell equation*, Computational algebra and number theory (Sydney, 1992), Mathematics and Its Applications, vol. 325, Kluwer Academic Publishers, Dordrecht, 1995, pp. 187–200.

[Ste96a]   A. Stein, *Algorithmen in reell-quadratischen Kongruenzfunktionenkörpern*, Ph.D. thesis, Universität des Saarlandes, Saarbrücken, Germany, 1996.

[Ste96b]   P. Stevenhagen, *On a problem of Eisenstein*, Acta. Arith. **74** (1996), no. 3, 259–268.

[Sti05]   D. Stinson, *Cryptography: Theory and Practice*, 3rd ed., Chapman and Hall/CRC Press, Boca Raton, FL, 2005.

[Sti93]   H. Stichtenoth, *Algebraic Function Fields and Codes*, Springer-Verlag, Berlin, 1993.

[Sto00]   A. Storjohann, *Algorithms for matrix canonical forms*, Ph.D. thesis, ETH Zurich, Zurich, Austria, 2000.

[Sto52]   B. Stolt, *On the Diophantine equation $u^2 - Dv^2 = \pm 4N$, Parts I,II,III*, Ark. Math. **2** (1952), 1–23, 251–268; **3** (1955), 117–132.

[Stö97]   C. Störmer, *Quelques théorèmes sur l'équation de Pell $x^2 - Dy^2 = \pm 1$ et leurs applications*, Skrifter Videnskabs-selskabet (Christiania), Mat.-Naturv. Kl. **I** (1897), no. 2, 48 pp.

[Str61]   Strabo, *The Geography of Strabo, Vol. 3*, Loeb Classical Library, No. 182, Harvard University Press, Cambridge, MA, 1961.

[Sut07]   A. V. Sutherland, *Order Computations in Generic Groups*, Ph.D. thesis, Department of Mathmatics, Massachusetts Institute of Technology, Cambridge, MA, 2007.

[SV05]   A. Schmidt and U. Vollmer, *Polynomial time quantum algorithm for the computation of the unit group of a number field (extended abstract)*, STOC '05: Proceedings of the Thirty-Seventh Annual ACM Symposium on Theory of Computing, 2005, pp. 475–480.

[SvdV91]   R. Schoof and M. van der Vlugt, *Hecke operators and the weight distributions of certain codes*, Journal of Combinatorial Theory, Series A **57** (1991), 163–186.

[SW88a]   A. J. Stephens and H. C. Williams, *Computation of real quadratic fields with class number one*, Math. Comp. **51** (1988), no. 184, 809–824.

[SW88b]   _____, *Some computational results on a problem concerning powerful numbers*, Math. Comp. **50** (1988), 619–632.

[SW89]   _____, *Some computational results on a problem of Eisenstein*, Théorie des nombres/Number Theory, Walter de Gruyter, Berlin, 1989, pp. 869–886.

[SW99]   A. Stein and H. C. Williams, *Some methods for evaluating the regulator of a real quadratic function field*, Experimental Mathematics **8** (1999), 119–133.

[SZ04]   D. Stehlé and P. Zimmermann, *A binary recursive GCD algorithm*, Sixth International Algorithmic Number Theory Symposium (D. Buell, ed.), Lecture Notes in Computer Science, Vol. 3076, Springer, Berlin, 2004, pp. 411–425.

[SZ91]   A. Stein and H. G. Zimmer, *An algorithm for determining the regulator and the fundamental unit of a hyperelliptic congruence function field*, International Symposium on Symbolic and Algebraic Computation (ISSAC'91) (Bonn, Germany), ACM Press, New York, 1991, pp. 183–184.

[Tan37]   P. Tannery, *Sur la mesure de circle d'Archimède*, Mémoires Scientifiques, T. I., Toulouse 1912–1937, pp. 226–253.

[Tan84]   _____, *La perte de sept livres de Diophante*, Bull. des Sciences Math. **8** (1884), 192–206.

[Tat51]   T. Tatuzawa, *On a theorem of Siegel*, Japan J. Math. **21** (1951), 163–178.

[Ter00]   D. C. Terr, *A modificiation of Shanks' baby-step giant-step algorithm*, Math. Comp. **69** (2000), no. 230, 767–773.

[Tes98]   E. Teske, *A space efficient algorithm for group structure computation*, Math. Comp. **67** (1998), no. 224, 1637–1663.

[Thi95]   C. Thiel, *On the complexity of some problems in algorithmic algebraic number theory*, Ph.D. thesis, Universität des Saarlandes, Saarbrücken, Germany, 1995.

[Tho80]   I. Thomas, *Greek Mathematical Works*, Loeb Classical Library, Vol. 335, 362, Harvard University Press, Cambridge, MA, 1980.

[Thu92]   Thucydides, *History of the Peloponnesian War, Books V and VI*, C. F. Smith (trans.), Loeb Classical Library, No. 110, Harvard University Press, Cambridge, MA, 1992.

[tRW03]   H. J. te Riele and H. C. Williams, *New computations concerning the Cohen-Lenstra heuristics*, Experimental Mathematics **12** (2003), no. 1, 99–113.

[TVW05]   A. Togbe, P. M. Voutier, and P. G. Walsh, *Solving a family of Thue equations with an application to the equation $x^2 - dy^4 = 1$*, Acta Arith. **120** (2005), 39–58.

[TW06]   W. Trappe and L. C. Washington, *Introduction to Cryptography with Coding Theory*, 2nd ed., Prentice-Hall, Eaglewood Cliffs, NJ, 2006.

[TW95]   R. Taylor and A. Wiles, *Ring-theoretic properties of certain Hecke algebras*, Ann. Math. **141** (1995), 553–572.

[TW99]   E. Teske and H. C. Williams, *A problem concerning a character sum*, Experimental Mathematics **8** (1999), no. 1, 63–72.

[Var98]   I. Vardi, *Archimedes' Cattle Problem*, Amer. Math. Monthly **105** (1998), 305–319.

[vdP03]   A. van der Poorten, *A note on NUCOMP*, Math. Comp. **72** (2003), 1935–1946.

[vdP94]   A. J. van der Poorten, *Explicit formulas for units in certain quadratic number fields*, Algorithmic Number Theory Symposium, ANTS I, (Ithaca, NY, 1994), Lecture Notes in Computer Science, vol. 887, Springer, Berlin, 1994, pp. 194–208.

[vdP96]   ———, *Notes on Fermat's Last Theorem*, Wiley, New York, 1996.

[vdP99a]   ———, *Beer and continued fractions with periodic periods*, Number Theory (Ottawa, ON, 1996), CRM Proceedings Lecture Notes, Vol. 19, American Mathematical Society, Providence, RI, 1999, pp. 309–314.

[vdP99b]   ———, *Reduction of continued fractions of formal power series*, Continued fractions: from analytic Numbe Theory to Constructive Approximations, (Columbia, MO, 1998), Contemporary Mathematics, No. 236, American Mathematical Society, Providence, RI, 1999, pp. 343–355.

[vdPT00]   A. J. van der Poorten and X. C. Tran, *Quasi-elliptic integrals and periodic continued fractions*, Mohatshefte Math. **131** (2000), 155–169.

[vdPT02]   ———, *Periodic continued fractions in elliptic function fields*, Algebric Number Theory, proceedings of ANTS-V, Lecture Notes in Computer Science, Vol. 2369, Springer, Berlin, 2002, pp. 390–404.

[vdPtRW01] A. J. van der Poorten, H. te Riele, and H. C. Williams, *Computer verification of the Ankeney-Artin-Chowla conjecture for all primes less than 100000000000*, Math. Comp. **70** (2001), no. 235, 1311–1328.

[vdPW99]   A. J. van der Poorten and H. C. Williams, *On certain continued fraction expansions of fixed period length*, Acta. Arith. **89** (1999), no. 1, 23–25.

[vdW54]   B. L. van der Waerden, *Science Awakening*, P. Noordhoff, Groningen, The Netherlands, 1954.

[Vin54]   I. M. Vinogradov, *Elements of Number Theory*, Dover Publications, New York, 1954.

[Vol00]   U. Vollmer, *Asymptotically fast discrete logarithms in quadratic number fields*, Algorithmic Number Theory — ANTS-IV, Lecture Notes in Computer Science, Vol. 1838, Springer, Berlin, 2000, pp. 581–594.

[Vol02]   ———, *An accelerated Buchmann algorithm for regulator computation in real quadratic fields*, Algorithmic Number Theory — ANTS-V, Lecture Notes in Computer Science, Vol. 2369, Springer, Berlin, 2002, pp. 148–162.

[Wag96]   C. Wagner, *Class number 5, 6, and 7*, Math. Comp. **65** (1996), no. 214, 785–800.

[Wal48]   H. S. Wall, *Analytic Theory of Continued Fractions*, Chelsea Publishing Company, Bronx, NY, 1948.

[Wat04]   M. Watkins, *Class numbers of imaginary quadratic fields*, Math. Comp. **73** (2004), no. 246, 907–938.

[Wat95]   W. Waterhouse, *On the Cattle Problem of Archimedes*, Historia Math. **22** (1995), 186–187.

[WB76]   H. C. Williams and J. Broere, *A computational technique for evaluating $L(1, \chi)$ and the class number of a real quadratic field*, Math. Comp. **30** (1976), 887–893.

[WB79]   H. C. Williams and P. Buhr, *Calculation of the regulator of $\mathbb{Q}(\sqrt{D})$ by use of the nearest integer continued fraction algorithm*, Math. Comp. **33** (1979), 369–381.

[WCS80]   H. C. Williams, G. V. Cormack, and E. Seah, *Calculation of the regulator of a pure cubic field*, Math. Comp. **34** (1980), 567–611.

[WDS83]   H. C. Williams, G. W. Dueck, and B. K. Schmid, *A rapid method of evaluating the regulator and class number of a pure cubic field*, Math. Comp. **41** (1983), 235–286.

[Web95]   K. Weber, *The accelerated integer GCD algorithm*, ACM Transactions on Mathematical Software **21** (1995), no. 1, 111–122.

[Wei04]   D. Weimer, *An adaptation of the NICE cryptosystem to real quadratic orders*, Master's thesis, Technische Universität Darmstadt, Darmstadt, Germany, 2004, Available at http://www.cdc.informatik.tu-darmstadt.de/reports/reports/DanielWeimer.diplom.pdf.

[Wei06]   A. Weilert, *Two efficient algorithms for the computation of ideal sums in quadratic orders*, Math. Comp. **75** (2006), 941–981.

[Wei79]   A. Weil, *Fermat et l'équation de Pell*, Collected Papers, Vol. III, Springer, Berlin, 1979, pp. 413–419.

[Wei84]   ———, *Number Theory. An Approach Through History*, Birkhäuser, Boston, 1984.

[Wer02]   G. Wertheim, *Die Algebra des Johann Heinrich Rahn (1659) und die englische übersetzung derselben*, Bibliotheca Math. **3** (1902), no. 3, 113–126.

[WGZ65]   H. C. Williams, R. A. German, and C. R. Zarnke, *Solution of the Cattle Problem of Archimedes*, Math. Comp. **19** (1965), 671–674.

488   References

[Whi12]   E. E. Whitford, *The Pell Equation*, College of the City of New York, New York, 1912.

[Wil00]   H. C. Williams, *A number theoretic function arising from continued fractions*, Fibonacci Quart. **38** (2000), no. 3, 201–211.

[Wil02]   _____, *Solving the Pell equation*, Proc. Millennial Conference on Number Theory, A. K. Peters, Natick, MA, 2002, pp. 397–435.

[Wil80a]   _____, *A modification of the RSA public-key encryption procedure*, IEEE Transactions on Information Theory **26** (1980), no. 6, 726–729.

[Wil80b]   _____, *Some results concerning the nearest integer continued fraction expansion of $\sqrt{D}$*, J. Reine. Agnew. Math. **315** (1980), 1–15.

[Wil81]   _____, *A numerical investigation into the length of the period of the continued fraction of $\sqrt{D}$*, Math. Comp. **36** (1981), 593–601.

[Wil85a]   _____, *A note on the period length of the continued fraction expansion of certain $\sqrt{D}$*, Util. Math. **28** (1985), 201–209.

[Wil85b]   _____, *On mid period criteria for the nearest integer continued fraction expansion of $\sqrt{D}$*, Utilitas Math. **27** (1985), 169–185.

[Wil85c]   _____, *Some public-key crypto-functions as intractable as factorization*, Cryptologia (1985), no. 9, 223–237.

[Wil85d]   _____, *Some public-key crypto-functions as intractable as factorization, extended abstract*, Advances in Cryptology — CRYPTO '84, Lecture Notes in Computer Science, Vol. 196, Springer-Verlag, Berlin, 1985.

[Wil90]   _____, *Eisenstein's problem and continued fractions*, Utilitas Math. **37** (1990), 145–158.

[Wil95a]   A. Wiles, *Modular elliptic-curves and Fermat's last theorem*, Ann. Math. **141** (1995), 443–551.

[Wil95b]   H. C. Williams, *Some generalisations of the S_n sequence of Shanks*, Acta. Arith. **69** (1995), no. 3, 199–215.

[Wil98]   _____, *Edouard Lucas and Primality Testing*, Wiley-Interscience, New York, 1998.

[Wor81]   R. T. Worley, *Estimating $|\alpha - p/q|$*, J. Austral. Math. Soc. Ser. A **31** (1981), 202–206.

[Wur30]   J. F. Wurm, *Review of J. G. Hermann's pamphlet: De Archimedis Problemate Bovino, Leipzig, 1828*, Jahrbücher für Philologie und Pädagogik **14** (1830), 194–202, (in German).

[WW87]   H. C. Williams and M. C. Wunderlich, *On the parallel generation of the residues for the continued fraction factoring algorithm*, Math. Comp. **48** (1987), 405–423.

[WY02]   W. A. Webb and H. Yokota, *Polynomial Pell's equation*, Proc. Amer. Math. Soc. **131** (2002), 993–1006.

[WZ72]   H. C. Williams and C. R. Zarnke, *Computation of the solutions of the Diophantine equation $x^2 - dy^4 = 1$*, Proc. 3rd S-E Conference, Combinatorics, Graph Theory and Computing, Utilitas Math., Winnipeg, 1972, pp. 463–483.

[Yam70]   Y. Yamamoto, *Real quadratic number fields with large fundamental units*, Osaka Math. J. **7** (1970), 57–76.

[Zuc97]   R. J. Zuccherato, *The continued fraction algorithm and regulator for quadratic function fields of characteristic 2*, Journal of Algebra **190** (1997), 563–587.

Index